INTRODUCTION TO ENGINEERING MATERIALS

Introduction to Engineering Materials

Vernon John

Bsc, MSc, CEng, MIM, MIMM

Formerly Senior Lecturer in Materials
The Polytechnic of Central London

Third Edition

Industrial Press Inc.
200 Madison Avenue
New York, N.Y. 10016

Library of Congress Cataloging-in-Publication Data
John, Vernon.
    Introduction to engineering materials / Vernon John. — 3rd ed.
        536p   24.6 x 18.9 cm.
    Includes index.
    ISBN 0–8311–3043–1
    1. Materials.   I. Title.
TA403.J54   1992
620.1'1—dc20   91–44159
                                                        CIP

Industrial Press Inc.
200 Madison Avenue
New York, N.Y. 10016

Published in Great Britain by
THE MACMILLAN PRESS LTD

First edition 1972
Reprinted 1973, 1974, 1977, 1978, 1979, 1981, 1982
Second edition 1983
Reprinted 1986, 1988, 1990
Third edition 1992

Printed in Hong Kong

Contents

PART I
INTRODUCTION

1.1 Introduction; 1.2 The range of materials; 1.3 Properties of engineering materials; 1.4 Cost and availability; 1.5 Possibilities for the future.

PART II
MATERIALS SCIENCE

2.1 Introduction; 2.2 Elementary particles; 2.3 Atomic number and atomic mass number; 2.4 Isotopes and isotones; 2.5 The gramme atom, gramme molecule and Avogadro's number; 2.6 The structure of the atom and quantum numbers; 2.7 The Pauli exclusion principle; 2.8 The periodic table; 2.9 The nucleus and radioactivity; 2.10 Artificial radioactive materials; 2.11 Interatomic and intermolecular bonding; 2.12 The ionic bond; 2.13 The covalent bond; 2.14 The co-ordinate bond; 2.15 The metallic bond; 2.16 Secondary bonds; 2.17 Mixed bonds; 2.18 Questions.

3.1 Introduction; 3.2 Structure and density; 3.3 Stability and melting point; 3.4 Stiffness; 3.5 Electrical properties.

4.1 Introduction; 4.2 Addition polymerisation; 4.3 Condensation polymerisation; 4.4 Linear and non-linear polymers; 4.5 Branching; 4.6 Cross-linking; 4.7 Stereoregularity; 4.8 Degree of polymerisation; 4.9 Polymerisation methods; 4.10 Questions.

PART IV
FORMING AND FABRICATION OF MATERIALS

PART V
BEHAVIOUR IN SERVICE

PART VI
EVALUATION OF MATERIALS

Preface to the First Edition

Not many years ago it was possible to obtain a first degree or other qualification in engineering with no knowledge whatsoever of metallurgy or the other materials of engineering. Today it is fully accepted that a sound knowledge of the science and technology of materials is very necessary to the engineer if he is to be able efficiently to translate a design into functional 'hardware'. The selection of materials and manufacturing route is an integral part of design procedure. Because of this, materials' science and technology now feature prominently in the educational programme for all engineering technologists and technicians. Engineering, including materials' science, is also beginning to appear in the curricula of some schools.

This book was conceived against the general background of the changing patterns in engineering education, and the aim was to produce a text which dealt with the basic principles of materials science and technology in a simple, yet meaningful manner.

It is my belief that no aspect of engineering should be studied in a vacuum, but that academic studies be related to our real cost-conscious world. It is for this reason that I have included some materials' costs and a short note on the selection of materials. Although actual costs will vary from year to year, this section should be of value as it indicates the principle that any work or processing performed on a material is reflected in an increase in the cost of the material.

I wish to acknowledge the help and advice received from various individuals and organisations, and in particular to thank Professor Bob Fergusson for providing the initial stimulus for this work. I am indebted to some of my colleagues at PCL, especially Mr C. J. Beesley, Mr G. E. Drabble, and Mr M. D. Munro Mackenzie, for their assistance in reading the manuscript and suggesting improvements. I also wish to acknowledge the assistance, in the way of photographs and information, which I received from the following firms and organisations: British Aluminium Company Ltd, British Metals Sinterings Association, Bound Brook Ltd, Copper Development Association, I.C.I. Ltd (Plastics Division), International Nickel Ltd, Raleigh Industries Ltd, Sintered Products Ltd, and the Zinc Development Association. Several colleges and polytechnics provided me with question papers, and I wish to record my appreciation of their help. Finally, I am very grateful to my wife for not merely putting up with me while I was struggling with the manuscript, but for helping considerably by sustaining me with refreshment and also typing some of the manuscript.

Vernon John   *1972*

Preface to the Second Edition

Ten years have elapsed since this book was first published and after such a period of time an updating is necessary for any technical text. In the preparation of this edition I have attempted to satisfy two objectives; firstly to keep a basic and straightforward approach so that the text can be readily understood by students beginning their studies, and secondly to broaden the scope of the volume and so increase its suitability for degree and diploma students of both mechanical and production engineering in the later years of their courses.

When preparing a new edition of a book an author is torn between the desire to tear everything up and start again with a ream of blank paper or merely to make small cosmetic changes. This book was generally well received when it was first published and students appeared to like the general style and content. I have, therefore, retained much of the original text and confined myself to meeting most of the criticisms (fortunately, these were not too numerous) levelled at the first edition and also to making changes that reflect the development in my own teaching at PCL.

Some of the major differences between this text and the first edition are: (1) the inclusion of more worked examples within the text, (2) an increase in the sections dealing with dislocations and plastic deformation within metals, (3) a substantial enlargement of the chapter on material-forming processes, (4) a major increase in the chapters devoted to plastics materials, (5) a more detailed treatment of oxidation and corrosion and (6) the development of a small section on welding, formerly placed as an Appendix, into a full chapter within the book. In addition to these changes, an introduction to the principles of fracture mechanics has been included as an Appendix to the main text. It is my hope that these changes will find favour with both students and teachers.

I wish to thank my colleagues in the Materials section at PCL for the helpful comments freely given during the preparation of this revision and to record my indebtedness to my wife for her great help and for the typing of the draft.

Vernon John   *1983*

Preface to the Third Edition

There is a process of continuous development for curricula for courses in engineering. One significant development of the last decade is that the subject of engineering materials has moved into a centre-stage position, especially as a subject field within mechanical and manufacturing engineering courses. There has also been debate on the subject of a 'core curriculum'. During the preparation of this revised and enlarged edition, I have attempted to cater for current curriculum developments and changes in emphasis within syllabi. One possible failing of the last edition was, perhaps, an overemphasis on metals. I have attempted to redress this with increased coverage of non-metallic materials. My aim has been to present that body of core knowledge in materials science and technology which, in my opinion, should be a necessary part of the educational programme for any engineer.

The earlier editions of this book were well received and feedback from students indicated that the style of writing and presentation was clear and understandable. In consequence, I have retained much of the original text and illustrations in this new edition, even though the general structure of the book has been changed significantly and the number of chapters increased. The principal changes from the second edition include an introductory chapter outlining the range of materials, their basic characteristics, costs and availability to set the scene. The materials science section has been enlarged with elasticity and plasticity separated into two chapters and with new chapters dealing with polymer formation, visco-elasticity, toughness and fracture, phase transformations, and optical and thermal properties. Worked examples have been included in the text where appropriate. There has been a major increase in the textual matter covering polymers, ceramics and glasses. The formation and fabrication of materials has been reorganised with separate chapters for each class of material and a new chapter for material removal processes. The importance of non-destructive testing has been recognised by including a new chapter devoted to this subject. Material which appeared as appendices in the last edition has now been incorporated into the main text. It is hoped that these changes will find general acceptance by both teachers and students.

I hope that those engineering students who read and use this book will be stimulated and encouraged in their studies. Also, that they will develop an understanding and a feel for materials so that, as practising engineers, they will be able to select and use materials effectively and efficiently.

Vernon John   *1991*

Symbols Used in Text

Quantity	Symbol	
Atomic mass number (atomic weight)	M	
Atomic number	Z	
Avogadro's number	N_0	$N_0 = 6.023 \times 10^{23}$ molecules per mole
Bulk modulus of elasticity	K	
Density	ρ	
Direct strain	ε	
Direct stress	σ	
Force	F	
Fracture toughness	K_e	
Glass transition temperature	T_g	
Melting temperature	T_m	
Modulus of rigidity	G	
Poisson's ratio	v	(Greek nu)
Shear strain	γ	
Shear stress	τ	
Temperature	T	
Time	t	
Universal constant	R_0	$R_0 = 8.314\,\mathrm{kJ/kmol\,K}$
Viscosity	η	
Young's modulus of elasticity	E	

Units

The units used throughout this book conform to the SI system. The principal units that are quoted in the text are given below. Preferred SI units are printed in bold type.

Quantity	Unit	Symbol
mass	**kilogramme**	**kg** $(1\,\text{kg} = 2.205\,\text{lb})$
	gramme	g
	tonne	$\text{t}\,(\text{Mg})(1\,\text{t} = 1000\,\text{kg}$ $= 0.984\,\text{ton})$
length	**metre**	**m**$(1\,\text{m} = 39.37\,\text{in})$
	millimetre	mm
time	**second**	**s**
	minute	min
	hour	h
temperature	degree Kelvin	K
	degree Celsius	°C
amount of substance	mole	mol
	kilomole	**kmol**
area	**square metre**	**m^2**
	square millimetre	mm^2
volume	**cubic metre**	**m^3** $(1\,\text{m}^3 = 35.315\,\text{ft}^3)$
	cubic millimetre	mm^3
density	**kilogramme per cubic metre**	**kg/m^3** $(1\,\text{kg/m}^3$ $= 10^{-3}\,\text{g/cm}^3$ $= 0.062\,\text{lb/ft}^3)$
force	**newton**	**N** $(1\,\text{N} = 0.225\,\text{lbf})$
	kilonewton	kN
	meganewton	MN
stress (pressure)	**newton per square metre**	**N/m^2** $(1\,\text{N/m}^2$ $= 0.000145\,\text{lbf/in}^2)$
	meganewton per square metre	MN/m^2 $(1\,\text{MN/m}^2$ $= 0.0648\,\text{tonf/in}^2)$
	giganewton per square metre	GN/m^2
	pascal	**Pa** $(1\,\text{Pa} = 1\,\text{N/m}^2)$
	bar	bar or b $(1\,\text{bar} = 10^5\,\text{N/m}^2)$
energy	**joule**	**J** (Nm)
	electron volt	eV $(1\,\text{eV}$ $= 1.602 \times 10^{-19}\,\text{J})$

electric current	**ampere**	**A**
voltage	**volt**	**V**
quantity of electricity	**coulomb**	**C** (A s)
electrical resistance	**ohm**	Ω (V/A)
electrical resistivity	**ohm metre**	**Ωm**
magnetic flux	**weber**	**Wb** (V s)
magnetic flux density	**tesla**	**T** (Wb/m^2)
		(1 T = 10^4 gauss)
frequency	**hertz**	**Hz** (s^{-1})

PART I
INTRODUCTION

1

The Materials of Engineering

1.1 Introduction

Today's engineers have a vast range, comprising of several thousand materials available to them. Some of these, timbers, stone and clay products, some cast irons and copper alloys have been in use as constructional materials for centuries. At the other end of the scale, many polymers, high temperature superalloys, industrial ceramics and fibre reinforced composite materials have come into use in recent decades, while other materials of great potential interest, including new alloy compositions, new polymers, metallic glasses and metal-matrix composites are currently in the development stage. Running parallel to the invention of new and improved materials there have been equally important developments in materials processing including vacuum melting and casting, new moulding techniques for polymers, ceramics and composites, and new joining technologies.

In addition to the need for an increased knowledge of materials and processing technology, other challenges are having to be met by our design engineers. In earlier times, with a much smaller number of materials available, engineers often produced their designs and products by a process of trial and error, in many cases using far more material than was really necessary. Today there is a requirement to use materials more effectively and efficiently in order to manufacture quality products which can compete in world markets and to minimise cost. Also, the advent of product liability legislation places an increased burden on design and materials engineers and they now need to foresee and cater for possible misuse of the product in addition to normal usage by customers. For example, the materials and corrosion protection treatments specified and used in the production of a small car may be adequate for the normal user but a few vehicles may be purchased by inshore fishermen who will leave them for long periods on quaysides exposed to salt water spray.

There is a complex inter-dependence between design, material and manufacture and the design engineer, materials engineer and manufacturing engineer need to function as a close-knit team. Many factors have to be considered when selecting possible materials to fit a design and manufacturing requirement.

Does the material possess the necessary mechanical, electrical and thermal properties?
Can the material be formed to the desired shape?
Will the properties of the material alter with time during service?
Will the material be adversely affected by the environmental conditions and resist corrosion and other forms of attack?
Will the material be acceptable on aesthetic grounds?
Will the material give sufficient degree of reliability and quality? And, of course:
Can the product be made at an acceptable cost?

1.2 The range of materials

The complete range of materials can be classified into the categories:

METALS, POLYMERS, CERAMICS AND INORGANIC GLASSES,
and COMPOSITES

The classification, composites, contains materials with constituents from any two of the first three categories, for example, fibre reinforced polymers. A broad comparison of the properties of metals, ceramics and polymers is given in Table 1.1.

Composite materials have been developed to overcome some of the deficiencies of members of a particular class of materials and there are examples of ceramic/metal, polymer/ceramic and metal/polymer composites in current use. Ceramics, though strong in compression, generally are weak in tension, but metals tend to have equal strengths in both tension and compression. Reinforced and prestressed concretes are composites designed to improve the tensile characteristics of concrete structural members. Polymers have low densities but also have low strength and stiffness. The use of glass, carbon or other fibre reinforcement gives greatly increased strength and stiffness without adding an excessive weight penalty. The load bearing characteristics of metals and the low friction characteristics of a polymer such as PTFE are combined to good effect in metal particle/PTFE composites developed as bearing materials.

1.3 Properties of engineering materials

Very many properties, or qualities, of materials have to be considered when choosing a material to meet a design requirement (see Table 1.2). These include a wide range of physical, chemical and mechanical properties together with forming, or manufacturing characteristics, cost and

Table 1.1 *Comparison of properties of metals, ceramics and polymers*

Property	Metals	Ceramics	Polymers
Density (kg/m^3 × 10^{-3})	2–16 (average 8)	2–17 (average 5)	1–2
Melting points	Low to high Sn 232°C, W 3400°C	High, up to 4000°C	Low
Hardness	Medium	High	Low
Machineability	Good	Poor	Good
Tensile strength (MPa)	Up to 2500	Up to 400	Up to 120
Compressive strength (MPa)	Up to 2500	Up to 5000	Up to 350
Young's Modulus (GPa)	40–400	150–450	0.001–3.5
High temperature creep resistance	Poor	Excellent	—
Thermal expansion	Medium to high	Low to medium	Very high
Thermal conductivity	Medium	Medium but often decreases rapidly with temperature	Very low
Thermal shock resistance	Good	Generally poor	—
Electrical properties	Conductors	Insulators	Insulators
Chemical resistance	Low to medium	Excellent	Generally good
Oxidation resistance at high temperatures	Poor, except for rare metals	Oxides excellent SiC and Si$_3$N$_4$ good	—

Table 1.2 *Material properties and qualities*

Physical properties	Density, melting point, hardness. Elastic moduli. Damping capacity.
Mechanical properties	Yield, tensile, compressive and torsional strengths. Ductility. Fatigue strength. Creep strength. Fracture toughness.
Manufacturing properties	Ability to be shaped by: moulding and casting, plastic deformation, powder processing, machining. Ability to be joined by adhesives, welding, etc.
Chemical properties	Resistance to oxidation, corrosion, solvents and environmental factors.
Other non-mechanical properties	Electrical, magnetic, optical and thermal properties.
Economic properties	Raw material and processing costs. Availability.
Aesthetic properties	Appearance, texture and ability to accept special finishes.

availability data and, in addition, more subjective aesthetic qualities such as appearance and texture.

Values of E, tensile yield strength, tensile strength, fracture toughness and density for some common groups of materials are given in Table 1.3, and the physical properties of some pure metals are given in Table 1.4.

Table 1.3 *Properties (at 25°C) of some groups of materials*

Material	E (GPa)	Yield strength (MPa)	Tensile strength (MPa)	Fracture toughness (MPa m$^{\frac{1}{2}}$)	Density (kg m^{-3} × 10^{-3})
Steels	200–220	200–1800	350–2300	80–170	7.8–7.9
Cast irons	150–180	100–500	300–1000	6–20	7.2–7.6
Aluminium alloys	70	25–500	70–600	5–70	2.7–2.8
Copper alloys	90–130	70–1000	220–1400	30–120	8.4–8.9
Magnesium alloys	40–50	30–250	60–300		1.7–1.8
Nickel alloys	180–220	60–1200	200–1400	>100	7.9–8.9
Titanium alloys	100–120	180–1400	350–1500	50–100	4.4–4.5
Zinc alloys	70–90	50–300	150–350		6.7–7.1
Polyethylene (LDPE)	0.12–0.25		1–16	1–2	0.91–0.94
Polyethylene (HDPE)	0.45–1.4		20–38	2–5	0.95–0.97
Polypropylene (PP)	0.5–1.9		20–40	3.5	0.90–0.91
PTFE	0.35–0.6		17–28		2.1–2.25
Polystyrene (PS)	2.8–3.5		35–85	2	1.0–1.1
Rigid PVC	2.4–4.0		24–60	2.4	1.4–1.5
Acrylic (PMMA)	2.7–3.5		50–80	1.6	1.2
Nylons (PA)	2.0–3.5		60–100	3–5	1.05–1.15
PF resins	5–8		35–55		1.25
Polyester resins	1.3–4.5		45–85	0.5	1.1–1.4
Epoxy resins	2.1–5.5		40–85	0.3–0.5	1.2–1.4
GFRP	10–45		100–300	20–60	1.55–2.0
CFRP	70–200		70–650	30–45	1.40–1.75
Soda glass	74		50*	0.7	2.5
Alumina	380		300–400*	3–5	3.9
Silicon carbide	410		200–500*		3.2
Silicon nitride	310		300–850*	4	3.2
Concrete	30–50		7*	0.2	2.4–2.5

* Modulus of rupture value.

Table 1.4 *Physical properties of some pure metals*

Metal	Symbol	Melting point (°C)	Density (kg/m³ ×10⁻³)	Crystal structure*	Elastic constants (GPa) E	G	Poisson's ratio	Transformation temperature (°C)
Aluminium	Al	660	2.7	f.c.c.	70.5	27.0	0.34	
Antimony	Sb	630	6.67	r.				
Beryllium	Be	1280	1.85	c.p.h.	313	160	0.28	
Bismuth	Bi	271	9.8	r.	32	12.5	0.33	
Chromium	Cr	1888	7.1	b.c.c.	238	88.5	0.3	
Cobalt	Co	1492	8.7	αc.p.h.	203	75	0.31	
				βf.c.c.				430† α → β
Copper	Cu	1083	8.9	f.c.c.	122.5	45.6	0.35	
Gold	Au	1063	19.3	f.c.c.	80	28.3	0.42	
Iron	Fe	1535	7.87	αb.c.c.	215	84.8	0.29	
				γf.c.c.				908 α → γ
				δb.c.c.				1388 γ → δ
Lead	Pb	327	11.3	f.c.c.	16.5	5.6	0.44	
Magnesium	Mg	649	1.74	c.p.h.	44	17.6	0.28	
Manganese	Mn	1244	7.4	αc.cub.	200	77.6	0.24	
				β c.cub.				700† α → β
				γf.c.t.				1100† β → γ
				δb.c.c.				1140† γ → δ
Molybdenum	Mo	2620	10.2	b.c.c.	338	120	0.3	
Nickel	Ni	1453	8.9	f.c.c.	208	78.7	0.31	
Niobium	Nb	2420	8.57	b.c.c.	104	36.7	0.39	
Platinum	Pt	1769	21.65	f.c.c.	173	61.2	0.39	
Silicon	Si	1412	2.34	d.				
Silver	Ag	961	10.5	f.c.c.	79	29	0.37	
Tin	Sn	232	7.3	αd.	40.8	13.6	0.33	
				β b.c.t.				18 α → β
Titanium	Ti	1660	4.51	αc.p.h.	106.4	40	0.36	
				β b.c.c.				880† α → β
Tungsten	W	3380	19.3	b.c.c.	393	152	0.30	
Uranium	U	1130	19.05	αorth.				
				β t.				668 α → β
				γb.c.c.				774 β → γ
Vanadium	V	1920	6.15	b.c.c.	127	46.8	0.37	
Zinc	Zn	419	7.14	c.p.h.	92	37.3	0.25	
Zirconium	Zr	1860	6.4	αc.p.h.	95.5	36.4	0.33	
				βb.c.c.				

*Key: b.c.c. – body centred cubic
      c.cub. – complex cubic
      d. – diamond
      f.c.t. – face centred tetragonal
      r. – rhombohedral

      b.c.t. – body centred tetragonal
      c.p.h. – close packed hexagonal
      f.c.c. – face centred cubic
      orth. – orthorhombic
      t. – tetragonal

†Approximate value.

1.4 Cost and availability

One of the most important aspects affecting the selection and use of materials is cost and availability and the costs of some materials (July 1991) are given in Table 1.5. In many cases purchase cost of materials accounts for about one-half of the total works cost of the finished product. It follows from this that the use of a cheaper raw material should have a significant effect on the final product cost. This is not always true as, in some cases, the choice of an expensive material may permit the use of relatively simple and low cost processing methods whereas a cheaper material may require lengthy, complex and expensive production methods.

It is usual to see the cost of materials quoted per unit mass, for example £100 per tonne. This may give a misleading picture as often it is the volume of material which is important rather than its mass. The relative position of a material in a league table of costs may change when the criterion is altered from £($) per tonne to £($) per unit volume. This can be seen in Table 1.5. Much of the data for metals in the table are for refined metal in ingot form and it should be realised that

Table 1.5 *Costs of some materials*, by mass and by volume*

Material	Cost		Material	Cost	
	(£/kg)	($/kg)		(£/100 cm³)	($/100 cm³)
Germanium	209	365.75	Germanium	122	213.50
Silver	93.7	163.98	Silver	98.4	172.20
Cobalt	18.02	31.54	Cobalt	15.67	27.42
PTFE	7.0	12.25	Nickel	4.54	7.95
Nickel	5.10	8.93	Chromium	3.37	5.90
Chromium	4.75	8.31	Tin	2.53	4.43
Tin	3.46	6.06	Brass (sheet)	2.38	4.17
Titanium	3.09	5.41	Beryllium-copper	1.97	3.45
Brass (sheet)	2.87	5.02	Cadmium	1.68	2.94
Al/Cu alloy sheet	2.50	4.38	Phosphor bronze (ingot)	1.63	2.85
Beryllium-copper	2.23	3.90	18/8 stainless (sheet)	1.55	2.71
Nylon 66 (PA 66)	2.2	3.85	PTFE	1.50	2.63
18/8 stainless (sheet)	2.0	3.50	Copper (tubing)	1.40	2.45
Cadmium	1.94	3.40	Titanium	1.39	2.43
Phosphor bronze (ingot)	1.85	3.24	Copper (grade A ingot)	1.21	2.12
Magnesium (ingot)	1.65	2.89	Manganese	1.07	1.87
Acrylic (PMMA)	1.6	2.80	Brass (ingot)	1.05	1.84
Copper (tubing)	1.57	2.75	Al/Cu alloy sheet	0.74	1.30
ABS	1.5	2.63	Zinc (ingot)	0.46	0.81
Manganese	1.44	2.52	Lead (ingot)	0.38	0.67
Copper (grade A ingot)	1.36	2.38	Magnesium (ingot)	0.29	0.51
Brass (ingot)	1.25	2.18	Mild steel (sheet)	0.27	0.47
Amino resin thermoset	0.85	1.49	Nylon 66 (PA 66)	0.25	0.44
Aluminium (ingot)	0.78	1.37	Aluminium (ingot)	0.21	0.37
P–F thermoset	0.75	1.31	Acrylic (PMMA)	0.19	0.33
Silicon	0.71	1.24	Silicon	0.17	0.30
Polystyrene	0.64	1.12	ABS	0.16	0.28
Zinc (ingot)	0.64	1.12	Mild steel (ingot)	0.14	0.25
Polyethylene (HDPE)	0.62	1.09	Amino resin thermoset	0.13	0.23
Polypropylene (PP)	0.57	1.00	Cast iron	0.11	0.19
Natural rubber	0.56	0.98	P–F thermoset	0.09	0.16
Polyethylene (LDPE)	0.42	0.74	Polystyrene	0.07	0.12
Rigid PVC	0.41	0.72	Natural rubber	0.07	0.12
Mild steel (sheet)	0.34	0.60	Polyethylene (HDPE)	0.06	0.11
Lead (ingot)	0.34	0.60	Rigid PVC	0.06	0.11
Mild steel (ingot)	0.18	0.32	Polypropylene (PP)	0.05	0.09
Cast iron	0.15	0.26	Polyethylene (LDPE)	0.04	0.07
Portland cement	0.05	0.09	Portland cement	0.016	0.03
Common brick	0.04	0.07	Common brick	0.008	0.01
Concrete (ready mixed)	0.02	0.04	Concrete (ready mixed)	0.006	0.01

* The costs are based on information obtained from *European Chemical News, Metal Bulletin, Procurement Weekly* and private sources and are for bulk quantities quoted in July 1991. The data are quoted with the kind permission of the proprietors of those journals. The dollar translation is based on $1.75/£.

the costs of processed products, such as sheet, plate, sections and forgings will be much higher. Every process and every heat treatment will give added value and increase the final material cost. Also, the process of alloying will mean that, generally, the costs of alloys will be higher than those for unalloyed metals. For example, the cost of mild steel cold rolled strip material is approximately twice that of mild steel ingot while stainless steel sheet is almost six times the cost of mild steel sheet (Table 1.6). Similarly, in the aluminium industry, an alloy of aluminium with 5 per cent magnesium is about 50 per cent more expensive than commercial purity aluminium (Table 1.7).

Availability also influences the choice of a material. In the last century, during the growth period of railways, most railway bridges in Britain were constructed of masonry or wrought iron. In the same period when tracks were being pushed westwards in the United States and Canada, bridges and viaducts were usually of timber trestle construction owing to the ready availability of

Table 1.6   *Cost build-up (steel products)* *

Material	Cost	
	(£ per tonne)	($ per tonne)
Iron from blast furnace	120	210
Mild steel (ingot)	180	315
Mild steel (black bar)	280	490
Mild steel (cold drawn bright bar)	380	665
Mild steel (hot rolled sections)	285	498.75
Mild steel (hot rolled strip coil)	272	476
Mild steel (cold rolled strip coil)	339	593.25
Mild steel (galvanised sheet)	394	689.50
Austenitic stainless steel (cold rolled sheet)	2000	3500

* These cost figures applied in July 1991 and are based on information in *Procurement Weekly* and *Metal Bulletin*. They are reproduced by kind permission of the proprietors of those journals. The dollar translation is based on $1.75/£.

Table 1.7   *Cost build-up (aluminium products)* *

Material		Cost	
		(£/kg)	($/kg)
Aluminium 99.7 per cent (ingot)		780	1365
Aluminium commercial purity	sheet	1660	2905
	strip coil	1820	3185
	simple extruded sections	2100	3675
Aluminium alloy 5251 (2% Mg)	sheet	1770	3097.5
Aluminium alloy 5056A (5% Mg)	plate	2500	4375
Aluminium alloy 6082 (0.9% Mg, 1% Si, 0.7% Mn)	extruded sections (T6 temper)	2300	4025

* These cost figures applied in July 1991 and are based on information in *Metal Bulletin*. They are reproduced by kind permission of the proprietors. The dollar translation is based on $1.75/£.

Table 1.8 *Cost build-up (plastics industry)* *

Stage	Cost	
	(p/kg)	($/kg)
Raw materials to the chemical industry	6	0.11
Intermediate products	11	0.19
Monomer	29	0.51
Polymer	59	1.03
Compounded moulding material	72	1.26
Simple moulded components	121	2.12

* These cost figures applied to HDPE in July 1991 and are based on information in *European Chemical News*. They are reproduced by kind permission of the proprietors. The dollar translation is based on $1.75/£.

suitable timber close to the point of use. The same principle holds today and often a material or source of material supply will be chosen on the basis of proximity or availability.

1.5 Possibilities for the future

The world's mineral resources are non-renewable and finite and, according to present day estimates, the present known reserves of some metals of economic importance, including copper, lead, silver, tin and zinc, could be exhausted in the early part of the 21st century, well within the lifetimes of today's young engineers. These estimates on the size of reserves are based on knowledge of the availability of the sources being worked at present. This does not necessarily mean that these metals would not be available at all. These metallic elements are present in small quantities, lower than in current workable ores, in other rock formations. If it becomes necessary to extract the metals from such low level deposits then the cost of extraction and, more importantly, the energy consumption for extraction, would rise considerably. A high proportion of the metal products made today is derived from primary metal, that is metal produced direct from ores. Greater emphasis will need to be placed on use of secondary metal sources, that is metal obtained by recovery from recycled scrap. Some problems may be overcome by substitution of one material for another. An example of this is the replacement of tin-plated steel by aluminium for can manufacture. Much of the substitution which has taken place, however, is substitution of polymers for metals. The future supply position for polymers may also be uncertain as they are largely derived from petroleum. The supply situation for ceramic materials is much different as the raw materials for ceramic manufacture are present in great abundance. The earth's crust is composed mainly of silicates and alumino-silicates.

Changes in the relative quantities, costs, and availabilities of materials during the next century will provide a series of major challenges to engineers and designers.

Now, having tried a little crystal gazing, it is time to get back to basics. The next few chapters present the background science necessary for a proper understanding of the properties and behaviour of the various classes of engineering materials.

PART II
MATERIALS SCIENCE

2
Atomic Structure and Bonding

2.1 Introduction

It is necessary for the proper understanding of materials technology to have some knowledge of the underlying science and the fundamental principles that control the properties of materials. There are many thousands of materials available for use by today's engineers and, clearly, it is impossible for one person to have a full and detailed knowledge of all of them. However, as we have seen in Chapter 1, materials can be classified within several well-defined groups. There is a further common feature—all substances are composed of atoms. There is a comparatively limited number of different atomic species or chemical elements (the total number of elements, including the artificially created transuranic elements, is 103) and all these elements are composed of the same basic sub-atomic particles. The manner in which these sub-atomic particles are assembled into atoms and the ways in which the various atoms are bonded to one another in the make-up of bulk materials play a major part in determining the final properties of any material. For example, it is the different nature of the interatomic bonding in metals, as compared with polymer materials, which allows for electrical conduction in the former while making the latter dielectrics or insulators.

2.2 Elementary particles

A large number of sub-atomic particles have been identified by physicists but, in terms of our general understanding of atoms and atomic structure, it is convenient to consider three types of sub-atomic particle only. These are the *proton, neutron* and *electron*. The proton is a particle that has a positive electrical charge and a mass of 1.672×10^{-27} kg. The neutron has a mass which is almost identical to that of a proton, and is 1.675×10^{-27} kg but possesses no electrical charge. The mass of an electron is extremely small, being 1/1836 of that of a proton, and it possesses an electrical charge which is equal in magnitude but opposite in sign to that of a proton. The characteristics of the sub-atomic particles are summarised in Table 2.1.

Table 2.1   *Sub-atomic particles*

Particle	Mass (kg)	Charge (C)	Relative mass	Relative charge
Proton	1.672×10^{-27}	1.602×10^{-19}	1	$+1$
Neutron	1.675×10^{-27}	0	1	0
Electron	0.910×10^{-30}	1.602×10^{-19}	0	-1

Within an atom the protons and neutrons together form a small compact *nucleus* and the electrons are positioned around this nucleus in a series of energy shells (refer to Section 2.6). The general term *nucleon* may be used to describe both nuclear protons and neutrons. Generally, atoms contain an equal number of protons and electrons and, thus, are electrically neutral. When, as a result of excitation or interactions with other atoms, there is an imbalance between the number of protons and electrons and the atom possesses a net positive or negative charge, this charged atom is termed an *ion*.

2.3 Atomic number and atomic mass number

There are 103 different chemical elements and each is characterised by having a different number of protons in its nucleus. The number of protons contained in the nucleus is termed the *atomic number*, Z. In an electrically neutral atom, therefore, there must be Z electrons.

The nucleus is composed of protons and neutrons, except in the case of the simplest atom, hydrogen, in which the nucleus comprises one proton only. It is the protons and neutrons which contribute to virtually all the mass of an atom and the *atomic mass number* (or *atomic weight*), M, indicates the total number of protons and neutrons in the nucleus. For the simplest atom, hydrogen, $Z = 1$ and $M = 1$, so that the nucleus is simply one proton but for oxygen, $Z = 8$ and $M = 16$, the nucleus contains eight protons and eight neutrons. The quoted values of atomic mass number are based on a unit termed the *atomic mass unit* (*a.m.u.*) and this is one-twelfth of the mass of an atom of the *isotope* of carbon which has a nucleus containing six protons and six neutrons. The a.m.u. has a measured mass of 1.6598×10^{-27} kg.

The quoted values for the atomic mass number of the chemical elements are not all integers. There are two reasons for this, *mass defect* and the existence of *isotopes*. The measured mass of an atom does not equate with the sum of the masses of the protons, neutrons and electrons present. For example, the mass of an oxygen atom containing eight protons, eight neutrons and eight electrons is 26.556×10^{-27} kg but the sum of the masses of eight protons, eight neutrons and eight electrons is 26.784×10^{-27} kg. This discrepancy is termed the *mass defect* and, generally, is expressed as mass defect per nucleon. From the above figures the mass defect per nucleon for oxygen is 14.205×10^{-30} kg. According to Einstein's theory of relativity, mass and energy are related according to the expression: $E = mc^2$ where m is mass (kg), E is energy (J) and c is the velocity of light (3×10^8 m s^{-1}). In the case of oxygen, the mass defect per nucleon is equivalent to an energy of 1.28×10^{-12} J, or about 8 MeV. (An electron volt (eV) is a unit of energy and is the energy change when an electron moves through a potential difference of one volt.) The energy equivalent of the mass defect per nucleon is referred to as the binding energy per nucleon. The value of mass defect per nucleon and, hence, nuclear binding energy varies with the atomic mass number as shown in Figure 2.1 and the elements with the greatest mass defect values are those with atomic mass numbers in the range from 40 to 100.

It can be seen from Figure 2.1 that when elements of high atomic mass number disintegrate into nuclei of lower mass number, that is, with higher mass defect values, mass will be converted into a large energy emission. This is the principle involved in nuclear fission reactions and is harnessed in nuclear reactors for power generation. It will be seen also that a nuclear fusion of hydrogen nuclei to form helium would result in a very large emission of energy. This is the principle involved in the 'hydrogen' bomb but as yet it has not proved possible to use nuclear fusion reactions for the controlled release of energy for power generation.

In chemical reactions, as opposed to nuclear reactions, the changes which occur are confined to the outer electron shells. Energy changes in chemical reactions are of an extremely small order

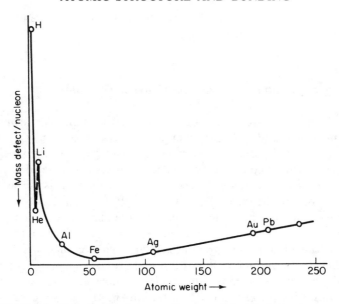

FIGURE 2.1   Relationship between the mass defect and atomic mass number

in comparison with the energy changes accompanying nuclear reactions, so small that any corresponding mass change is immeasurable.

2.4 Isotopes and isotones

Isotopes are atoms of the same element which possess different atomic mass numbers, that is there is a variation in the number of neutrons in the nuclei. Isotopes exist for the majority of chemical elements and the relative abundance of the various isotopes in any sample of an element is virtually constant, irrespective of the source. Some examples are given.

Two isotopes of hydrogen occur naturally, normal hydrogen ($Z = 1$, $M = 1$) and deuterium ($Z = 1$, $M = 2$) while a third isotope, tritium ($Z = 1$, $M = 3$), has been formed by the bombardment of lithium and boron by neutrons. The isotope tritium is radioactive. The proportion of deuterium in a sample of hydrogen is about 0.02%.

Isotopes may be symbolised in one of two ways, as for example $^{12}_{6}C$, where the superscript denotes atomic mass number and the subscript denotes atomic number, or as carbon-12 (C-12).

Any sample of oxygen contains three isotopes with mass numbers of 16, 17 and 18 while any sample of carbon has three isotopes with mass numbers of 12, 13 and 14, the last one, present in only very small quantities, being radioactive. It is the existence of the last radioactive isotope which permits use of the carbon dating technique for determining the age of ancient remains. Living matter absorbs carbon throughout its life and the proportion of C-14 atoms within the total carbon content is virtually constant. When the plant or animal dies no further carbon is absorbed and, as the C-14 atoms slowly decay radioactively, the proportion of C-14 atoms and, hence the total radioactivity, in the substance diminishes.

In addition to the isotopes of elements which occur naturally, other isotopes may be created by means of nuclear reactions (see Section 2.10) and these are almost invariably radioactive, disintegrating spontaneously with the emission of radiation.

Isotones, or *isobares*, are nuclei of different elements which possess the same atomic mass number. Examples of this are argon-40 (a nucleus containing 18 protons and 22 neutrons) and calcium-40 (a nucleus containing 20 protons and 20 neutrons).

2.5 The gramme atom, gramme molecule and Avogadro's number

Some substances are termed atomic while others are termed molecular, a *molecule* being a number of atoms in combination as, for example in gaseous oxygen, O_2 (two atoms per molecule), carbon dioxide, CO_2 (one carbon atom and two oxygen atoms per molecule), or calcium chloride, $CaCl_2$ (one calcium and two chlorine atoms per molecule). In the case of a molecular substance the *molecular weight*, or molecular mass number, is the sum of the atomic mass numbers of the atoms in the molecule.

A gramme atom of an element is the amount of the element with a mass in grammes numerically equal to its atomic mass number (atomic weight) and similarly, a gramme molecule is the amount of substance with a mass in grammes equal to its molecular weight. The SI unit of molar mass is the *mole* (abbreviation mol) and this is the alternative name for both a gramme atom of a monatomic element and a gramme molecule of a molecular substance. There is a larger unit of molar mass, this being the kilomole (abbreviation kmol), and being either a kilogramme atom or kilogramme molecule.

Equal molar masses of any substance will always contain the same number of molecules (the same number of atoms in the case of monatomic substances). The number of molecules in a mole or kilomole is given by the universal constant, Avogadro's number, N_0. The numerical value of Avogadro's number is 6.023×10^{23}/mol or 6.023×10^{26}/kmol.

2.6 The structure of the atom and quantum numbers

Rutherford in 1911 performed a series of experiments in which he bombarded thin gold foil with α particles (helium nuclei, which are positively charged particles with an atomic mass number of 4). Very many α particles passed through the foil without being deflected, but some particles were deflected. The inference was that much of the atom was void space allowing most of the α particles to pass straight through, but with some particles being deflected owing to their passing close to, or colliding with, a central nucleus of very small dimensions. The effective diameter of a gold atom is approximately 3×10^{-10} m and the diameter of the nucleus is about 1/10 000 of this size.

One of the first postulations about electrons was that they are in orbit around the central nucleus in much the same way as planets are in orbit around the sun. On the basis of classical mechanics this picture cannot be valid. The electron is a charged particle and in following a circular or elliptical path will be subject to angular acceleration. But from electromagnetic theory an electrical charge will radiate energy when it is accelerated. (It is the acceleration of charges in an aerial that is responsible for the transmission of energy as radio waves). This would not allow stability in atoms. If orbiting electrons were continually losing energy, the orbits would not remain constant. The electrons would follow spiral paths and eventually collapse into the nucleus. Electrons, therefore, cannot simply be regarded as corpuscular particles. They exhibit some of the characteristics of waves—an electron beam may be diffracted—and the motion of electrons can be described not by classical mechanics, but by quantum mechanics and wave mechanics.

In 1913, Bohr suggested a quantum model for an atom, making the following assumptions:

The electrons exist in stable circular orbits of fixed energy, with the angular momentum of an electron in an orbit being an integral multiple of $h/2\pi$, where h is Planck's constant.
An electron can emit or absorb energy only when making a transition from one possible orbit to another.

According to Bohr, when an atom is excited by the input of energy an electron moves to an orbit of greater radius. Subsequently, when the electron moves back to the 'ground state' of energy, energy will be released at some specific frequency giving rise to a spectral line. Planck postulated that energy is not absorbed or emitted in a continuous manner, but in discrete packets termed *quanta*, with one quantum of energy being hf, where h is Planck's universal constant ($h = 6.62 \times 10^{-34}$ Js) and f is the frequency of the radiation.

The modern concept is that the electrons exist in a series of energy levels, or shells, surrounding the nucleus. The quantum theory of atoms requires that four quantum numbers be defined to give a complete description of the energy state of an electron. The first, or principal, quantum number, n, which may have positive integral values, 1, 2, 3, 4, etc is a description of the general energy level of the electron in terms of distance from the nucleus, that is the layer or shell containing the electron. These principal shells may be denoted by letters: $K = 1$, $L = 2$, $M = 3$, $N = 4$, $O = 5$, $P = 6$ and $Q = 7$.

The secondary quantum number, l, is a measure of the angular momentum of the electron. l may have integral values from 0 to $n - 1$ and the values of l are denoted by the letter symbols s, p, d, f, where $s = 0$, $p = 1$, $d = 2$, $f = 3$. A value of $l = 0$ does not mean that the electron is stationary, but rather that there is an equal probability of finding an s state electron at any point on the surface of a spherical orbital. The probability envelope for a p state electron, that is when $l = 1$, has a dumb-bell shape (Figure 2.2(b)).

The third quantum number, m_l, describes the magnetic moment of an electron. A moving electron will produce a magnetic field and this, in turn, will be affected by an external magnetic field. The quantum number m_l may have any integral value from $-l$ to $+l$, including 0. Consider the p state ($l = 1$) where m_l may have three possible values, -1, 0 and $+1$. Three

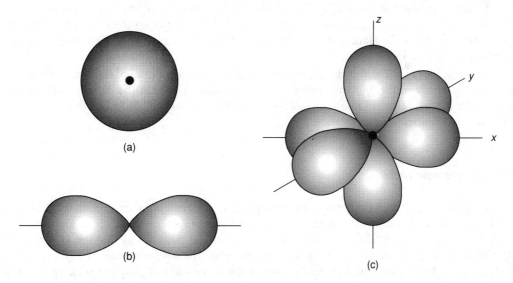

FIGURE 2.2 (a) Spherical probability envelope for the s state. (b) Probability envelope for the p state. (c) Three mutually perpendicular p state envelopes for m_l values of -1, 0 and $+1$

mutually perpendicular dumb-bell shaped orbitals, corresponding to the three values of m_l are possible (Figure 2.2(c)).

The fourth quantum number considers motion within the electron. It is assumed that the electron is spinning about an axis as well as being in orbit around a nucleus. This spin may be deemed either clockwise or counter-clockwise. The spin quantum number, m_s may have the values $\pm \frac{1}{2}$.

The four quantum numbers, n, l, m_l and m_s, must be defined to fix the energy level of an electron. In the 'ground state' of energy, the electrons will have the lowest quantum number values possible. For example, for hydrogen in its ground state, the one electron would be defined as: $n = 1$, $l = 0$, $m_l = 0$, $m_s = \pm \frac{1}{2}$.

2.7 The Pauli exclusion principle

In the case of an atom possessing more than one electron, the 'ground state' of energy does not mean that all electrons can be described by the quantum numbers $n = 1$, $l = 0$, $m_l = 0$, and $m_s = \pm \frac{1}{2}$. It is true that the atom possesses the minimum possible energy when it is in the ground state, but no two electrons in the same atom can be assigned the same set of quantum numbers. This important principle was first deduced by Pauli. If two electrons possessed the same values of n, l, and m_l, they could technically be regarded as being 'in the same place', and they would tend to repel one another strongly. This would be the case if the value for spin, m_s, was the same. On the other hand, if two electrons possessed opposite spins, the repulsive force would not be strong and they could possess the same values of n, l, and m_l.

We are now in a position to see how the electronic structure of atoms can be built up.

For a principal quantum number of $n = 1$, (that is, the first, or K shell), l may only have the value of 0 (s). Similarly, m_l can only be 0. The spin quantum number may be $+\frac{1}{2}$ or $-\frac{1}{2}$. Therefore only two electrons can be accommodated in the K shell, these being in the $1s$ state.

In the L shell, $n = 2$, therefore l may have values of 0 or 1 (s or p). When $l = 0$, m_l can only be 0. m_s may be $+\frac{1}{2}$ or $-\frac{1}{2}$, so that only two s state electrons can be placed in the L shell. But with $n = 2$, l can also have the value of 1. This allows the possibility of m_l having values of -1, 0, or $+1$. With two values of m_s possible in each case a total number of six electrons can exist in this, the p state of the L shell. Hence, the L shell can contain a total number of eight electrons, two in the $2s$ state and six in the $2p$ state.

Similarly, we can show that the M, or third, shell can contain a total of eighteen electrons, two in the $3s$ state, six in the $3p$ state, and ten in the $3d$ state. Also that the N, or fourth, shell can contain thirty-two electrons made up as follows: two in the $4s$ state, six in the $4p$ state, ten in the $4d$ state, and fourteen in the $4f$ state.

2.8 The periodic table

Table 2.2 gives the first 38 elements in ascending order of atomic number, and shows the electron configuration of each element. As the atomic number increases, at least up to argon ($Z = 18$), each additional electron takes up a place in the appropriate shell, filling first the $1s$, then $2s$ and $2p$, followed by $3s$ and $3p$. When the $3p$ shell is completed, the nineteenth electron (for potassium) goes not into the $3d$ state, but into the $4s$ state, as this state is at a slightly lower energy level than $3d$. Also the $3d$ state is at a slightly lower energy level than $4p$. This means that after element 20 (calcium), when the $4s$ state is full, additional electrons go into the $3d$ state. In passing from elements 21 to 30 (scandium to zinc) the $3d$ sub-shell is completely filled, and with element 31 (gallium) we find that the $4p$ sub-shell is commenced.

For all states above the $4s$ state, the energy levels are fairly close together and the general order in which the shells fill up is

$$4s,\ 3d,\ 4p,\ 5s,\ 4d,\ 5p,\ 6s,\ 4f,\ 5d,\ 6p,\ 7s,\ 6d,\ \text{and}\ 5f.$$

The number of electrons in the outermost shell of any atom has an important bearing on the properties of the element. Some elements, those with atomic numbers of 2, 10, 18, 36, 54, and 86, are extremely stable, and inert chemically. These, the inert or noble gases, are characterised by having all their energy levels completely filled with electrons. Apart from helium, which possesses only two electrons, these elements all have a complement of eight electrons in the outermost principal shell.

All other elements, which have at least one incompletely filled shell or sub-shell, will take part in chemical reactions. It is the outermost shell electrons, the valence electrons, that are responsible for the chemical reactivity of elements, and elements in combination attempt to achieve an outer shell configuration of eight electrons.

It was noticed more than a hundred years ago, long before deductions had been made about the inner structure of the atom, that there existed certain groups of atoms showing some similarity in properties. The alkali metals, lithium, sodium, and potassium, formed one such group, and the halogens, fluorine, chlorine, bromine, and iodine, formed another. When the various elements known at that time were placed in order of ascending atomic weight, a certain periodicity of elements with specific properties, such as the alkali metals and the halogens, was observed. In 1870, Mendeléev arranged the elements in a table in a manner that highlighted this effect. This was the forerunner of the modern periodic table of elements. With the discovery of the electron and the build-up of electrons in shells and sub-shells, as outlined above, came a fuller understanding of the properties of the elements.

Table 2.3 gives the full periodic table of the elements. The figures in front of each row of elements, for example $3p$ in front of elements 13 to 18, indicate the particular sub-shell being filled with electrons. The electrons in the outermost shell are the valence electrons. For the vertical columns IA, IIA, IIIB, IVB, VB, VIB, VIIB, and 0, within each column, the elements possess the same number of outer shell electrons, and hence, possess similar properties. But, considerable property differences occur between the elements of one column and those of another.

In the transition series, the first of which is the horizontal row containing the elements 21 to 30, the sub-shell being filled is not in the outer shell. This means that all the elements in the row possess somewhat similar properties. However, because the $4s$, $4p$, and $3d$ energy levels are close to one another, there is a tendency for electrons to move from one sub-shell to another. This causes these elements to possess variable valency characteristics. For example, iron (element 26) has a valency of three in the ferric (or iron III) state, and a valency of two in the ferrous (or Iron II) state.

In the cases of the two long series, the lanthanides, elements 58 to 71, and the actinides, elements 90 to 103, the sub-shell being filled is well in towards the core of the atom and so the chemical differences between each member of these two groups is very small.

2.9 The nucleus and radioactivity

It is known that the nucleus is extremely small in size, and that its diameter is only about 1/10 000 of the effective diameter of the atom. Particles of like electrical charge should repel one another

Table 2.2   *Electronic configuration of the elements*

Element	Symbol	Atomic Number Z	Atomic Weight M	K	L		M			N				O				P				Q
				1s	2s	2p	3s	3p	3d	4s	4p	4d	4f	5s	5p	5d	5f	6s	6p	6d	6f	7s
Hydrogen	H	1	1.008	1																		
Helium	He	2	4.003	2																		
Lithium	Li	3	6.940	2	1																	
Beryllium	Be	4	9.013	2	2																	
Boron	B	5	10.82	2	2	1																
Carbon	C	6	12.011	2	2	2																
Nitrogen	N	7	14.008	2	2	3																
Oxygen	O	8	16.000	2	2	4																
Fluorine	F	9	19.000	2	2	5																
Neon	Ne	10	20.183	2	2	6																
Sodium	Na	11	22.991	2	2	6	1															
Magnesium	Mg	12	24.32	2	2	6	2															
Aluminium	Al	13	26.98	2	2	6	2	1														
Silicon	Si	14	28.09	2	2	6	2	2														
Phosphorus	P	15	30.975	2	2	6	2	3														
Sulphur	S	16	32.066	2	2	6	2	4														
Chlorine	Cl	17	35.457	2	2	6	2	5														
Argon	Ar	18	39.944	2	2	6	2	6														

Number of electrons in each shell and sub-shell

Table 2.2—*continued*

Element	Symbol	Z	At. wt.	1s	2s	2p	3s	3p	3d	4s	4p	5s
Potassium	K	19	39.10	2	2	6	2	6		1		
Calcium	Ca	20	40.08	2	2	6	2	6		2		
Scandium	Sc	21	44.96	2	2	6	2	6	1	2		
Titanium	Ti	22	47.90	2	2	6	2	6	2	2		
Vanadium	V	23	50.95	2	2	6	2	6	3	2		
Chromium	Cr	24	52.01	2	2	6	2	6	5	1		
Manganese	Mn	25	54.94	2	2	6	2	6	5	2		
Iron	Fe	26	55.85	2	2	6	2	6	6	2		
Cobalt	Co	27	58.94	2	2	6	2	6	7	2		
Nickel	Ni	28	58.69	2	2	6	2	6	8	2		
Copper	Cu	29	63.54	2	2	6	2	6	10	1		
Zinc	Zn	30	65.38	2	2	6	2	6	10	2		
Gallium	Ga	31	69.72	All K, L, and M shells are completely filled						2	1	
Germanium	Ge	32	72.60							2	2	
Arsenic	As	33	74.91							2	3	
Selenium	Se	34	78.96							2	4	
Bromine	Br	35	79.916							2	5	
Krypton	Kr	36	83.80							2	6	
Rubidium	Rb	37	85.48							2	6	1
Strontium	Sr	38	87.63							2	6	2

Table 2.3 *Periodic table of the elements*

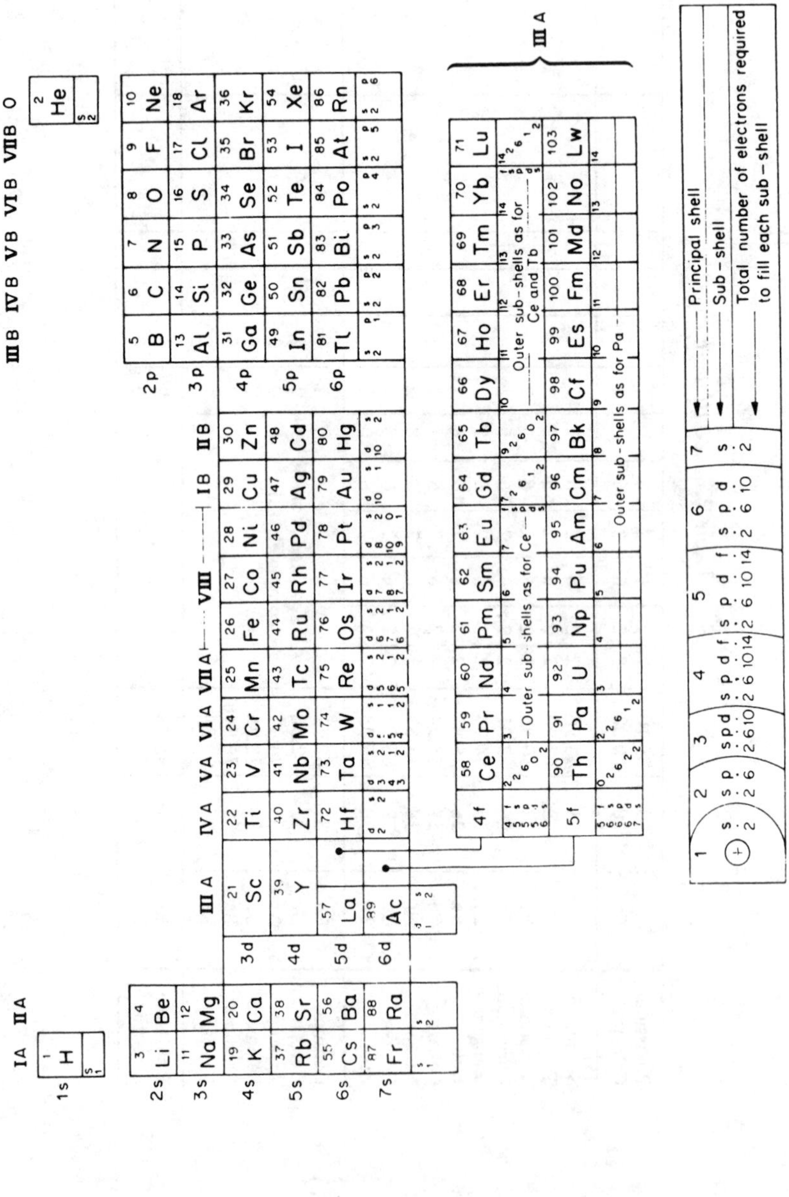

strongly, and yet the nuclei of most elements are extremely stable. The presence of neutrons seems to play a major role in ensuring the stability of an atomic nucleus, but the nature of the nuclear binding is imperfectly understood. The lighter elements have roughly equal numbers of protons and neutrons in their nuclei. As the atomic number increases, so the ratio of neutrons to protons increases from 1:1 to about 1.5:1 in stable nuclei. Elements with atomic numbers greater than 83 do not possess stable nuclei. These large nuclei disintegrate spontaneously, emitting energy and particles, these emissions being termed *radioactivity*.

Natural radioactivity was first discovered in 1896 by Becquerel. Subsequent research workers found that all elements of high atomic number were radioactive, emitting small particles and very high frequency electromagnetic radiation, similar to X-radiation. Two types of particle emission were identified, these being termed α and β particles. The α particle had a mass number of 4 and possessed two unit positive charges (that is, a helium nucleus), while the β particle was identified as an electron. The very high frequency radiation emitted was termed γ-radiation.

The activity of a radioactive substance decreases exponentially with time. It is impossible to predict when any individual atom will disintegrate but, statistically, the number of atoms that will disintegrate in a given time can be accurately forecast, The rate of radioactive decay at some particular time is proportional to the activity at that instant. If I is the activity at some time t then

$$\frac{\mathrm{d}I}{\mathrm{d}t} = \lambda I \text{ where } \lambda \text{ is a constant}$$

Integration gives $I = I_0 \exp(-\lambda t)$ where I_0 is the initial activity at zero time. The constant λ is known as the radioactive constant.

An important characteristic of any radioactive substance is its *half life*. This is the time required for half of the radioactive nuclei originally present to disintegrate, or in other words, the time needed for the activity to reduce to half of its original value. If the half life is T then

$$I_T = \frac{I_0}{2} = I_0 \exp(-\lambda T)$$

Therefore

$$\lambda T = \ln 2$$

$$\lambda = \frac{0.693}{T}$$

Example

The radioactive isotope of cobalt, cobalt-60, has a half-life period of 5.3 years. What is the radioactive constant for cobalt-60 and how long would it take for the activity of a sample of cobalt-60 to reduce to 85 per cent of its original value?

$$T \text{ for cobalt-60} = 5.3 \text{ years} = 1.67 \times 10^8 \text{ s}$$

$$\lambda = \frac{0.693}{T} = 4.14 \times 10^{-9} \text{ s}^{-1}$$

$$I_t = 0.85I_0 = I_0 \exp(-\lambda t)$$

$$\ln 0.85 = -\lambda t$$

$$t = \frac{-\ln 0.85}{\lambda}$$

$$= 3.92 \times 10^7 \text{ s}$$

$$= 1.24 \text{ years}$$

It is customary to classify radioactive substances according to their half life. The half-life periods for radioactive elements vary considerably, and range from about 10^{10} years for uranium and thorium to less than a second for some isotopes. The eventual product of natural radioactive disintegration will be some stable nucleus. In fact, in all cases this end product is an isotope of lead. Table 2.4 gives the disintegration sequence for thorium. The loss of an α particle from the nucleus must mean the loss of two protons, and, hence, a decrease of two in the atomic number, so producing a different element. Similarly, the emission of a β particle from the nucleus means a gain of one in the number of unit positive charges remaining in the nucleus, or an increase of one in the atomic number.

Table 2.4   *Radiation disintegration sequence for thorium*

Nucleus		Isotope symbol	Particle emitted	Half life
Thorium		$^{232}_{90}\text{Th}$	α	1.39×10^{10} years
Mesothorium I	(radium)	$^{228}_{88}\text{Ra}$	β	6.7 years
Mesothorium II	(actinium)	$^{228}_{89}\text{Ac}$	β	6.13 hours
Radiothorium I	(thorium)	$^{228}_{90}\text{Th}$	α	1.90 years
Thorium X	(radium)	$^{224}_{88}\text{Ra}$	α	3.64 days
Thorium emanation	(radon)	$^{220}_{86}\text{Rn}$	α	54.5 seconds
Thorium A	(polonium)	$^{216}_{84}\text{Po}$	α	0.16 second
Thorium B	(lead)	$^{212}_{82}\text{Pb}$	β	10.6 hours
Thorium C	(bismuth)	$^{212}_{83}\text{Bi}$	(a)β, (b)α	47 minutes
(a) Thorium C'	(polonium)	$^{212}_{84}\text{Po}$	α	3×10^{-7} second
Thorium D	(lead)	$^{208}_{82}\text{Pb}$	stable	
(b) Thorium C''	(thallium)	$^{208}_{81}\text{Tl}$	β	2.1 minutes
Thorium D	(lead)	$^{208}_{82}\text{Pb}$	stable	

2.10 Artificial radioactive materials

The naturally occurring elements with atomic numbers of less than 84 possess stable nuclei, but unstable nuclei of low atomic number elements may be made artificially. The natural form of phosphorus has an atomic mass number of 31, but in 1933 Joliot–Curie produced a radioactive isotope of phosphorus, with an atomic mass number of 30, by bombarding aluminium with α particles. The reaction may be written as follows

$$^{27}_{13}\text{Al} + ^{4}_{2}\text{He} \text{ (α particle)} = ^{30}_{15}\text{P} + ^{1}_{0}\text{n}$$

Subscripts denote atomic number, Z, and superscripts denote atomic mass number, M. Thus $_0^1 n$ represents the neutron formed in the reaction.

α particle bombardment is not a very efficient means of producing radioactive isotopes, as the positively charged atomic nuclei repel the bombarding particles of equal electrical sign, so limiting the number of collisions. The use of neutrons for the bombardment of nuclei is a much better technique, and it is this system that is used for the production of the radioactive isotopes required by industry. A typical reaction is the neutron bombardment of aluminium to produce radioactive sodium. An α particle is emitted in this reaction.

$$_{13}^{27}\text{Al} + _0^1\text{n} \rightarrow _{11}^{24}\text{Na} + _2^4\text{He} \ (\alpha \text{ particle})$$

The $_{11}^{24}$Na nucleus produced in the above reaction is an unstable nucleus and this decays radioactively into a stable isotope of magnesium with the emission of a β-particle (electron):

$$_{11}^{24}\text{Na} \rightarrow _{12}^{24}\text{Mg} + _{-1}^0\text{e} \ (\text{electron})$$

Radioactive isotopes are used in a number of engineering applications (refer to Table 2.5). Examples are:

the use of γ-radiography for defect detection
the use of β-particle emitters for the continuous monitoring of the thickness of rolled strip
the use of radioactive tracers to investigate wear in components that are in sliding or rolling contact
the use of radioactive markers to follow fluid flow in pipe-lines and pumping systems

Table 2.5 *Some radioactive isotopes used in engineering*

Element	Symbol	Radiation emitted	Half life	Application
Argon	$_{18}^{41}$Ar	β, γ	109 minutes	Leak detection
Caesium	$_{55}^{137}$Cs	β, γ	30 years	Radiography
Cobalt	$_{27}^{60}$Co	β, γ	5.3 years	Radiography
Iridium	$_{77}^{192}$Ir	β, γ	74.4 days	Radiography
Iron	$_{26}^{59}$Fe	β, γ	45 days	Engine wear
Strontium	$_{38}^{90}$Sr	β	28 years	Thickness gauges
Tungsten	$_{74}^{187}$W	β, γ	24.1 hours	Tool wear

2.11 Interatomic and intermolecular bonding

The fluid and solid substances that we are familiar with are composed of very large aggregates of atoms and the properties of these materials derive in part from the manner in which the individual atoms are bonded together and the strengths of these bonds. The bonding which exists between atoms is not the same for all materials as there are several types of interatomic bond possible. Generally, the bonding involves some degree of interaction between the outer shell, or *valence*, electrons and is, therefore, dependent on the number and distribution of electrons within the atom. It has been stated already (Section 2.8) that a completely filled outer electron shell confers a very high degree of stability to an atom, this being the electronic structure of the inert or noble gases. In the formation of interatomic bonds, atoms of elements with incomplete outer electron shells attempt, in combination with other atoms, to achieve filled outer electron shells, thus satisfying this condition for stability.

Interatomic bonds are relatively strong and are termed *primary* bonds. There are other types of bonds, weaker than the primary bonds, which can occur within substances. Generally, these weaker *secondary* bonds exist between molecules. The principal types of interatomic bond are the ionic bond, the covalent bond and the metallic bond. Types of intermolecular bond are the hydrogen bond and van der Waal's bonds.

2.12 *The ionic bond*

Elements in the lower groups of the periodic table would expose a completely filled electron shell if they could lose the electrons present in their incomplete outer shell. Group IA elements, the alkali metals, would need to shed only one electron, while Group IIA elements would need to lose two electrons from each atom. Conversely, atoms of the elements in Groups VIB and VIIB can complete their unfilled outer electron shell by capturing one or two electrons. An element of Group I, therefore, could react chemically with an element of Group VIIB, resulting in an electron transfer from one atom to the other. For example, sodium ($Z = 11$) would transfer its one electron from the third principal shell, so exposing a completely filled second shell, to chlorine ($Z = 17$), so giving the chlorine atom the complement of eight electrons in its outer layer (Figure 2.3).

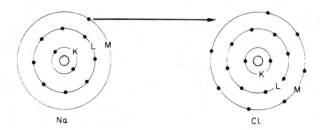

FIGURE 2.3   Electron transfer from sodium to chlorine

By losing an electron, the sodium atom has become out of balance electrically, and is said to be in an ionised state

$$Na \rightarrow Na^+ + e^-$$

Similarly, by gaining an additional electron, the chlorine atom has become ionised

$$Cl + e^- \rightarrow Cl^-$$

An ion is a charged particle and the net charge may be positive or negative, depending on whether electrons are lost or gained. The magnitude of the net charge may be one, two, three, or in a few cases, even four units, a unit being the charge of a single electron. An ion is symbolised by using the symbol for the chemical element with a superscript indicating both the number and sign of the electrical charge.

The sodium and chlorine ions, being of opposing charge, will be strongly attracted to one another, the attractive force, F, being given by

$$F = q^2/4\pi\epsilon_0 r^2$$

where q is the charge on each ion, ϵ_0 is the permittivity of a vacuum and r is the separation distance between the ions. There is also a gravitational attractive force between the nuclear masses of the two ions.

If these were the only forces acting we could expect the ions to continue moving closer together until they completely coalesced. This does not happen. As the two ions close to one another their outer electron shells begin to overlap and there is a strong force of repulsion and there will be one particular separation distance at which the forces of attraction and repulsion will be equal. At this separation distance, when the net force on the ions is zero, the potential energy of the ions will be a minimum. This is the stable separation distance between the ions as an input of energy would be necessary to move the ions either closer to one another or further apart.

The potential energy of the ions, in relation to their separation distance is given by the expression

$$P.E. = -\frac{A}{r} + \frac{B}{r^n}$$

where A is a function of ionic charge and the value of B is a function of atomic number and atomic mass number. The first term in the expression is the potential energy due to attraction and the second term is that due to repulsion. For the ionic bond, n is a power with a value of order of 9. Usually, this equation, termed the Mie question is written in the general form

$$P.E. = -\frac{A}{r^m} + \frac{B}{r^n}$$

In this general form the equation can be applied to all types of bond. The variation of potential energy with separation distance is shown diagrammatically in Figure 2.4.

In the interaction of Na^+ and Cl^- ions to form sodium chloride the transfer of only one electron is involved in the formation of one ionic molecule. In the case of a Group IIA element reacting with a Group VIB element, two electrons would be transferred to make one ionic molecule. For example, calcium and sulphur

$$Ca \rightarrow Ca^{2+} + 2e^-$$

$$S + 2e^- \rightarrow S^{2-}$$

or we may simply write

$$Ca + S \rightarrow CaS$$

It is also possible for elements of Group IIA to combine with elements of Group VIIB. In this case, the Group II element sheds two electrons, which will satisfy the requirements of two Group VIIB atoms. For example, calcium and chlorine (see Figure 2.5)

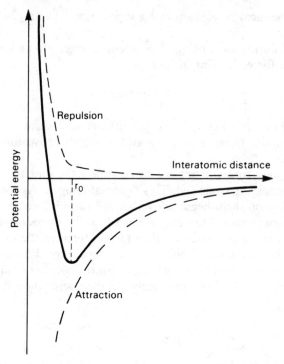

FIGURE 2.4   Variation of potential energy with interatomic distance

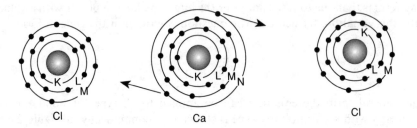

FIGURE 2.5   Electron transfer from calcium to two chlorine atoms

$$Ca \rightarrow Ca^{2+} + 2e^-$$

$$2Cl + 2e^- \rightarrow 2Cl^-$$

giving one molecule $CaCl_2$.

Similarly a Group I element could combine with a Group VIB element as

$$2Na \rightarrow 2Na^+ + 2e^-$$

$$O + 2e^- \rightarrow O^{2-}$$

giving one molecule, Na_2O.

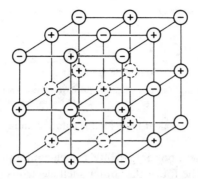

FIGURE 2.6   Symmetrical arrangement of ions in NaCl structure

The ionic bond is a strong bond and, in the solid state, individual ionic molecules do not exist. Instead, the ions pack into symmetrical crystalline arrays. In the case of a sodium chloride crystal, for example, the ions pack into a cubic array with each Na^+ ion surrounded by six Cl^- ions, and vice versa. The cubic crystalline pattern for sodium chloride is shown in Figure 2.6. The charged ions occupy fixed sites within the crystal lattice and little, if any, movement of ions is possible, even in a strong electrical field, rendering the ionically bonded crystal an electrical insulator. Most ionic crystals, however, are soluble in a polar solvent such as water and in solution the charge carrying ions have mobility. Thus, in an electrical field they will move preferentially constituting an electrical current. Ionic conductive solutions are termed *electrolytes*. Similarly, when an ionic crystalline substance melts the ions have mobility and molten salts are electrolytically conductive.

2.13 *The covalent bond*

In covalent bonding, the stable arrangement of eight electrons in an outer shell is achieved by a process of electron sharing rather than electron transfer. In the case of chlorine, for example, individual atoms combine to form diatomic molecules. The reaction can be written

$$2Cl \rightarrow Cl_2$$

The bond is achieved by the sharing of a pair of electrons. One electron from each atom enters into joint orbit around both nuclei, so giving both nuclei an effective complement of eight outer shell electrons (Figure 2.7(a)). This may be represented symbolically as

Cl:Cl or Cl–Cl

where a pair of dots, or a hyphen represents a pair of electrons shared between adjacent atoms, namely then, one covalent bond.

In an oxygen molecule, two pairs of electrons are shared between two adjacent atoms (Figure 2.7(b)) to give each atom a complement of eight outer shell electrons.

O:O or O=O

In a similar manner to that described for the ionic bond, covalently bonded atoms are subject to both attractive and repulsive forces and the energy of the bond can be represented by the Mie equation with $m > n$.

Covalent bonds tend to be highly directional. The shape of the water molecule, H_2O, is shown

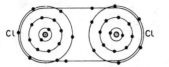

(a) Chlorine molecule Cl_2. One pair
of electrons shared

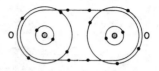

(b) Oxygen molecule O_2. Two pairs
of electrons shared

FIGURE 2.7   Covalent bonds in (a) chlorine and (b) oxygen

in Figure 2.8. There are two covalent bonding orbitals, each containing two electrons, linking the hydrogen nuclei to the oxygen. The four other outer shell electrons of oxygen occupy two non-bonding orbitals, two electrons in each, and these are much smaller than the two bonding orbitals. Within the two covalent bonds the electron sharing is not equal and the shared electrons spend a greater portion of the time in the region of the oxygen atom. This means that the molecule is polarised with the two hydrogen ends being slightly positive relative to the two non-bonding orbitals. Each H_2O molecule is a small dipole and secondary bonds, *hydrogen bonds*, based on electrostatic attraction, occur between these polar molecules (see Section 2.16).

Carbon, with four outer shell electrons, is capable of forming four covalent bonds, and the shared orbitals point towards the corners of a regular tetrahedron, as shown in Figure 2.9. The diamond form of carbon shows this structure with each carbon atom being covalently bonded to four other carbon atoms (Figure 2.10).

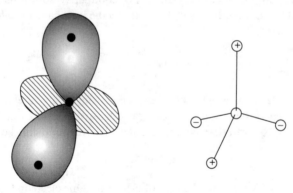

FIGURE 2.8   Representation of a water molecule

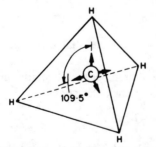

(a) Tetrahedral form of methane, CH_4

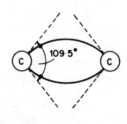

(b) Double covalent bond between carbon
atoms. Bonds under strain

FIGURE 2.9   Symmetrical arrangement of the four covalent bonds in carbon. The angle between bonds is
109.5°

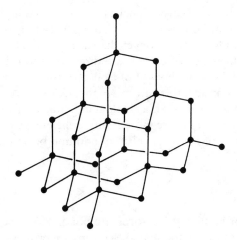

FIGURE 2.10   Structure of diamond

The covalent bond is the form of bonding in organic molecules. ('Organic chemistry' is the chemistry of carbon and its compounds. Organic molecules are composed principally of carbon and hydrogen.)

The most simple compound of this type is methane, CH_4, in which the carbon atom is bonded covalently to four hydrogen atoms thus

$$
\begin{array}{ccc}
 & & H \\
 & & | \\
H:\ddot{C}:H & or & H-C-H \\
 & & | \\
H & & H
\end{array}
$$

The molecule is tetrahedral in shape with the carbon nucleus at the centre of a regular tetrahedron (Figure 2.9). The angle between adjacent bonds is 109.5°.

Not all bonds in organic molecules need be of the single covalent type. Multiple bonding is possible also. Consider the three compounds ethane, C_2H_6, ethene (or ethylene), C_2H_4 and ethine (or acetylene), C_2H_2.

Ethane has single covalent bonds only

$$
\begin{array}{cc}
H & H \\
| & | \\
H-C-C-H \\
| & | \\
H & H
\end{array}
$$

Ethene has a double covalent bond linking the two carbon atoms

$$
\begin{array}{c}
H \diagdown \quad \diagup H \\
\quad C=C \\
H \diagup \quad \diagdown H
\end{array}
$$

and ethine has a triple covalent bond between the two carbon atoms

$$
H-C\equiv C-H
$$

The separation distance between carbon nuclei is smaller in multiple covalent bonds than in a single covalent bond. The separation distance in ethane is 0.154 nm, but is 0.134 nm in ethene and 0.120 in ethine. Multiple bonds are not stronger than a single covalent link. In fact the reverse is true. Normally, the covalent bonds of carbon are directed outwards towards the corners of a tetrahedron. A multiple bond can be thought of as two or three small spring linkages which are curved and, hence, under considerable strain (Figure 2.9(b)). A multiple bond in organic molecules can be broken more easily than a single bond. Stated in another way this means that compounds such as ethene and ethine are more reactive chemically than ethane. The splitting of double covalent bonds is part of the mechanism of addition polymerisation (see Chapter 5).

2.14 The co-ordinate bond

A variation on the covalent bond is the co-ordinate bond. This type of bond also involves the sharing of electrons, but in this case, the shared electrons are provided by one atom. An example is the sulphate group SO_4^{2-} (Figure 2.11). In this group, the sulphur atom provides all the electrons required for covalent linkages with three oxygen atoms. The two additional electrons required for the covalent link with the fourth oxygen atom are obtained from an ionic donor, for example, calcium, so making the SO_4 group an ion with a double negative charge.

$$Ca^{2+}[SO_4]^{2-}$$

In the above example, calcium sulphate, it will be seen that two types of inter-atomic bond exist within the one molecule. There are very many other substances with interatomic bonding that is in part covalent, and in part ionic.

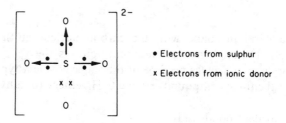

FIGURE 2.11   Sulphate group SO_4^{2-}

2.15 The metallic bond

In the bond types discussed so far, the electron distribution is rearranged to provide each nucleus with an external shell containing eight electrons. In the case of metals, where the number of valency electrons to each atom is small, it is not possible to fully satisfy the requirement of eight outer shell electrons per atom, and there is only a partial satisfying of this condition. In an assembly of metal atoms, the principal forces acting will be an attraction due to gravity and a repulsion due to negatively charged electron shells in close proximity. For two atoms, the variation in energy with separation distance will be similar to that for ions (Figure 2.4) discussed earlier. (The separation distance, r_0, for minimum energy can be regarded as an atomic diameter, and for the discussion of many properties of metals, it is feasible to consider metal atoms as hard spheres of finite radius.) At the equilibrium separation distance, the outer, or valence, electron shells of the atoms can be regarded as in contact or slightly overlapping. The outer shell electrons, at certain points in their orbits, are attracted as much by one nucleus as by another, and the valence electrons follow complex paths around many nuclei. Thus, all valence electrons are shared by all the atoms in the assembly. This has similarities to a covalent bond, but the strength of the bond is weaker than that of a true covalent bond. One can liken the metallic state to an arrangement of positive ions permeated by an electron cloud or gas (Figure 2.12.). The extreme mobility of the valence electrons accounts for the high electrical conductivity, and other properties, of metals.

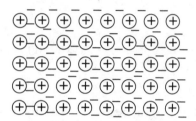

FIGURE 2.12   The metallic state, positive ions in an electron gas

2.16 Secondary bonds

In addition to the types of primary bond discussed in the preceding sections, there also exist weaker secondary bonds. These are the hydrogen bond and van der Waal's bonds. The bond energies of the secondary bonds, again, are given by the Mie equation.

$$\text{P.E.} = -\frac{A}{r^m} + \frac{B}{r^n}$$

but for these bonds the values of the indices are $m \simeq 6$ and $n \simeq 12$ and typical bond energies are of the order of 0.2 eV for a hydrogen bond and between 0.002 and 0.1 eV for van der Waal's bonds, as compared with 6.5 eV for the ionic bond in sodium chloride.

The bonding in the H_2O molecule was described in Section 2.13 and it was shown that the molecule is polar with the two hydrogen atoms being positive relative to the two non-bonding orbitals of the oxygen atom. There is quite a strong force of attraction between the hydrogens and the negative ends of adjacent molecules (see Figure 2.13). The hydrogen bond can be written thus: H–O–H...O

The hydrogen bond does not occur only in water and ice but in a number of polymer materials. In general terms, a hydrogen bond can be formed between an electronegative atom and a hydrogen atom which is already covalently bonded to another electronegative atom. Examples of hydrogen bonding in polymers are –N–H...O bonds between polyamide (nylon) molecules and –O–H...O bonds in cellulose and polyvinyl alcohols.

Many molecular compounds are polarised to some extent and electrostatic attractive forces exist between the molecular dipoles. These weak electrostatic attractive forces are termed van der Waal's bonds. Van der Waal's bonds can occur also between atoms. The monatomic inert gases, with full outer electron shells will condense into liquids and solids at extremely low temperatures. This indicates the existence of weak bonding forces. Within such atoms, owing to the continual

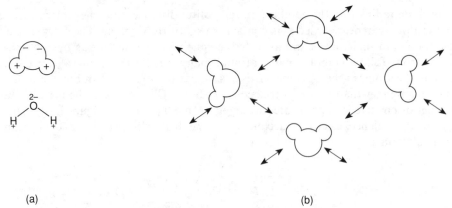

(a)   (b)

FIGURE 2.13   (a) Representation of a polar H_2O molecule; (b) attraction between neighbouring H_2O molecules—the hydrogen bond

movement of electrons, at any instant the centroid of negative charge need not coincide with the centre of positive charge, that is, the nucleus (Figure 2.14). The atom becomes slightly polarised and may be weakly attracted to a similar polarised atom.

At very low temperatures, the kinetic energy of the atoms will be insufficient to overcome even these very weak attractions and the gas will condense. The energies of the van der Waal's bonds in the inert gases is of the order of 0.002 eV but in some molecular substances with strong dipoles the magnitude of the bond energy may approach that of the hydrogen bond.

The influences of the type of bonding on the properties of materials is considered in the next chapter.

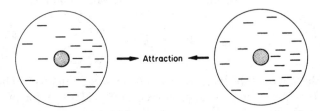

FIGURE 2.14   Momentary uneven electron distribution in atoms giving weak dipoles and weak interatomic attraction

2.17 Mixed bonds

The main types of primary and secondary bonds have been described in the preceding sections but, in very many cases, the bonding in substances cannot be described as purely ionic or purely covalent. There are a large number of compounds in which a mixture of bond types occurs. In many ceramics and glasses, the bonding is part ionic and part covalent. Dipoles occur in a number of ceramic materials and permit the creation of van der Waal's bonds, examples being clay minerals and graphite. The interatomic bonding within polymers is predominantly covalent but van der Waal's intermolecular bonds occur in most of these and, as has been mentioned, hydrogen bonds also occur in some polymers. Intermetallic compounds are formed in some metallic alloy systems and this is an indication that some alloys contain bonds which are not entirely metallic.

2.18 Questions

2.1 Copper contains two isotopes, with mass numbers of 63 and 65. The atomic mass number of copper is 63.54. Estimate the relative proportions of the two isotopes.

2.2 Silver, $Z = 47$, contains 56.5 and 43.5 per cent, respectively, of the isotopes with mass numbers of 107 and 109. Estimate the atomic mass number of silver.

2.3 Determine the number of molecules in 1 kg of $MgCl_2$. Atomic mass numbers are: Mg—24.32, Cl—35.46.

2.4 The nucleus of an atom contains 13 protons and 14 neutrons. State

    (a) the values of Z and M for the element,
    (b) the electronic structure for atoms of this element, and
    (c) the valence value.

2.5 Iron, $Z = 26$, may show a valency of 2 or 3. State the electron configurations for the two valence states of iron.

2.6 Complete the following disintegration table.

Nucleus	Symbol	Emission
Uranium	$^{235}_{92}U$	α
Thorium	?	?
Protoactinium	?	α
Actinium	?	?
Thorium	$^{227}_{90}Th$	α
Radium	?	?
Radon	$^{219}_{86}Rn$	α
Polonium	?	β
Astatine	$^{215}_{85}At$	?
Bismuth	?	?
Polonium	$^{211}_{84}Po$	?
Lead	$^{207}_{82}Pb$	stable

2.7 Write down the values of M and Z for

    (a) the element X formed from thorium-232 after six α and four β particle emissions, and
    (b) the element Y formed from uranium-238 after six α and two β particle emissions.
    ($Z = 90$ for thorium, $Z = 92$ for uranium.)

2.8 The radioisotope iridium-192, which has a half-life period of 74.4 days is used as a source material for industrial radiography. Determine the source intensity, as a percentage of the initial intensity, after (a) 25 days, (b) 50 days and (c) 100 days.

2.9 The remains of wooden artifacts were excavated from an archaeological site and were found to be emitting β particles at a rate of 210 per second per kg of wood. A living tree absorbs the radioactive isotope carbon-14 and is mildly radioactive, emitting 255 β particles per second per kg of wood. Estimate the age of the remains, assuming that the half-life period of carbon-14 is 5000 years.

2.10 The elements A, B, X, Y and Z have atoms with outer electron shell configurations containing 1, 2, 4, 7 and 8 electrons respectively. State and describe the type of bonding which is likely to occur in the following cases:

    (a) between a large number of atoms of A,
    (b) between a large number of atoms of X,
    (c) between atoms of Y,
    (d) between atoms of Z,

(e) between a mixture of A and B atoms,

(f) between equal numbers of A and Y atoms,

(g) between one atom of X and four atoms of Y.

2.11   Two elements X and Y which have atomic numbers of 12 and 17 respectively combine to form a compound. What type of bonding would exist in such a compound and what would be the chemical formula of the compound formed?

3

Influence of Bond Type on Structure and Properties

3.1 Introduction

The various types of interatomic and intermolecular bonds were described in Chapter 2 and it was stated that the bond energies, that is the strengths of bonds, varied from one type to another. It was noted also that covalent bonds were directional whereas the ionic and metallic bonds were not directional in nature. These characteristics exert influences on both the structures and properties of materials. An attempt is made in this chapter to indicate the general ways in which the bond type helps to determine the properties of a material. It should be remembered though that bond type is only one of many disparate factors which influence the final properties of any material.

3.2 Structure and density

Neither the metallic nor the ionic bond is directional and this allows metal atoms to pack closely together in a regular crystalline array. The majority of metals form closely packed arrangements with *coordination numbers* of either 12 (close packed hexagonal or face centred cubic systems) or 8 (body centred cubic systems). The coordination number is the number of other atoms or ions with which a particular atom or ion is in direct contact. The coordination number for ionic crystals is frequently 4, 6 or 8 and so, on average, ionic crystals possess lower packing densities than metals. The reasons for this are: (i) an ionic compound contains at least two types of ion which may be of different size, compared with a pure metal where all atoms are identical, and (ii) the ions possess opposing charges.

Consider the two compounds caesium chloride, CsCl, and sodium chloride, NaCl. The ions of caesium and chlorine are comparable in size and eight chlorine ions can pack around a caesium ion whereas a sodium ion is small in relation to a chlorine ion and only six chlorine ions can pack around a sodium ion (Figure 3.1).

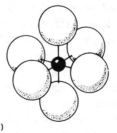

(a)   (b)

FIGURE 3.1   Caesium and sodium chloride structures: (a) caesium chloride; (b) sodium chloride

Why cannot we achieve a coordination number of 12 in ionic crystals? This is best illustrated by looking at a two-dimensional representation of the packing (Figure 3.2). In both caesium chloride and sodium chloride, the negatively charged chlorine ions, while in contact with a positive ion, are not in direct contact with one another, but negative ions would need to be in contact with one another to obtain a coordination number of 12. Such a structure would not be stable as the ions of like sign would repel one another strongly.

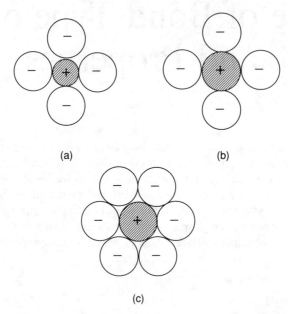

FIGURE 3.2 Two dimensional representations of packing arrangements with coordination numbers of (a) 6, (b) 8, and (c) 12. In (c), negative ions would be in contact and, hence, it is an unstable arrangement

Covalent bonds are directional and this will affect the structures of covalent crystals. It was shown (Section 2.13) that the coordination number for diamond is four with each carbon atom bonded to four other atoms of carbon. A similar tetrahedral arrangement occurs in silicon and in silica, SiO_2, where each silicon atom is bonded to four oxygen atoms with each oxygen being bonded to two silicon atoms.

Metals and ceramics are fully crystalline but polymers and glasses are either partially crystalline or completely amorphous. In amorphous structures, the atoms or molecules are not packed in a regular and symmetrical pattern and the amount of void space is greater than within a crystal. Polymer molecules are very large and, in many instances, of complex shape so that void spaces of considerable size can occur within the structure. (See Chapters 4 and 6 for details of polymer structures.)

The densities of materials are related to structure. The range of densities of metals and ceramics is large, being from about 2000 to 16 000 kg/m³ in each case, but because of the higher coordination numbers in metal crystals the average density of metals (8000 kg/m³) is considerably higher than the average density of ceramics (5000 kg/m³). Inorganic glasses have lower densities than crystalline forms of the same material and the densities of polymers are much lower than those of metals, ceramics and inorganic glasses, ranging from about 900 to 3000 kg/m³.

3.3 Stability and melting point

The stability of a substance is related to the bond strength. One indicator of bond strength and stability is the melting point of a substance. When a substance is heated the absorption of heat energy causes an increase in the vibrational amplitudes of the atoms. A point will be reached when the vibrational energy is sufficient to overcome the interatomic bonding energy and the atoms become mobile, in other words the substance melts.

Both ionic bonds and covalent bonds are strong bonds possessing a similar range of bond energies. However, within both ionically and covalently bonded substances there are major variations in melting point. For example, consider the two substances sodium chloride, NaCl, and magnesium oxide, MgO. The bonding in MgO is much stronger than in NaCl as force of attraction between the ions Mg^{2+} and O^{2-} is much greater than that between Na^+ and Cl^- ions, by virtue of the increased ionic charge. The melting point of MgO is about 2800°C as compared with about 800°C for NaCl.

In covalently bound crystals, the strength of the bonds lessens as the atomic number of the element increases and the distance of the outer shell, or valence, electrons from the nucleus becomes greater. This is illustrated by comparing the melting points of the Group IV elements carbon ($Z = 6$), silicon ($Z = 14$), germanium ($Z = 32$) and tin ($Z = 50$). The melting points are: carbon (diamond) 3800°C, silicon 1420°C, germanium 937°C and tin 232°C. The melting points of the major ceramic materials, generally oxides, carbides and nitrides of the lower atomic number elements such as aluminium, boron, magnesium, silicon, are all high being in the general range 1000–3000°C.

It has been said that bonding in polymer molecules is of the covalent type but, generally, polymers have low melting points in the range 100–400°C. Although the bonding within molecules is covalent, only weak van der Waal's bonds exist between molecules and it is these secondary bonds that influence melting temperatures. The polyamides (nylons) have higher melting temperatures than other polymers such as polyethylene because of the presence of hydrogen bonds in the former (see Section 2.16).

The melting temperatures of molecular compounds are also influenced by molecular weight and increase as the molecular weight of the substance increases.

In metals, although the strength of the metallic bond tends to be less than that of ionic and covalent bonds, the close packing of atoms leads towards relatively high melting points. Some general tendencies are observable. One is that melting points tend to increase as the number of valence electrons increase, for example, sodium (Group I of the periodic table) melting point, $T_m = 98$°C, magnesium (Group II) $T_m = 649$°C, aluminium (Group III) $T_m = 660$°C. Another trend is for the melting point to decrease as the atomic number increases for metals within the same group in the periodic table, for example, lithium ($Z = 3$) $T_m = 180$°C, sodium ($Z = 11$) $T_m = 98$°C, potassium ($Z = 19$) $T_m = 64$°C, and caesium ($Z = 55$) $T_m = 30$°C. The metals of highest melting temperature are those of the three transition series and here the bonding may not be purely metallic but have some aspects of the covalent type.

3.4 Stiffness

The modulus of elasticity, E, or Young's modulus, is a measure of the stiffness of a material, and is defined as the ratio of the stress applied to the elastic strain produced. Crystalline structures, metals and ceramics, have high values of E, generally within the range 40–450 GPa while polymer materials have low stiffnesses, usually in the range 0.7–3.5 GPa, but one group, the elastomers, have elastic moduli which are extremely small in the range 0.01–0.1 GPa.

The bonds between atoms can be regarded as having the properties of tiny springs and when an external force is applied these springs will either extend, compress or twist, according to the type of force system. The effect of a tensile force on atoms in a crystal is illustrated in Figure 3.3.

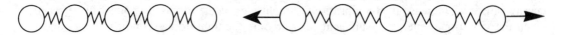

FIGURE 3.3   Effect of tensile force is to extend the bond between atoms giving a small total extension

In crystals, the strain produced per unit of stress is small and so the value of Young's modulus is high.

In polymer materials, the basic structure of many polymer molecules is a linear chain of carbon atoms. The angle between adjacent covalent bonds in carbon is 109.5°. When a tensile force is applied the effect of the force is not only to extend the interatomic bonds but also to tend to straighten them. This alone would cause a greater amount of extension per unit of stress than in crystals but, in addition, the bonding between molecules consists of weak secondary bonds, or to use the spring analogy, bonding springs of lower stiffness than the primary bond springs. These effects give the materials low values of Young's modulus. This is illustrated in Figure 3.4.

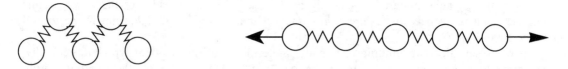

FIGURE 3.4   Effect of a tensile force on a polymer. Covalent bonds are both extended and straightened

The group of materials termed *elastomers*, of which natural rubber is one, consists of polymers at a temperature above their *glass transition temperature*. The significance of the glass transition temperature is discussed in Chapter 6. It is sufficient to say at this stage that this temperature is sufficiently high to break the van der Waal's bonds. Many elastomers possess light cross-linking, that is a few covalent links between adjacent polymer molecules, and also, in many cases, the molecules twist into spirals and behave in the manner of miniature helical springs. The amount of extension per unit stress is extremely high for elastomers giving them extremely low values of Young's modulus. The stiffness of elastomers can be increased by increasing the number of covalent cross-linkages, and if carried to the ultimate this will produce a strong rigid material. Natural rubber in which the maximum number of cross-links is created converts into the rigid material Ebonite. Elastomers lose their very low elastic moduli if cooled to a temperature below their glass transition allowing the van der Waal's bonds to reform.

3.5 Electrical properties

In both covalent and ionic bonding, the valency electrons are localised in orbitals around one or a pair of atoms but in the metallic bond the valency electrons are not localised and a metal crystal can be likened to an array of positive ions permeated by a cloud of electrons, the '*electron gas*'. In an electric field, the electron gas will move preferentially in one direction, thus constituting an electric current. Metals are electrically conductive. Ionic and covalent solids are not conductive and are insulators or dielectrics because their valency electrons are localised. However, the ions in ionic crystals become mobile when the substance is dissolved in a solvent or is melted and these mobile ions can move preferentially when influenced by an electric field. This ionic conduction is termed *electrolytic conduction* and the solution or molten salt is termed an *electrolyte*. It may be possible for ions of small size to migrate, or diffuse, through a solid in a strong electric field giving the solid a small, though measurable, conductivity.

Some crystalline dielectric materials show a special effect, the *piezo-electric effect*. These materials polarise when strained mechanically creating an electrical voltage or field. This phenomenon is reversible and such a material will strain mechanically when subjected to an electrical field. The piezo-electric effect is used in *transducers* which convert electrical fields into mechanical vibrations and vice versa. This is dealt with in more detail in Chapter 13.

Although most materials can be regarded as either conductors or insulators there is a small but very important group of materials possessing semiconductivity, notably silicon and germanium. The mechanism of semiconduction is covered in Chapter 13.

4

The Formation of Polymers

4.1 Introduction

Living organisms have the capacity for producing large and complex molecules from simple ingredients. These large molecules are principally composed of carbon and hydrogen atoms. The long fibrous molecules produced by their respective organisms help to determine the properties possessed by naturally occurring materials such as timbers, wool, cotton and natural rubber. The whole range of plastics materials is also based on very large molecular structures, but in this case the large molecules are often produced synthetically. These substances containing large molecules are termed *polymers*, the name derived from *poly*, meaning many, and *mer*, unit. The chemical compounds which are reacted together to create polymer molecules are referred to as *monomers*. These monomers, the building blocks of plastics materials, can be reacted together, or polymerised, in several ways and these are described in subsequent sections of this chapter. The polymers produced may possess *linear* or *non-linear* structures.

The plastics materials produced by polymerisation can be classified under three headings: as *thermoplastics*, *thermosets* and *elastomers*. A thermoplastic material is one which becomes more plastic when heated and becomes rigid again on cooling. A thermoset is a material which, on first heating, becomes plastic and can be moulded and then 'cures' at temperature, setting into a hard rigid structure. Once curing and setting has occurred the material cannot again be made plastic. Some materials are *cold-setting*. This type usually involves mixing a resin with a 'hardener' which causes curing to occur at ordinary temperature to produce a hard, rigid material. Elastomers are a group of polymer materials of very low elastic moduli giving them great flexibility and the ability to suffer major elastic deformations.

4.2 Addition polymerisation

Many covalently bonded compounds of carbon and hydrogen exist in nature. The simplest group of hydrocarbon compounds is the paraffin series, many members of which are found in petroleum. The first few compounds in the paraffin series are

$$\text{methane, } CH_4 \quad \text{or} \quad \begin{array}{c} H \\ | \\ H-C-H \\ | \\ H \end{array}$$

ethane, C_2H_6 or

$$H-\overset{\overset{\displaystyle H}{|}}{\underset{\underset{\displaystyle H}{|}}{C}}-\overset{\overset{\displaystyle H}{|}}{\underset{\underset{\displaystyle H}{|}}{C}}-H$$

propane, C_3H_8 or

$$H-\overset{\overset{\displaystyle H}{|}}{\underset{\underset{\displaystyle H}{|}}{C}}-\overset{\overset{\displaystyle H}{|}}{\underset{\underset{\displaystyle H}{|}}{C}}-\overset{\overset{\displaystyle H}{|}}{\underset{\underset{\displaystyle H}{|}}{C}}-H$$

The paraffins are known as saturated hydrocarbons as every linkage is a single covalent bond, and the carbon chain is fully saturated with hydrogen. Other hydrocarbons may be unsaturated and contain double or triple covalent bonds. An example is ethylene, or ethene,

$$C_2H_4 \quad \text{or} \quad \overset{H}{\underset{H}{>}}C=C\overset{H}{\underset{H}{<}}$$

An unsaturated compound may be made to react with itself, or with other compounds, to produce large complex molecules. This is termed *polymerisation*. In the case of ethylene, the double bonds can split allowing the separate molecules to link up, forming polyethylene (polythene)

$$\overset{H\ \ H}{\underset{H\ \ H}{C=C}} + \overset{H\ \ H}{\underset{H\ \ H}{C=C}} + \overset{H\ \ H}{\underset{H\ \ H}{C=C}} \rightarrow -\overset{H\ H\ H\ H\ H\ H}{\underset{H\ H\ H\ H\ H\ H}{C-C-C-C-C-C}}-$$

$$\text{or } n\left(\overset{H\ \ H}{\underset{H\ \ H}{C=C}}\right) \rightarrow \left(-\overset{H\ \ H}{\underset{H\ \ H}{C-C}}-\right)_n \quad \text{where } n \text{ is a very large number.}$$

This is the process termed *addition polymerisation*. According to the reaction equation shown above there is a free valence at each end of the molecular chain. This is not allowable and there must be a small quantity of some other reactant present to provide a logical ending to the chain, for example chlorine, giving

$$----\overset{H\ H\ H\ H}{\underset{H\ H\ H\ H}{C-C-C-C}}-Cl$$

The reactant used to provide a chain ending is termed the *initiator*. Certain peroxides are frequently used as initiators. The amount of initiator added will affect the average molecular weight of the resultant polymer and this, in turn, will influence the properties of the polymer. In polymers of the addition type, the number of atoms in each molecular chain will be extremely large running into thousands.

There are many unsaturated compounds that can be polymerised by an addition reaction and these include:

$$\text{vinyl chloride} \qquad \begin{matrix} H & H \\ | & | \\ C & = & C \\ | & | \\ H & Cl \end{matrix} \qquad \text{giving polyvinyl chloride (PVC)}$$

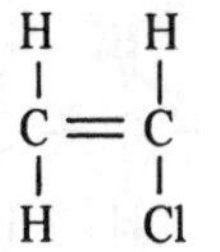

$$\text{tetrafluoroethylene} \qquad \begin{matrix} F & F \\ | & | \\ C & = & C \\ | & | \\ F & F \end{matrix} \qquad \text{giving PTFE}$$

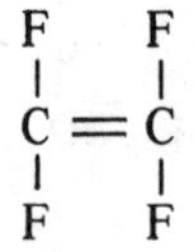

$$\text{styrene} \qquad \begin{matrix} H & H \\ | & | \\ C & = & C \\ | & | \\ H & C_6H_5 \end{matrix} \qquad \text{giving polystyrene (PS)}$$

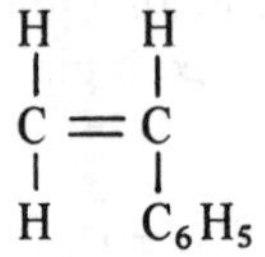

$$\text{propylene} \qquad \begin{matrix} H & H \\ | & | \\ C & = & C \\ | & | \\ H & CH_3 \end{matrix} \qquad \text{giving polypropylene (PP)}$$

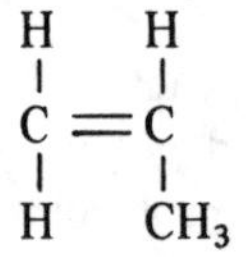

It will be seen in the above examples that unsaturated monomers produce polymers that are fully saturated. In some addition polymerisation reactions, however, the resultant polymer is unsaturated. An example of this is the polymerisation of butadiene. Butadiene has two double bonds per molecule and the polymerisation reaction can be written

$$n(CH_2=CH-CH=CH_2) \rightarrow (-CH_2-CH=CH-CH_2-)_n$$

The product, polybutadiene, is a rubber-like material and, being unsaturated, is capable of being further reacted. The type of reaction that can be brought about in unsaturated polymers will create cross-links between molecules. A good example of this is the vulcanisation of rubbers. Refer to Section 4.6.

The addition polymerisation of a single monomer produces a *homopolymer* but it is also possible to polymerise a mixture containing two or more monomers. *Copolymerisation* is the term used for the addition polymerisation of two or more monomers. Two monomers, represented by A and B, when polymerised together, may produce a copolymer with either a random or ordered linear structure, as shown:

–A–A–B–A–B–B–B–A–A–B–A–B–B–A–A–A–B–A–B–B– (random)
–A–A–B–B–B–A–A–B–B–B–A–A–B–B–B–A–A–B–B–B– (ordered)

It is also possible to produce a *block copolymer* in which the structure is composed of large blocks of A alternating with large blocks of B, as:

$-(-A-)_m-(-B-)_n-(-A-)_m-(-B-)_n-$, where m and n are large integers.

Examples of copolymers are SAN (styrene/acrylonitrile) and ABS (a terpolymer of acrylonitrile, butadiene and styrene).

4.3 Condensation polymerisation

Another type of polymerisation reaction is condensation polymerisation. A condensation reaction is one in which two molecules combine to form a larger, more complex molecule, together with a small molecular substance as a by-product. Frequently, the by-product in condensation reactions is water, H_2O. The reaction between an alcohol and an organic acid to form an ester is an example of a condensation reaction. Ethylene glycol, a di-ol, and teraphthalic acid, a di-acid, can be reacted together to form a linear polyester, as shown in the equation below:

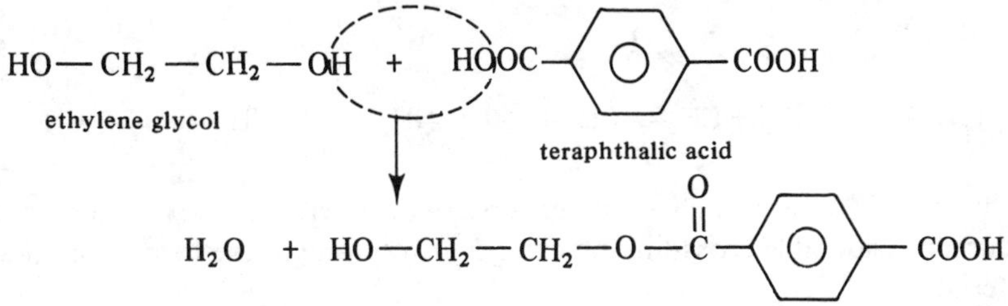

The product of the above reaction is used as a synthetic fibre, one trade name for which is *terylene*. Linear polyamides, also known as *nylons*, are produced by the condensation of a di-acid and a di-amine.

If one of the reactants in condensation polymerisation has more than two functional groups, or react at more than two points, then a non-linear polymer structure can be created. A good example of this is the condensation between phenol, C_6H_5OH, and formaldehyde, HCHO, creating –CH_2– linkages between adjacent phenol molecules, as shown below. The carbon atoms in the phenol structure shown are numbered and condensation reactions can occur involving the hydrogen atoms at the 2, 4 and 6 positions. Only one –CH_2– linkage is shown in the equation but they can form from all three reactive positions in each phenol to create a three-dimensional network molecular structure. This type of structure is shown schematically in Figure 4.1.

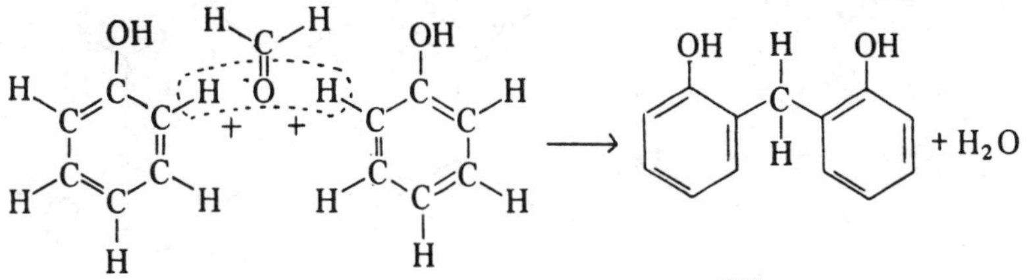

Within a cyclic carbon compound the sites may be numbered thus:

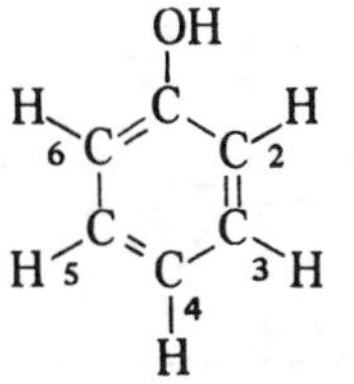

Other condensation reactions of importance in the formation of non-linear network polymers are those between formaldehyde and amines (giving urea-formaldehyde and melamine-formaldehyde materials), and between acids and alcohols to give non-linear polyesters.

Epoxides are another valuable group of polymer materials with non-linear network structures but the epoxy reaction is not a condensation reaction as there is no reaction by-product evolved. Uncured epoxy resin consists of polymer molecules with a reactive epoxy group at both ends of each molecular chain.

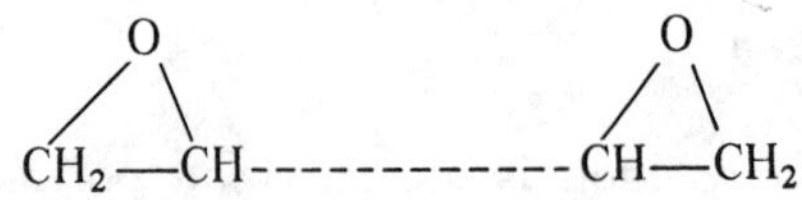

$$
\underset{CH_2-CH}{\overset{O}{\triangle}}\text{-----------}\underset{CH-CH_2}{\overset{O}{\triangle}}
$$

It is necessary to mix a hardener compound with an epoxy resin to convert, or cure, it into a fully rigid network structure. Several types of hardener may be used, but the more commonly used types contain at least two amino, $-NH_2$, groups. The amine hardener will form cross-links between the epoxy molecules. This cross-linking reaction is shown below.

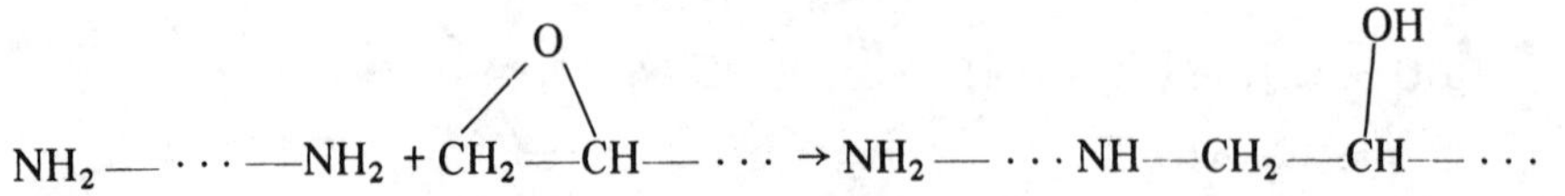

$$
NH_2 -\cdots-NH_2 + \underset{CH_2-CH}{\overset{O}{\triangle}}-\cdots \rightarrow NH_2-\cdots NH-CH_2-\underset{}{\overset{OH}{CH}}-\cdots
$$

Both hydrogen atoms in each amino group can react with an epoxy group in this way and, in consequence, considerable cross-linking can occur resulting in the formation of a very hard and rigid material.

4.4 Linear and non-linear polymers

In general, the polymers produced by addition polymerisation are linear polymers with long fibrous molecules. As has been stated, some polymers produced by condensation polymerisation are also linear in structure. Several types of non-linear structures are also possible in polymers. These are branched, cross-linked and full network (see Figure 4.1).

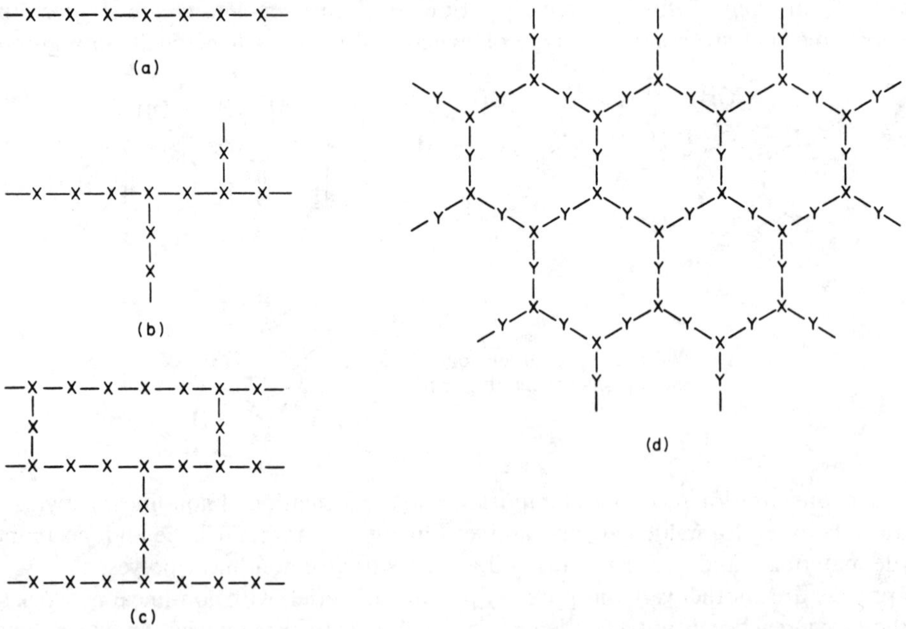

FIGURE 4.1   (a) linear polymer, (b) branched polymer, (c) cross-linked polymer, (d) network polymer

Linear, branched and lightly cross-linked polymers are thermoplastic, while heavily cross-linked polymers and those with a full network structure are thermosets. Most elastomers are linear polymers with light cross-linking and it is the presence of these cross-links which causes the elastomer to return to its original dimensions after the removal of a deforming stress.

4.5 Branching

Branching is the linking together of some polymer molecules to create side branches to the long chain molecules. Branching reactions require a certain input of energy (activation energy) before they can occur and the necessary energy can be provided by high-energy photons. Irradiation of a polymer such as polyethylene with X- or γ-radiation can cause branching. The formation of branch chains stiffens the polymer and increases its melting point.

Branching reactions can be used to good effect in the manufacture of some products. For example, polyethylene, with low softening and melting points, can be processed and moulded with ease but the moulded product is flexible and can soften and lose its shape in boiling water or steam. Branching reactions can be initiated in the already moulded products to give stronger and more rigid articles and which possess higher softening points than untreated polyethylene. Branched polyethylene products can withstand steam sterilisation without losing their shape and, thus, this relatively low cost material can be used for the manufacture of items of medical equipment.

Some copolymers can be produced by branching. These, known as *graft copolymers*, are made by initiating branching reactions in a mix of two linear polymers. Chains of polymer A can be grafted on at points along the linear chains of polymer B.

4.6 Cross-linking

A good example of a cross-linking reaction is the vulcanisation of natural rubber using sulphur. Natural rubber, cis-polyisoprene, is an unsaturated polymer with the formula $(-CH_2-C(CH_3)=CH-CH_2-)_n$. During vulcanising flexible cross-links are formed between adjacent polyisoprene chains with a consequent increase in the hardness and stiffness of the material. Each link may be either a single atom of sulphur or a short chain of sulphur atoms.

Generally, between 1 and 5 parts sulphur per 100 parts of rubber are used for vulcanisation, the number of cross-links formed and, hence, the hardness of the vulcanised rubber increasing as the amount of sulphur is increased. If the sulphur content is increased to about 40 parts per 100 parts rubber, complete vulcanisation, with all possible reaction sites used, will take place creating a three-dimensional network structure. The product of this full vulcanisation of natural rubber is the hard, rigid, elastic solid, *ebonite*.

4.7 Stereoregularity

A linear homopolymer which has side groups attached to the linear chain, for example, $-CH_3$ groups in polypropylene or $-C_6H_5$ groups in polystyrene, can show structural variations. These variations are termed *stereoregularity* or *tacticity*. When all the attached groups are on the same side of the carbon chain the arrangement is termed *isotactic*. A *syndiotactic* configuration is one in which the attached groups alternate in a regular manner along both sides of the linear chain. The third type of tacticity is the *atactic* structure in which there is a random arrangement of the attached groups. These three configurations are shown schematically in Figure 4.2 for the general case of a vinyl polymer with the repeating unit $-CH_2-CH(R)-$.

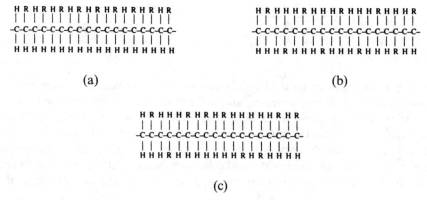

(a)   (b)

(c)

FIGURE 4.2   (a) Isotactic structure. (b) Syndiotactic structure. (c) Atactic structure

The tacticity of polymers has implications for the microstructures and properties of polymers as it affects the way in which adjacent molecules, or parts of molecules, may fit together. A regular isotactic homopolymer can have a strong tendency to crystallise whereas atactic polymers possess insufficient regularity to form crystals and, instead, give glassy polymers (see Section 6.6). Some syndiotactic polymers possess sufficient regularity to give some crystallisation. Commercial polypropylene is isotactic while commercial polystyrene is atactic.

4.8 Degree of polymerisation

The molecular chain lengths within many linear polymers is very large, each molecule containing thousands of carbon atoms. Within any polymer the chain lengths of the constituent molecules are not equal and, frequently, a figure is quoted for the average molecular size. The average number of repeating units, or mers, in molecules of a linear polymer is referred to as the *degree of polymerisation*.

It was stated in Section 4.2 that the process of addition polymerisation requires an initiator. The function of an initiator is to provide terminal groups for the polymer molecules. Therefore, the amount of initiator used will determine the degree of polymerisation and, hence, affect some of the properties of the resulting material.

Example.

Calculate the degree of polymerisation and the average molecular weight of a sample of polyethylene made using hydrogen peroxide, H_2O_2, as initiator if the amount of hydrogen peroxide used was 500 g per 1000 kg ethylene. Assume that all of the peroxide is used to create terminal groups. (Take atomic mass numbers as: $H = 1$, $C = 12$, $O = 16$)

One molecule of hydrogen peroxide, H_2O_2, will provide two hydroxyl, –OH, terminal groups, so that each peroxide molecule will generate one polymer molecule.

Molecular weight $H_2O_2 = (1 \times 2) + (16 \times 2) = 34$
500 g $H_2O_2 = 500 \div 34 = 14.71$ moles
Molecular weight ethylene, $C_2H_4 = (12 \times 2) + (1 \times 4) = 28$
1000 kg $C_2H_4 = (1000 \times 10^3) \div 28 = 35.71 \times 10^3$ moles
Degree of polymerisation $= 35.71 \times 10^3 \div 14.71 = 2428$
The average molecular weight of the polyethylene $= 2428 \times 28 = 67\,984$

For any particular polymer an increase in the degree of polymerisation will cause an increase in its melting temperature and also increase strength and hardness. The variation of melting temperature, T_m (in K), with the degree of polymerisation, n, is given by an empirical relationship of the form

$$T_m = (A + B/n)^{-1} \text{ where } A \text{ and } B \text{ are constants.}$$

4.9 Polymerisation methods

There are four basic methods used for the production of polymers from one or more monomers. These are:

(a) mass, or bulk, polymerisation,
(b) solution polymerisation,
(c) emulsion polymerisation,
(d) suspension polymerisation.

The choice of polymerisation method used for any particular material will depend on either the nature of the reaction chemistry or the nature of the resultant polymer, whether it is a low or high viscosity liquid or a flexible, friable or rigid solid.

Mass polymerisation is the simplest method. Monomer, with the appropriate amount of initiator added, is allowed to polymerise at a predetermined temperature. The process rate may be accelerated by means of a catalyst. The main disadvantage of this process is that, as polymerisation proceeds, the viscosity of the mass increases and stirring of the mass so as to maintain a uniform temperature becomes more difficult. Polymerisation is an exothermic process and the heat evolved during the reaction can cause an uneven distribution of temperature in the reactive mass and render it difficult to achieve full control of process temperature. A consequence of this could be a product of variable quality. Polystyrene and PVC may be made by mass polymerisation methods.

Solution polymerisation, as the name implies, is polymerisation with the monomer dissolved in a suitable solvent. Control of process temperature is much easier in this process than in mass polymerisation. However, there are disadvantages. Additional processes are required to separate solvent and any unreacted monomer from the polymer to give a material of marketable quality. These additional purification procedures may be complex, expensive and also hazardous. Polystyrene and SAN copolymer may be made by solution polymerisation, a suitable solvent being ethyl benzene.

A variation on solvent polymerisation is interfacial condensation and this can be used for the condensation polymerisation of an organic acid and an amine in the manufacture of a polyamide (nylon). The two reactants are dissolved in different solvents. The diamine is dissolved in water and the diacid dissolved in a chlorinated solvent giving a solution of high density. The aqueous amine solution is carefully poured over the other solution to form two immiscible layers. The condensation reaction occurs at the interface between the two liquid layers and the resulting polymer is drawn off continuously as nylon fibre.

Some monomers can be highly dispersed in water, forming an emulsion, and it may be possible to polymerise from such an emulsion. A water emulsion may possess a viscosity sufficiently low to permit good control of process temperature. In the same way as for solution polymerisation, it is necessary to have additional procedures for the separation of pure polymer from water and unreacted monomer. The advantage offered over solvent polymerisation is that the purification processes are less hazardous. Polyvinyl acetate and PVC may be produced by emulsion polymerisation.

Suspension polymerisation is a process in which the monomer is broken up into small particles or droplets and maintained in suspension in water by continuous agitation, but without an emulsion being formed. Polymer particles form and these tend to coalesce into larger globules which are removed from the suspension by screening or centrifuging. Styrene is almost insoluble in water and some polystyrene is produced by the polymerisation of a suspension of one part styrene in two parts water.

4.10 Questions

4.1 What is meant by the term *polymerisation* and distinguish clearly between addition and condensation polymerisation?

The structural formulae of some chemical compounds are given below. For each of the compounds, state whether or not it can be polymerised and by what type of reaction.

4.2 Natural rubber is polyisoprene with the formula $(-CH_2-C(CH_3)=CH-CH_2-)_n$. What mass of sulphur per kg of rubber is needed to lightly vulcanise by cross-linking at one-fifth of the available sites assuming that each cross-link is composed of a short chain of two sulphur atoms? (Atomic mass numbers are: H = 1.008, C = 12.01, S = 32.066)

4.3 Determine the average molecular mass of polypropylene if the degree of polymerisation is 1000. (H = 1.008, C = 12.01)

4.4 The mean molecular mass of a sample of PTFE, $(-CF_2-CF_2-)_n$, is 3500. What is the degree of polymerisation of this sample? (C = 12.01, F = 19.00)

4.5 Calculate the degree of polymerisation and the average molecular mass of a sample of polyvinyl acetate, $(-CH_2-CH(COO.CH_3)-)_n$, polymerised using 100 g hydrogen peroxide, H_2O_2, as an initiator per 1000 kg vinyl acetate. Assume that all of the peroxide is used to provide terminal groups. (H = 1.008, C = 12.01, O = 16.00)

4.6 What mass of hydrogen peroxide, H_2O_2, would be needed as an initiator for the polymerisation of 100 kg of propylene, $CH_2=CH.CH_3$, to give an average degree of polymerisation of 750? (H = 1.008, C = 12.01, O = 16.00)

4.7 Determine the ratio of styrene, $CH_2=CH.C_6H_5$, to acrylonitrile, $CH_2=CHCN$, in the copolymer SAN having a degree of polymerisation of 600 and an average molecular mass of 50 000 kg/kmol. (H = 1.008, C = 12.01, N = 14.008)

4.8 The melting temperature, T_m, of polyethylene, in K, is given by the expression: $T_m = \{0.0024 + (0.034/n)\}^{-1}$, where n is the degree of polymerisation. Estimate the melting temperatures for degrees of polymerisation of 50, 500 and 1000.

5

Crystalline Structures

5.1 Introduction

The terms crystalline, glassy or amorphous may be used to describe solid materials. What do these terms mean? The crystalline state is that in which the constituent atoms or molecules of the substance are arranged in a regular, repetitive and symmetrical pattern. The regularity of a structure may be termed long range, meaning that the same symmetrical pattern of atoms or ions exists over large distances within the material, or short range, in which local groupings of atoms or ions may be in a symmetrical pattern but the relationships between these local groupings may not be regular. This short range order is found in many inorganic glasses. Amorphous, meaning literally without form, is the term used to describe non-crystalline structures, even though these may possess a short range order.

Many solids are fully crystalline, some are partially crystalline, and others are amorphous. Metals and ceramics are crystalline solids, but in many polycrystalline ceramics there is frequently a glassy phase in the spaces between crystal grains and a new range of materials, metallic glasses is under development. Many polymer materials show a greater or lesser degree of crystallinity but others are amorphous. Inorganic glasses may have the same, or similar, compositions as some ceramic materials yet are amorphous rather than being crystalline.

If a material is allowed to cool slowly from the liquid state the shape of the cooling curve obtained may be either continuous (Figure 5.1 (a)), or show discontinuities (Figure 5.1 (b), (c)). As the temperature falls, there is a steady decrease in the kinetic energy of the atoms or molecules that make up the liquid, and hence a steady increase in the viscosity of the fluid. Materials that show cooling curves of the type shown in Figure 5.1(a) change during cooling from true fluids, through stages of steadily increasing viscosity, into an apparently solid condition. The atoms or molecules present still have the same type of random arrangement that existed in the true fluid state. Such materials are amorphous solids or glasses.

Other solid materials are crystalline in nature, that is their constituent atoms or molecules are arranged in a definite symmetrical pattern. Figure 5.1(b) shows the type of cooling curve obtained for the freezing of a pure crystalline substance. There is a very definite freezing or solidification point, shown by a marked arrest on the cooling curve. At this temperature the atoms cease their random movement and tend to 'stick' together in relatively fixed positions in a regular pattern. Atomic motion does not cease abruptly upon solidification, and the atoms or molecules in a crystalline solid vibrate about fixed positions.

The change from random fluid motion to vibration about a point within a crystal is a change to a much lower energy state. Energy, the latent heat, is emitted from the material at the freezing temperature. Figure 5.1(c) shows the type of cooling curve applicable to some crystalline mixtures. Freezing is not completed at constant temperature, but takes place over a definite temperature range. The latent heat energy emitted during freezing shows up as a less steep gradient on the cooling curve within the freezing range.

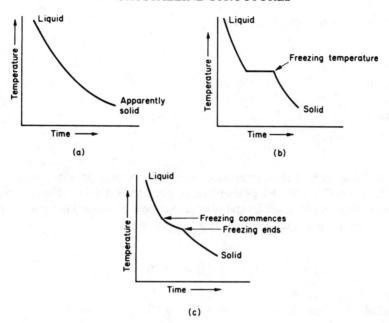

FIGURE 5.1   (a) Cooling curve for a glassy material—no definite solidification temperature. (b) Cooling curve for a pure crystalline material. (c) Solidification and crystallisation occurring over a temperature range

The remainder of this chapter contains details of the various classes of crystals and analysis of crystal structures. The glass state and crystallinity in polymers is dealt with in Chapter 6.

5.2 Crystal classes

A shape is said to be symmetrical if it possesses one or more of the elements of symmetry. The elements of symmetry are planes, axes, and points of symmetry. A shape of low symmetry may possess only one plane of symmetry, whereas a highly symmetrical shape, such as the cube, will contain several planes and axes of symmetry (Figure 5.2). A shape is said to be symmetrical about a plane if the plane divides the shape into either two identical halves, or into two halves that are mirror images of one another. If a shape can be rotated about an axis so that the shape occupies the same relative position in space more than once in a complete revolution then such an axis is termed an axis of symmetry. Such axes may be either 2, 3, 4, or 6-fold.

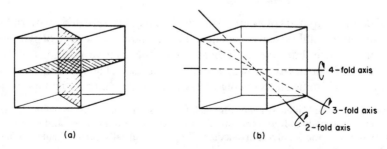

FIGURE 5.2   Elements of cubic symmetry: (a) planes; (b) axes

There are seven crystal systems. These are

triclinic
monoclinic
rhombohedral
hexagonal
orthorhombic
tetragonal
cubic

It is possible to define each of these systems by reference to three principal axes.

Consider three axes OA, OB, OC, of lengths a, b, and c respectively, inclined to one another at angles α, β, and γ (Figure 5.3). Each crystal system may now be defined in terms of the lengths of the principal axes, and the angles between them, as follows:

triclinic	$a \overset{\neq}{=} b \overset{\neq}{=} c$	$\alpha \neq \beta \neq \gamma \neq 90°$
monoclinic	$a \overset{\neq}{=} b \overset{\neq}{=} c$	$\alpha = \beta = 90° \neq \gamma$
rhombohedral	$a = b = c$	$\alpha = \beta = \gamma \neq 90°$
hexagonal*	$a = b \neq c$	$\alpha = \beta = 90°,\ \gamma = 60°$
orthorhombic	$a \neq b \neq c$	$\alpha = \beta = \gamma = 90°$
tetragonal	$a = b \neq c$	$\alpha = \beta = \gamma = 90°$
cubic	$a = b = c$	$\alpha = \beta = \gamma = 90°$

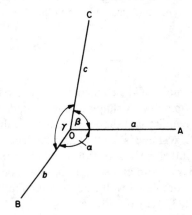

FIGURE 5.3   Principal axes of reference

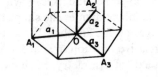

FIGURE 5.4   Miller–Bravais reference axes

A crystal contains myriads of atoms arranged in a regular, repetitive pattern known as a *space lattice*. The various crystal systems, as defined above, indicate the type of symmetry to be found within the space lattice, but give no indications of the relative positions of atoms within the crystal. To show this the device known as a unit cell is used. The unit cell can be regarded as the smallest grouping of atoms still showing the symmetry elements of the type (Figure 5.5).

* Another system, the Miller–Bravais nomenclature, is sometimes used to define the hexagonal form. In this system reference is made to four axes, namely, three coplanar axes, a_1, a_2, and a_3, all of equal length and inclined at 120° to each other, and a fourth axis, of length c, mutually perpendicular to the other three axes. (See Figure 5.4.)

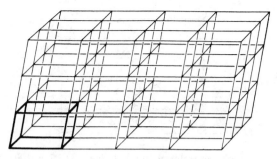

FIGURE 5.5   Representation of part of a space lattice with a unit cell outlined

5.3 Miller notation

As will be seen later, it is sometimes necessary to refer to specific planes and directions within a crystal lattice. The notation used is the Miller indexing system.

Referring to the principal axes OA, OB, and OC of the unit cell (Figure 5.6), the plane PQR can be described by the Miller indices h, k, and l, where these are the reciprocals of the intercepts of the plane with the principal axes, in terms of the lengths of the axes.

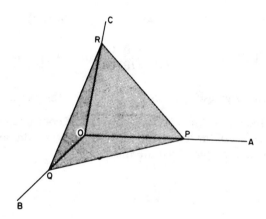

FIGURE 5.6   Plane PQR in relation to crystal axes

$$h = \frac{OA}{OP}, \; k = \frac{OB}{OQ}, \; \text{and} \; l = \frac{OC}{OR}$$

The describing indices are enclosed in brackets, thus (h, k, l). Any fractions are cleared to give the smallest integers.

Figure 5.7 shows three planes in the cubic system. These planes would be described as follows:

Plane ADEF intercepts the principal axes at A, ∞, and ∞.

Indices of plane ADEF are $\qquad \dfrac{OA}{OA}, \dfrac{OB}{\infty}, \dfrac{OC}{\infty}$ or $(1,0,0)$

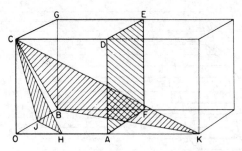

FIGURE 5.7   Identification of planes (refer to text). Plane ADEF = (100), plane CJH = (231), plane BCK = (122)

On the same basis plane CDEG would be (0,0,1) and plane BFEG would be (0,1,0). All face planes of a cube are similar and are termed a family of planes. The use of curly brackets, thus, {1,0,0} means reference to the whole family of face planes rather than to one specific plane.

Plane CJH intercepts the principal axes at H, J, and C.

Indices of plane CJH are $\dfrac{OA}{OH}, \dfrac{OB}{OJ}, \dfrac{OC}{OC}$ or (2,3,1)

Plane BCK intercepts the principal axes at K, B, and C.

Indices of plane BCK are $\dfrac{OA}{OK}, \dfrac{OB}{OB}, \dfrac{OC}{OC}$ or $(\tfrac{1}{2},1,1)$

Removal of fractions gives (1,2,2) as the indices for this plane.

To describe a direction within a crystal, subtract the coordinates of the start of the line from the coordinates of the end of the line (Figure 5.8). Again, fractions are cleared to give the smallest integers. The following examples illustrate this.

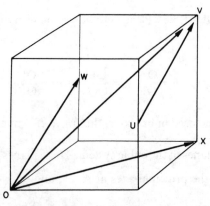

FIGURE 5.8   Directions in crystals (refer to text). Direction UV = direction OW = [021], direction OV = [111], direction OX = [110]

For direction UV: coordinates of $V = 1,1,1$, coordinates of $U = 1,0,\frac{1}{2}$. So direction $UV = 1,1,1 - 1,0,\frac{1}{2} = 0,1,\frac{1}{2}$. Removal of fractions gives the indices as $0,2,1$. Indices for crystallographic directions are enclosed in square brackets, thus, [021]. It will be noticed that line OW is parallel to UV and has the same direction indices. The reciprocal direction, WO, would be $0,0,0 - 0,1,\frac{1}{2} = 0,-1,-\frac{1}{2}$.

Removal of fractions would make this $0,-2,-1$, which would be written as [0$\bar{2}\bar{1}$]. Direction OV becomes [111] and direction OX becomes [110].

5.4 Metallic crystals

The concept of the unit cell can be used to indicate the relative positions of atoms within a space lattice. Figure 5.9 shows three of the variations on the cubic theme, with the circles representing the positions of atom centres.

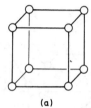 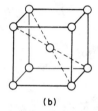 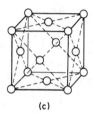

(a) (b) (c)

FIGURE 5.9   Unit cells in the cubic system (circles represent atomic centres). (a) Simple cubic. (b) Body centred cubic (b.c.c.). (c) Face centred cubic (f.c.c.)

As stated earlier (Section 2.15), atoms and ions in the bound state can sometimes be regarded as solid spheres in contact. In the case of the crystalline form of a pure element all the spheres will be identical. However, when considering a chemical compound or a metallic alloy, atoms of two or more elements will be present, and the different elements will possess differing atomic, or ionic, diameters. In the former case, with identical atoms, the structures tend to have high symmetry. Consequently very many metals crystallise in the cubic or hexagonal forms.

Consider uniform spheres. These may be packed together in more than one way. For example, the packing of spheres within a single plane could be square, or hexagonal (Figure 5.10), this latter being the closest possible packing for uniform spheres. A crystal lattice is three dimensional, so that the planes of the type shown in Figure 5.10 must be stacked above one another.

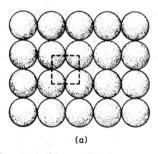 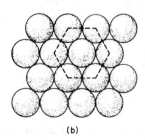

(a) (b)

FIGURE 5.10   Two possible packing arrangements for spheres in a plane: (a) square packing; (b) hexagonal packing

A series of planes with square packing, as in Figure 5.10(a), if stacked upon one another so that the centres of atoms in successive layers all lie on a line normal to the base, would give a three-dimensional space lattice in the simple cubic pattern. In this system, each atom is in contact with six other atoms, that is, a co-ordination number of six. The spheres in a simple cubic pattern occupy 52 per cent of the available space.

In the body centred cubic system, the atoms are packed more closely than in the simple cubic form. The most densely packed planes within the body centred cubic system are those of the (110) type (Figure 5.11). The three-dimensional crystal is built up by stacking a series of identical planes of the (110) type on one another in the manner shown in Figure 5.11(c). It will be seen that the atoms in alternate planes, or layers, are all in line, giving an ABABAB stacking pattern. Figure 5.14(a) shows the unit cell of the body centred cubic system as a series of spheres in contact. It can be seen that the co-ordination number for this type of packing is 8. This is a higher density of packing than for simple cubic, and it can be calculated that the spheres occupy 68 per cent of the available space. A number of metals, including iron and chromium, crystallise in this form.

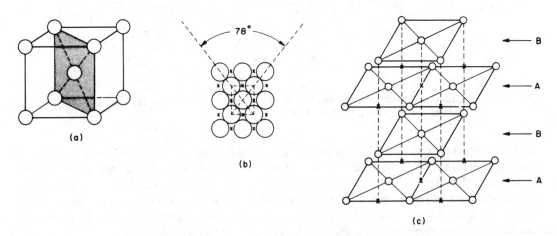

FIGURE 5.11   Body centred cubic system: (a) unit cell with (110) plane shaded; (b) plan view of a (110) plane, the most densely packed in the B.C.C. system. The positions of atomic centres in the second layer are marked x. (c) ABABAB type stacking sequence of 110 planes

The closest possible packing of uniform spheres in a plane is hexagonal packing, and the stacking of planes of this type upon one another gives rise to the closest possible packing in three dimensions. It is, however, possible to build up a symmetrical three-dimensional structure from close packed planes in two different ways. Figure 5.12 shows the centres of atoms (plane A) arranged in a hexagonal plane pattern. When a second plane of atoms is stacked adjacent to the first, or A, plane, the centres of atoms in the second layer may be in either the position marked B or that marked C. If three relative spatial positions exist for close packed layers then symmetry in three dimensions can be obtained if the layers are stacked in either an ABABAB sequence (Figure 5.12(a)), or else in an ABCABC sequence (Figure 5.12(b)).

In both cases, the density of packing is the same, with 74 per cent of the total space filled with spheres, and in both cases the co-ordination number is 12. The first type of stacking sequence produces the crystal form known as close packed hexagonal, and the second sequence gives rise to the face centred cubic type of crystal (Figure 5.12(c), (d)).

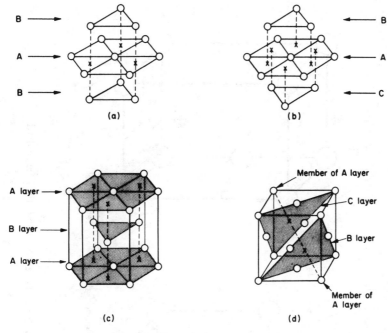

FIGURE 5.12   Close packed lattices: (a) ABAB type stacking sequence; (b) ABC type stacking sequence; (c) close packed hexagonal crystal structure: ABAB stacking; (d) face centred cubic crystal system: ABCABC stacking

Many metals crystallise in these close packed systems, including magnesium and zinc (close packed hexagonal), and aluminium and copper (face centred cubic).

5.5 Interstitial sites

In the structures considered so far we have considered that all the atoms have been spheres of uniform diameter. In such a build-up there are gaps, or interstices, between these uniform spheres into which may be fitted atoms or ions of smaller diameter. The co-ordination number of the various types of interstice or space in crystals, that is the number of atoms that an inserted atom will be in contact with, may be 4, 6 or 8 and the names given to the three types of interstitial site are tetrahedral, octahedral and cubic, respectively. In the first type the interstice is the central space between four atoms forming a tetrahedron, in the second the space is at the centre of six atoms which occupy the corners of an octahedral figure, and in the third type the space is the central space of a simple cubic structure. The locations of these spaces in some crystal systems is shown in Figure 5.13.

5.6 Ceramic crystals

With the exception of diamond, all ceramic materials are compounds and, therefore, ceramic crystals contain atoms, and/or ions, of more than one type. Consider the case of compounds of the alkali metals (Group IA of the periodic table) and the halogen elements (Group VIIB). These

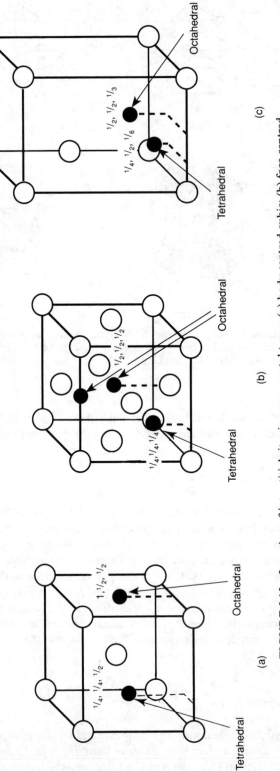

FIGURE 5.13   Location of interstitial sites in some crystal types: (a) body centred cubic; (b) face centred cubic; (c) hexagonal close packed

are all ionic compounds between monovalent elements, for example LiF, NaCl, CsCl, KBr. These compounds possess symmetrical crystalline structures but, although all similar chemically, they do not all form in the same crystal pattern. In the case of caesium chloride, CsCl, the diameters of the caesium and chlorine ions are comparable in size and the caesium chloride structure is a body centred cubic pattern (Figure 5.14(a),(b)) with each caesium ion in contact with eight chlorine ions, and vice versa. (This is the highest possible co-ordination number for an ionic crystal, as explained in Section 3.2.)

In the case of sodium chloride, NaCl, the sodium ion is small in comparison with the chlorine ion and there is room for only six chlorine ions to surround and be in contact with a sodium ion, as in Figure 5.14(c). This arrangement also builds up into a symmetrical cubic pattern (Figure 5.14(d)), but of a different type from caesium chloride.

Calcium fluoride, known also as fluorite or fluorspar, with the formula CaF_2, is an example of an ionic compound between a divalent metallic ion and a halogen ion.

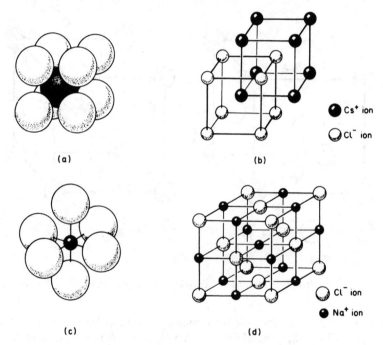

FIGURE 5.14 Caesium chloride and sodium chloride structures. (a) Caesium chloride unit cell showing Cs^+ ion in contact with eight Cl^- ions. (b) Interpenetration of cubes showing Cl^- ion in contact with eight Cs^+ ions. (c) Sodium chloride showing Na^+ ion surrounded by six Cl^- ions. (d) Cubic crystal of sodium chloride

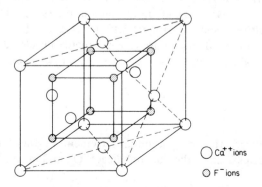

FIGURE 5.15 The structure of calcium fluoride, CaF_2

Many compounds with a composition of the type AB_3 crystallise in a face centred cubic structure with A atoms occupying cell corner sites and the B atoms in the face centre sites. One such compound is tungsten trioxide, WO_3 (Figure 5.16(a)). The *perovskite* structure is similar but in this case the central site in the unit cell is occupied by a small third ion. The mineral perovskite, calcium titanate, has the formula $CaTiO_3$ and the crystal structure is shown in Figure 5.16(b). A number of ceramic materials possess the perovskite structure and one important one is barium titanate. The structure of the barium titanate crystal, at temperatures below 120°C, is distorted. The TiO_6 octahedra are not perfectly regular as the Ti ion is displaced from the body centre position by a small amount, about 0.01 nm. This causes a spontaneous polarisation of the crystal. Barium titanate is a *ferroelectric* material in which the polarisation of the crystal can be affected by an external electric field, giving it piezoelectric characteristics. It is used as a piezo crystal in transducers, for example ultrasonic inspection probes. Some materials which possess the perovskite structure are also ferromagnetic.

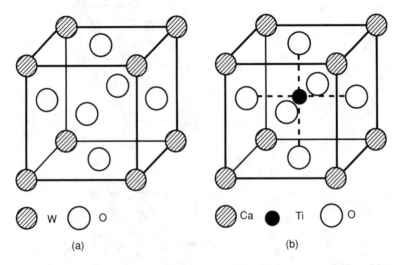

FIGURE 5.16   (a) Structure of tungsten trioxide, WO_3. (b) Structure of perovskite, $CaTiO_3$

There are some oxide ceramics with more complex crystal structures and an important group is the *spinels*. Spinels have the general formula AB_2O_4 where A is a divalent metal ion and B is a trivalent metal ion. The normal spinel structure has a structure in which oxygen ions are arranged in a face centred cubic structure with A^{2+} ions in tetrahedral interstices and B^{3+} ions in octahedral interstices. Not all the available interstitial sites are occupied. A related structure is that of *inverse spinels*. In an inverse spinel half of the B^{3+} ions occupy octahedral sites while the other half, together with the A^{2+} ions, are in tetrahedral interstices. The arrangement for spinel and inverse spinel structures is shown in Figure 5.17. Inverse spinels in which B is the Fe^{3+} ion are termed *ferrites*, ceramic materials which are ferromagnetic but also dielectrics and so electrical eddy currents are not established in them when they are subject to an alternating magnetic field.

5.7 *Silica and silicates*

Silicon is, next to oxygen, the most abundant element in the earth's crust. It occurs in combination with oxygen as silicon dioxide, or silica, SiO_2, and in combination with oxygen and

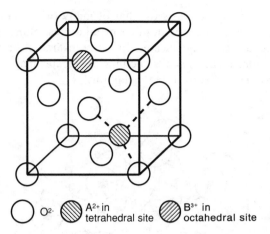

O^{2-} A^{2+} in tetrahedral site B^{3+} in octahedral site

FIGURE 5.17   Spinel structure. Only a small number of the available tetrahedral and octahedral sites are occupied

other metallic elements in a wide range of silicates. Silica and silicates are the basis of many ceramic materials and glasses.

The silicon atom is tetravalent and forms bonds with four oxygen atoms in a tetrahedral structure (Figure 5.18). The SiO_4 unit tetrahedron is the basis of silica and silicate structures. Bonds are formed between the corner oxygen atoms and other silicon atoms to give silica crystals (or with atoms of other metals in silicates). The size of the silicon atom is small in comparison with oxygen and the silicon atom fits easily into the space between oxygen atoms giving a comparatively open structure. There are several allotropic forms of silica as the SiO_4 tetrahedra may add together in more than one way. In one form of silica, tridymite, the tetrahedra build up into a hexagonal array while in another form, cristobalite, the tetrahedra form into a cubic array. The build-up of SiO_4 tetrahedra gives rise to a fairly open network. One possible build-up of tetrahedra giving a three-dimensional structure is shown in Figure 5.19. The bonding within silica is principally of the covalent type and this means that silica is very hard and possesses a high melting point (1710°C).

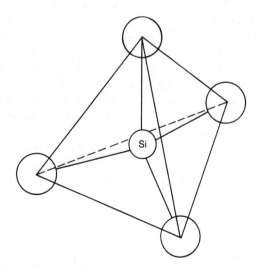

FIGURE 5.18 SiO_4 tetrahedron

The structure of silica forms the basis for that of many other ceramics, the silicates. It is possible for some substitution to occur with other metal atoms taking the place of some of the silicon atoms in the structure forming silicates. In many instances the substituted atoms bond ionically to the oxygen atoms. The crystal structures of silicates may or may not be distorted, depending on the relative sizes of substituted metal ions.

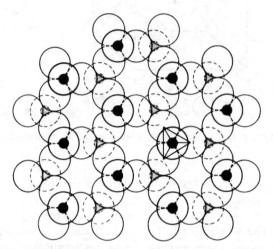

FIGURE 5.19   Plan view of a silica structure

Some silicates possess chain structures with the tetrahedra adding in one plane only, and in some cases the chains could be linked forming a sheet or layer structure (Figure 5.20). Sheet or layer lattices are found in clays, talc and mica. In layer lattice structures, there is strong covalent and ionic bonding within the layer, but comparatively weak van der Waal's secondary bonds between layers. This allows for easy cleavage of the mineral into thin sheets, as is the case with mica. Kaolin, the chief mineral in china clay, is a hydrated aluminium silicate and is a good example of a mineral with a layer lattice (Figure 5.21). Because the structure of kaolin is not symmetrical, the crystal layers become highly polarised. The polarisation of layer crystals allows the clay to absorb water molecules, which are also polar, between the layers. This gives good intercrystalline lubrication and contributes significantly to the plasticity of the clay.

Silica and many silicates can form glasses. The structures discussed so far exhibit crystal symmetry, that is, the same geometrical arrangements of groups of atoms or ions occur throughout the structure. In other words, the structure possesses long range order. Glasses are not crystalline, in other words they do not possess a full and regular symmetrical structure, but they cannot be said to be entirely structureless. Glasses based on silica and silicates have the same composition as the ceramic from which they are derived and, therefore, will still contain SiO_4 tetrahedra. However, these tetrahedra are not assembled in an ordered and symmetrical manner. This is termed short range order as, in silica glass for example, each silicon atom is bonded to four oxygen atoms. Glasses are dealt with in greater detail in Chapter 6.

5.8 Analysis of crystals

The lattice constants, or parameters, are the lengths of the principal axes of the unit cell. For cubic crystals, therefore, there is only one constant to be quoted. The lattice constants of pure metal crystals can be directly related to the atomic diameter of the element, as below.

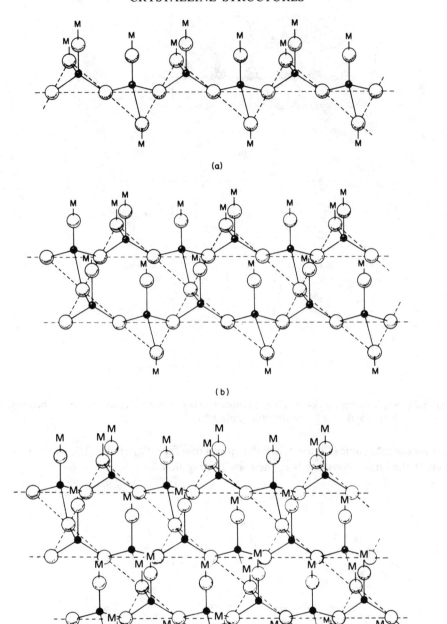

FIGURE 5.20   Silicate chain and sheet structures: (a) single chain; (b) double chain; (c) sheet structure.
Metal ions may be attached at points marked M

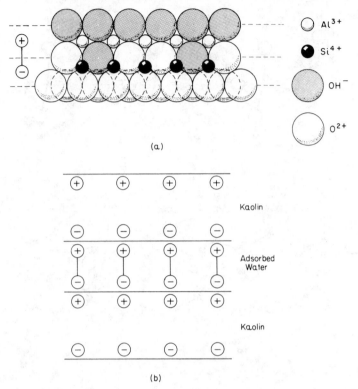

(a)

(b)

FIGURE 5.21 (a) Structure of kaolin. The asymmetric layers are polarised. (b) Adsorbed water between layers acts as a lubricant aiding plasticity

For face centred cubic (Figure 5.22): the length of a face diagonal $= 2D$, where $D =$ an atomic diameter. If the lattice constant is a, then, by Pythagoras

$$(2D)^2 = a^2 + a^2$$

$$4D^2 = 2a^2$$

$$a = \frac{2D}{\sqrt{2}} = D\sqrt{2}$$

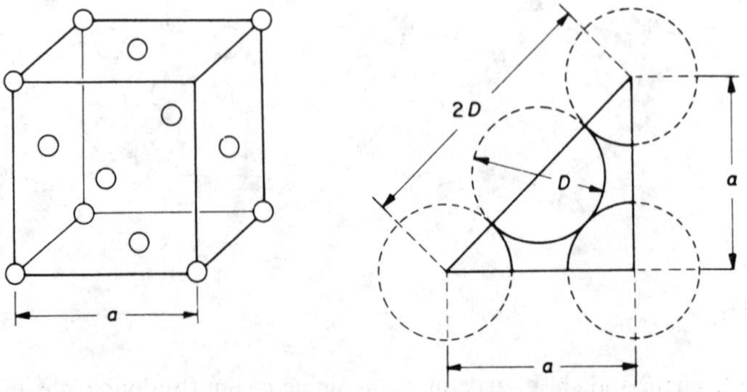

FIGURE 5.22 Face centred cubic unit cell. Relation between a and D

Similarly, for body centred cubic (Figure 5.23): the length of the cube diagonal $= 2D$. In this case, by Pythagoras

$$(2D)^2 = a^2 + 2a^2$$

$$4D^2 = 3a^2$$

$$a = \frac{2D}{\sqrt{3}}$$

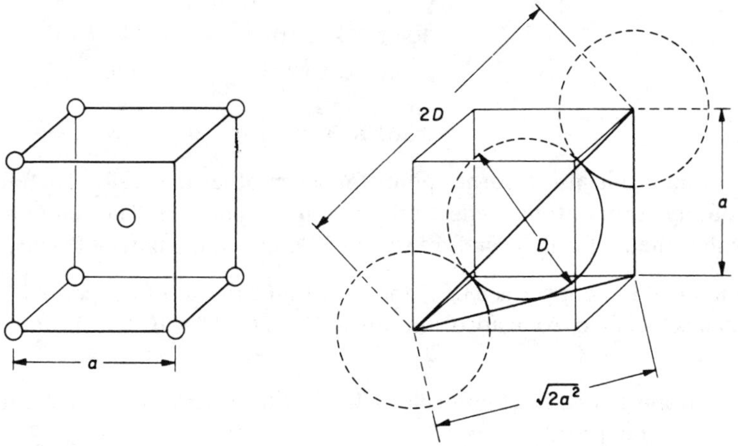

FIGURE 5.23 Body centred cubic unit cell. Relation between a and D

In the close packed hexagonal system there are two lattice constants, a and c (Figure 5.24). From Figure 5.24 it will be seen that parameter a is equal to one atomic diameter D. It can be calculated that the parameter c is equal to $2h$, where h is the height of the regular tetrahedron of edge length a.

$$c = 2h = \frac{D2\sqrt{2}}{\sqrt{3}}$$

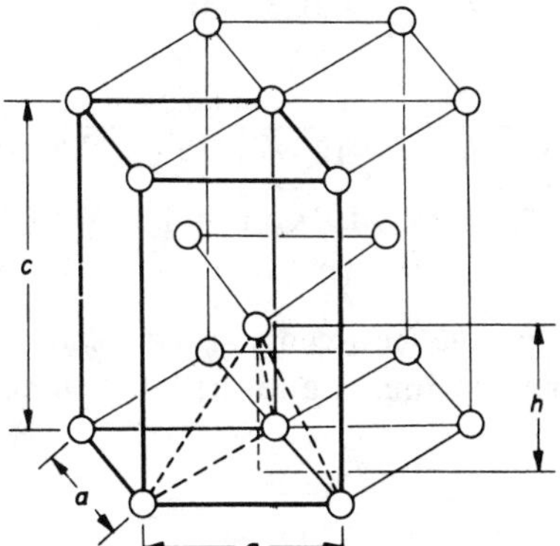

FIGURE 5.24 Close packed hexagonal structure with unit cell outlined, and showing relationship between a, c and h, the height of a tetrahedron

Examples

1. The atomic radius of an iron atom is 1.238×10^{-10} m. Iron crystallises as b.c.c. Calculate

the lattice parameter of the unit cell, a. How many atoms are contained within the b.c.c. unit cell?

For the b.c.c. unit cell

$$a = \frac{4r}{\sqrt{3}}$$

and for iron $r = 1.238 \times 10^{-10}$ m.
Therefore

$$a = \frac{4 \times 1.238 \times 10^{-10}}{\sqrt{3}}$$

$$= 2.861 \times 10^{-10} \text{ m}$$

A corner of a unit cell is also a corner point for seven other unit cells. Therefore, only one-eighth part of any corner atom is effectively within the unit cell. The centre atom is wholly contained within the unit cell. Therefore the b.c.c. unit cell contains $8 \times 1/8 + 1 = 2$ atoms.

2. The atomic weight of copper is 63.54, and the atomic radius of copper is 1.276×10^{-10} m. Copper crystallises as f.c.c. Avogadro's number, N_0, is 6.023×10^{23}. Calculate the density of copper.

Number of atoms in one f.c.c. unit cell $= 4(1/8$ of 8 corner atoms $+ 1/2$ of 6 face atoms).
Lattice parameter of copper, a, $= 2r\sqrt{2}$

$$= 2 \times 1.276 \times \sqrt{2} \times 10^{-10} \text{ m}$$

Volume of unit cell, $a^3 = (2 \times 1.276 \times \sqrt{2} \times 10^{-10})^3 \text{ m}^3$

$$\equiv 4 \text{ atoms of copper}$$

1 mole of copper (63.54 g) contains N_0 atoms.

$$\text{Mass of 4 copper atoms} = \frac{63.54 \times 4}{N_0} = \frac{63.54 \times 4}{6.023 \times 10^{23}} \text{ g}$$

$$\text{Density of copper} = \frac{\text{Mass}}{\text{Volume}} = \frac{63.54 \times 4}{6.023 \times (2 \times 1.276 \times \sqrt{2} \times 10^{-10})^3 \times 10^{23}}$$

$$= 8.98 \times 10^6 \text{ g/m}^3 = 8.98 \times 10^3 \text{ kg/m}^3$$

Within the cubic system the spacing between parallel planes can be readily calculated. The relationship between interplanar spacing, d, and the lattice constant, a, is

$$d = \frac{a}{\sqrt{(h^2 + k^2 + l^2)}}$$

where h, k, and l are the Miller indices of the plane.

The wavelengths of X radiation are of the same order of magnitude as interplanar spacings in crystals. Consequently, when X radiation is directed at a crystalline material there will be some diffraction effects. The radiation will be reflected by atomic planes, including both surface and sub-surface planes. This is illustrated in Figure 5.25. For radiation striking atomic planes at some

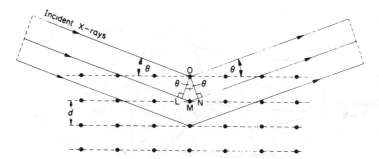

FIGURE 5.25   Reflection of X-rays by crystal planes

angle, θ, there will be effective reflection only when the various ray paths differ by an amount that is equal to one wavelength, or an integral number of wavelengths. Under these conditions the reflected rays will be exactly in phase. From the diagram this condition will be satisfied when distance LMN is equal to $n\lambda$, where λ is the wavelength of the radiation, and n is a small integer.

	OM = d, the interplanar spacing
therefore	MN = $d \sin\theta$
and	LMN = $2d \sin\theta$

The condition for reflection is, therefore,

$$n\lambda = 2d \sin\theta$$

This is the Bragg equation.

By using monochromatic X radiation and measuring the angles at which Bragg reflections occur it is possible to determine the lattice constants for crystals. In the case of cubic crystals of metals the planes that give Bragg reflections are:

body centred cubic—(110), (200), (211), (220), (310), and (222) planes
face centred cubic —(111), (200), (220), (311), and (222) planes

It will be seen from the above that different types of crystal give different diffraction patterns. For the less symmetrical crystal types, and for those containing more than one type of atom, the analysis is similar, but considerably more complex.

One method which is used to obtain X-ray diffraction patterns, and hence determine crystal interplanar spacings, is the Debye–Scherrer method. A powdered metal sample is placed in the centre of a circular diffraction camera and monochromatic X-radiation is directed at it. A strip of film, sensitive to X-radiation, is mounted around the circumference of the camera. The incident X-ray beam is diffracted by crystal planes, according to Bragg's law, giving a series of emerging cones of radiation and a series of lines are recorded on the film. The included angle of each radiation cone is equal to $4\theta_n$, where n is an integer denoting the number of the Bragg diffraction. After exposure and development of the film the strip of film is held flat and the distances between diffraction lines measured. If the distance between a pair of lines is x then $x/R = 4\theta$ (in radians), where R is the radius of the camera (see Figure 5.26).

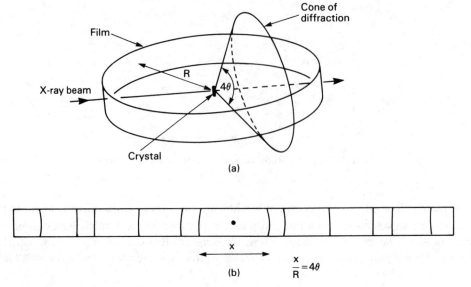

FIGURE 5.26   (a) Principle of Debye–Scherrer camera. (b) Diffraction pattern

Examples

1. Copper atoms are 2.552×10^{-10}m diameter and form an f.c.c. structure. X-radiation of wavelength 1.52×10^{-10}m is used for the analysis of two samples of copper. For sample A the first-order Bragg reflection from (111) planes occurred at an angle of 21°00′, while for sample B the first-order Bragg reflection from (111) planes was at 21°23′.
Give an explanation for the difference between the samples.

For an f.c.c. crystal the lattice parameter $a = D\sqrt{2}$

Lattice parameter for pure copper, $a = 2.552 \times 10^{-10} \times \sqrt{2}$

$$= 3.609 \times 10^{-10} \, m$$

For sample A
$$d_{(111)} = \frac{\lambda}{2 \sin \theta} = \frac{1.52 \times 10^{-10}}{2 \sin 21°00'}$$

$$= 2.12 \times 10^{-10} \, m$$

For sample B
$$a = d\sqrt{(h^2 + k^2 + l^2)} = 2.12 \times 10^{-10} \times \sqrt{3}$$
$$= 3.673 \times 10^{-10} \, m$$

$$d_{(111)} = \frac{1.52 \times 10^{-10}}{2 \sin 21°23'} = 2.08 \times 10^{-10} \, m$$

$$a = 2.08 \times 10^{-10} \times \sqrt{3}$$

$$= 3.61 \times 10^{-10} \, m$$

The lattice parameter for sample B is the same as that calculated for pure copper. Therefore sample B is high-purity copper. The lattice parameter for sample A is 1.75 per cent greater than that for pure copper. Sample B is not pure and the presence of impurity atoms has imposed strain in the crystal lattice of the copper.

2. X-radiation of wavelength of 1.71×10^{-10} m is directed at a cubic crystalline metal. The first two Bragg reflections occur at angles of 30°00′ and 35°17′ respectively.

Determine: (a) whether the crystal type is b.c.c. or f.c.c.,
(b) the lattice parameter of the unit cell, and
(c) the atomic diameter (assuming that all atoms are identical).

If the metal is b.c.c. the first two Bragg reflections will be from the (110) and (200) planes respectively, whereas if the metal is f.c.c. the first two reflecting planes are the (111) and (200). Assuming that the metal is b.c.c.

$$d_{(110)} = \frac{1.71 \times 10^{-10}}{2 \sin 30°00'} \qquad \text{and} \qquad d_{(200)} = \frac{1.71 \times 10^{-10}}{2 \sin 35°17'}$$

$$= 1.71 \times 10^{-10} \text{ m} \qquad\qquad = 1.48 \times 10^{-10} \text{ m}$$

The lattice parameter, $a = d\sqrt{h^2 + k^2 + l^2}$
Therefore

$$a = 1.71 \times 10^{-10} \times \sqrt{2} \text{ or} \qquad\qquad a = 1.48 \times 10^{-10} \times 2$$

$$= 2.42 \times 10^{-10} \text{ m} \qquad\qquad\qquad = 2.96 \times 10^{-10} \text{ m}$$

The two values of a are not the same, therefore the metal is not b.c.c.
Assuming that the metal is f.c.c.

$$d_{(111)} = 1.71 \times 10^{-10} \text{ m} \qquad \text{and} \qquad d_{(200)} = 1.48 \times 10^{-10} \text{ m}$$

giving

$$a = 1.71 \times 10^{-10} \times \sqrt{3} \text{ and} \qquad\qquad a = 1.48 \times 10^{-10} \times 2$$

$$= 2.96 \times 10^{-10} \text{ m} \qquad\qquad\qquad = 2.96 \times 10^{-10} \text{ m}$$

These two values of a are the same showing that the metal is f.c.c. with a lattice parameter of 2.96×10^{-10} m. In the f.c.c. system the relationship between atomic diameter, D, and the lattice parameter is given by $D = a/\sqrt{2}$.
Therefore

$$D = \frac{2.96 \times 10^{-10}}{\sqrt{2}} = 2.093 \times 10^{-10} \text{ m}$$

(In producing the above solution the author deliberately made an initial assumption that the metal was b.c.c. However, from inspection of the data such an assumption was unlikely. The first two Bragg angles were 30°00′ and 35°17′. The relative closeness of these two angles is an indication that they are for planes 3 (111) and 4 (200) rather than for planes 2 (110) and 4 (200). An angle of 30° for plane 2 (110) would mean an angle of about 45° for plane 4 (200).)

3. Bragg reflections may occur from the following planes in b.c.c. crystals: (110), (200), (211), (220), (310), (222), (321), (400), (411), (420), (332), (422).

Which of these planes will give reflections when X-radiation of wavelength 1.54×10^{-10} m is directed at a sample of chromium? Assume that the atomic diameter of chromium is 2.494×10^{-10} m.

$$\lambda = 2d \sin \theta \text{ or } \frac{\lambda}{2d} = \sin \theta$$

$$\sin \theta \not> 1 \text{ therefore } d \not< \frac{\lambda}{2}$$

$$d_{(hkl)} \not< \frac{1.54 \times 10^{-10}}{2} \text{ m}$$

But

$$d_{(hkl)} = \frac{a}{\sqrt{(h^2 + k^2 + l^2)}}$$

Therefore

$$\sqrt{(h^2 + k^2 + l^2)} \not> \frac{2a}{1.54 \times 10^{-10}}$$

In b.c.c. crystals $a = \frac{2D}{\sqrt{3}}$

Therefore lattice parameter for chromium, $a = \dfrac{2 \times 2.494 \times 10^{-10}}{\sqrt{3}}$

$$= 2.88 \times 10^{-10} \text{ m}$$

Therefore

$$\sqrt{(h^2 + k^2 + l^2)} \not> \frac{2 \times 2.88 \times 10^{-10}}{1.54 \times 10^{-10}}$$

$$\not> 3.74$$

Therefore

$$h^2 + k^2 + l^2 \quad \not> 13.98$$

The highest possible values of (hkl) are therefore (222).
Bragg reflections will occur from the first six reflecting planes, that is, up to and including plane (222).

5.9 Questions

5.1 Give the Miller indices for the planes A, B, C, D and E in Figure 5.27.
5.2 Give the Miller indices for the directions A, B, C, D and E in Figure 5.28.
5.3 (a) Sketch a cubic unit cell and mark the positions of the following planes: (220), (010) and ($\bar{1}$11), (b) What is the separation distance between adjacent planes of each type, in terms of the lattice parameter a?
5.4 Considering that metal atoms in a single plane are represented as discs of uniform diameter, show, by calculation, that the packing density in FCC (111) planes is greater than in BCC (110) planes.
5.5 (a) Estimate the theoretical density of nickel, given that nickel crystallises in the FCC form and its atoms have a radius of 0.124 nm. (Avogadro's number, $N_0 = 6.023 \times 10^{26}$/kmol; atomic

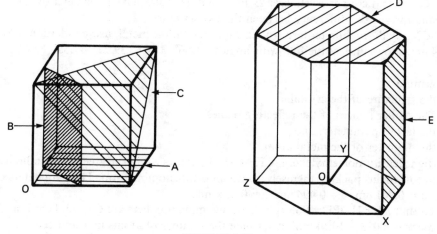

FIGURE 5.27

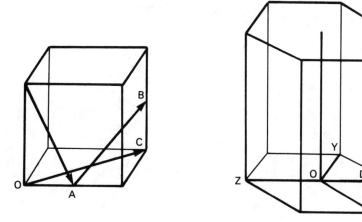

FIGURE 5.28

mass number of nickel is 58.69.) (b) Explain why this theoretical density is greater than the actual density of 8900 kg/m³.

5.6 Determine the diameter of (a) the octahedral site in BCC iron, and (b) the tetrahedral site in FCC iron, assuming the radii of the iron atoms in the two types of lattice are 0.1258 nm and 0.1292 nm respectively.

5.7 Sodium chloride, NaCl, crystallises in a cubic pattern and the unit cell can be regarded as Cl^- ions in a FCC arrangement with Na^+ ions occupying octahedral sites (see Figure 2.6). Given that the ionic radii of sodium and chlorine ions are 0.097 nm and 0.181 nm respectively, calculate (a) the lattice parameter of the unit cell and (b) the packing factor for the crystal.

5.8 X-ray diffraction analysis is conducted on two samples of copper using X-radiation of wavelength 0.152 nm and a Debye–Scherrer camera with a film diameter of 180 mm. For Sample A, the measured distance between the first pair of lines was 132.0 mm while for sample B the distance between the first pair of lines was 134.4 mm. Determine the lattice parameter, a, for each sample. Given that copper crystallises as FCC and its atoms have a diameter of 0.2552 nm, give an explanation for the difference between the two samples.

·**5.9** The X-ray diffraction pattern from a cubic crystalline metal, using radiation of wavelength of 0.154 nm, shows lines at the following angles: 21.66°, 31.47°, 39.74°, 47.58°, 55.63°, 64.71° and 77.59°.

Determine

(a) the structure of the crystal,

(b) The Miller indices of the reflecting planes,

(c) the lattice parameter, and

(d) the diameter of the metal atoms.

5.10 High density polyethylene, $(-CH_2-CH_2-)_n$, shows a high degree of crystallinity with the hydrocarbon chains packing themselves into an orthorhombic form. The parameters of the unit cell are: $a = 0.741$ nm, $b = 0.494$ nm, $c = 0.255$ nm.

Given that ρ for HDPE $= 965$ kg/m^3, atomic mass numbers are C $= 12$, H $= 1$, and Avogadro's number is 6.023×10^{26}/kmol, determine the number of atoms in a unit cell.

6

Glasses and Partial Crystallinity

6.1 Introduction

Glasses are non-crystalline or amorphous elastic solids formed by cooling from the fully molten state at a rate sufficiently rapid that crystallisation is prevented. The glass, or *vitreous*, state can be achieved in all classes of materials. Inorganic glasses have compositions which are the same as or similar to those of ceramics. Many polymeric materials exist in the glass state and, recently, metallic glasses have been developed. A glass is not necessarily structureless. For example, in vitreous silica the bonding between atoms is the same as in the crystalline forms of silica with each silicon atom bonded to four oxygen atoms forming an SiO_4 tetrahedron. The difference is that the tetrahedra are linked together into a large symmetrical array in the crystalline forms whereas, in vitreous silica, there is no long range order (see Figure 6.2).

6.2 Formation of glasses

When a substance is cooled from the fully molten state to, or just below, the temperature T_m, the melting point, the random fluid atoms or molecules tend to rearrange themselves into an ordered crystalline structure. This process of freezing into a crystalline solid does not occur instantaneously but by a process of nucleation and growth. Small crystals grow from nuclei through the liquid. Outward growth of a crystal ceases when it meets the advancing boundary of another crystal growing from a neighbouring nucleus and the result, when the freezing process is complete, is a polycrystalline solid. Generally, the change from a random fluid into an ordered solid is accompanied by an increase in density and a consequent decrease in volume. The rate at which this crystallisation process occurs in a liquid at a temperature just below T_m depends on the viscosity of the liquid and this parameter is dependent on the size and complexity of the molecules that compose the substance. Liquid metals tend to be of low viscosity and the atoms form into a crystalline structure readily. Many inorganic ceramic materials contain complex molecular arrangements but can still crystallise with relative ease. Polymeric materials tend to have very large complex molecules and have relatively high viscosities at T_m and, generally, are more slow to crystallise than other classes of materials.

When a liquid is cooled rapidly to below T_m, the viscosity of the liquid in the supercooled state may be so high that crystallisation is prevented. As the temperature falls further the viscosity steadily increases until, eventually, it is so high that the material is effectively solid but is amorphous. It is then termed a glass. A material is generally classed as being a solid when the viscosity increases beyond 10^{16} Pa s. There is no sharp change in density at T_m during the formation of a glass and, on cooling of the supercooled liquid below T_m, the volume continues to reduce with the same rate of thermal contraction as applied above T_m. At some lower temperature, there is a change in slope of the volume-temperature curve and this temperature is termed the *glass transition* or *fictive temperature*, symbolised as T_g (see Figure 6.1). T_g is considered to be the temperature at which, on cooling, the substance changes from a supercooled liquid to a glass.

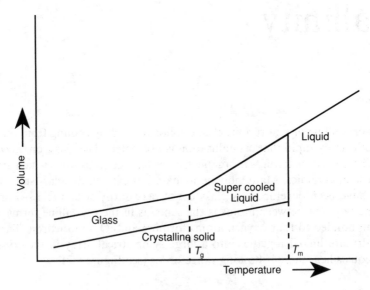

FIGURE 6.1   Change in volume with temperature for both a crystalline solid and a glass

Polymers and some inorganic ceramic materials can be cooled sufficiently rapidly from the molten state to produce a supercooled liquid and a glass by rates of cooling which are readily achieved during normal commercial processing but to achieve the glass state in metals extremely high cooling rates, of the order of 10^6 °C/s, are required.

6.3 Inorganic glasses

Many inorganic glasses are based on silica, SiO_2, and the basic structural unit is the SiO_4 tetrahedron. The glass form of silica, *vitreous silica*, consists of a three-dimensional network of SiO_4 tetrahedra but with these tetrahedra arranged at random with no long range order. Figure 6.2 is a schematic representation of the structural difference between crystalline silica and silica glass. For simplicity in this two-dimensional representation each silicon atom is shown bonded to three oxygen atoms only.

Some other oxides can form glass networks, the principal ones being B_2O_3 and P_2O_5. These oxides, together with silica, are referred to as *glass formers* or *network formers*. Alumina, Al_2O_3, can form a joint glass network with silica. Silica has a high melting point, > 1700°C, and the major commercial inorganic glasses contain other oxides, particularly the alkaline oxides CaO, Na_2O and K_2O, to lower the melting temperature. These alkaline oxides are not capable of

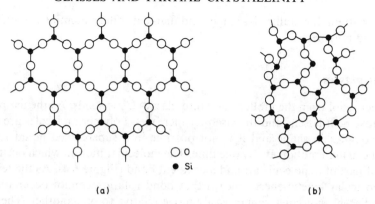

FIGURE 6.2 Schematic representation of (a) crystalline silica, (b) silica glass

forming glasses on their own as they cannot form continuous networks. They are referred to as *glass* or *network modifiers*. The oxygen of the oxide enters the silica network but, as there are now insufficient silicon atoms to combine with the extra oxygen, the network begins to break up. The metal cations enter holes within the network. This is shown schematically in Figure 6.3.

The main commercial inorganic glasses, apart from silica glass, are common glass (soda-lime-silica), leaded glasses, borosilicate glasses and alumino-silicate glasses.

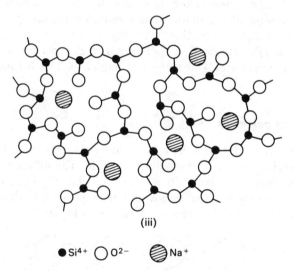

FIGURE 6.3 Schematic diagram of soda glass structure. Network begins to break up and cations enter holes

6.4 Metallic glasses

A metallic glass may be made by cooling a liquid metal extremely rapidly. Cooling rates of about 10^6 °C/s or greater are required. This can be achieved by melt spinning from a chilled disc to form very thin tapes or filaments. Typical tape thicknesses are of the order of 40 μm. Some metallic glasses will devitrify on heating to moderate temperatures (350°C+) and become brittle, but for some a controlled crystallisation to form a ductile, high strength microcrystalline structure can be achieved. High strength metallic glass tapes and filaments offer potential as reinforcing elements in ceramics and polymers.

6.5 Polymer glasses

Many polymers cool from the molten state into glassy elastic solids. As the temperature reduces molecular movement lessens and the viscosity increases. Polymer molecules are extremely large and, often, complex in shape and it is not possible in a supercooled liquid for the complete molecule to be in fluid motion at any one time. The molecular motion which occurs is largely due to rotation of part of a molecule around a covalent bond (Figure 6.4). As the temperature falls, the lower molecular kinetic energy means that bond rotation cannot occur to any significant extent and adjacent molecules cannot easily move relative to one another. The *glass transition temperature*, T_g, is the temperature at which bond rotation largely ceases.

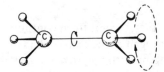

FIGURE 6.4   A single covalent bond permits bond rotation

The value of T_g varies considerably from one polymer to another and is much lower for simple flexible chain molecules than for those with bulky side groups attached to the chain. For example, T_g is of the order of $-90°C$ for polyethylene while for polystyrene, with bulky $-C_6H_5$ groups attached, it is 95°C. Below T_g the polymer is an elastic solid but at temperatures between T_g and T_m the material is viscoelastic (refer to Chapter 9). This is sometimes termed the *rubber* state.

6.6 Crystallinity in polymers

X-ray analysis of polymer materials show that many polymer materials give diffraction patterns which indicate a regular array, or crystalline structure. The diffraction patterns obtained also show diffuse regions indicating that there is not full crystallinity but a mixture of crystalline and amorphous zones. It is easy to visualise how in polyethylene, for example, one can readily have regular arrangements of the $-CH_2-$ chains, as shown in Figure 6.5.

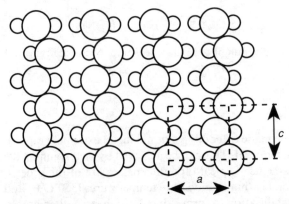

FIGURE 6.5   Crystalline packing within polyethylene, with unit cell outlined

An early theory was that the polymer structure contained small high density areas, termed *crystallites*, where sections of molecular chains were aligned together, and other, less dense amorphous areas (Figure 6.6(a)). In this model, one individual polymer chain could form part of both crystalline and amorphous zones. At a later stage, single crystals of polymer were separated out from solution and analysis of these showed that they were small platelets with a thickness of about 10 nm and a plate size of about 10 μm. Within these platelets the linear polymer chain was repeatedly folded (Figure 6.6(b)).

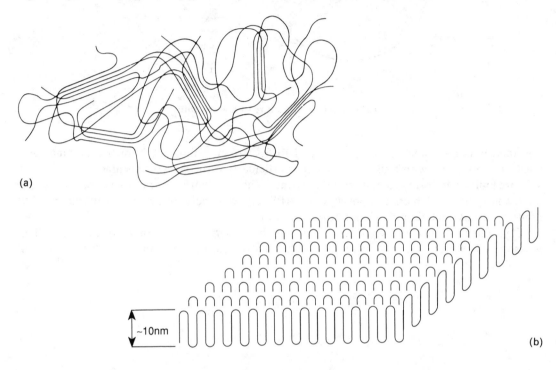

(a)

(b)

~10nm

FIGURE 6.6   (a) Crystallites. (b) Crystal platelets formed by folding

Many crystalline polymers show a structure containing areas termed *spherulites* when examined under a polarising microscope. These spherulites may vary in size from several millimetres across to less than a micrometre. A spherulite is not a single crystal but is believed to consist of many crystalline platelets which grow from a nucleus, and that these platelets are twisted (see Figure 6.7). Within the spherulite there are both platelets and amorphous regions.

The structure of a crystalline polymer can also be affected by deformation. During the cold drawing of a polymer there can be a major alignment, or orientation, of the structure in the direction of drawing. This orientation produces an anisotropic material which is much stronger in the direction of the draw than in a transverse direction. A similar orientation takes place during film blowing.

The degree of crystallinity in polymers varies. It is about 50% in commercial polyethylene and may be considerably less than this in some other materials. Nevertheless, polymers which show partial crystallinity are termed *crystalline polymers*. The main crystalline polymers are linear homopolymers in which there are strong van de Waal's intermolecular forces, and some block copolymers. Frequently polymers with isotactic and syndiotactic structures (refer to Section 4.7) possess sufficient regularity for them to form crystals. Polyethylene, polypropylene, PTFE,

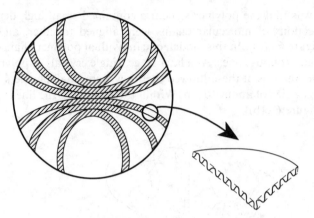

FIGURE 6.7   Structure of a spherulite

polyamides, polycarbonate, PEO, POM and PEEK are crystalline. A branched polymer or a linear homopolymer with bulky side groups, particularly if the structure is atactic (see Section 4.7), are unlikely to form ordered crystalline zones. This is also true for random copolymers. The main amorphous polymers are polystyrene, ABS, PMMA, polyurethanes and the thermosets. Some materials, including PVC, are borderline cases.

Crystalline polymers show viscoelasticity and possess a glass transition temperature, T_g. This is associated with the amorphous regions of the structure, which may make up 50% or more of the volume.

7

Elastic Behaviour

7.1 Stress and strain

When a load is applied to a material a balancing force is set up within the material, and this internally acting force is termed a *stress*. The stress acting upon a material is defined as the force exerted per unit area.

For a force of 10 newton acting on a surface of 10 mm square the stress is:

$$\frac{\text{force}}{\text{area}} = \frac{10}{10 \times 10 \times 10^{-6}} = 10^5 \text{ newton/metre}^2 \text{ (N/m}^2\text{)} = 10^5 \text{ pascal (Pa)}$$

The types of stress normally considered are tensile, compressive and shear stresses. When a material is in a state of stress its dimensions will be changed. A tensile stress will cause an extension of the length of the material, while a compressive stress will shorten the length. Tensile and compressive stresses are termed *direct stresses*. A shear stress imparts a twist to the material. The dimensional change caused by stress is termed *strain*. In direct tension or compression the strain is the ratio of the change in length to the original length. As a ratio strain has no units and is simply a numerical value.

A bar of length l and cross-sectional area A, when subjected to a tensile force F, will increase in length by an amount δl (Figure 7.1(a)). The direct stress, σ, in the material is given by $\sigma = F/A$. The direct strain, ε, developed is given by $\varepsilon = \delta l/l$. Similarly, a compressive force F will cause a reduction in length of δl, so that a direct compressive stress of $\sigma = F/A$ will cause a strain $\varepsilon = -\delta l/l$ (Figure 7.1(b)). The action of a shear force F will cause a block of material to twist through a small angle ϕ. Referring to Figure 7.1(c), the shear force F acting on a plane of area A creates a shear stress, τ, given by $\tau = F/A$. The shear strain, γ, which is generated is given by $\gamma = y/x = \tan \phi$.

In elastic behaviour the strain developed in a material, when the material is subjected to a stress, is fully recovered immediately the stress is removed. Some materials, notably ceramics, glasses and many thermosets, show elastic properties up to their fracture stress. Metals and alloys exhibit both elasticity and plasticity, with non-recoverable or plastic strain occurring when the applied stress exceeds a critical level, the *elastic limit* or *yield stress*. Many thermoplastic materials possess little, if any, true elasticity but show viscoelasticity (see Chapter 9) and much of the strain developed is not fully recoverable.

Robert Hooke, in 1678, enunciated his law stating that the strain developed in a material is directly proportional to the stress producing it. This law holds, at least within certain limits, for many materials.

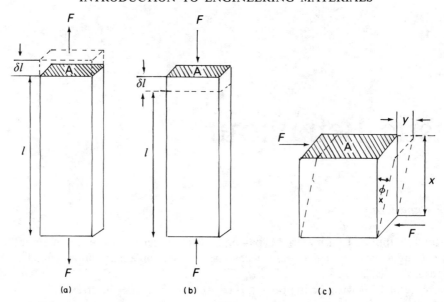

FIGURE 7.1 Stresses and strains: (a) tensile force F, cross-sectional area A, tensile stress $\sigma = F/A$, tensile strain $\varepsilon = \delta l/l$; (b) compressive force F, cross-sectional area A, compressive stress $\sigma = F/A$, compressive strain $\varepsilon = -\delta l/l$; (c) shear force F, cross-sectional area A, shear stress $\tau = F/A$, shear strain $\gamma = \tan \phi = y/x$

Figure 7.2 is a force–extension diagram for a metal stressed in tension. The first portion of the curve, OA, shows that the length of the specimen increases in direct proportion to the applied load, hence strain will be proportional to stress. Hooke's law does not hold beyond point A. Behaviour within the region OA will be elastic. Beyond point A the extension of the material ceases to be wholly elastic and some permanent strain is developed. The non-elastic permanent strain is termed *plastic strain*. If the material were loaded from O to B and the load then removed the material would not revert to its original length, but follow the unloading path BC showing a permanent extension, or plastic deformation, of OC when completely unstressed. Point A is known as the *limit of proportionality* or *elastic limit*.

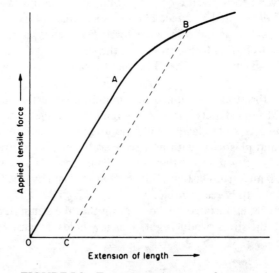

FIGURE 7.2 Force–extension curve for a metal

7.2 Elastic constants

For an elastic material, namely one that obeys Hooke's law, the ratio of stress to strain will be a constant of the material. For direct stresses acting in tension or compression:

$$\frac{\text{direct stress }(\sigma)}{\text{direct strain }(\varepsilon)} = E \text{ (a constant for the material)}$$

Strain, being the ratio of change in dimension to original dimension, has no dimension, therefore the constant E has the dimensions of stress, namely pascals (Pa) (or newtons/metre^2 (N/m^2)). E is termed the *modulus of elasticity*, or *Young's modulus*, and is a measure of the stiffness of a material.

There is a similar relationship between shear stress and shear strain within the elastic limit.

$$\frac{\text{shear stress }(\tau)}{\text{shear strain }(\gamma)} = G$$

G is a second constant for a material and is termed the *modulus of rigidity* or *shear modulus*.

The modulus of elasticity can be determined in a tensile or compressive test (refer to Chapter 29) and the modulus of rigidity may be determined from the results of a torsion test but vibration test methods (refer to Chapter 31) are used for the most accurate evaluation of these elastic constants.

There is a third elastic modulus for a material, this being the *bulk modulus of elasticity, K*. This is the elastic response of the material to equilateral tension or compression.

$$\frac{\text{stress}}{\text{volume strain}} = K$$

If a material is subjected to a direct longitudinal stress, a longitudinal strain will be developed extending the length of the specimen. This will be accompanied by a certain lateral contraction, compensating for the increase in length (Figure 7.3). The ratio of lateral to axial strain is termed *Poisson's ratio, v*. The value of Poisson's ratio for many materials lies between 0.28 and 0.35.

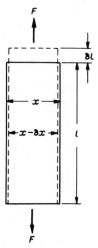

FIGURE 7.3   Poisson's ratio. Longitudinal strain = $\delta l/l$, lateral strain = $\delta x/x$, Poisson's ratio v = lateral strain/longitudinal strain

The various elastic constants are related to one another as follows:

$$G = \frac{E}{2(1 + v)} \text{ or } E = 2G(1 + v)$$

$$K = \frac{E}{3(1 - 2v)} \text{ or } E = 3K(1 - 2v)$$

The values of the elastic constants for a material will be independent of the direction of stressing for some materials. Such materials are termed *isotropic* (possessing the same properties in all directions). However, in an anisotropic material the values of the elastic constants differ according to the directions in which they are measured. Crystals show anisotropy and a single crystal will possess differing values of E, G and Poisson's ratio, depending on the direction of stressing relative to the crystal axes. However, most commercial ceramics and metals are polycrystalline, their structures consisting of many small crystals with a random orientation. In consequence, they approximate in behaviour to isotropic materials.

Some bulk materials, for example, timbers and fibre reinforced composites, particularly composites with uniaxial fibre reinforcement, exhibit marked anisotropy and possess different elastic constants in different directions. In timber, for example, the values of E and G are different in the longitudinal, radial and tangential directions relative to the grain. Thus, for a timber, there are three values of Young's modulus, three of shear modulus and six values of Poisson's ratio.

7.3 Fibre reinforced composites

Fibre reinforced composites are a very important group of engineering materials. Many of those in current use are composed of a thermoset or thermoplastic matrix material reinforced with thin glass, carbon or aramid fibres. The greatest reinforcing effect is obtained when the fibres are continuous and parallel to one another and the maximum strength is obtained when such a composite is stressed in tension in a direction parallel to the fibre axis.

Consider a uniaxial continuous fibre composite subjected to an axial tensile force F (Figure 7.4). The force will be divided between fibres and matrix such that:

$$F = F_f + F_m \tag{1}$$

where F_f and F_m are the forces carried by fibres and matrix respectively.

The same longitudinal strain, ε, will be developed in both fibres and matrix so that:

$$\varepsilon_c = \varepsilon_f = \varepsilon_m \tag{2}$$

Stress, σ, = force F/area A, giving $F = \sigma A$ so, from Equation (1):

$$\sigma_c A_c = \sigma_f A_f + \sigma_m A_m$$

giving

$$\sigma_c = \sigma_f \frac{A_f}{A_c} + \sigma_m \frac{A_m}{A_c} \tag{3}$$

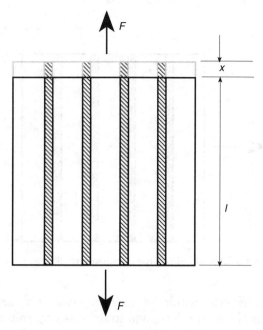

FIGURE 7.4   Strain in composite, ε_c, strain in fibres, ε_f, and strain in matrix, ε_m, are all equal to x/l

where A_f, A_m and A_c are the cross-sectional areas of fibre, matrix and composite respectively. For fibres of uniform cross-section, the area fractions A_f/A_c and A_m/A_c will equal the fibre and matrix volume fractions V_f and V_m so that, from Equation (3) we have:

$$\sigma_c = \sigma_f V_f + \sigma_m V_m \tag{4}$$

From Hooke's law $\sigma = \varepsilon E$ giving, from Equation (4)

$$\varepsilon_c E_c = \varepsilon_f E_f V_f + \varepsilon_m E_m V_m$$

As, in Equation (2) there is equal strain in composite, fibre and matrix, the value of E for the composite, E_c, is given by:

$$E_c = E_f V_f + E_m V_m \tag{5}$$

where E_f and E_m are the values of Young's modulus for fibre and matrix respectively.

Equations (4) and (5) show that the properties of the composite are proportional to the volume fractions of the components of the composite. This is the *Rule of Mixtures* and this rule can also be used to calculate other properties such as the density or the thermal conductivity of a composite in the fibre direction.

The values of E and tensile strength are greatest in the fibre axial direction and will be very much less in the transverse direction. When subjected to a force, F, normal to the fibre axis (see Figure 7.5) the effective cross-sectional areas of fibre and matrix can be assumed to be equal and so the stress developed in both fibres and matrix can be assumed as equal, giving:

$$\sigma_c = \sigma_f = \sigma_m \tag{6}$$

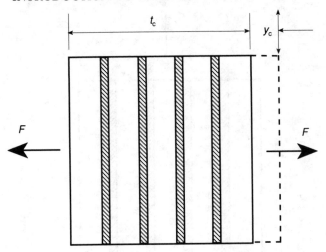

FIGURE 7.5   A transverse force, F, causes the composite thickness, t_c, to be increased by an amount y_c

The total transverse deformation of the composite is y_c and will be the sum of the deformations of fibres and matrix. The total strain, $\varepsilon_c = y_c/t_c$, so that

$$y_c = \varepsilon_c t_c$$

Therefore

$$\varepsilon_c t_c = \varepsilon_f t_f + \varepsilon_m t_m \tag{7}$$

where t_f and t_m are the respective thicknesses of the fibre and matrix elements.
From Equation (7)

$$\varepsilon_c = \varepsilon_f \frac{t_f}{t_c} + \varepsilon_m \frac{t_m}{t_c}$$

But the thickness ratios t_f/t_c and t_m/t_c equate to the respective volume fractions, so:

$$\varepsilon_c = \varepsilon_f V_f + \varepsilon_m V_m \tag{8}$$

From Hooke's law $\qquad\qquad\qquad \varepsilon = \sigma/E$

so

$$\frac{\sigma_c}{E_c} = \frac{\sigma_f V_f}{E_f} + \frac{\sigma_m V_m}{E_m}$$

But, from Equation (3) $\qquad\qquad \sigma_c = \sigma_f = \sigma_m$

so

$$\frac{1}{E_c} = \frac{V_f}{E_f} + \frac{V_m}{E_m}$$

or

$$E_c = \frac{E_f E_m}{E_m V_f + E_f V_m} \tag{9}$$

Example 1

Calculate the values of E in both the longitudinal and transverse directions for an epoxy composite reinforced with continuous, parallel glass fibres, the volume fraction of glass in the composite being 0.35. Assume E for epoxy is 4 GPa and E for glass fibre is 70 GPa.

$$E_{cL} = E_f V_f + E_m V_m = (70 \times 0.35) + (4 \times 0.65) = 27.1 \text{ GPa}$$

$$E_{cT} = \frac{E_f E_m}{E_m V_f + E_f V_m} = \frac{70 \times 4}{(4 \times 0.35) + (70 \times 0.65)} = 5.97 \text{ GPa}$$

In practice, the use of Equation (9) does not give an accurate value for E in the transverse direction. One reason for this is that there may not be an equal distribution of force between the fibres and matrix. Several equations of an empirical nature have been proposed to reduce inaccuracies. Two such equations are the Halpin–Tsai and the Brintrup equations.

Halpin–Tsai equation $\quad E_{cT} = E_m \dfrac{1 + 2\beta V_f}{1 - \beta V_f} \quad$ where $\beta = \dfrac{(E_f/E_m) - 1}{(E_f/E_m) + 2}$

Brintrup equation $\quad E_{cT} = \dfrac{E_f E_m'}{E_f V_m + V_f E_m'} \quad$ where $E_m' = \dfrac{E_m}{(1 - v_m^2)}$

Example 2

Determine values of E_{cT} for the glass-filled epoxy composite quoted in Example 1 on the basis of both the Halpin–Tsai and Brintrup equations. Assume Poisson's ratio for the epoxy matrix = 0.35.

Using the Halpin–Tsai equation

$$\beta = \frac{(E_f/E_m) - 1}{(E_f/E_m) + 2} = \frac{70/4 - 1}{70/4 + 2} = \frac{16.5}{19.5} = 0.846$$

$$E_{cT} = E_m \frac{1 + 2\beta V_f}{1 - \beta V_f} = 4\left[\frac{1 + (2 \times 0.846 \times 0.35)}{1 - (0.846 \times 0.35)}\right] = 4\left[\frac{1.592}{0.704}\right] = 9.05 \text{ GPa}$$

Using the Brintrup equation

$$E_m' = \frac{E_m}{(1 - v_m^2)} = \frac{(4}{(1 - 0.35^2)} = 4.56 \text{ GPa}$$

$$E_{cT} = \frac{E_f E_m'}{E_f V_m + V_f E_m'} = \frac{70 \times 4.56}{(70 \times 0.65) + (4.56 \times 0.35)} = 6.78 \text{ GPa}$$

As stated earlier, a composite with uniaxial fibre reinforcement will be highly anisotropic and have a low modulus value and be relatively weak in a direction transverse to the fibre direction. In order to obtain some degree of isotropy in, say, flat sheet material, reinforcing fibres can be arranged in the form of a two-dimensional mat. This gives a greater uniformity of properties in the plane of the sheet but at the expense of the maximum strength obtainable. The maximum strength in any direction in such sheet is only about one-third of the maximum strength parallel to the fibre direction in a composite with uniaxial fibre reinforcement. If a fibre-composite has reinforcing fibres arranged in a random three-dimensional manner then the maximum strength in any direction is only about one-sixth of that for a uniaxial composite.

Another method for reducing anisotropy effects in sheet material is to compile a series of thin uniaxially reinforced laminae into a laminate. Generally, these laminates are symmetrical in build-up and contain an odd number of laminae with each lamina orientated at some angle $\pm \phi$ to a particular direction (see Figure 7.6). Plywood is a timber product with reduced anisotropy produced in similar manner by bonding an odd number of laminae, or plies, together. The analysis of laminates is beyond the scope of this volume.

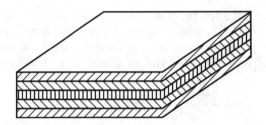

FIGURE 7.6   Build-up of a laminate to reduce anisotropy effects

7.4 Thermal stresses

Under certain circumstances stresses can be induced in materials and components through a change in temperature. The cooling of a casting, forging or moulding from the processing temperature may cause residual thermal stress to be generated and, in some cases, the level of internal stress developed may be so high as to exceed the fracture stress of the material.

When a material is heated it will, if completely unrestrained, expand. The change in length, δl, can be calculated using the expression $\delta l = l_0 \alpha \delta \theta$ where

l_0 = original length,
α = the coefficient of linear expansion, and
$\delta \theta$ = the temperature interval.

If the material is so constrained that its dimensions cannot change and it is then heated or cooled through a temperature interval of $\delta \theta$ a stress will be set up within the material. The length of the sample should have been changed by an amount $l_0 \alpha \delta \theta$. The length is forced to remain as l_0 and this is equivalent to a direct strain of $l_0 \alpha \delta \theta / l_0 = \alpha \delta \theta$. If Young's modulus, E, for the material is known then the stress induced in the material may be calculated, the stress, σ, being given by $\sigma = E \alpha \delta \theta$.

Example 3

A 1000 m length of welded steel railway track is heated to 85°C before being firmly secured to the track bed. Calculate the stress in the track at a temperature of 15°C. E for steel is 208 GPa, α for steel is 12×10^{-6}/°C.

The tensile strain developed in the track on cooling from 85°C to 15°C is given by $\alpha \delta \theta = 12 \times 10^{-6} \times (85 - 15)$.
The tensile stress induced in the rail is:
$$12 \times 10^{-6} \times 70 \times 208 \times 10^9 = 174.7 \times 10^6 \, \text{Pa} = 174.7 \, \text{MPa}.$$

Thermal stresses can be generated in composite structures through the unequal expansion or contraction of each element. The bimetallic strip is a good example showing the effects of differing expansion coefficients. Consider a bimetallic strip made of copper and steel. Copper has a larger coefficient of expansion than steel and so, if the initially flat strip is heated through a temperature interval of $\delta\theta$, the copper, if unrestrained, would expand to a greater extent than the steel. However, the two elements of the strip are fully bonded to one another and the greater expansion of the copper forces the strip into a curved profile. Full expansion of the copper is prevented by the restraining presence of the steel with the result that the copper will be in a state of compression while the steel contains a tensile stress.

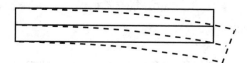

FIGURE 7.7   Bimetallic strip. The upper element, of higher expansion coefficient, is not allowed to expand to the full extent, and the strip is forced to adopt a curved shape

If the various elements in a composite construction are arranged in a symmetrical manner bending or twisting of the composite will not occur when thermal stresses are induced. An example of such a symmetrical construction is a high voltage electrical transmission cable consisting of an aluminium conducting element surrounding a reinforcing steel core.

Example 4

A high voltage electrical conducting cable with an external diameter of 25 mm consists of a steel core of 5 mm diameter surrounded by aluminium. Assuming that there is no relative movement between the steel and aluminium, calculate the thermal stresses developed in both elements if the conductor is heated through a temperature interval of 40°C.
Assume: α for steel = $12 \times 10^{-6}/°C$; α for aluminium = $26 \times 10^{-6}/°C$; E for steel = 208 GPa; E for aluminium = 70 GPa.

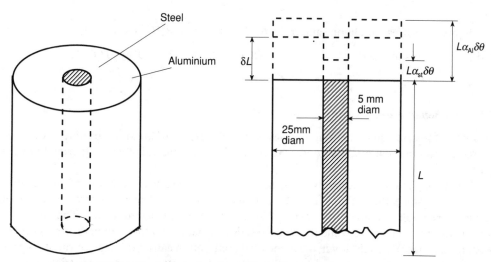

FIGURE 7.8   Thermal strains induced in an aluminium/steel composite

Refer to Figure 7.8. When heated through some temperature interval $\delta\theta$, if completely unrestrained, the aluminium would expand by an amount $L\alpha_{Al}\delta\theta$ and the steel expand by an amount $L\alpha_{st}\delta\theta$. But both of the elements are constrained to move together so that they will both expand by some amount δL. This means that the aluminium is effectively subject to a compressive strain of $(\alpha_{Al}\delta\theta - \delta L/L)$ at the higher temperature while the steel is subject to a tensile strain of $(\delta L/L - \alpha_{st}\delta\theta)$. The assembly is in equilibrium at the higher temperature, therefore the compressive force in the aluminium must be equal to the tensile force in the steel.

$$\text{Force} = \text{stress} \times \text{cross-sectional area,}$$

Therefore

$$F = \sigma_{Al}A_{Al} = \sigma_{st}A_{st}$$

Stress, $\sigma = \varepsilon E$ so

$$\varepsilon_{Al}E_{Al}A_{Al} = \varepsilon_{st}E_{st}A_{st}$$

Substituting for strain gives:

$$(\alpha_{Al}\delta\theta - \delta L/L)E_{Al}A_{Al} = (\delta L/L - \alpha_{st}\delta\theta)E_{st}A_{st}$$

$$\alpha_{Al}\delta\theta E_{Al}A_{Al} - (\delta L/L)E_{Al}A_{Al} = (\delta L/L)E_{st}A_{st} - \alpha_{st}\delta\theta E_{st}A_{st}$$

$$\frac{\delta L}{L} = \frac{\alpha_{Al}\delta\theta E_{Al}A_{Al} + \alpha_{st}\delta\theta E_{st}A_{st}}{E_{st}A_{st} + E_{Al}A_{Al}}$$

$$\frac{\delta L}{L} = \frac{\{26 \times 10^{-6} \times 40 \times 70 \times 10^9 \times ((25 \times 10^{-3})^2 - (5 \times 10^{-3})^2) \times \pi/4\} + \{12 \times 10^{-6} \times 40 \times 208 \times 10^9 \times (5 \times 10^{-3})^2 \times \pi/4\}}{\{208 \times 10^9 \times (5 \times 10^{-3})^2 \times \pi/4\} + \{70 \times 10^9 \times ((25 \times 10^{-3})^2 - (5 \times 10^{-3})^2) \times \pi/4\}}$$

$$= 9.783 \times 10^{-4}$$

So

$$\varepsilon_{Al} = (\alpha_{Al}\delta\theta - \delta L/L) = (26 \times 10^{-6} \times 40) - 9.783 \times 10^{-4} = 61.7 \times 10^{-6}$$

and

$$\varepsilon_{st} = (\delta L/L - \alpha_{st}\delta\theta) = 9.783 \times 10^{-4} - (12 \times 10^{-6} \times 40) = 498.3 \times 10^{-6}$$

$$\sigma = \varepsilon E$$

so compressive stress in aluminium, $\sigma_{Al} = 61.7 \times 10^{-6} \times 70 \times 10^9 = 4.32 \, \text{MPa}$

and tensile stress in steel, $\sigma_{st} = 498.3 \times 10^{-6} \times 208 \times 10^9 = 103.6 \, \text{MPa}$

7.5 Toughened glass

In some cases, residual thermal stresses are deliberately induced to produce some specific characteristics in a material. A good example of this is the manufacture of thermally toughened glass.

A plate of glass to be thermally toughened is heated to a temperature which is sufficiently high to allow for some thermal movement of molecules and adjustment of stresses and then the surfaces are cooled quickly by means of an air blast or quenching in oil. The surface layers contract and become rigid while the centre section remains at a high temperature. As the centre section subsequently cools down, the associated contraction causes the generation of compressive stresses in the surface layers and tensile stresses in the centre. The stress distribution across a section of the glass plate is shown in Figure 7.9. Glasses are stronger in compression than in

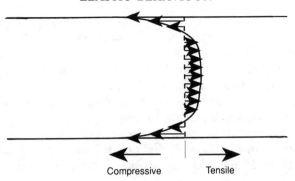

FIGURE 7.9   Stress distribution across a section of thermally toughened glass

tension and the residual stress distribution in toughened glass means that a considerable tensile force must be applied before the residual compressive stress in the surface layers is overcome and the net stress moves into the tensile mode. The major disadvantage of thermally toughened glass is that if the surface is damaged and a surface chip or scratch goes through the compressive layer and into the tension zone then failure may be sudden and catastrophic. This type of toughened glass was used to a considerable extent at one time for the manufacture of car windscreens. However, impact with a stone or other object thrown up by the wheels of a preceding vehicle could be sufficient to penetrate the compressive skin leading to immediate total failure of the screen. Today, the majority of vehicle windscreens are made of laminated glass rather than of thermally toughened glass. Catastrophic failure does not occur in glass which is strengthened by lamination.

7.6 *Questions*

7.1   When a steel bar, of 12 mm × 12 mm cross-section is subjected to a tensile force of 11 kN the measured elastic extension over a length of 150 mm is 5.5×10^{-2} mm. Determine: (a) the stress in the bar, (b) the elastic strain developed and (c) the value of Young's modulus for steel.
7.2   Determine the shear stress in the coupling pin when the coupling (shown in Figure 7.10) is subject to a tensile force of 5 kN.

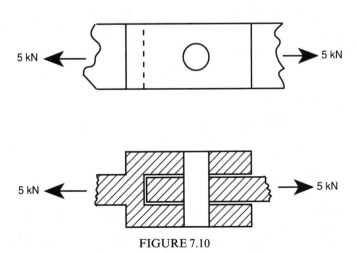

FIGURE 7.10

7.3 (a) Estimate the shear modulus of bronze, given that its modulus of elasticity is 80 GPa and Poisson's ratio is 0.34. (b) Estimate the value of Poisson's ratio for nickel, given that the values for E and G for nickel are 208 GPa and 79 GPa respectively.

7.4 A fibre composite rod of 10 mm diameter, with a 0.3 volume fraction of continuous axially aligned glass fibres, is subject to an axial force of 1.8 kN.

Calculate: (a) the force carried by the fibres, (b) the stress in the fibres, and (c) the stress in the matrix. Assume E for matrix is 3 GPa and E for fibres is 85 GPa.

7.5 A fibre-reinforced composite is made from a polyester resin and long uniaxially aligned glass fibres with a fibre volume fraction of 45 per cent. Estimate the maximum strength of the composite and the values of E for the composite in both the longitudinal and transverse directions.

7.6 A uniaxially aligned carbon fibre/epoxy composite with a fibre volume fraction of 0.4 is in the completely unstressed condition at a temperature of 15°C. Calculate the stresses in both fibre and matrix when the temperature is raised to 50°C.

Coefficients of thermal expansion are: epoxy 60×10^{-6}/°C; carbon 5.4×10^{-6}/°C. E for epoxy = 3 GPa; E for carbon fibre = 200 GPa.

7.7 A component is to be manufactured from a uniaxially-aligned continuous fibre composite material. Two composites are available, one containing 40 per cent, by volume, of glass fibre and the other containing 40 per cent, by volume, of carbon fibre. A minimum axial tensile strength of 700 MPa and a minimum value of E of 50 GPa parallel to the fibre direction is required. Using the property values in the table below, which of the two composite materials is the most suitable?

Material	Tensile strength (MPa)	E (GPa)
Matrix for glass composite	60	3
Matrix for carbon composite	15	3
Glass fibre	3500	62
Carbon fibre	2000	415

8

Dislocations and Plasticity in Metals

8.1 Plastic flow in metals

Many materials possess an elastic limit and when stressed they strain in an elastic manner up to a certain point. Beyond this point the strain developed is no longer directly proportional to the applied stress, and also, the strain developed is no longer fully recoverable. If the stress is removed elastic strain is recovered but the material will be left in a state of permanent, or plastic, strain. The mechanism of plastic deformation is not the same for all classes of materials and it is necessary to consider the various materials groups separately.

Metals, in general, are characterised by possessing high elastic modulus values, and also the ability to be strained in a plastic manner. Some metals will begin to deform plastically at very low values of stress and will yield to a very considerable extent before fracture occurs. Other metals and alloys show little plastic yielding before fracture. These latter materials are termed brittle. Plastic deformation in metals may take place by the process of slip, or by twinning. Slip is the more common deformation process encountered and will be dealt with first.

An early theory evolved to explain plastic deformation in metals was the 'block slip' theory. In this it was postulated that when the yield stress of the metal was exceeded plastic deformation took place by the movement of large blocks of atoms sliding relative to one another across certain planes—slip planes—within the crystal (Figure 8.1).

The 'block slip' theory accounted for many of the observed phenomena but possessed a number of drawbacks. The theoretical shear yield strength of a metal, τ_c, calculated on the basis of this theory is given by $\tau_c = G/2\pi$, where G is the modulus of rigidity. This calculated value of τ_c is many times greater than the experimentally observed yield strengths of pure metals. Present day theories of plastic flow in metals are based on the existence of small imperfections, or defects, within crystals. These are structural defects, termed dislocations, and plastic deformation is due to the movement of dislocations across the slip planes of a crystal under the action of an applied stress. The calculated stress required to bring about the movement of dislocations is of the same order of magnitude as observed yield stresses in metals. In recent years, it has been possible to produce small single crystals that are virtually defect free. These small perfect crystals, termed whiskers, possess properties close to the very high theoretical strengths predicted for perfectly crystalline metals.

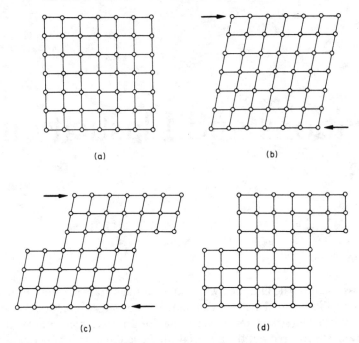

FIGURE 8.1 Plastic deformation according to the 'block slip' theory. (a) Original unstressed condition. (b) Elastic strain only. Specimen will revert to (a) when the stress is removed. (c) Elastic and plastic deformation. When stress is removed the specimen will appear as in (d). (d) Unstressed. No elastic strain but permanent plastic strain remains

8.2 Slip planes

The crystal planes on which slip generally takes place are those that possess the highest degree of atomic packing. The direction of slip within a crystal plane is the direction of greatest atomic line density. The majority of metals crystallise as close packed hexagonal, face centred cubic, or body centred cubic, and slip within these three systems will be discussed.

In the close packed hexagonal system, the planes with the greatest density of atomic packing are the basal planes (001) and within planes of this type there are three possible slip directions (Figure 8.2). Within any one hexagonal crystal space lattice there is only one series of parallel planes of this type.

The results of research work in the 1930s into the effects of tensile stresses on single crystal

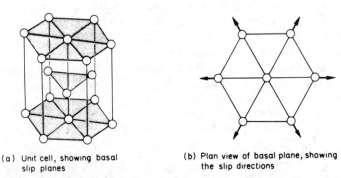

(a) Unit cell, showing basal slip planes

(b) Plan view of basal plane, showing the slip directions

FIGURE 8.2   Slip planes and directions in the close packed hexagonal system

specimens of zinc, a close packed hexagonal metal, showed that the behaviour of the sample was dependent on the angle at which the slip planes were inclined to the axis of the applied stress. If the slip planes were aligned either normal to, or parallel to, the stress axis, failure occurred in a brittle manner with negligible plastic deformation (Figure 8.3). If however, the slip planes were inclined at some angle, θ, other than 0° or 90°, plastic yielding took place before failure. The amount of plastic yielding and the value of the direct tensile stress necessary to commence yielding were not constant but varied with angle θ (Figure 8.5).

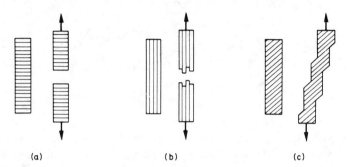

(a)   (b)   (c)

FIGURE 8.3   Effect of stress on single crystals of a hexagonal metal. (a) Slip planes normal to the applied stress; brittle fracture; no slip. (b) Slip planes in line with the applied stress; brittle fracture; no slip. (c) Slip planes inclined to the applied stress; slip and plastic deformation caused by shear force acting on slip planes

Analysis of the results indicated that, for single crystals of zinc, plastic deformation was due to slip initiated by the action of a shearing stress on the slip plane and also that the magnitude of the critical shearing stress (namely, that necessary to initiate slip), was constant. This has been enunciated as Schmid's law: *slip in a metal crystal begins when the stress resolved on the slip plane in the slip direction reaches a certain value termed the critical resolved shear stress.*

The magnitude of a resolved shear stress, τ, across a plane inclined at some angle θ to the axis of direct stress, σ, (see Figure 8.4) can be determined as follows.

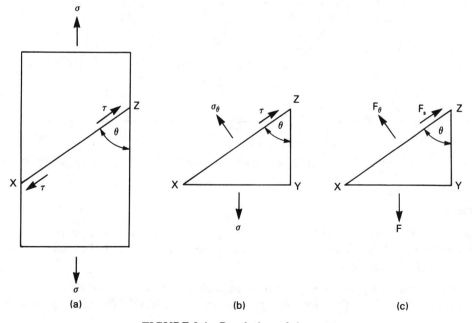

(a)   (b)   (c)

FIGURE 8.4   Resolution of shear stress

Consider the prism XYZ. Figure (8.4(b)) shows the stresses acting on this prism. The forces F, F_θ and F_s (Figure 8.4(c)) must be in equilibrium. Force is stress times area, so that if the thickness of the prism XYZ is w then:

$$F = \sigma \times XY \times w, \quad F_\theta = \sigma_\theta \times XZ \times w, \quad \text{and} \quad F_s = \tau \times XZ \times w.$$

Resolving along XZ: $\tau \times XZ \times w = \sigma \times XY \times w \times \cos \theta$

$$\tau = \sigma \times (XY/XZ) \times \cos \theta$$

$$= \sigma \times \sin \theta \cos \theta$$

$$= \tfrac{1}{2}\sigma \sin 2\theta$$

For a constant value of critical shearing stress, τ_c, the applied tensile stress necessary to initiate slip will be at a minimum when $\theta = 45°$, that is, when $\sin 2\theta = 1$, and infinite when $\theta = 0°$ or $90°$, that is when $\sin 2\theta = 0$.

There is a similar expression, namely $\tau = \sigma \times \cos \theta \cos \varphi$, where θ is the angle between the axis of direct stress and the slip direction within a slip plane and φ is the angle between the axis of direct stress and the normal to the slip plane.

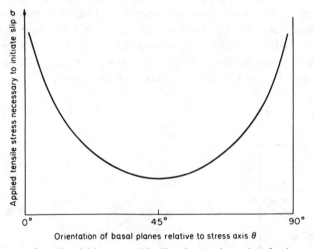

FIGURE 8.5   Variation of tensile yield stress with slip plane orientation for hexagonal single crystal

Example

The critical shear stress, τ_c, for the $<110>\{111\}$ slip system of a pure face centred cubic metal is found to be 1.6 MPa. What direct stress, σ, must be applied in the [001] direction to produce slip in the [101] direction on the $(\bar{1}11)$ plane?

Refer to Figure 8.6. It will be seen that the angle θ between the direction of direct stress [001] (line AB) and the direction of slip [101] (line OB), is $45°$. The angle ϕ between the

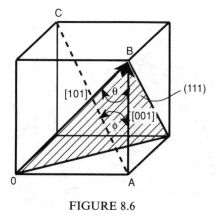

FIGURE 8.6

direction [001] and the normal to the slip plane (line AC) is $\sin^{-1}$ (BC/AC). But BC = $a\sqrt{2}$ and AC = $a\sqrt{3}$, where a is the lattice parameter, so that $\phi = \sin^{-1} (2/3)^{\frac{1}{2}} = 54.74°$.

From the expression $\tau_c = \sigma \times \cos \theta \cos \phi$

$$\sigma = 1.6 \div (\cos 45° \cos 54.74°)$$

$$= 3.92 \text{ MPa}$$

For the face centred cubic system the most densely packed planes of atoms are those of the {111} family and within each (111) plane there are three possible directions of slip (Figure 8.7). The combination of a slip plane and a slip direction is termed a *slip system*. There are four sets of planes of this type, each set occurring at different inclinations, within the face centred cubic structure. Consequently, no matter from what direction relative to the crystal a direct stress is applied there will be resolved shearing stresses acting on several slip planes and at least one slip system will be inclined in such a way that plastic deformation can occur. This means that face centred cubic crystals are comparatively soft and ductile.

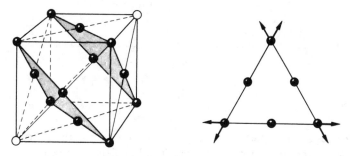

FIGURE 8.7   Slip plane and directions in the f.c.c. system

In the body centred system, the most densely packed planes are those of the {110} type. Within this type of plane there are two possible directions (Figure 8.8).

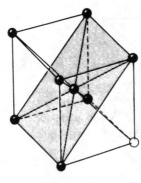

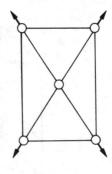

(a) Unit cell, showing a (110) slip plane

(b) Plan view of (110) plane showing the slip directions

FIGURE 8.8   Slip plane and directions in the b.c.c. system

Slip in body centred cubic metals is more complex than in the other two systems already discussed. In addition to slip taking place on (110) planes, slip may also occur on planes of the (112) and (123) types. Although there are numerous possible slip directions within body centred cubic crystals, metals crystallising in this form are generally harder and less ductile than face centred cubic metals. This is thought to be due to the body centred cubic lattice being less densely packed than the face centred cubic lattice. An increase in temperature will bring more slip planes into action, and also reduce the value of critical shearing stress. This latter statement also applies to other forms of metallic crystal.

The main slip planes and directions for the three principal types of crystal found in metals are summarised in Table 8.1

Table 8.1   *Slip planes and directions in metal crystals*

Crystal type	Slip planes	No. of slip planes	Slip directions	No. of slip directions per plane	Total No. of slip systems
h.c.p.	(001)	1	$(100)(010)(\bar{1}10)$	3	3
f.c.c.	{111}	4	<110>	3	12
b.c.c.	{110}	6	<111>	2	12

8.3 Dislocations

Crystal space lattices are not perfectly crystalline but contain certain structural defects. These may be classified as either line or point defects. The line defects are known as *dislocations*, and these may be classified as either *edge-type* or *screw-type*. In practice, dislocation lines are rarely of the pure-edge or pure-screw type but are mixed dislocations, that is, lines of dislocation containing both an edge and a screw component. An important property of a dislocation is its *Burger's vector, b,* which indicates the extent of lattice displacement caused by the dislocation. The Burger's vector also indicates the direction in which slip will occur. Figure 8.9 is a representation of an edge dislocation. It can be considered that the edge dislocation is due to the presence of an additional half-row of atoms within the lattice. If an atom-to-atom circuit is described within a portion of regular lattice, as shown at the bottom left of Figure 8.9(a), it will be a complete closed circuit. If, however, a similar circuit is described around the dislocated portion of lattice the start and finish will not be coincident. The distance SF in Figure 8.9(a) will be the Burger's vector, b. In an edge dislocation, the Burger's vector is normal to the dislocation line.

When a shearing stress is applied to a section of crystal lattice, and this stress is beyond the elastic limit, minor atomic movements will occur causing the dislocation line to move through the lattice (see Figure 8.9(b)). The magnitude of the stress necessary to initiate the movement of a dislocation across a slip plane is considerably less than that which would be required to bring about block slip. It will be seen that plastic flow occurs in the direction of the Burger's vector.

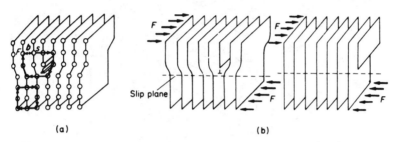

FIGURE 8.9 Edge dislocation: (a) representation of a portion of crystal lattice containing an edge dislocation, dislocation shown as ⊥ and Burger's vector as b: (b) application of shear stress F causes the dislocation to move along the slip plane until it leaves this section of lattice causing an increment of plastic deformation

A diagrammatic representation of a screw dislocation is shown in Figure 8.10. In this case, a Burger's circuit describes a helical, or screw, path and the Burger's vector, b, is in the same direction as the line of the dislocation. It should be noted, however, that plastic flow under the action of a shearing stress still occurs in the direction of the Burger's vector.

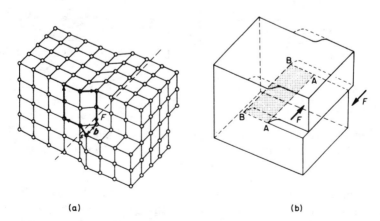

FIGURE 8.10 Screw dislocation: (a) representation of a lattice containing a screw dislocation; (b) application of a shear stress F will cause the dislocation to move from AA to BB (slipped area is shaded)

As stated earlier, many dislocations are mixed dislocations with both an edge and a screw component. The magnitude and direction of the Burger's vector are constant for all points on any one dislocation line. Referring to Figure 8.11 it will be seen that the dislocation line XY which is pure screw at X and pure edge at Y has the same vector, b, at each end. When a shear stress, F, is applied to such a curved dislocation line lying within a slip plane ABCD it will cause slip to occur, as shown in Figure 8.12. All slip will be in the direction of the Burger's vector, b, but it will be noted that the curved dislocation line XY has now moved to position X'Y', the pure screw portion at X having moved normal to the applied force and the pure edge portion at Y having moved parallel to the applied force (see Figure 8.12(b)).

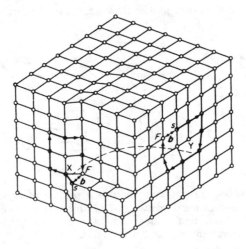

FIGURE 8.11   Line of dislocation that is screw at X and edge at Y

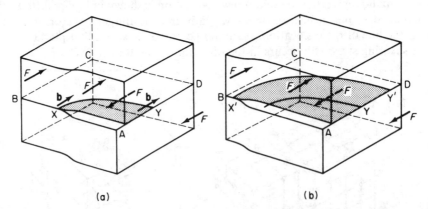

(a)   (b)

FIGURE 8.12   Slip caused by movement of curved dislocation line across a slip plane ABCD. The dislocation moves from XY to X'Y' (slipped region is shown shaded)

It should be appreciated that there will be strain fields associated with dislocations. In the example of an edge-type dislocation, considered as an extra half-row of atoms inserted into the structure, the crystal lattice in the region of the dislocation will be strained. There will be a compressive strain above the dislocation and a tensile strain below it, as shown in Figure 8.13(a).

It can be said that dislocations of like sign repel one another. Consider two similar

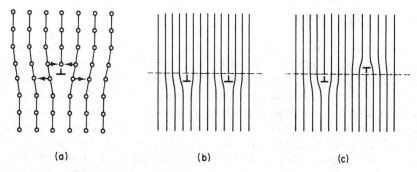

(a)   (b)   (c)

FIGURE 8.13   (a) Strain field surrounding a dislocation. (b) Dislocations of like sign tend to repel one another. (c) Dislocations of opposite sign tend to attract one another

dislocations on the same slip plane (Figure 8.13(b)). These will tend to repel one another in an effort to reduce the strain in that portion of the lattice, and it would require an increase in the level of applied stress to move them closer together. Conversely, dislocations of opposite sign attract one another. If the two dislocations shown in Figure 8.12(c) were brought together they would cancel one another out leaving a perfect unstrained section of lattice.

Crystals invariably contain numerous point defects, in addition to the structural defects mentioned above. The principal types of point defect which occur in metal crystals are the vacancy, the substitutional defect, and the interstitial type defect (Figure 8.14). The latter two types are due to the presence of 'stranger' atoms of other elements which are present either as impurities or are deliberately added as alloying elements. The interstitial type occurs when the stranger atoms are very small in comparison with the atoms of the parent metal and fit into interstitial sites (see Section 5.5), while the substitutional type occurs when parent and stranger atoms are comparable in size. It will be noticed from Figure 8.14 that strain is developed within a crystal lattice in the neighbourhood of point defects.

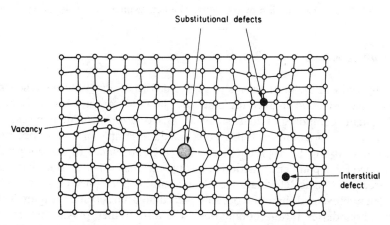

FIGURE 8.14   Lattice containing several types of point defect. Note that substitutional atoms may be larger or smaller than the parent atoms causing positive or negative lattice strain

8.4 Deformation by twinning

Another mechanism of plastic deformation, namely deformation by twin formation, may sometimes occur in close packed hexagonal, body centred cubic or tetragonal metal crystals. This type of deformation may occur if the slip planes within the crystals are not aligned to the axis of stress in a favourable manner. Twin formation in these crystals often occurs as the result of impact or shock loading conditions. In this type of deformation, the atoms in each plane undergo some displacement and a revised lattice arrangement is formed which will produce planes more favourably placed for slip to occur. The new lattice arrangement is a mirror image of the original in the twin, or shear, plane (Figure 8.15). When a sample of the metal tin, a body centred tetragonal structure, is deformed plastically it is sometimes possible to hear high pitched sounds emanating from the metal. These distinctive sounds, known as *tin cry*, are emitted during mechanical twin formation within the material. Deformation by mechanical twinning does not occur in crystals of the face centred cubic type but twinned crystals, known as annealing twins, commonly occur in face centred cubic metals that have been annealed following cold deformation.

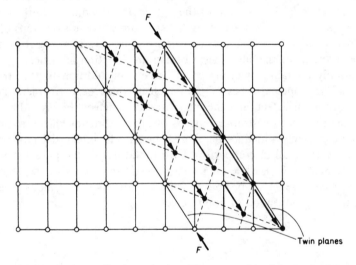

FIGURE 8.15   Deformation by twinning. The new twinned lattice is shown dotted

8.5 Polycrystalline metals

In the foregoing sections, the mechanisms of plastic deformation within metallic single crystals have been discussed. Metallic engineering materials, however, are not normally in the form of single crystals, but are composed of many small crystals, or grains, and in many cases these small crystals have a random orientation. For a polycrystalline sample of a close packed hexagonal metal, such as zinc, it will be apparent that while the slip planes in some crystals might be favourably inclined for slip under the action of an applied stress other crystals may not be aligned in a suitable direction. Plastic deformation of the favourably positioned crystals will be hindered, or even completely prevented, by unfavourably placed adjacent crystals (Figure 8.16(a)).

The crystal boundaries (grain boundaries) will also hinder plastic deformation. These boundaries are not simple planes but are transition zones between adjacent crystals of differing orientation (Figure 8.16(b)). These transition zones, which may be of several atoms in thickness,

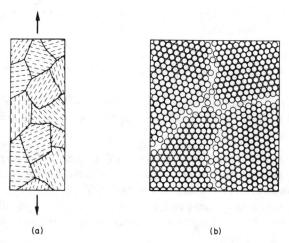

FIGURE 8.16 Polycrystalline metals. (a) Random orientation of slip planes (shown dashed). Some crystals are more favourably placed than others for plastic deformation. (b) Representation of grains and boundaries

do not possess regular crystal planes and consequently act as barriers to the movement of dislocations. From this it follows that the grain boundaries in a pure metal tend to be stronger than the crystal grains and that, consequently, a sample composed of a large number of very small crystals will be stronger than another sample of the same metal containing a smaller number of larger crystal grains. These statements do not hold true at all times. For example, in some alloys and impure metals there may be segration of alloy constituents, or impurities, to grain boundaries. In some cases this renders the crystal boundaries weak; for example, copper is embrittled by the segregation of traces of bismuth impurity to the grain boundaries. At very high temperatures close to the melting temperature, the grain boundaries become weaker than the crystals and plastic deformation may occur by a process of viscous flow of the grain boundary material. The turbine blades for aircraft gas turbines, which operate in a very high temperature environment, are manufactured by a special process which produces a single crystal structure with a specific orientation.

The effect of grain size on tensile yield strength has been quantified by Petch. The Petch relationship relates the yield strength, σ_y, of a polycrystalline metal to the crystal grain size and is: $\sigma_y = \sigma_0 + K\, d^{-\frac{1}{2}}$, where d is the average grain diameter in mm and σ_0 and K are constants for the material.

Example

Annealed samples of a steel with varying grain sizes were tested and the results obtained are shown in the Table below.

	Average grain size (grains/mm)	Tensile yield strength, σ_y (MPa)
Sample A	13	250
Sample B	34	300
Sample C	45	320

Using the Petch relationship, estimate the grain size which would be necessary to give the steel a tensile yield strength of 360 MPa.

The average grain sizes, d, for the three samples are:
Sample A—$1/13=0.077$ mm; B—$1/34=0.030$ mm; C—$1/45=0.022$ mm, from which $d^{-\frac{1}{2}}$ values are: Sample A—3.6; B—5.8; C—6.75.

A plot of σ_y againsy $d^{-\frac{1}{2}}$ yields a straight line (Figure 8.17). From the graph $\sigma_y = 360$ MPa corresponds to a value of $d^{-\frac{1}{2}} = 8.5$ from which $d = 0.014$ giving an average grain size of 72 grains/mm.

A single crystal will exhibit anisotropy, that is, it will possess differing properties in different directions. This is very evident in the case of hexagonal metals that possess only one series of parallel slip planes. However, a polycrystalline material in which the orientation of the individual crystals is a purely random arrangement will, on a macro scale, be virtually isotropic.

The deformation behaviour of the principal crystal types in polycrystalline samples of pure metals may be summarised as follows:
(a) Close packed hexagonal metals are generally brittle, or of low ductility only. If ductility occurs, it usually exists only within a narrow range of temperatures.
(b) Face centred cubic metals are invariably highly ductile over a wide range of temperatures.
(c) The body centred cubic system is less favourable to ductility than the face centred cubic system. The body centred cubic system contains both ductile and brittle metals, but even the ductile metals in this system often show good ductility only over a fairly small range of temperature.

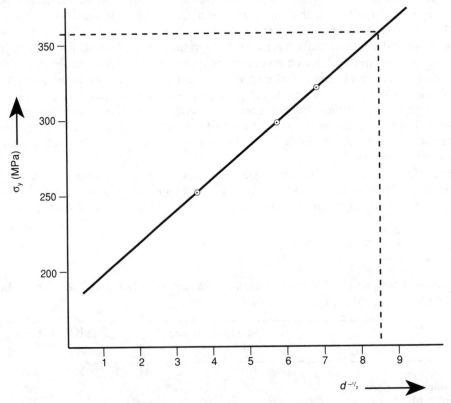

FIGURE 8.17   Graph of σ_y against $d^{-\frac{1}{2}}$

8.6 *Plastic deformation and strain hardening*

When a metal is stressed beyond its elastic limit, dislocations within the metal begin to move and plastic deformation occurs. The movement of one dislocation through a lattice will give one increment of plastic deformation, namely a displacement of the order of one atomic spacing. The movement of extremely large numbers of dislocations would be necessary to give a visible amount of plastic slip and the number of dislocations that would need to be present to account for the large amount of plastic deformation that takes place in commercial metal-working processes such as rolling or forging would be so great as to mean that there would be virtually no regular sections of lattice at all. In fact, the number of dislocations present in a fully annealed metal is comparatively small and additional dislocations are generated during plastic deformation. The dislocation density in a metal can be considered as N, the number of dislocations that intersect a unit area of 1 mm². For fully annealed metals the dislocation density, N, is usually between 10^4 and 10^6/mm². The number of atoms per unit area in a typical metal crystal is of the order of 10^{13}/mm² and so it can be seen that the dislocations in an annealed metal are widely separated. There is a much higher dislocation density in plastically deformed metals and typical values for heavily cold worked material would be between 10^9 and 10^{10}/mm².

One suggested mechanism for the generation of dislocations is the Frank–Read source. This supposes a length of dislocation line firmly anchored at each end. Application of a shear force will cause the dislocation line to bow. The successive stages of movement of a dislocation line AB, fixed at both A and B, is shown in Figure 8.18. As movement and growth continues a kidney-shaped loop (Figure 8.18(iv)) develops. The two sections of the loop advancing towards one another are of opposing sign and will cancel out when they meet (Figure 8.18(v)) forming a complete loop and a new dislocation line between A and B, thus allowing the process to be repeated.

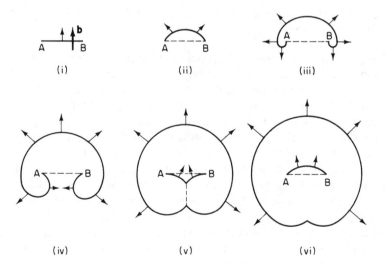

FIGURE 8.18 Dislocation generation from a Frank–Read source

Frank–Read sources and dislocation generation have been observed experimentally by means of transmission electron microscopy of thin metal foils. The effect of expanding dislocation loops from a Frank–Read source giving plastic deformation is shown in Figure 8.19.

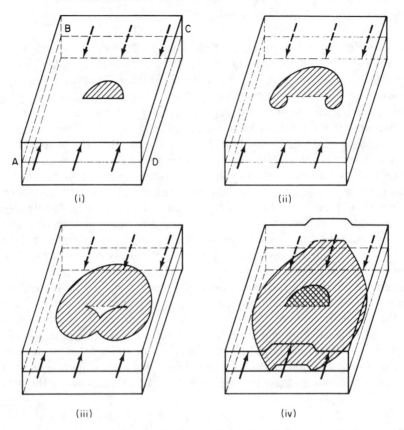

FIGURE 8.19   Dislocation loops developing from a Frank–Read source on slip plane ABCD and causing plastic deformation

As plastic deformation occurs, the level of applied stress must be continually increased if the process is to be continued. In other words, as more plastic deformation is given to the metal, the greater will be the force required to continue deforming the material, and the metal can be said to become strengthened. Dislocations cannot move freely throughout the whole of a crystalline material. Their progress will be impeded by barriers such as grain boundaries. A continued movement of the dislocations, that is, further plastic deformation, can occur only if the applied stress is raised so that the dislocations can move across the potential barriers. Also, in cubic crystals, where slip may be occurring simultaneously on several intersecting slip-planes, the dislocations may interfere with one another's movement. After severe cold working, the array of dislocations within the metal will be in a highly tangled state, but there will be some areas that are almost dislocation free while other areas will have a very high dislocation density. Figure 8.20 is a diagrammatic representation of the dislocation distribution in a sample of heavily cold worked metal. This type of pattern can be seen using transmission electron microscopy.

There is a high degree of structural disorder in those areas with a high dislocation density and these eventually become new crystal grain boundaries. One of the results of cold working is that the crystal grains of the original annealed metal become distorted and fragmented.

This phenomenon of strain, or work, hardening is utilised in practice for the strengthening of metals. By performing cold deformation operations, such as rolling or drawing, the strength of ductile metals may be increased considerably (see Table 8.2). Strain hardening is accompanied by a decrease in ductility and an increase in the electrical resistivity of the material.

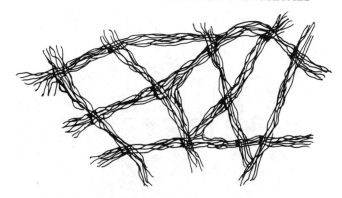

FIGURE 8.20   Sketch showing formation of dislocation tangles in heavily cold worked metal (from an electron micrograph). The dislocation-free areas become small crystals

Table 8.2   *Effect of cold rolling on the properties of commercial purity aluminium*

Reduction in thickness (per cent)	Condition	Harness (V.P.N.)	Tensile strength (MN/m²)	Elongation (per cent)
0	Annealed	20	92	40
15	Quarter hard	28	107	15
30	Half hard	33	125	8
40	Three-quarters hard	38	140	5
60	Fully work hardened	43	155	3

If attempts are made to continue cold working beyond the point where maximum hardness is achieved, cracks will develop within the material and failure will occur. The start of failure may be due to a number of dislocations of like sign on the same slip plane being forced together at a major barrier, such as a grain boundary, by a large applied stress, thus forming a void or internal crack (see Figure 8.21).

8.7 Recrystallisation

A metallic material that has been subjected to straining beyond the elastic limit becomes harder and stronger, but more brittle, and its crystal structure becomes deformed. When the external deforming stress is removed, the material will still be in a state of some internal stress. Each individual crystal of the original material will have been strained both elastically and plastically. Removal of the external force should allow for the recovery of all elastic strain, but for any individual crystal complete elastic recovery will be hindered by the surrounding rigid crystals and there will be some locked-in elastic strain.

When a material in this condition is heated, changes will begin to take place. These changes may be classified under three headings:

(a) stress relief,
(b) recrystallisation,
(c) grain growth.

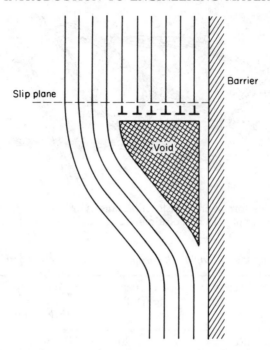

FIGURE 8.21   The pile-up of several dislocations of like sign can form a void which can be a starting point for total failure

As the temperature of the material is raised so the vibrational energies of the individual atoms are increased and atomic movements can occur. Comparatively minor atomic movements result in the removal of the residual stresses associated with the locked-in elastic strains. This change, which occurs at comparatively low temperatures, has a negligible effect on the strength and hardness of the material, and the microstructure of the metal is unchanged in its appearance. When the temperature is raised further the process of recrystallisation begins. New unstressed crystals begin to form and grow from nuclei until the whole of the material has a structure of unstressed polygonal crystals (Figure 8.22). This change in structure is accompanied by a reduction in hardness, strength, and brittleness to the original values prior to plastic deformation. The driving force for the recrystallisation process is the release of the strain energy stored in the zones of high dislocation density. The temperature at which recrystallistion occurs is, for a pure metal, within the range from one-third to one-half of the melting temperature (K). The recrystallisation temperature is not, however, constant for any material as its value is affected by the amount of plastic deformation prior to heating, and it is lower for very heavily cold worked metals than for samples of the same material which have received small amounts of plastic deformation. The recrystallisation temperature is also affected by composition. The presence of impurities or alloying elements will increase the recrystallisation temperature of the material. If the temperature is raised further, grain growth may occur following the completion of recrystallisation, with some crystal grains growing in size at the expense of others by a process of grain boundary migration. Again, the driving force for grain growth is the release of boundary surface energy as the amount of total grain boundary surface is reduced.

The industrial process of *annealing* is a heat treatment allowing recrystallisation to take place with consequent softening of work hardened materials. Some metals, such as lead, have recrystallisation temperatures below room temperature. This means that they cannot be work

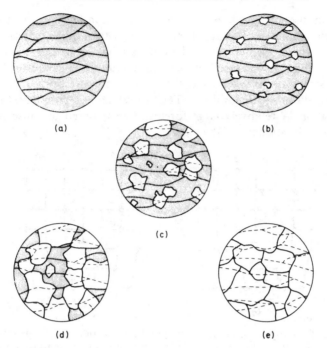

FIGURE 8.22   Recrystallisation: (a) original cold worked structure; (b) nuclei forming; (c) and (d) new grains growing from nuclei; (e) recrystallisation complete

hardened at ordinary temperatures. Note that many of the various annealing treatments for steels, while still being softening treatments, are based on a different principle (refer to Chapter 16).

Plastic deformation processes carried out at temperatures above the recrystallisation temperature are referred to as hot working. There is no strain hardening effect due to hot working as there is almost immediate recrystallisation. Hot working processes such as hot rolling, forging and extrusion are used for the initial major plastic deformation of large ingots in industrial metal processing.

8.8 Solution hardening

As mentioned in Section 8.3, the presence of interstitial and substitutional point defects introduces strain in the lattice and disturbs the regularity of crystal planes. This has the effect of increasing the stress necessary to cause movement of dislocations and, therefore, strengthens the metal. The presence of stranger atoms within the parent metal space lattice also affects the movement of valency electrons and increases the electrical resistivity of the metal.

Atoms can move, or diffuse, through a crystal lattice because there are vacancies within the lattice. An atom can move from a position adjacent to a vacancy into that vacant site. Conversely, it could be considered that the vacancy had moved in the opposite direction to the atom jump. There is a tendency for solute atoms within a crystal lattice to be associated with vacancies. In this way the extent of the strain associated with a solute atom will be reduced to a minimum. Movement of solute atom/vacancy pairs can result in diffusion, a movement of solute atoms through the crystal lattice. (There is further coverage of diffusion in Chapter 12.) Solute atoms will tend to diffuse or migrate to sites near dislocation lines as a means of reducing the lattice strain to a minimum. When solute atoms are larger than the parent metal atoms they will diffuse to sites below the dislocation line, where the strains are tensile (Figure 8.23(a)). Solute metal atoms smaller than parent metal atoms will migrate to sites above the dislocation line where compressive strains occur (Figure 8.23(b)). These concentrations of solute atoms close to dislocations are termed *Cottrell atmospheres*. The Cottrell atmospheres have the effect of pinning dislocations in position. A relatively large force will be required to cause the dislocations to move, thus increasing the strength of the alloy.

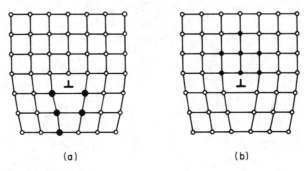

(a)   (b)

FIGURE 8.23   Cottrell atmospheres. These tend to pin the dislocations. (a) Large solute atoms migrate to sites below the dislocation. (b) Small solute atoms migrate to sites above the dislocation

Interstitial solute atoms also tend to migrate to sites near dislocations (see also Section 8.10). The alloying of metals to produce interstitial or substitutional solid solutions is a useful technique for increasing the strengths of metals (see also Section 11.5). Table 8.3 shows the hardness and strength increases obtained in copper–nickel solid solutions.

Table 8.3   *Solid solution hardening. Properties of annealed copper-nickel solid solutions*

Composition (per cent)		Hardness (V.P.N.)	Tensile strength (MN/m^2)	Elongation (per cent)
Cu	Ni			
100	—	60	210	65
95	5	65	230	50
90	10	70	250	45
80	20	75	315	45
70	30	80	370	45
60	40	90	430	45
30	70	120	520	45
—	100	95	450	50

8.9 Dispersion hardening

A third principle for the strengthening of metallic materials is termed *dispersion hardening*. A very finely dispersed second phase distributed throughout the crystal lattice will provide a series of barriers to the movement of dislocations, and considerably higher stresses will be necessary to cause yielding and plastic deformation than would be the case for a single-phase material. This dispersed second phase may be metallic or non-metallic. An example of a non-metallic dispersion is the use of thorium oxide to increase the high-temperature strength of tungsten wire in electric lamp filaments. Sintered aluminium powder (SAP) is another example of an oxide dispersion strengthened material. The compacting and sintering of finely powdered aluminium produces aluminium containing a very fine dispersion of aluminium oxide.

Precipitation hardening (age hardening) is another dispersion hardening process, but in this case the finely divided second phase is produced by giving an alloy a specific type of heat treatment. The types of alloys that may respond to this type of treatment are those in which partial solid solubility occurs (refer to Section 11.7). The first stage of the heat treatment produces a supersaturated solid solution (Figure 8.24(a)). Such a supersaturated solution is metastable and there will be a tendency for diffusion of solute atoms to occur leading to the formation of a precipitate. Such diffusion will be accelerated if the temperature is increased. The first results of the diffusion of solute atoms is to increase their concentration in some localised

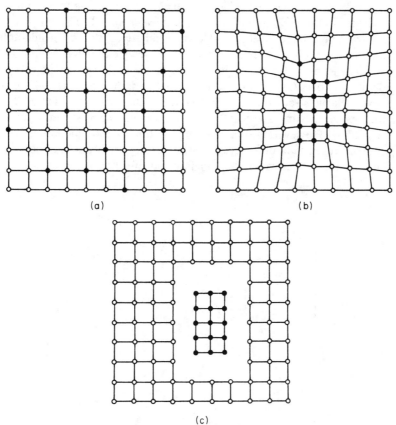

FIGURE 8.24   Formation of a precipitate. (a) Supersaturated solid solution with solute atoms uniformly distributed through the lattice. (b) Diffusion of solute atoms has occurred producing severe strain in area around concentration of solute element. The solute atoms are still coherent with the parent metal lattice, and termed a coherent precipitate. (c) Solute has now separated as a true non-coherent precipitate. Both parent metal and precipitate lattice are un-strained

areas of the crystal lattice. At this stage the solute atom aggregates are termed a *coherent* precipitate (Figure 8.24(b)). As the process continues the solute metal atoms separate from the parent metal lattice to form small crystals with their own lattice structure. At this stage the precipitate is a true precipitate, non-coherent with the parent metal (Figure 8.24(c)).

During the diffusion stage leading up to the formation of a coherent precipitate, areas of high strain are created within the parent lattice. These high-strain areas will hinder the movement of dislocations, thus strengthening the metal. Once true precipitation begins the strain in the parent lattice is released and the alloy loses the increased strength brought about by diffusion.

Certain commercial alloy systems, of which one of the best known is aluminium–copper, may be strengthened in this way. The dispersion strengthening of aluminium–copper alloys, also known as age hardening or precipitation hardening, is discussed in Sections 12.6 and 15.3.

8.10 Yield point in mild steel

In most metallic materials, there is a gradual transition from elastic to plastic strain with increasing stress, but in the case of mild steel there is a sharp discontinuity in the tensile stress–strain curve (Figure 8.25). The material behaves in an elastic manner up to a certain point, A, and then suddenly yields. Point A is termed the *upper yield point*. The stress level necessary to continue plastic straining falls to level B, the lower yield stress, and a considerable amount of plastic strain takes place at this lower level. After this sudden yielding has occurred, the stress has to be increased again to bring about further plastic strain and, beyond point C on the curve normal strain hardening occurs. If the metal is stressed to some point, D, on the curve and the stress then removed, elastic strain will be recovered along the path DO′. Immediate reloading of the test-piece will give elastic strain to point D with continued plastic strain beyond point D, but with no sudden yielding. However, if after stressing to point D and unloading, the sample is left at ordinary temperatures for about a week before retesting (or heated at 100°C for about one hour) it is found that the sudden yield phenomenon returns. This is termed strain ageing.

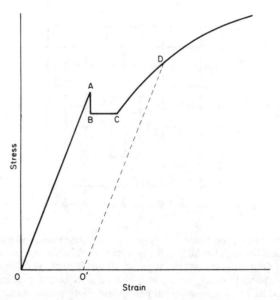

FIGURE 8.25   Tensile stress–strain curve for mild steel showing sharp yield point

Mild steel contains a very small percentage of carbon present in interstitial solid solution within the iron space lattice. The carbon atoms, small though they are, are somewhat larger than the interstices between iron atoms, and so the lattice is in a state of strain. The interstices along a dislocation line are slightly larger than the normal lattice interstitial sites and carbon atoms tend to migrate by diffusion through the lattice to positions along dislocations. This has the effect of partially locking the dislocation in position and requiring a relatively high stress, the upper yield stress, to initiate movement of the dislocation from the neighbourhood of carbon atoms.

Once the dislocations have been freed, their movement can be sustained at a lower stress level, the lower yield stress. During the ageing period, after stressing beyond the yield point and unloading, the interstitial atoms have a chance to diffuse through the lattice and return to positions on lines of dislocation. The diffusion process is speeded up if the temperature is raised. Pure iron completely free from interstitial impurities does not show a pronounced yield point.

8.11 Diffusion and dislocation climb

As stated earlier (Section 8.3) crystal lattices contain vacancies and it is possible, through the medium of vacancies, for a certain amount of atomic movement, or diffusion, to take place. An atom occupying a position next to a vacancy in a crystal lattice may jump from its original position into the vacant site. Alternatively, it could be said that the vacancy has moved one atomic position in the opposite direction. In the case of a pure metal with all atoms identical, this atomic movement cannot be observed, but observations and measurements have been made of diffusion in pure metals containing a small number of atoms of a radioactive isotope. The movements of these 'marked' atoms can be registered. Diffusion becomes of consequence in an alloy, or in a metal containing impurities, as the atoms of an alloying element or impurity may migrate through the parent lattice. Any individual atom, in moving into a vacant site, has to cross a potential barrier, and must possess an energy equal to, or greater than, q, in order to break away from the attraction of its neighbours. q is the activation energy for the movement of one atom. Diffusion is a thermally activated process and is governed by the Arrhenius rate law. The Arrhenius law and the laws of diffusion are dealt with in Chapter 12.

The migration of vacancies through a crystal lattice can lead to interaction with dislocations. When a vacancy reaches a point at a dislocation line, there will be a collapse of the vacancy at that point and the dislocation will move to the next lattice plane (Figure 8.26). The collapse of

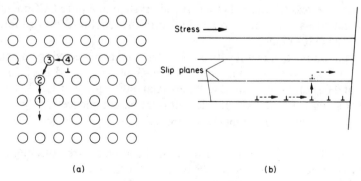

(a)   (b)

FIGURE 8.26   Dislocation climb. (a) Lattice containing vacancy and dislocation. If numbered atoms diffuse into the vacancy the vacancy will move to position 4 causing dislocation to climb. (b) Dislocation pile-up at barrier. The climb of a dislocation will allow further dislocation movement

one vacancy will only cause the dislocation line to be moved to the next plane at one point. This is termed a *jog*. If many vacancies migrate to the same area then a large portion, or the whole, of a dislocation line will move to the next plane. This is known as the climb of dislocations. This has significance in connection with the creep of metals. Metallic creep is the continued slow straining of the material under constant stress. In most metals, creep occurs at significant rates only at elevated temperatures, although lead will creep at ordinary temperatures. Figure 8.26(b) shows a series of dislocations, that have been moved by the action of a stress, piled up at a barrier. No further movement (plastic strain) can take place unless the level of stress is increased. However, if thermally activated diffusion leading to dislocation climb occurs, then some further dislocation movement may take place at constant stress. This has been postulated as a possible mechanism of metallic creep (see also Chapter 26).

8.12 Superplasticity

Plastic deformation through the medium of dislocation movement has been discussed in earlier sections but some alloys, under certain circumstances may be plastically deformed to an extremely large extent (elongation values of the order of a thousand per cent) under low deformation forces. This phenomenon is termed *superplasticity*. The superplastic alloys possess an extremely fine grain structure with grain diameters of less than 5 μm and, generally, possess a duplex microstructure. The strain rate must be closely controlled for superplastic deformation to occur. Neither very low nor high rates of strain will permit superplastic deformation. Superplastic behaviour occurs at high temperatures and a temperature between 0.5–0.7 T_m (K) is generally required. The mechanism of deformation is largely by grain boundary sliding. A good example of a superplastic alloy is the zinc–aluminium alloy of eutectoid composition (22 per cent Al) (see Section 15.11) which will show very large deformations when formed at 275°C. Other superplastic alloys of commercial importance include the titanium alloy containing 6% Al and 4% V, which can be deformed at about 930°C, and some of the Ni–Cr–Fe superalloys.

8.13 Questions

8.1   A direct stress of 100 MPa acts on a sample of material. Calculate the shear stress acting within planes at the following angles to the axis of direct stress: (a) 20°, (b) 40°, (c) 60°, (d) 80°.

8.2   The critical shear stress for the $<111>\{110\}$ slip system of a pure BCC metal is determined to be 23 MPa. What stress is required to be applied in the [00$\bar{1}$] direction to initiate slip in the [11$\bar{1}$] direction on the (110) plane?

8.3   A single crystal sample of a metal with a HCP structure is subjected to an axial tensile force. The sample is of circular cross-section, with a diameter of 4 mm, and is of 50 mm length. An elastic extension of 0.047 mm is measured when an axial tensile force of 1.25 kN is applied. When the tensile force is increased to 2.8 kN plastic deformation commences.

(a) Given that Poisson's ratio for the metal is 0.35, estimate the theoretical shear strength of the material.

(b) Given that the [0001] slip direction and the (0001) slip plane are both aligned at 35° to the tensile force axis, calculate the critical shear stress for the metal.

(c) Why do the values from (a) and (b) differ by so much?

8.4   Given that slip in a FCC metal is across a (111) plane in a [110] direction, calculate the length of the Burger's vector, b, in a sample of silver. (The lattice parameter for silver is 0.420 nm).

8.5   The density of dislocations in soft iron is approximately $10^5/mm^3$. Estimate the total length of dislocations in a 1 kg sample of iron. (Density of iron = $7870 \, kg/m^3$.)

8.6   Samples of cold worked and annealed commercial purity aluminium were subjected to tensile testing and the average grain size in each sample determined. The results obtained were:

Sample A   Tensile strength—97.5 MPa; grain size—800 grains/mm²
Sample B   Tensile strength—85.0 MPa; grain size—190 grains/mm²
Sample C   Tensile strength—95.0 MPa; grain size—600 grains/mm²

(a) Estimate the maximum grain size allowable if the minimum tensile strength required in the aluminium product is 87.5 MPa.

(b) Determine the value of σ_0 in the Petch relationship and explain its significance.

8.7   Copper and nickel, when alloyed, form a continuous range of FCC solid solution. The radii of copper and nickel atoms are 0.1276 nm and 0.1243 nm respectively. Calculate the proportion of nickel atoms in a solid solution if the lattice parameter of the alloy is determined as 0.358 nm.

8.8   Estimate the extent of lattice strain in the copper lattice when it contains in solid solution:
(a) 5 per cent nickel, (b) 5 per cent tin, (c) 5 per cent zinc.
   Atomic radii are: copper 0.1276 nm, nickel 0.1243 nm, tin 0.1508 nm, zinc 0.1329 nm.

9

Viscoelastic Behaviour

9.1 Introduction

The behaviour of polymeric materials, when subjected to stress, varies very considerably from one material to another and also is greatly affected by temperature. Thermoplastic materials, at temperatures below their *glass transition* temperatures, and thermosets are brittle elastic solids but thermoplastics, above T_g, show *viscoelastic* behaviour. In a viscoelastic material the deformation under stress is time-dependent and the phenomenon of *creep* can occur. Also, in such a material, there can be *relaxation*, that is, a reduction in the level of stress with time at constant strain. A consequence of viscoelasticity is that parameters such as modulus, strength and ductility, are functions of time and rate of strain.

9.2 Viscoelasticity

For an ideal elastic material the strain developed, ε, is directly proportional to the stress applied, σ, and this may be expressed as $\varepsilon = \sigma \times$ constant. For this Hookean elasticity, this relation is usually rewritten as $\sigma = \varepsilon \times E$, the constant E being the *modulus of elasticity*.

In the case of a Newtonian fluid, an ideal viscous fluid, the rate of strain, $d\gamma/dt$, is directly proportional to the magnitude of the shear stress, τ, and this may be expressed as $\tau = d\gamma/dt \times$ constant. In this case, the constant is the *viscosity*, η.

Thermoplastic materials at temperatures above T_g possess properties which lie somewhere between these two ideal cases and show some aspects of elasticity and also some viscosity, hence the term *viscoelastic solids*. In a viscoelastic solid, stress, σ, is a function of both strain, ε, and time, t, and can be expressed by an equation: $\sigma = f(\varepsilon, t)$ or $\sigma = \varepsilon f(t)$. It is also temperature dependent as viscosity decreases with increase in temperature according to the Arrhenius law (see Chapter 12).

Viscoelastic behaviour can be simulated using a series of models. These models are composed of two simple elements, a Hookean spring, representing ideal elasticity for which strain $\varepsilon = \sigma/E$, and a dashpot, representing Newtonian viscosity for which strain rate $d\varepsilon/dt = \sigma/\eta$, combined in various ways. Several models have been proposed and some of these are considered below.

9.3 The Maxwell model

The Maxwell model for viscoelasticity consists of a Hookean spring and a dashpot combined in series (Figure 9.1).

FIGURE 9.1   Maxwell model

Consider the behaviour of this model when subjected to some constant stress, σ_c. There will be an instantaneous strain, ε_1, developed in the spring, such that:

$$\varepsilon_1 = \sigma_c / E \tag{1}$$

and a continuous straining at a constant rate of the dashpot occurs according to the equation:

$$d\varepsilon / dt = \sigma_c / \eta \tag{2}$$

The total strain developed after some time t, ε_t, is given by:

$$\varepsilon_t = \sigma_c \left(\frac{1}{E} + \frac{t}{\eta} \right) \tag{3}$$

This behaviour is shown graphically in Figure 9.2(a) (the section between zero time and t_1). If the strain is then maintained constant until some time t_2 the phenomenon of stress relaxation will occur (Figure 9.2(b)), the stress reducing in this time interval from σ_c to σ_1.

From Equation (1), the strain rate is

$$\frac{d\varepsilon}{dt} = \frac{1}{E} \times \frac{d\sigma}{dt}$$

Combining this with Equation (2), we have

$$\frac{d\varepsilon}{dt} = \left(\frac{1}{E} \times \frac{d\sigma}{dt} \right) + \left(\frac{1}{\eta} \times \sigma \right) \tag{4}$$

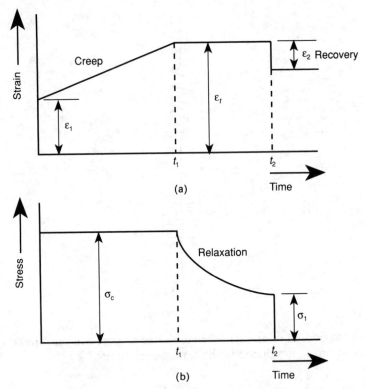

FIGURE 9.2   (a) Strain response of a Maxwell model. Stress is maintained constant until time t_2. (b) Stress response of a Maxwell model showing relaxation between t_1 and t_2

At constant strain $d\varepsilon/dt = 0$. The solution of the differential Equation (4) when $d\varepsilon/dt = 0$ and the condition that when $t = 0$ then $\sigma = \sigma_c$ becomes:

$$\sigma_t = \sigma_c \exp \left(\frac{-E}{\eta}\right)t \text{ where } \sigma_t \text{ is the stress at some time } t \qquad (5)$$

In other words stress relaxation is exponential, as shown in Figure 9.2(b) with a time constant of η/E.

If the stress relaxes from σ_c to σ_1 by time t_2 and the stress is then removed completely, there will be a corresponding *recovery* of elastic strain of amount ε_2, as shown in Figure 9.2(a), such that $\varepsilon_2 = \sigma_1/E$.

The Maxwell model gives a reasonable approximation to the stress relaxation phenomenon in thermoplastics but does not correspond particularly well to either the creep behaviour or the elastic recovery of these materials.

9.4 Voigt–Kelvin model

The Voigt–Kelvin model for viscoelasticity comprises a Hookean spring and a dashpot arranged in parallel (Figure 9.3).

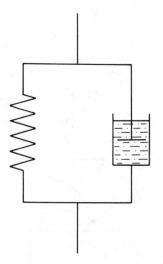

FIGURE 9.3   Voigt–Kelvin model

Consider the response of this model when subject to a stress, σ. As the two elements of the model are arranged in parallel the stress will be distributed between the elements such that:

$$\sigma = \sigma_s + \sigma_d$$

where σ_s is the stress in the spring and σ_d is the stress in the dashpot, so that, from Equations (1) and (2):

$$\sigma = E\varepsilon + \eta\,d\varepsilon/dt \tag{6}$$

Which may be rewritten as:

$$\frac{dt}{\eta} = \left(\frac{d\varepsilon}{\sigma - E\varepsilon}\right)$$

Integration gives

$$\frac{t}{\eta} = -\frac{1}{E}\ln(\sigma - E\varepsilon) \quad \text{or} \quad \exp\left(\frac{-Et}{\eta}\right) = \sigma - E\varepsilon$$

From which

$$\varepsilon = \frac{\sigma}{E}\left\{1 - \exp\left(\frac{-Et}{\eta}\right)\right\} \tag{7}$$

This equation indicates that when a stress, σ, is applied to the model strain will increase exponentially from zero to a value of σ/E, that is, the instantaneous strain which would have been developed in the spring alone in the absence of a dashpot. This behaviour is illustrated in Figure 9.4(a).

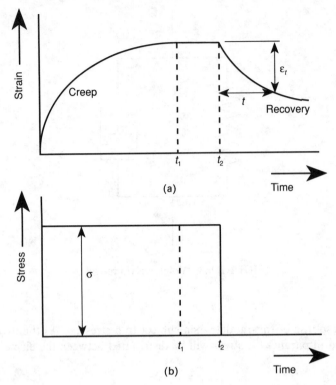

FIGURE 9.4   (a) Strain response of a Voigt–Kelvin model. Strain is maintained constant until t_2, when stress is removed. (b) Stress response of a Voigt–Kelvin model. No relaxation occurs between t_1 and t_2

The Voigt–Kelvin model does not show any stress relaxation as when the maximum strain has been developed, at time t_1, the dashpot is fully extended and the elastic spring element bears all the stress. Maintaining constant strain for a time interval between t_1 and t_2 causes the stress to remain constant also. (See Figure 9.4.)

If the stress is removed there will be strain recovery. With zero stress, Equation (6) becomes:

$$E\varepsilon + \eta \, d\varepsilon/dt = 0$$

and the solution of this differential equation gives:

$$\varepsilon_t = \varepsilon \exp\left(\frac{-E}{\eta}\right)t$$

where ε_t is the reduced strain at some time t after the removal of stress. This exponential recovery is shown in Figure 9.4(a).

The Voight–Kelvin model gives a response which corresponds to both the creep and recovery behaviour of thermoplastics but is deficient in terms of relaxation.

9.5 Other models

In the preceding sections it was shown that the Maxwell model was realistic in terms of relaxation but does not accurately relate to creep and recovery while the Voigt–Kelvin model approximates to the creep and recovery of many thermoplastics but shows no relaxation phenomenon. Various complex models comprising combinations of these two simple models have been proposed. One such is the Maxwell–Kelvin model in which a Maxwell model and a Voigt–Kelvin model are combined in series as shown in Figure 9.5. The response of this model shows creep, recovery and stress relaxation and is a closer approximation to the actual behaviour of a thermoplastic than either model singly. The response of a Maxwell–Kelvin model is shown in Figure 9.6.

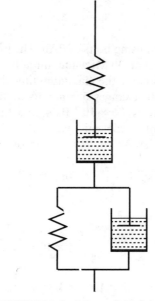

FIGURE 9.5   Maxwell–Kelvin model

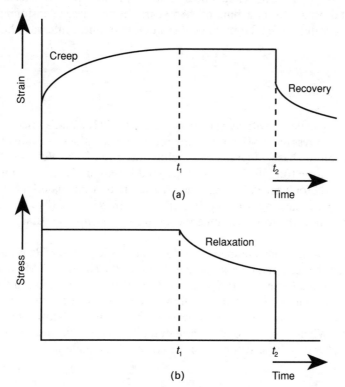

FIGURE 9.6   (a) Strain response of a Maxwell–Kelvin model. (b) Stress response of a Maxwell–Kelvin model

The Maxwell–Kelvin model, while showing creep, recovery and relaxation, is only an approximation and does not truly reflect the complex responses of real thermoplastic materials. More accurate simulations can be achieved by making other models with more Maxwell and Kelvin elements added in various combinations. However, in these more complex models the mathematical analysis becomes very much more involved and complicated.

Example

A coupling in a fluid line is fastened using bolts of PA6. The joint may be subject to leakage if the tensile stress in the bolts is less than 30 MPa and, initially, the bolts are tightened until the tensile stress in them is 45 MPa. What is the maximum time that should be allowed to elapse before the bolts are retightened if leakages are not to occur? Assume that E for PA6 is 2.5 GPa, that the viscosity of PA6 is 7.5×10^{17} Pa s, and that the relaxation of PA6 is as predicted by the Maxwell model of viscoelasticity.

According to the Maxwell model, stress relaxation is according to Equation (5).

$$\sigma_t = \sigma_c \exp\left(\frac{-E}{\eta}\right) t \quad \text{or} \quad \ln\left(\frac{\sigma_c}{\sigma_t}\right) = \frac{Et}{\eta}$$

Substituting

$$\ln\left(\frac{45 \times 10^6}{30 \times 10^6}\right) = 0.4055 = \frac{2.5 \times 10^9 t}{7.5 \times 10^{17}} = 3.333 \times 10^{-9} t$$

from which

$$t = 1.22 \times 10^8 \text{ s} \simeq 3.85 \text{ years.}$$

According to this, if the Maxwell model accurately reflects the stress relaxation of PA6, no leakages should occur within a time of 3.85 years. In practice, retensioning of the holding bolts would occur during regular maintenance at more frequent intervals than this maximum time.

9.6 Questions

9.1   A thermoplastic material with an elastic modulus of 1.5 GPa and a viscosity of 100 GPa s is subject to a constant stress of 15 MPa. Calculate the predicted values of strain after (a) 2 minutes and (b) 5 minutes on the basis of (i) the Maxwell model and (ii) the Voigt–Kelvin model.

9.2   A cylindrical plastic rod, of 12 mm diameter, and length of 300 mm is subjected to an axial tensile force of 1400 N for a time of 20 minutes. Assuming that the material behaves as a Voigt–Kelvin model, estimate the length of the rod after: (a) 10 minutes, (b) 20 minutes and (c) 20 minutes after removal of the force. Assume the spring and dashpot constants are 900 MPa and 300 GPa s.

9.3   A rubber band with an instantaneous modulus of elasticity of 4 MPa is stretched around a sheaf of papers and the instantaneous stress developed in the rubber is 2 MPa. The stress within the rubber is determined to be 1.9 MPa after 40 days. Estimate the stress in the rubber after a period of 1 year, assuming that relaxation follows the pattern of the Maxwell model.

9.4   A thermoplastic material which can be represented by a Maxwell model has a tensile modulus of 1 GPa. When in a state of constant strain it is found that the stress reduces to 0.6 of its original value after 300 days. Estimate the viscosity of the material.

10

Toughness and Fracture of Materials

10.1 Introduction

Not all materials fail in the same manner and the terms *tough*, *ductile* and *brittle* are frequently used to describe the fracture behaviour of a material. In a tough, or ductile fracture, failure is preceded by a considerable amount of plastic deformation and considerable energy is required to bring about fracture. On the other hand, in a brittle, or non-ductile, fracture there is little or no plastic deformation prior to failure and the energy input necessary to bring about fracture is relatively low. The type of fracture which occurs is largely dependent on the nature and condition of the material but it is also affected by other factors, including the type of stress applied, the rate of application of stress, temperature and environmental conditions. Other factors that have an effect on the failure behaviour of a material are the component geometry and surface condition. Stress concentration effects will occur at a change in section, a surface imperfection or internal flaw.

In some circumstances, components made from materials which would be expected to yield and plastically deform if overstressed may suddenly fracture in a brittle manner, the fracture growing from some existing crack or flaw at sonic speed, even though the average level of stress within the material is well below the design stress. The resistance of a material to fast crack propagation is termed *fracture toughness*.

10.2 Crack propagation in materials

Most commercial materials contain internal cracks and other defects. Cracks within a material subject to a stress propagate by one of two methods, either by ductile tearing or by cleavage. In the former process, which applies to ductile metals and thermoplastic materials at temperatures higher than T_g, considerable energy is absorbed in plastically deforming the material just ahead of an advancing crack. The plastic deformation tends to blunt the tip of the crack with a consequent reduction in the stress concentrating effect of the crack. However, the severe plastic deformation tends to create small cavities in the deformed area ahead of the crack tip and these eventually link up, so extending the crack length. In non-ductile or brittle materials little, if any, plastic deformation can occur ahead of a crack tip to blunt the crack and the local stress developed at the crack tip exceeds the fracture stress of the material and ruptures the interatomic bonds. The crack then develops catastrophically as a fast, brittle fracture. Materials which show this behaviour are ceramics, glasses, brittle metals, thermosets and thermoplastics at temperatures below T_g.

Many fibre reinforced composite materials possess higher toughness than either the matrix or reinforcing material on their own. The growth of a crack through the matrix in a direction normal to the line of reinforcing fibres can be stopped at a fibre. The high stress concentrated at the crack tip can cause a rupture of the bonding between matrix and fibre, thus blunting the crack tip and reducing the local stress so that fracture cannot proceed into the fibre. This crack stopping effect only applies to crack growth in a direction normal to the fibre direction. Crack growth in a direction parallel to the fibre direction can proceed without hindrance. A good example of this is the fracture characteristics of timber, a fibrous material. Timbers split easily along the grain but give a tough fracture across the grain.

The theoretical strength of a material, σ_t, is given by: $\sigma_t = (E\gamma/b)^{\frac{1}{2}}$ where E is the modulus of elasticity, γ is the surface energy and b is the interatomic spacing. The calculated value of σ_t for a material is approximately one-tenth of the value of E. The actual strengths of most materials are very much lower than this theoretical value, differing by some factor between 10^3 and 10^5. Real materials contain structural imperfections and small cracks. The presence of a crack will cause a high local concentration of stress. The classical equation which may be used for calculating the stress concentrating effect of an elliptical defect lying normal to the stress axis is given by the expression: $\sigma_{loc} = \sigma(1 + 2(a/r)^{\frac{1}{2}})$, where σ_{loc} is the local stress at the defect tip, σ is the nominal stress, $2a$ is the size of the defect and r is the radius at the defect tip (refer to Figure 10.1). (Note that $2a$ is the size of a wholly enclosed flaw while for a defect which breaks the surface the size is quoted as a.)

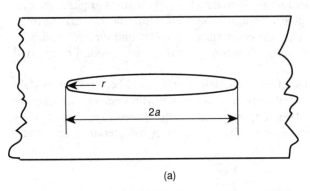

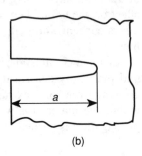

(a)

(b)

FIGURE 10.1   (a) Enclosed elliptical defect. (b) Edge defect

For a defect of high aspect ratio with a crack tip radius which is small in relation to the defect size, then

$$\sigma_{loc} = \sigma(1 + 2(a/r)^{\frac{1}{2}}) \simeq 2\sigma(a/r)^{\frac{1}{2}}.$$

The quantity $(1 + 2(a/r)^{\frac{1}{2}})$ or $2(a/r)^{\frac{1}{2}}$ is termed the *stress concentration factor*, K_t. Consider the relative stress concentrating effects of (i) a circular hole and (ii) small enclosed crack of length 2 mm with a crack tip radius of 1 μm.

For (i) $a = r$ and so the stress around the periphery of the hole is: $\sigma_{loc} = \sigma(1 + 2) = 3\sigma$.
For (ii) $\sigma_{loc} = \sigma(2(1 \times 10^{-3}/1 \times 10^{-6})^{\frac{1}{2}}) \simeq 63\sigma$.

10.3 Fracture mechanics

Fracture mechanics is the study of the relationships between crack geometry, material strength and toughness, and stress systems, as they affect the fracture characteristics of a material. The aim of fracture mechanics is to determine the critical size of a crack or other defect necessary for the occurrence of fast fracture, that is, catastrophic crack propagation and failure, under service loading conditions.

The analysis of the critical stress to cause crack propagation in brittle materials was made by Griffith who proposed the relationships:

$$\sigma_c = \left(\frac{2\gamma E}{\pi a}\right)^{\frac{1}{2}} \text{ (for plane stress)}$$

$$\sigma_c = \left\{\frac{2\gamma E}{\pi(1-v^2)a}\right\}^{\frac{1}{2}} \text{ (for plane strain)}$$

where σ_c is the critical stress for fracture, γ is the fracture surface energy per unit area, E is the modulus of elasticity, a is one half the crack length, and v is Poisson's ratio. The plane stress version is used for crack propagation in thin materials.

Example 10.1

(a) Determine the fracture stress of a glass if it contains a series of microcracks with a maximum dimension of 10 μm. Assume that E for glass is 70 GPa and the surface energy is 0.6 J/m².

(b) What would be the likely fracture stress for the glass if a fine scratch with a depth of 0.3 mm was made on the surface?

(a) Using the Griffith equation for plane stress, $\sigma_c = (2\gamma E/\pi a)^{\frac{1}{2}}$ and submitting the values for γ, E and a ($2a = 10$ μm, $a = 5 \times 10^{-6}$m) we have:

$$\sigma_c = \left(\frac{2 \times 0.6 \times 70 \times 10^9}{\pi \times 5 \times 10^{-6}}\right)^{\frac{1}{2}} = 73.1 \times 10^6 \text{ Pa} = 73.1 \text{ MPa}$$

(b) For a surface scratch the depth is taken as a, so $a = 3 \times 10^{-4}$ m

$$\sigma_c = \left(\frac{2 \times 0.6 \times 70 \times 10^9}{\pi \times 3 \times 10^{-4}}\right)^{\frac{1}{2}} = 9.44 \times 10^6 \text{ Pa} = 9.44 \text{ MPa}$$

The presence of a small surface scratch causes a major reduction in the fracture stress of the glass.

The Griffith relationship only applies to brittle materials and was modified by Orowan and Irwin to take account of the plastic flow which occurs at a crack tip at the beginning of crack development. With the assumption that the size of the plastic zone at the crack tip is very small, the Orowan–Irwin modification becomes:

$$\sigma_c = \left(\frac{2(\gamma+\gamma_p)E}{\pi(1-v^2)a}\right)^{\frac{1}{2}}$$

where γ_p is the work done in plastic deformation per unit area of crack extension. The equation is generally written as:

$$\sigma_c = \left(\frac{G_cE}{\pi(1-v^2)a}\right)^{\frac{1}{2}} \text{ or } \sigma_c = \left(\frac{G_cE}{\pi a}\right)^{\frac{1}{2}} \text{ for plane stress}$$

The term G_c ($G_c = 2(\gamma + \gamma_p)$) is a property of the material and is termed the *toughness*, or *critical strain energy release rate*, and has the units J/m^2.

There are three possible ways to apply stress to a crack so as to cause a crack extension. These are in plane strain (opening mode I), in plane shear (opening mode II), and in anti-plane shear (opening mode III). These modes are shown in Figure 10.2. The respective toughnesses for each opening mode are written as G_{Ic}, G_{IIc} and G_{IIIc}.

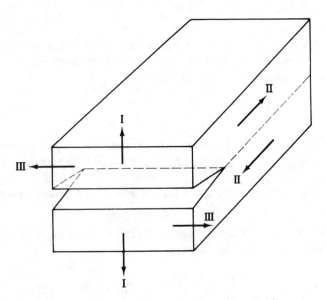

FIGURE 10.2   Crack opening modes

The toughness value G_{Ic} is the value generally quoted for materials as mode I opening is the easiest mode for crack extension and, hence, represents the lowest toughness value. Values of toughness, G_{Ic}, for a range of materials are given in Table 10.1.

The plane stress equation above can be rewritten as $\sigma_c(\pi a)^{\frac{1}{2}} = (G_cE)^{\frac{1}{2}}$. Both G_c and E are material properties, that is, constants for the material. It follows that the particular combination of stress and defect size at which fracture will occur must also be a material constant.

As stated above, the aim of fracture mechanics is to determine the critical values of stress and crack size necessary for fast fracture. The following examples should aid understanding.

Example 10.2

If the fracture stress of a sheet of high-tensile steel containing a central crack of length 30 mm is 500 MPa, calculate the fracture stress in the case of a 50 mm crack in a similar sheet.

$$\sigma_c = (G_{Ic}E/\pi a)^{\frac{1}{2}} \text{ which may be written as } \sigma_c = \text{constant} \times a^{-\frac{1}{2}}$$

When $2a = 30$ mm (0.03 m) $\sigma_c = 500$ MPa.

Table 10.1 *Toughness and fracture toughness values for some materials*

Material	Toughness G_{Ic}(kJ/m²)	Fracture toughness K_{Ic}(MPa m$^{\frac{1}{2}}$)	Material	Toughness G_{Ic}(kJ/m²)	Fracture toughness K_{Ic}(MPa m$^{\frac{1}{2}}$)
Steels	30–135	80–170	acrylic (PMMA)	0.4–1.6	0.9–1.6
Cast irons	0.2–3	6–20	nylons (PA)	1–4	3–5
Aluminium alloys	0.4–70	5–70	rigid PVC		
			(UPVC)	1.3–1.4	1–4
Copper alloys	10–100	30–120	polyester resins	0.1–0.2	0.5–1
Nickel alloys	50–110	100–150	epoxy resins	0.1–0.3	0.3–0.5
Titanium alloys	20–100	50–100	GFRP	20–100	20–60
Polyethylene (LDPE)	6–10	1–2	CFRP	5–30	30–45
Polyethylene (HDPE)	3–7	1–5	soda glass	0.01	0.7
Polypropylene (PP)	7–10	3–4	alumina	0.02	3–5
Polycarbonate (PC)	0.4–5	1–2.5	silicon nitride	0.1	4–5
Polystyrene (PS)	0.4–0.8	0.7–1.1	concrete	0.03	0.2

So: $$\text{constant} = \sigma_c \times a^{\frac{1}{2}} = 500 \times 0.015^{\frac{1}{2}} \text{ MPa m}^{\frac{1}{2}}$$

For a 50 mm crack $a = 25$ mm (0.025 m)

Therefore
$$\sigma_c = 500 \times (0.015/0.025)^{\frac{1}{2}}$$
$$= 387 \text{ MPa}$$

For a crack of length 50 mm the critical fracture stress will be 387 MPa.

Example 10.3

Determine the maximum crack size that can be tolerated in a sheet of high-tensile material with a tensile yield stress of 1100 MPa and a toughness of 25 kJ/m², if the design stress safety factor is 1.5. Assume that E for the material is 208 GPa.

The maximum design stress = yield stress $\div$ safety factor = 1100/1.5 = 733.3 MPa

Since
$$\sigma = \left(\frac{G_{Ic}E}{\pi a}\right)^{\frac{1}{2}}$$

So
$$733.3 \times 10^6 = \left(\frac{25 \times 10^3 \times 208 \times 10^9}{\pi a}\right)^{\frac{1}{2}}$$

and hence
$$a = \frac{25 \times 208 \times 10^{12}}{(733.3 \times 10^6)^2 \, \pi} = 3.08 \times 10^{-3} \text{ m} = 3.08 \text{ mm}$$

The maximum crack size that can be tolerated is a crack of length $2a$, namely 6.16 mm.

Irwin noted that stresses at a crack tip were proportional to $(\pi a)^{-\frac{1}{2}}$ and proposed a parameter, K, the *stress intensity factor*, where $K = \sigma(\pi a)^{\frac{1}{2}}$. K has the units MPa m$^{\frac{1}{2}}$. It should be noted that this expression is based on an elliptical flaw under plane stress in a plate of infinite size. The general form of the equation can be written as $K = \sigma M(\pi a)^{\frac{1}{2}}$, where M is a *geometry factor* to take account of varying flaw shapes and finite boundaries. Stress intensity factors for some geometries are given in Table 10.2. This factor, K, while it is associated with stress distribution close to the tip of a crack, should not be confused with the *stress concentration factor*, K_t, described in Section 10.2.

Table 10.2   *Stress intensity factors for some geometries*

Type of crack	Stress intensity factor, K
Centre-crack of length $2a$ in infinite plate	$K = \sigma(\pi a)^{\frac{1}{2}}$
Centre-crack of length $2a$ in plate of width W*	$K = \sigma(\pi a)^{\frac{1}{2}}\{W/\pi a \times \tan(\pi a/W)\}^{\frac{1}{2}}$ or $K = \sigma(\pi a)^{\frac{1}{2}}\{\sec(\pi a/W)\}^{\frac{1}{2}}$
Single edge crack of length a in semi-infinite plate	$K = 1.12\,\sigma(\pi a)^{\frac{1}{2}}$
Two symmetrical edge cracks, each of length a in plate of width W	$K = \sigma(\pi a)^{\frac{1}{2}}\{[W/\pi a \times \tan(\pi a/W)] + [0.2W/\pi a \times \sin(\pi a/W)]\}^{\frac{1}{2}}$
Central penny-shaped crack of diameter $2a$ in infinite body	$K = \sigma(\pi a)^{\frac{1}{2}}(2/\pi)$

* Of the two functions listed the sec function is probably the more accurate.

Sudden fast fracture will occur when K reaches some critical value, K_c. The parameter $K_c = (G_c E)^{\frac{1}{2}}$ and, hence, is a constant for the material. K_c, the *critical stress intensity factor*, is generally termed the *fracture toughness* of the material and has the units MPa m$^{\frac{1}{2}}$. The fracture toughness value generally quoted for a material is that for K_{Ic}, the value for mode I crack opening. Values of K_{Ic} for a range of materials are given in Table 10.1.

10.4 Applications of K_{Ic}

No material is wholly free from defects but it is essential that any crack-like defects that are present are relatively harmless. Using K_{Ic} values of fracture toughness it is possible to calculate the defect size necessary to initiate fast fracture failure at a given value of applied stress, or the value of applied stress required to cause failure in a component containing a defect of a known size. A high value of fracture toughness, K_{Ic}, is an indication that the resistance to fracture will be high. Generally, as the hardness and tensile yield stress of a material increases so its fracture toughness decreases (see Figure 10.3). This is because materials with high yield strengths tend to have low values of ductility and their ability to plastically deform ahead of a crack with a consequent blunting of the crack tip is negligible.

It can be seen from the general expression $\sigma_c = K_{Ic}/(\pi a)^{\frac{1}{2}}$ that σ_c, the critical stress for fast fracture, reduces as the size of a flaw, denoted by a, increases. This relationship is shown graphically in Figure 10.4. Many materials, particularly metals and alloys, can plastically deform when stressed beyond a certain level, the yield stress. The value of yield stress, σ_y, is marked in the Figure showing that there is a critical flaw size, below which plastic yielding of a material will occur. If the material contains a flaw larger than this critical size it will fail by fast fracture before the yield stress is reached. Quantitative non-destructive testing methods can be used to determine the location and sizes of flaws in many materials. Thus it is possible to decide whether the dimensions of a discovered defect are below the critical value and thus 'safe', or above the critical value and fail catastrophically in service. Flaws in components which are subject to fluctuating

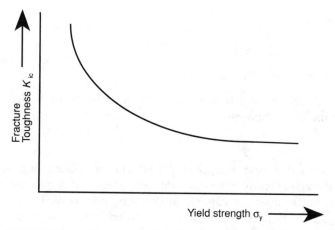

FIGURE 10.3   General relationship between yield strength and K_{Ic}

stresses in service can grow in size due to *fatigue*. Similarly, corrosion effects could produce crack growth with time. Thus, a flaw which was considered 'safe' initially may become unstable and reach critical dimensions after a period of time in service. Many engineering fabrications, such as pressure vessels, aircraft structures and bridges, are examined at regular intervals, using quantitative non-destructive testing techniques, and components taken out of service or the damage rectified before a crack has reached the critical size.

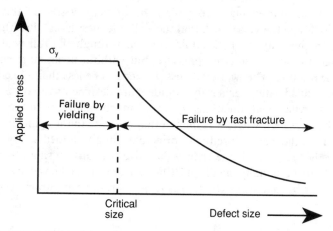

FIGURE 10.4   Plot of σ_c and σ_y for a material against defect size

Example 10.4

A component manufactured from a steel with a tensile yield strength of 1200 MPa and a fracture toughness of 150 MPa m$^{\frac{1}{2}}$ is found to contain a small crack of length 4.5 mm. Determine whether or not the component is safe in terms of its resistance to fast fracture assuming that the maximum applied stress is limited by the yield stress.

From the expression $\sigma_c = K_{Ic}/(\pi a)^{\frac{1}{2}}$, $a = (K_{Ic}/\sigma_c)^2 \div \pi$

Making $\sigma_c = \sigma_y$ we have:

$$a = \left(\frac{150 \times 10^6}{1200 \times 10^6}\right)^2 \div \pi = 4.97 \times 10^{-3}\,\text{m} = 4.97\,\text{mm}$$

from which the critical crack size, $2a = 9.94\,\text{mm}$.

The critical crack size for fast fracture is just over twice the size of the detected crack and it would be safe to use the component.

Example 10.5

An aluminium alloy with a yield strength of 475 MPa and a fracture toughness of 21 MPa m$^{\frac{1}{2}}$ is suggested for the manufacture of a structural component with a maximum design stress of $\sigma_y/2$. The inspection facilities are capable of detecting defects of 3 mm length or greater. Determine whether the inspection capability is sufficiently sensitive to detect internal flaws before they reach the critical dimensions for fast fracture.

Using the expression: $a = (K_{Ic}/\sigma_c)^2 \div \pi$ and making $\sigma_c = \sigma_y/2$ we have:

$$a = \left(\frac{2 \times 21 \times 10^6}{475 \times 10^6}\right)^2 \div \pi = 2.49 \times 10^{-3}\,\text{m} = 2.49\,\text{mm}$$

from which the critical crack size, $2a = 4.98\,\text{mm}$, which is well within the capabilities of the inspection system.

The failure of a pressure vessel in service by the mechanism of fast fracture because of the unmonitored slow growth of a flaw could be extremely dangerous. One measure which can be adopted to prevent this happening is to design the pressure vessel on the *leak before break* principle, that is, to design the vessel such that the wall thickness is less than the critical crack size for fast fracture. In this way, a crack could develop through the full wall thickness, with a consequent leakage and loss of internal pressure, but without being large enough to reach the critical size for fast fracture. The wall thickness, t, needs to be less than the critical crack size $2a$ but in practice, for added safety, generally cylinders are designed so that $t \simeq a$.

For a cylinder in which wall thickness, t, is very much smaller than the radius, r, the hoop stress, σ, is given by: $\sigma = Pr/t$, where P is the internal pressure in the cylinder. It will be seen that although designing on the leak before break principle will give added safety there may be some disadvantages in other respects. If the critical flaw size is relatively small then the cylinder wall thickness will have to be equally small and this can only be achieved by either reducing the internal pressure P or by increasing the radius r, or altering both parameters.

Example 10.6

A pressure cylinder is to be constructed from a steel with a yield strength of 1100 MPa and a fracture toughness of 110 MPa m$^{\frac{1}{2}}$. Safety factors of 2 on the yield strength and ratio of wall thickness to critical flaw size are to be used. If the maximum working pressure is to be 150 bar (1 bar $= 10^5$ Pa), calculate the maximum wall thickness and minimum cylinder diameter required if failure by leak before break is to occur.

Maximum design stress $= \sigma_y/2 = 550\,\text{MPa}$

Using $a = (K_{Ic}/\sigma)^2/\pi$

we have

$$a = \left(\frac{110}{550}\right)^2 \div \pi = 12.7 \times 10^{-3}\,\text{m} = 12.7\,\text{mm}.$$

Using $\sigma = Pr/t$, $P = 150$ bar $= 15\,$MPa and making $t = a = 12.7\,$mm

we have

$$r = \frac{550 \times 10^6 \times 12.7 \times 10^{-3}}{15 \times 10^6} = 0.466\,\text{m}$$

To satisfy the leak before break criteria with a maximum working pressure the cylinder should have a maximum wall thickness of 12.7 mm and a minimum diameter of $(2 \times 0.466) = 0.932$ m.

10.5 Effect of temperature

A change in temperature can have a significant effect on the fracture toughness of a material, a reduction in temperature increasing the liability to fail by brittle fracture. In the case of thermoplastic materials, there is a major change in the fracture toughness as the temperature falls through the glass transition temperature, T_g. There is also a major change in the behaviour of many metals.

An increase in temperature will render dislocations in metals more mobile and, hence, will cause a reduction in the yield strength. Conversely, a decrease in temperature will be reflected in an increased yield strength. Metals with face centred cubic structures remain ductile at all temperatures but metals with body centred cubic and hexagonal crystal structures undergo a change from ductile to brittle behaviour as the temperature is lowered. For example, zinc (HCP structure) is brittle at ordinary temperatures but will deform plastically at temperatures in excess of 100°C. Low carbon steel is ductile at low strain rates at all temperatures above a value of about -170°C, but when a low carbon steel is subjected to impact loading conditions the transition from a tough, ductile fracture to a brittle cleavage fracture occurs at about 0°C. The energy absorbed in fracturing a low carbon steel test-piece in a Charpy impact test may reduce from about $2\,\text{MJ/m}^2$ at $+15$°C to less than $200\,\text{kJ/m}^2$ at -10°C.

The cleavage fracture stress, σ_f, of a metal does not vary as much with temperature as does the yield stress, σ_y. In Figure 10.5(a), curve (i) shows the variation of σ_f with temperature, while curves (ii) and (iii) show the variation of σ_y with temperature for a body centred cubic metal. Curve (ii) is the effect for slow rates of strain and curve (iii) applies under conditions of high rates

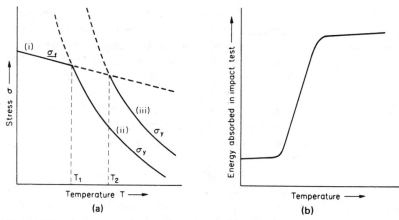

FIGURE 10.5 (a) Variation of σ_f and σ_y with temperature for a BCC metal: (i) $\sigma_f - T$, (ii) $\sigma_y - T$ for slow strain rate, (iii) $\sigma_y - T$ for fast strain rate; (b) Notch-impact energy against temperature showing ductile–brittle transition

of strain (impact loading conditions). At low rates of strain $\sigma_f < \sigma_y$ at all temperatures below T_1, while for impact conditions $\sigma_f < \sigma_y$ at temperatures below T_2. In other words, at temperatures between T_1 and T_2, the metal will show ductile behaviour at low rates of strain but will be brittle under impact conditions.

Figure 10.5(b) shows the relationship between impact strength and temperature, as measured in a Charpy notch-impact test, for body centred cubic metal. The transition from tough to ductile behaviour occurs over a narrow range of temperature but, generally, a specific transition temperature is quoted. In the case of the transition in steels, the transition temperature is that at which the energy for fracture of a standard size Charpy test-piece is 40 J (200 kJ/m²). The ductile–brittle transition temperature, which is around 0°C for a low carbon steel, is affected by alloying and impurity elements. It is reduced by manganese and nickel but increased by carbon, nitrogen and phosphorus.

It should be noted that the parameter of toughness, as measured in a notch-impact test of the Charpy or Izod type, is not the same as the fracture toughness of a material. The units of notch-impact toughness are J/m^2 while those of fracture toughness are MPa $m^{\frac{1}{2}}$.

10.6 Determination of fracture toughness

Standard tests for fracture toughness are designed to determine reproducible values for K_{Ic} (mode I crack opening). Two types of standard test-piece may be used. These are the single edge-notched bend test-piece (SEN) and the compact tensile test-piece (CTS) (see Figures 10.6 and 10.7). Both types of test-piece are prepared with a notch conforming to standard dimensions and with a crack emanating from the base of the notch. The crack is generally developed from the notch by a fatigue process to a pre-determined size. During a fracture toughness test the values of applied force required to cause given amounts of crack extension are measured. The full details of the test procedures and the manner in which K_{Ic} can be determined from the force-displacement results is given in BS 5447 'Methods of Test for Plane-Strain Fracture Toughness (K_{Ic}) Testing of Metallic Materials' and in ASTM E 399-83.

10.7 Yielding fracture mechanics

The linear elastic fracture mechanics principles outlined earlier in this chapter are used in connection with high-strength materials where valid K_{Ic} values can be obtained using relatively small test-pieces. However, with low-strength, high fracture toughness materials such as low-carbon plain carbon steels, it would be necessary to use an extremely large test-piece in order to satisfy the standard conditions for the determination of K_{Ic} and this would be impracticable.

If the influence of plasticity on the stress field at a crack tip is considered then, for a plastic zone of length $2r_y$ at the tip of a crack of length $2a$, the relationship $K_{Ic} = \sigma\{\pi(a + r_y)\}^{\frac{1}{2}}$ is valid, where r_y is given by $\frac{1}{2}a(\sigma_c/\sigma_y)^2$. The use of this modified relationship is not appropriate if the ratio of applied stress to yield stress is large and the size of the plastic zone becomes large, but it does permit the use of a smaller sized test-piece.

In the case of materials that show a high degree of plasticity, a small test-piece could show general yielding before fracture, even though a large structure could fail by fast fracture before general yield takes place. In these circumstances the plane strain fracture toughness is estimated from *crack opening displacement* (COD) measurements. A test-piece of the SEN type (Figure 10.6) is used and displacement is measured by means of a clip gauge attached above the crack.

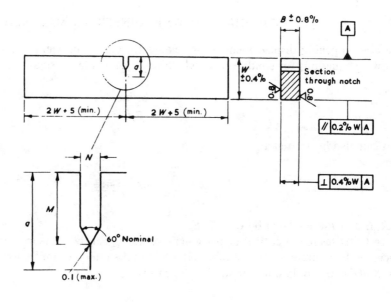

Width = W
Thickness = B = 0.5W
Half loading span L = 2W
Notch width N = 0.065W max. (if W is over 25 mm) or = 1.5 mm
 max. (if W is less than or equal to 25 mm)
Effective notch length M = 0.25W to 0.45W
Effective crack length a = 0.45W to 0.55W

All dimensions are in millimetres.

FIGURE 10.6   Details of standard bend test-piece (SEN) from BS 5447:1987

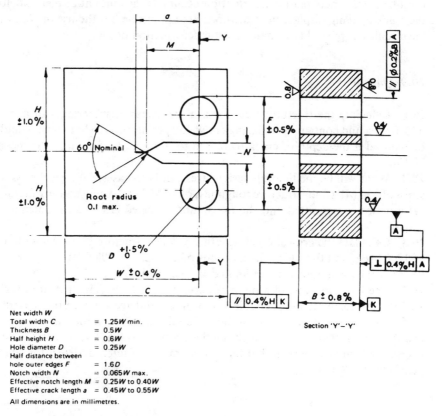

Net width W
Total width C = 1.25W min.
Thickness B = 0.5W
Half height H = 0.6W
Hole diameter D = 0.25W
Half distance between
hole outer edges F = 1.6D
Notch width N = 0.065W max.
Effective notch length M = 0.25W to 0.40W
Effective crack length a = 0.45W to 0.55W

All dimensions are in millimetres.

FIGURE 10.7   Details of standard tension test-piece (CTS) from BS 5447:1987

The crack opening displacement, δ, is related to K and the yield stress, σ_y according to the following equations for conditions in which total yield is fairly small:

$$\delta = \frac{\sigma_a^2 \pi a}{\sigma_y E}$$

This can also be written as

$$\delta = \frac{K^2}{\sigma_y E} \quad \text{or} \quad \delta = \frac{G}{\sigma_y}$$

as K, G and E are related by $G = K^2/E$.

In COD testing, a critical displacement value, δ_c, can be measured and this can be used to calculate the fracture toughness of a material, but it should be emphasised that the relationships $\delta = K^2/\sigma E = G/\sigma$ hold only for small σ_a/σ_y ratios.

10.8 Conclusion

Failure of a material by fast or brittle fracture can have disastrous consequences. Fracture mechanics is useful for assessing the risk potential of cracks in structures and for determining a 'safe maximum' or permissible crack size in a component. This is invaluable when designing structures to be 'fracture safe.' The techniques can be used with confidence for very high strength materials where fracture occurs under conditions that are almost totally elastic, but greater care must be taken where plastic deformation occurs. If the amount of plastic deformation is limited the crack opening displacement methods are applicable but the use of K_{Ic} or δ_c factors does not have sufficient reliability if major plastic deformation occurs.

10.9 Questions

10.1   Calculate the fracture stress for a glazed ceramic material with a modulus of elasticity of 300 GPa and containing a glazing crack 150 μm deep. To what value would the fracture stress be reduced if the depth of crack increased to 0.4 mm? Assume that the mean surface energy is 20 J/m².

10.2   What will be the limiting design stress, in both cases, for a steel possessing a tensile yield stress of 550 MPa and a fracture toughness of 40 MPa m$^{\frac{1}{2}}$, using a safety factor of 1.5, if no plastic deformation is permitted and the maximum tolerated crack is (a) 3.0 mm, and (b) 5.0 mm in length?

10.3   Calculate the critical crack length for brittle fracture in a steel with a tensile yield stress of 1200 MPa, and for which the modulus of elasticity is 208 GPa, and the toughness, G_c, is 110 kJ/m², if the maximum stress is limited by the yield stress.

10.4   The fracture stress of a piece of perspex is 57 MPa. The fracture toughness of Perspex is 1.2 MPa m$^{\frac{1}{2}}$. Estimate the probable size of internal defects within the material.

10.5   Assess the relative merits of the materials described in the Table below to resist failure by fast fracture in structures in which the design stress is 0.5 × yield stress. The materials test and inspection facilities are such that the minimum size of internal defect for which detection can be guaranteed is 8 mm.

Material	Yield stress (MPa)	Fracture toughness (MPa m$^{\frac{1}{2}}$)
High strength steel	1730	110
Aluminium alloy	590	35
Titanium alloy	960	86

10.6 What should be the appropriate wall thickness and diameter for an aluminium alloy vessel if it is to operate at a maximum pressure of 100 bar and be designed to 'leak before break' criteria using safety factors of 2.5 on yield strength and 1.5 on ratio of wall thickness to critical flaw size. σ_y for aluminium alloy is 500 MPa and K_{Ic} is 23 MPa m$^{-\frac{1}{2}}$.

10.7 Charpy notched impact tests carried out on a steel over a range of temperatures gave the results in the Table below.

Test temperature (°C)	Energy absorbed (J)	Test temperature (°C)	Energy absorbed (J)
95	164	5	50
80	164	0	47
70	157	−10	34
60	149	−20	25
50	149	−30	18
40	123	−40	12
20	84	−50	12

Plot this data and estimate the ductile–brittle transition temperature for the steel.

11

Phase Diagrams and Alloy Formation

11.1 Introduction

Many of the engineering materials in use are not pure substances but contain two or more ingredients. There will be various interactive effects which can affect both the number of phases and the temperatures at which changes of phase occur. Firstly, what do we mean by the word *phase*? A phase may be defined as a portion of matter that is homogeneous. A pure substance may exist in three states, solid, liquid and vapour, each of these states being a single phase. The relationships between these states, or phases, for various conditions of temperature and pressure, can be shown on a simple diagram, a *p–t* diagram. Figure 11.1 is a *p–t* diagram, or phase diagram, for water. The bounding line between liquid and solid phases in this diagram is almost parallel with the ordinate, showing that an increase in pressure will have the effect of reducing slightly the temperature at which liquid and solid phases are in equilibrium with each other. Put in other words, the effect of a pressure change on the melting or freezing point is very small. The water phase diagram is a good example of a one-component system.

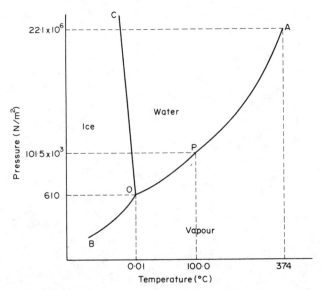

FIGURE 11.1   A one-component phase diagram—H_2O

Now, consider the effect of adding a second ingredient, or component, for example the addition of common salt, NaCl, to water. We now have an additional parameter, namely composition, to take into account. Salt will dissolve in water to a certain extent, and a salt solution is homogeneous. It is, therefore, a single phase. An effect of dissolving salt in water is to depress the freezing point of water. Fortunately, as the effect of pressure on the freezing point of water is so small, we can ignore pressure and show the phase relationships on a temperature–composition diagram. The temperature–composition phase diagram for the two-component system H_2O–NaCl is shown in Figure 11.2.

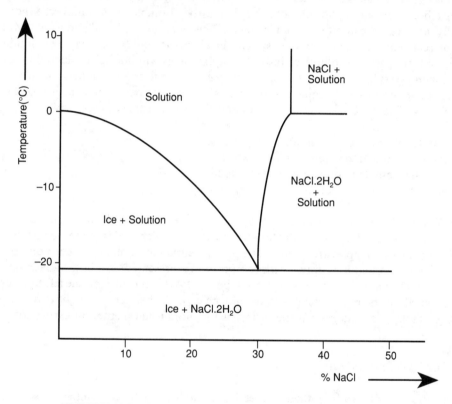

FIGURE 11.2   A two-component phase diagram—H_2O–NaCl

A phase diagram is a map of a system and it shows the phases which should exist under equilibrium conditions for any particular combination of composition and temperature. Phase diagrams are used widely and are particularly important in the understanding of metallic and ceramic alloy systems. For two-component, or binary, systems temperature–composition phase diagrams are used to show liquid–solid phase changes and the relationships between solid phases. If it is required to show phase relationships for a three-component, or ternary, system then it is necessary to construct a three-dimensional diagram. The composition of any three-component mixture can be shown by a point on an equilateral triangle. Temperature is represented on an axis orthogonal to such a composition plane. Refer to Section 11.12.

11.2 Alloy systems

The term *alloy* is used to describe a mixture of two or more metals or of two or more ceramic substances. Certain metallic alloys may be a mixture of a metal and a non-metallic element, a good example being alloys of iron and carbon, or steels. Many alloys are created by mixing substances in the liquid phase and it is convenient to consider the formation of alloy structures on the basis of the solidification of liquid alloys. Alloy structures and, hence, alloy properties, will depend on the nature of the solid phase or phases formed.

When two liquid substances are mixed they may dissolve in one another in all proportions, forming a homogeneous liquid solution, or they may be insoluble in one another, either partially or totally, and separate into two liquid layers on density grounds like oil and water. Most alloys of commercial importance are based on systems in which the liquids are completely soluble in one another and, so, partially soluble or insoluble liquid mixes will not be considered in this text.

Although two substances may be completely soluble in one another in the molten state it does not necessarily mean that the liquid solution will solidify to give an homogeneous solid phase, or solid solution. There are several possibilities and the two substances may:

(a) be totally insoluble in one another when solid,
(b) be totally soluble in one another when solid,
(c) be partially soluble in one another when solid, or
(d) combine with one another to form a compound.

The binary phase diagrams and the types of solid structure which can occur for each of the above possibilities are described in the following sections. Much of this relates to metallic systems but a few ceramics examples are quoted. The phase diagram for a binary system composed of the two components, A and B, is drawn with composition represented on a base and temperature on the ordinate. The base line will range from 100 per cent A, 0 per cent B on the left to 0 per cent A, 100 per cent B on the right. Compositions may be expressed either as weight percentages or as atomic, or molar, percentages but the former is the more usual method of expression.

11.3 Total solid insolubility

Consider the case of two hypothetical metals, A and B, which are completely soluble in one another in the liquid state, but are totally insoluble in one another in the solid state. If a composition base line and temperature scale is drawn (Figure 11.3), certain information can be plotted. The melting point of pure metal A can be marked off as point A on the left-hand temperature axis. Similarly point B on the right-hand axis represents the melting point of pure metal B. At high temperatures, any mixture of the two liquid metals will be a single-phase liquid solution.

In the same way as the presence of dissolved salt depresses the freezing point of water, so the freezing point of a liquid metal will normally be depressed if the liquid metal contains some other metal in solution. Line AL in Figure 11.3 indicates the depression of freezing point of pure metal A containing dissolved metal B. Similarly, line BM is the depression of the freezing-point curve for pure metal B containing dissolved metal A. It is important to note that at any point on line AL it is pure metal A that is freezing; that is, during cooling the solid that is forming is crystals of pure metal A. For example, X per cent of B dissolved in liquid A will depress the freezing point of A by an amount δT. The liquid solution will begin to solidify at a temperature of $A-\delta T$ (point S on curve AL in Figure 11.3) but it is crystals of pure A that will begin to solidify. The two curves

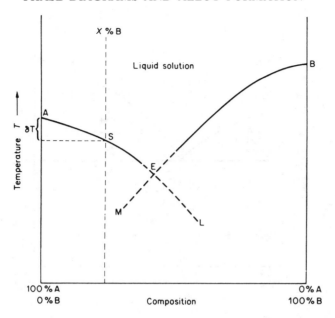

FIGURE 11.3   Freezing point curves for two metals insoluble in the solid state

AL and BM intersect at point E. The section of curves EL and EM are hypothetical as no liquid can exist at a temperature lower than that of point E.

Figure 11.4 shows the completed phase, or equilibrium, diagram for the alloy system of metals A and B. Consider again an alloy containing X per cent of B. At a high temperature, the alloy exists as a single-phase liquid solution. On cooling, the liquid will commence freezing at a point denoted by S on curve AE. Crystals of solid pure A will begin to form. If pure A is rejected from the solution, the composition of the remaining liquid must become enriched in metal B, that is, the composition of the liquid varies toward the right. This means that as the freezing of metal A continues, the temperature and composition of the liquid remaining follow the curve AE toward point E. Point E, which is the only point common to both freezing-point curves, represents the lowest temperature that a liquid solution can exist at, and at this point all remaining liquid solution solidifies, forming a fine-grained crystal mixture of both solid metals, A and B. Point E is termed the *eutectic* point and the fine-grained crystal mixture formed is termed the eutectic mixture. The final structure of the solid alloy containing X per cent of B will, therefore, be composed of large crystals of pure metal A (primary crystals) and a eutectic mixture of A and B. The primary crystals will probably be *dendritic* in form. (Many metals solidify from liquid in a dendritic manner. Solidification commences at a nucleus and outward growth from the nucleus occurs preferentially in three directions. Subsequently, secondary, and then tertiary arms grow, producing a skeleton-type crystal, as in Figure 11.5. Outward growth ceases when the advancing dendrite arms meet an adjacent crystal. When outward growth has ceased the dendrite arms thicken and eventually the whole mass is solid and no trace of the dendritic formation remains, except where shrinkage causes interdendritic porosity, or in alloy systems where the final liquid to solidify is of a different composition from the primary dendrites.)

If a liquid alloy containing Y per cent of metal B is allowed to solidify, solidification will follow a similar pattern, but in this case primary crystals of pure metal B will solidify first. It is important to note that the composition of the eutectic mixture remains constant.

In the phase diagram, line AEB is termed the *liquidus* and line CED is termed the *solidus*. At

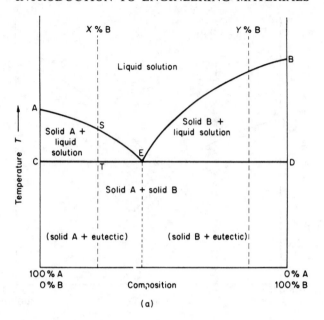

(a)

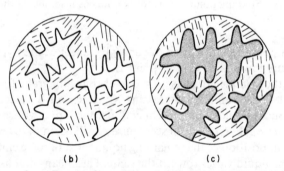

(b)   (c)

FIGURE 11.4   (a) Binary phase diagram for solid insolubility (simple eutectic). (b) Solid structure of alloy containing X per cent of B. Dendrites of A in eutectic. (c) Solid structure of alloy containing Y per cent of B. Dendrites of B in eutectic

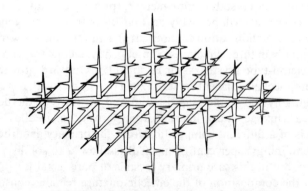

FIGURE 11.5   Representation of a dendrite

all points above the liquidus the alloy is always liquid, and below the solidus the alloy is always wholly solid. Between liquidus and solidus, in the solidification range, the alloy is in a pasty stage.

11.4 Interpretation of phase diagrams

It is appropriate to bring in at this stage some simple rules for the interpretation of binary phase diagrams.

(a) A phase diagram consists of lines that divide it into a number of areas, or fields. These fields may be single phase, as is the area above the liquidus in Figure 11.4, or two phase, as is area ACE in the same figure.

(b) Single-phase areas are always separated by a two-phase zone, and three phases can only coexist at a point, for example, the eutectic point.

(c) When a vertical line representing the composition of some alloy in the system crosses a line in the phase diagram, it indicates that some change is taking place in the alloy. For example, in Figure 11.4 the X per cent B line cuts two other lines, AE at S and CD at T, indicating that during cooling the alloy is beginning to freeze at point S and that solidification is complete at point T.

(d) For any point in a two-phase region the composition of the two phases in equilibrium with one another can be determined. If a horizontal tie-line is drawn through the point the intersections of this line with the phase boundary lines denote phase compositions. For point U in Figure 11.6 the intersections at x and y indicate that solid metal A is in equilibrium with a liquid solution containing y per cent of metal B.

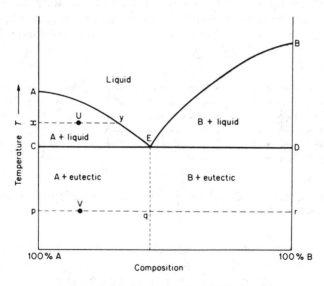

FIGURE 11.6   Application of lever rule

(e) The relative proportions of the phase present can be determined using the lever rule. The quantities of phases present are in proportion to the lengths of the lever lines, for example, for point U in Figure 11.6

$$\frac{\text{quantity of solid A}}{\text{quantity of liquid (composition y)}} = \frac{Uy}{Ux}$$

Similarly, at point V the phases present are solid metals A and B in the ratio

$$\frac{\text{quantity A}}{\text{quantity B}} = \frac{Vr}{Vp}$$

Alternatively it could be considered that the phases present are solid A plus eutectic mixture in the ratio

$$\frac{\text{quantity of A}}{\text{quantity of eutectic}} = \frac{Vq}{Vp}$$

or percentage of eutectic mixture in the solid alloy of compression V is

$$\frac{Vp}{pq} \times 100$$

11.5 Solid solubility

It is possible for a mixture of solid substances to form as a single phase; what is termed a *solid solution*. This concept may sound strange but it simply means that the atoms or molecules of the two substances have taken up positions in a common crystal lattice forming a single phase. Solid solutions may be of either the substitutional or interstitial type (refer also to Sections 5.5 and 8.3) and a schematic representation of substitutional and interstitial solid solutions is shown in Figure 11.7.

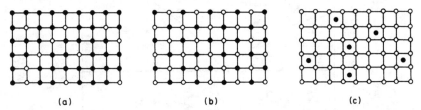

(a)   (b)   (c)

FIGURE 11.7   Schematic representation of solid solutions: (a) substitutional (random); (b) substitutional (ordered); (c) interstitial

Generally, the arrangement of solute in a substitutional solid solution is random, but in some instances solid solutions of an ordered type may be formed. An ordered solid solution (also known as *superlattice*) can exist only at either one fixed composition or over a narrow range of composition.

As stated previously (Section 8.3), atoms in interstitial or substitutional solid solution cause strain to be developed in the parent lattice. As there is an upper limit to the amount of strain that can be tolerated in a crystal lattice, it follows that there will be some restrictions to solid solution formation. The nature of solid solution in metal systems was studied extensively by Hume-Rothery and his work is summarised in the following rules.

(a) Relative size. If the sizes of the atoms of two metals do not differ by more than 14 per cent, conditions are favourable for the formation of substitutional solid solutions. If the relative sizes of atoms differ by more than 14 per cent, solid solution formation (if it occurs at all) will be extremely limited. Interstitial solid solutions may be formed if the atoms of the solute element are very small in comparison with those of the solvent metal.

(b) Chemical affinity. When two metals have a high affinity for one another the tendency is for solid solubility to be severely restricted and intermetallic compounds to be formed instead. This occurs when one element is electronegative and the other is electropositive.

(c) Relative valency. If a metal of one valency is added to a metal of another valency the number of valency electrons per atom, the electron ratio, will be altered. Crystal structures are very sensitive to a decrease in the electron ratio. Consequently, a metal of high valency can dissolve very little of a metal of low valency, although a metal of low valency might be able to dissolve an appreciable amount of a high-valency metal.

(d) Crystal type. If two metals are of the same crystal lattice type and all other factors are favourable, it is possible for complete solid solubility to occur over the whole composition range. (It is also necessary that the relative sizes of atoms differ by not more than 7 per cent for complete solid solubility).

Some examples of binary systems in which a continuous range of solid solutions is formed are Cu–Ni, Cu–Au, Al_2O_3–Cr_2O_3 and MgO–NiO.

11.6 Phase diagram for total solid solubility

The two possible types of phase diagram shapes for a binary alloy system where a continuous range of solid solution is formed, as exemplified by the Cu–Ni and Cu–Au systems, are shown in Figure 11.8. It should be noted that point C in the Cu–Au diagram (Figure 11.8(b)) is not an eutectic point.

An alloy containing X per cent of nickel (Figure 11.8(a)) would solidify in the following manner. Freezing of the liquid solution would commence at temperature t_1. At this temperature, liquid of composition 1 would be in equilibrium with a solid solution of a composition corresponding to point p on the solidus, so the first solid solution crystals to form are of composition p. Consequently, the composition of the remaining liquid becomes enriched in copper and the freezing temperature falls slightly. As the temperature falls so the composition of the solid solution tends to change by a diffusion process following the solidus line toward B. At some temperature t_2 liquid of composition m is in equilibrium with solid solution of composition q.

Solidification of this alloy will be complete at temperature t_3 when the last drops of liquid, of composition n, solidify, correcting the composition of solid solution crystals to r. If the solidification rate is very slow, allowing for the attainment of equilibrium at all stages during the cooling process, the final solid solution crystals will be uniform in composition. In practice however, solidification rates are too rapid for full equilibrium to be attained and the crystals will be cored. In a cored crystal the composition is not the same at all points. The crystal lattice is continuous but there will be a gradual change in composition across each crystal. The centre of a crystal will be rich in nickel while the outer edges will be rich in copper. In the Cu–Ni alloy system, the coring in crystals is clearly visible under microscopical examination of a prepared section as the alloy colour is dependent on composition. The centres of crystals are rich in nickel and silver in appearance while the outer edges of crystals are rich in copper and darker in colour. This colour shading clearly shows the dendritic manner of growth. Coring can occur in all solid solution alloys, even though it may not be visible.

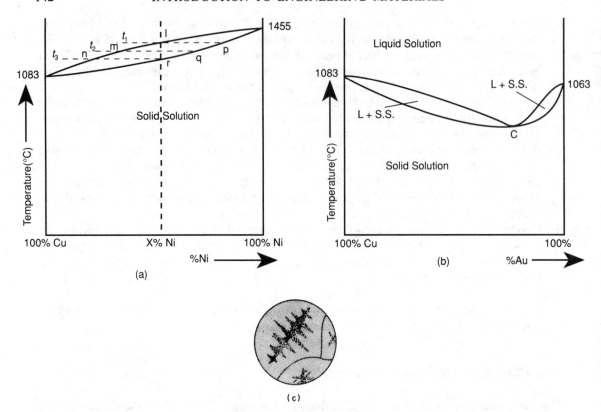

FIGURE 11.8   Phase diagrams for complete solid solubility: (a) Cu–Ni system; (b) Cu–Au system; (c) cored crystal structure

Coring in alloys may be subsequently removed by heating this material to a temperature just below the solidus. During this treatment, *annealing* or *homogenising*, diffusion takes place, evening out composition gradients within the crystals.

11.7 *Partial solid solubility*

It is far more common to find that solid metals are partially soluble in one another rather than totally insoluble or totally soluble. The phase diagram for the Pb–Sn system, which shows partial solid solubility, is given in Figure 11.9. This diagram is, in effect, a combination of the two previous types and shows solid solubility sections and also a eutectic. The liquidus is line AEB and the solidus is ACEDB. Lines FC and GD are *solvus* lines and denote the maximum solubility limits of tin in lead and of lead in tin respectively. As there are two separate solid solutions formed, the Greek letters α and β are used to identify them.

Consider the solidification of three alloy compositions in this system. For alloy composition (1) solidification begins at temperature t_1 with the formation of a tin-rich β solid solution of composition O. As cooling continues, the composition of the liquid varies along the liquidus towards point E and the composition of the solid β phase varies according to the solidus toward point D. When the eutectic temperature is reached, there will be primary cored crystals of β and liquid solution of the eutectic composition. This liquid then freezes to form a eutectic mixture of

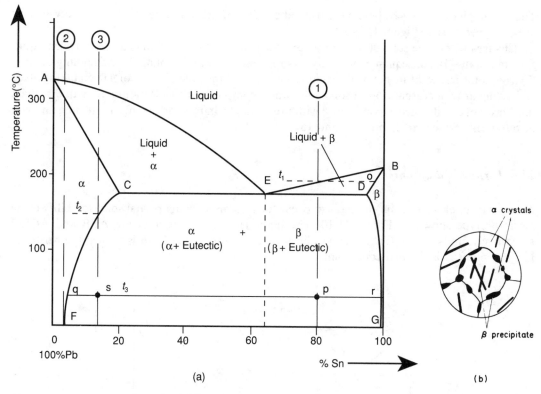

FIGURE 11.9 (a) Phase diagram for Pb–Sn (partial solid solubility with eutectic). (b) Structure of alloy (3)

two saturated solid solutions, α (lead-rich) of composition C and β of composition D. During further cooling the compositions of the α and β phases will adjust, following the solvus lines, until eventually at point p saturated α crystals of composition q will be in equilibrium with saturated β solid solution of composition r.

For alloy (2) solidification of the liquid solution takes place in the same manner as for a complete solid solution alloy (Section 11.6) and when solidification is complete the structure will be one of cored α (lead-rich) solid solution.

In the case of alloy (3), a new concept emerges, namely the possibility of structural changes occurring within the solid state. The liquid alloy will freeze on cooling to give a cored α solid solution. During further cooling below the solidus, the α solid solution will remain unchanged until temperature t_2 is reached. At this temperature, the composition line meets the solvus and the solid solution is fully saturated with dissolved tin. As the temperature of the alloy falls below t_2 the solubility limit is exceeded and excess tin is rejected from α solid solution as a precipitate. In this case, it is not pure tin that forms as a second solid phase, but rather, saturated β (tin-rich) solid solution. Eventually, at temperature t_3 the structure is composed of crystals of α of composition q with precipitated β particles of composition r. Applying the lever rule in this case the proportion of phases present would be in the ratio

$$\frac{\text{quantity of } \beta \text{ solid solution}}{\text{quantity of } \alpha \text{ solid solution}} = \frac{sq}{sr}$$

The second phase, β, may be precipitated either at the α grain boundaries, within the α crystals, or at both types of site (Figure 11.9(b)).

Changes within the solid state take place slowly in comparison with changes between liquid and solid states. In consequence they may be suppressed by rapid cooling. Rapid cooling of alloy (3) from some temperature below the solidus may prevent the precipitation of β from taking place and giving at temperature t_3 an α solid solution of composition s supersaturated with dissolved tin. This is of significance in connection with precipitation hardening and age hardening and will be discussed further in Chapters 12 and 15.

11.8 Peritectic diagram

Another form of phase diagram that can occur for systems showing partial solid solubility is the *peritectic* type shown in Figure 11.10. The liquidus and solidus lines are AEB and ACDB respectively, and FC and GD are solvus lines. The horizontal line CDE is termed the peritectic line and point D is the peritectic point.

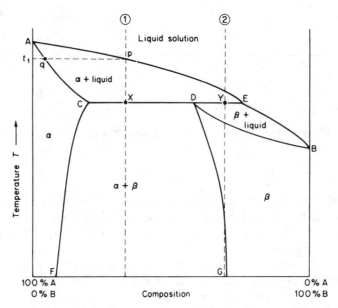

FIGURE 11.10   Phase diagram for partial solid solubility with peritectic

Consider the cooling of liquid alloy of composition (1). Solidification will commence at temperature t_1 with α solid solution of composition q forming. As freezing continues, the composition of the liquid follows the liquidus toward point E and the composition of the solid solution follows the solidus toward point C. When the peritectic temperature is reached liquid of composition E exists in equilibrium with α solid solution of composition C. At this temperature the two phases react together to form β solid solution according to the reaction

$$\alpha \text{ (composition C)} + \text{liquid (composition E)} \xrightarrow[\text{heating}]{\text{cooling}} \beta \text{ (composition D)}$$

If the reactants α and liquid were present in equivalent proportions, that is in the ratio

$$\frac{\text{amount } \alpha}{\text{amount liquid}} = \frac{DE}{CD}$$

they would both be totally consumed in the reaction producing β solid solution.

In the case of alloy (1) the reactants were present in the ratio

$$\frac{\text{amount } \alpha}{\text{amount liquid}} = \frac{XE}{CX} \text{ (by lever rule)}$$

where

$$\frac{XE}{CX} > \frac{DE}{CD}$$

so that the peritectic reaction will cease when all the liquid is consumed and there is some unreacted α remaining. The structure of this alloy below the peritectic temperature is, therefore, α and β. During further cooling, the compositions of both phases will vary according to the solvus lines. In the case of alloy (2) the ratio of reactants immediately before the peritectic reaction occurs is

$$\frac{\text{amount } \alpha}{\text{amount liquid}} = \frac{YE}{CY} < \frac{DE}{CD}$$

Consequently the reaction will cease when all the α has been consumed and there is some excess liquid remaining. During further cooling this liquid will solidify as β.

11.9 Compound formation

As stated earlier, in some binary systems the components may combine to form compounds. Often these are not compounds in the true chemical sense, but are ordered crystal space lattices with atoms of the various elements taking up specific positions within the lattice. Compound formation occurs in both ceramic and metallic alloy systems and in many cases more than one compound may be formed within a single binary alloy system. Compounds possess crystal structures that differ from those of either constituent component and they often possess unique melting points. In the case of intermetallic compounds, their melting points are often higher than the melting points of the constituent metals and they are generally of high hardness and quite brittle. An example of an intermetallic compound is the compound between magnesium and tin Mg_2Sn.

	Melting point (°C)	Crystal structure
Mg	649	close packed hexagonal
Sn	232	body centred tetragonal
Mg_2Sn	783	complex cubic

Compounds are formed in many ceramic alloy systems, examples being mullite, $Al_6Si_2O_{13}$, in the Al_2O_3–SiO_2 system and the four different compounds formed in the Al_2O_3–CaO system (see Figure 11.11 (a) and (b)). It will be seen from the figure that mullite and tricalcium aluminate

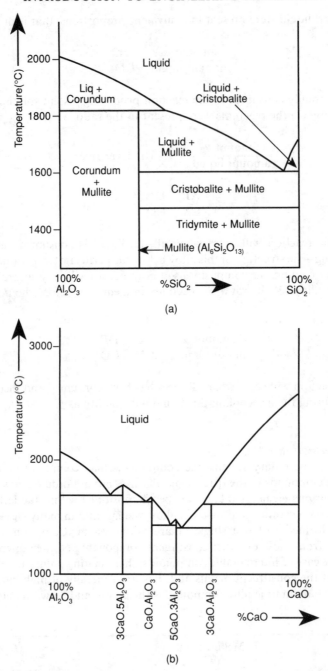

FIGURE 11.11   (a) Al₂O₃–SiO₂ phase diagram. (b) Al₂O₃–CaO phase diagram

($3CaO.Al_2O_3$) are formed as a result of peritectic reactions but the other compounds in the $Al_2O_3 - CaO$ system possess unique melting points.

From many points of view an intermetallic compound can be regarded as a pure metal. Frequently the binary phase diagram for an alloy system in which an intermetallic compound is formed is effectively two simple diagrams linked together. Figure 11.12(a) shows the phase diagram for two metals, A and B, which form one compound, A_xB_y, and assumes no solid solubility. It is simply two binary eutectic diagrams joined together, one between metal A and the compound and the other between metal B and the compound. Figure 11.12(b) shows a variation on the above in which there is some solid solubility. As there are now three separate solid solution zones three Greek letters, α, β, and γ, have to be used to describe these. Conventionally, these are used progressively, working from left to right. When, as in this case, a compound may exist over a small range of composition it is termed an *intermediate* phase.

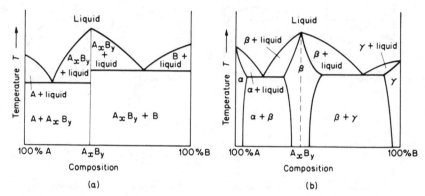

FIGURE 11.12   Binary diagrams showing intermetallic compound (a) with no solid solubility, (b) with partial solid solubility

In some binary alloy systems, several intermetallic compounds or intermediate phases may be formed. Although the complete phase diagram may look highly complex at first sight, it can usually be split into the small and comparatively simple elements discussed in earlier sections of this chapter, and it can be interpreted according to the same simple rules.

11.10 Effect of alloy type on properties in metallic systems

For systems in which the component metals are completely insoluble in one another in the solid state, the structure of the solid alloy is simply a mixture of two pure metals. Consequently, the variation of properties with alloy composition should obey the *rule of mixtures* and be linear. In actual practice, there is a departure from linearity owing to a grain-size effect. A eutectic is a finely divided mixture of two metals. The primary crystals that solidify first on either side of the eutectic point are much larger in size. A fine-grained crystal metal tends to be harder and stronger than a coarse-grained sample of the same material (refer to Section 8.5). Similarly, a fine grain size causes a reduction in electrical and thermal conductivities.

In Figure 11.13(a) the approximate relationship between two properties, hardness (H) and electrical conductivity (G), and alloy composition is shown for a simple eutectic alloy. The dotted lines show the expected property variation, neglecting the grain-size effect. Tensile yield and ultimate strengths follow a similar pattern to hardness.

In a solid solution alloy, the presence of the solute atoms imposes strain in the parent lattice, strengthening the alloy (see Section 8.8). Maximum strengthening occurs when the lattice is subjected to maximum strain, that is, when there are equal numbers of both types of atoms. 50 atomic per cent is not necessarily the same as 50 per cent by weight. Property variations with composition for solid solution alloys are shown in Figure 11.13(b).

Figure 11.13(c),(d) shows the relationships between properties and composition for the

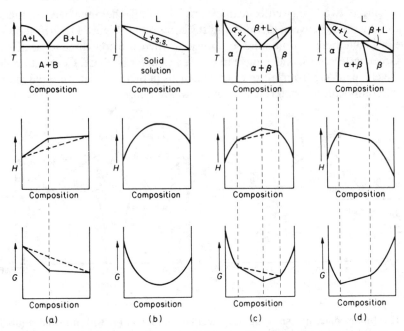

FIGURE 11.13 Relationships between alloy composition and hardness (H) and electrical conductivity (G) for (a) simple eutectic, (b) solid solubility, (c) partial solubility with eutectic and (d) peritectic

partial solid solubility cases. As the phase diagrams contain some features from each of the previous type of diagram, so the property diagrams are combinations of the former two types.

11.11 Allotropy

Certain elements and compounds can exist in more than one crystalline form.

The various separate forms are termed *allotropes* and the element or compound is said to exhibit allotropy. Diamond and graphite are two allotropic forms of the element carbon. A number of metals and ceramic compounds also exhibit allotropy and the presence of allotropes has an effect on the form of phase diagrams. In the Al_2O_3–SiO_2 diagram, Figure 11.11(a), a horizontal line below the solidus indicates the solid state transformation from the cristobalite form of silica to the tridymite form on cooling. The most important allotropic metal is iron, which can exist in two crystal forms, body centred cubic and face centred cubic. Pure iron is body centred cubic in structure (α iron) at all temperatures up to 908°C, and also between 1388°C and the melting point of iron at 1535°C. (The high temperature b.c.c. form is known as δ iron). Between 908°C and 1388°C the structure is face centred cubic and this form is termed γ iron. Some phase diagrams involving iron appear in Chapter 16.

If a metal is allotropic, this will have an effect on the shape of equilibrium diagrams for alloy systems involving the metal. Consider a hypothetical alloy system between two allotropic metals, A and B. Suppose that metal A is body centred cubic at low temperatures and face centred cubic at high temperatures while metal B is hexagonal at low temperatures and face centred cubic at high temperatures. If all factors are favourable, it is possible for a complete range of solid solutions to exist between the two metals at high temperatures. At lower temperatures only partial solid solubility can occur because the metals differ in crystal form. The complete phase diagram for the alloy system A–B could be as in Figure 11.14. It will be noticed in this diagram that, immediately below the solidus, there is a phase field containing one solid solution, β, and that below this there is a diagram apparently identical to the eutectic with partial solid solubility shown in Figure 11.9. This is a *eutectoid* diagram and point E is the eutectoid point. In a eutectic system it is a case of a single-phase liquid changing during cooling into two separate solid phases. A eutectoid is similar but it is a case of a single-phase solid solution changing during cooling into two differing solid phases. Line CED is termed the *liquidoid* and line CFEGD is termed the *solidoid*. The interpretation of a eutectoid diagram is fundamentally the same as the interpretation of a eutectic diagram. It must be remembered though, that reactions wholly within the solid state take place more slowly than liquid to solid changes. Variations in the rate of cooling through a eutectoid phase change can exert a profound effect on the structure and properties of the alloy. This will be discussed in greater detail in Section 12.7 and in connection with the heat treatment of steels in Chapter 16.

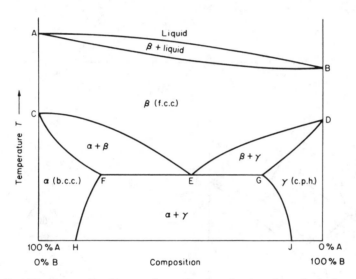

FIGURE 11.14 Possible form of a binary phase diagram for two hypothetical metals, both being allotropic

Eutectoids may also appear in alloy systems between non-allotropic metals if several intermediate phases occur. There are examples of this in the copper–aluminium and copper–tin systems (see Chapter 15).

11.12 Ternary diagrams

When three components are present in a system the composition of any mixture cannot be represented by a point on a line. The composition abscissa of the binary diagram becomes an equilateral triangle (Figure 11.15 (a)) in a ternary diagram. Temperature is represented on an axis orthogonal to the base triangle and the phase diagram becomes a three-dimensional solid figure. The ternary diagram for a hypothetical system of three components, A, B and C, all totally insoluble in one another in the solid state, is represented in Figure 11.15 (b). In this system, each pair of components would give a simple eutectic system, and the three components give a ternary eutectic. It will be seen that the liquidus curve of a binary system becomes a curved surface in a ternary system, and the eutectic point of each binary diagram becomes a eutectic line, with the three eutectic lines intersecting at point E, the ternary eutectic point. One method of representing this in two dimensions is to show the figure as a liquidus projection, with thermal contours marked. This form of representation is shown in Figure 11.15 (c).

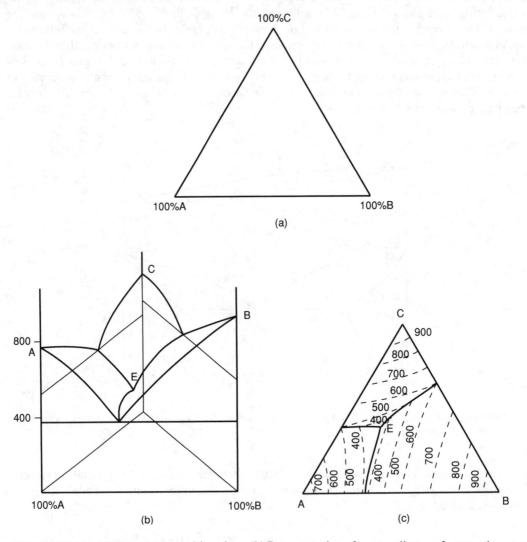

FIGURE 11.15   (a) Ternary composition plane. (b) Representation of ternary diagram for eutectic system.
(c) Liquidus projection of (b)

11.13 Questions

11.1   The phase diagram for a binary alloy system is shown in Figure 11.16.

(a) Label all the phase fields.
(b) Estimate the liquidus and solidus temperatures for the alloy containing 20 per cent B.

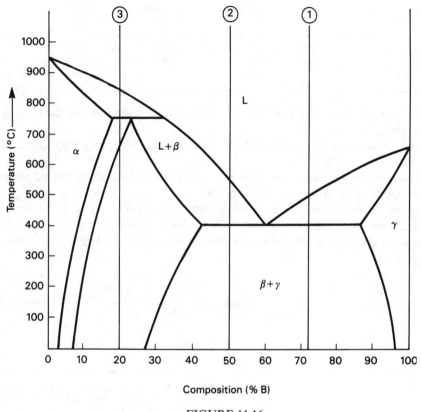

FIGURE 11.16

(c) For the alloy containing 40 per cent B, state what phases are present, and in what relative proportions, at (i) 600°C, (ii) 300°C and (iii) 100°C.

(d) What percentage of the microstructure is eutectic mixture in the alloy containing 70 per cent of B at room temperature?

11.2 A range of alloys is made from the two pure metals A, of melting point 750°C, and B, of melting point 1110°C, and the alloys are allowed to cool from the molten state to room temperature. During the cooling process, thermal arrest points are noted for each alloy. These are given in the table below.

Per cent B in alloy	8	30	45	55	78	90
1st arrest temperature (°C)	810	910	980	1010	1060	1090
2nd arrest temperature (°C)	760	830	910	910	940	1030
3rd arrest temperature (°C)	–	720	–	–	800	–

(a) Using these data draw and fully label the phase diagram for the alloy system of metals A and B.

(b) Describe the room temperature microstructure of an alloy containing 78 per cent of B in the following conditions: (a) sand cast, (b) cast, reheated to 900°C and quenched in water.

11.3 Two pure metals A and B, and a series of alloys of these two metals, were cooled from the liquid state and the information, in the Table below, was obtained.

Per cent A in alloy	100	95	85	60	30	10	5	0
1sr arrest point (°C)	600	575	535	425	300	400	425	450
2nd arrest point (°C)	–	500	300	300	–	300	350	–
3rd arrest point (°C)	–	40	–	–	–	–	200	–

(a) From the data, plot and fully label the thermal equilibrium diagram for the alloy system of metals A and B.

(b) For the alloy containing 75 per cent of A, what phases exist at the following temperatures: (a) 525°C, (b) 425°C, (c) 250°C?

(c) Explain how the properties of hardness and tensile strength of slowly cooled alloys would vary with composition from pure A to pure B.

11.4 Some data for the metallic elements absurdium, Ab, anglium, An, hypothetium, Hy, and Londonium, Ln, are given in the Table below. On the basis of this information, sketch the likely forms of the phase diagrams for the following binary systems: (i) Ab–Hy, (ii) Ab–Ln, (iii) An–Ln and (iv) Hy–Ln.

Metal	T_m(°C)	Atomic diameter (nm)	Structure	Valency	Electronegativity
Ab	855	0.255	HCP	2	1.4
An	720	0.271	FCC	3	1.8
Hy	1300	0.251	BCC	2	1.5
Ln	950	0.242	BCC	2	1.3

12

Phase Transformations and Diffusion

12.1 Introduction

The rate at which many processes occur is affected significantly by a change in temperature, reducing if the temperature is lowered and accelerated when the temperature is raised. Also, there is a certain minimum, or threshold, temperature below which the process will not be active. These principles apply to both physical processes, such as difusion and diffusion related phenomena, and chemical processes. The quantity of energy which is necessary to initiate a process is termed the *activation energy* for the process.

A simple mechanical analogy can be used to explain the significance of activation energy. Consider a tetragonal prism of mass m (see Figure 12.1). This prism may have two stable rest positions, marked as A and C in the figure, although the potential energy of the prism is greater in position A than in position C owing to the centroid of the prism being at a higher level above datum. If the prism is translated from position A to position C by pivoting about point O, it is seen that the centroid will follow the path GG′G″. When it is in an intermediate position B, the potential energy of the prism will be greater than at position A by an amount $mg\delta h$. This quantity of energy, $mg\delta h$, would be termed the activation energy for the process of translation of the prism from position A to position C.

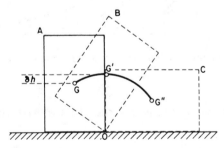

FIGURE 12.1   Activation of a process—mechanical analogy

12.2 Thermally activated processes

The atoms or molecules within any substance possess kinetic energy and are continually in motion. In a crystalline solid, the motion of atoms takes the form of vibration about fixed points.

The total energy content of a system is determined by temperature and rises as the temperature increases. However, at any particular temperature, the atoms do not all possess the same value of kinetic energy. At any instant in time, some atoms may be momentarily at rest while others are in motion. The distribution of energy between all the atoms or molecules in a system is given by the Maxwell–Boltzmann distribution law. This is illustrated in Figure 12.2. $N(E)$ is the number of atoms per unit volume possessing an energy in the range from E to $E + \delta E$, where E is some value of energy and δE is a small increment of energy. Figure 12.2 shows the $N(E)$ distribution at three different temperatures T_1, T_2 and T_3, where $T_1 < T_2 < T_3$. As the temperature is raised, the total energy content of the system increases and the number of atoms possessing high values of kinetic energy increases. But from the definition of $N(E)$:

$$\int_0^\infty N(E)\, dE = N$$

where N is the total number of atoms in the system. Therefore the areas under the three curves shown in Figure 12.2 are the same. If q is the activation energy of some process then the number of atoms per unit volume possessing an energy of q, or greater, will be given by $\int_q^\infty N(E)\, dE$. This function would have a value of zero at temperature T_1, but definite values at temperatures T_2 and T_3. It follows that the reaction would not take place at temperature T_1, but would occur at the temperature T_2 and T_3. Further, the rate at which the process would occur would be greater at T_3 than T_2 as a greater number of atoms possess energies above level q at this higher temperature.

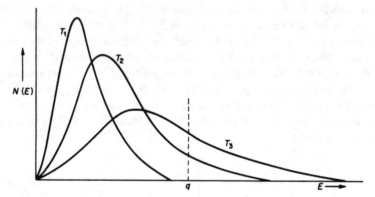

FIGURE 12.2   Maxwell–Boltzmann distribution of molecular energies in a system at three temperatures $(T_1 < T_2 < T_3)$

   The rate at which a process occurs is governed by the Arrhenius rate law, which can be written as

$$\text{Reaction rate} = v = A \exp\left(-\frac{q}{kT}\right)$$

where q is the activation energy for a single molecule, k is Boltzmann's constant, T is the temperature (K) and A is a constant. This may be rewritten as

$$v = A \exp\left(-\frac{Q}{R_0 T}\right)$$

where Q is the activation energy per kilomole, and R_0 is the universal gas constant (8.314 kJ/kmol K).

The Arrhenius expression can be written in logarithmic form as

$$\ln v = \ln A - \frac{Q}{R_0 T}$$

This is a linear equation of the form $y = c + mx$, the two variables being the logarithm of rate and $1/T$. A graph plot of $\ln v$ against $1/T$ will give a straight line. The slope of the line is equal to $-Q/R_0$ and $\ln A$ is given by the intercept with the ordinate. If preferred, the logarithmic equation can be written as

$$\log_{10} v = \log_{10} A - \frac{Q}{2.3026 R_0 T}$$

The value of Q is of the order of 100 to 400 MJ/kmol for many physical processes and in the range from 20 to 100 MJ/kmol for many chemical reactions. In industrial chemical processes, catalysts are frequently used to accelerate the rate of a reaction. Their presence effectively reduces the activation energy for the reaction. Conversely, a reaction may be retarded with the aid of a negative catalyst which effectively raises the activation energy. Catalysts and negative catalysts are not permanently consumed in the reaction.

Examples

1. The rate of a process varies with temperature and the following table gives experimentally determined values.

Temperature (°C)	175	250	355	500
Rate (mm/s)	1.13×10^{-5}	5.93×10^{-5}	3.10×10^{-4}	1.46×10^{-3}

Show graphically that these data are consistent with the Arrhenius law and calculate the activation energy for the process. Also determine the temperature at which the process rate will be 1.5×10^{-2} mm/s ($R_0 = 8.314 \times 10^3$ J/kmol).

The data should first be put into a form suitable for plotting as a graph. This is best done by producing a new table.

Temperature				
°C	K	$1/T$	v	$\ln v$
175	448	2.23×10^{-3}	1.13×10^{-5}	-11.39
250	523	1.91×10^{-3}	5.93×10^{-5}	-9.73
355	628	1.59×10^{-3}	3.10×10^{-4}	-8.08
500	773	1.29×10^{-3}	1.46×10^{-3}	-6.53
?	?	?	1.50×10^{-2}	-4.20

$\ln v$ can now be plotted against $1/T$ as shown in Figure 12.3.
The slope of the line is measured as $-5170 = -Q/R_0$.
Therefore the activation energy $Q = 5170 \times 8314 = 43 \times 10^6$ J/kmol.

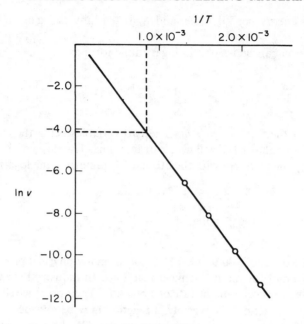

FIGURE 12.3   Plot of ln v against $1/T$

From the graph a value of ln $v = -4.20$ corresponds to $1/T = 0.83 \times 10^{-3}$.
Therefore the temperature at which v is 1.50×10^{-2} is 1205 K or 932°C.

2. The rates of linear growth of aluminium crystals during the recrystallisation of cold worked aluminium are given below

Temperature (°C)	200	250	300	400
Rate (mm/s)	5.62×10^{-10}	1.38×10^{-7}	1.35×10^{-5}	1.82×10^{-2}

Determine if this is an example of a reaction that obeys the Arrhenius law and if so, calculate the activation energy.

This problem can either be solved graphically as for example 1 or by a numerical method, as follows.
If the reaction rates of temperatures T_1, and T_2 are v_1 and v_2 respectively, then

$$v_1 = A \exp\left(-\frac{Q}{R_0 T_1}\right) \text{ and } v_2 = A \exp\left(-\frac{Q}{R_0 T_2}\right)$$

Dividing we have

$$\frac{v_2}{v_1} = \exp\left(-\frac{Q}{R_0 T_2} + \frac{Q}{R_0 T_1}\right) = \exp\left[+\frac{Q}{R_0}\left(\frac{T_2 - T_1}{T_1 T_2}\right)\right]$$

From which Q can be evaluated.
    If this calculation is completed for several pairs of values and each calculation gives the same result for Q then the process satisfies the Arrhenius law.

Let $v_1 = 5.62 \times 10^{-10}$ mm/s, $T_1 = 200°C = 473$ K

$v_2 = 1.38 \times 10^{-7}$ mm/s, $T_2 = 250°C = 523$ K

$v_3 = 1.35 \times 10^{-5}$ mm/s, $T_3 = 300°C = 573$ K

and $v_4 = 1.82 \times 10^{-2}$ mm/s, $T_4 = 400°C = 673$ K

$$\frac{v_2}{v_1} = \frac{1.38 \times 10^{-7}}{5.62 \times 10^{-10}} = \exp\left\{ \frac{Q}{8314} \left(\frac{523-473}{473 \times 523} \right) \right\}$$

$$245.6 = \exp(5.5) = \exp(24.3 \times 10^{-9} Q)$$

$$Q = 226.2 \text{ MJ/kmol}$$

$$\frac{v_3}{v_1} = \frac{1.35 \times 10^{-5}}{5.62 \times 10^{-10}} = \exp\left\{ \frac{Q}{8314} \left(\frac{573-473}{473 \times 573} \right) \right\}$$

$$24.02 \times 10^3 = \exp(10.09) = \exp(44.38 \times 10^{-9} Q)$$

$$Q = 227.4 \text{ MJ/kmol}$$

$$\frac{v_4}{v_1} = \frac{1.82 \times 10^{-2}}{5.62 \times 10^{-10}} = \exp\left\{ \frac{Q}{8314} \left(\frac{673-473}{473 \times 673} \right) \right\}$$

$$32.38 \times 10^6 = \exp(17.29) = \exp(75.57 \times 10^{-9} Q)$$

$$Q = 228.8 \text{ MJ/kmol}$$

The three calculated values of Q are in close agreement indicating that the process obeys the Arrhenius law. The average value of activation energy is 227.5 MJ/kmol.

3. A reaction is complete in 5 seconds at 600°C but requires 15 minutes for completion at 290°C. Calculate the time necessary for the reaction to be completed at 50°C.

If the reaction requires 5 s for completion at 600°C then the reaction rate is proportional to $1/5$ s^{-1}. At 600°C (873 K) $v = k/5$ and at 290°C (563 K) $v = k/(15 \times 60)$:

$$\frac{v_{873}}{v_{563}} = \frac{15 \times 60}{5} = \exp\left\{ -\frac{Q}{8314} \left(\frac{563-873}{873 \times 563} \right) \right\}$$

$$180 = \exp(5.19) = \exp(75.86 \times 10^{-9} Q)$$

$$Q = 68.4 \text{ MJ/kmol}$$

$$v_{563} = \frac{1}{5} = A \exp\left(-\frac{68.4 \times 10^6}{8314 \times 563} \right)$$

$$0.2 = A \exp(-14.62) = 4.49 \times 10^{-7} A$$

Therefore $A = 445 \times 10^3$

Let the time for the reaction at 50°C (323 K) be t s.

$$\frac{1}{t} = 445 \times 10^3 \exp\left(-\frac{68.4 \times 10^6}{8314 \times 323} \right) = 3.86 \times 10^{-6} \text{ s}^{-1}$$

$$t = 259 \times 10^3 \text{ s} = 71.94 \text{ hours}$$

12.3 Diffusion

Many processes of significance in the materials world involve solid state diffusion. How can material diffuse through a solid? It was stated in Chapter 8 that crystal lattices contain a number

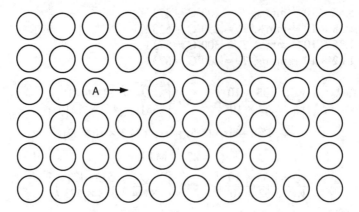

FIGURE 12.4   Section of crystal lattice containing vacancies

of point defects, one type of point defect being a vacancy. Consider a section of a crystal lattice containing some vacancies (Figure 12.4).

Atoms and vacancies can move through the lattice. For example, if the atom, marked A in Figure 12.4 and sited adjacent to a vacancy, possesses sufficient energy it can jump into the vacant site causing the vacancy to move one incremental distance in the opposing direction. As temperature rises not only is there an increase in the mean kinetic energy of atoms but also there is an increase in the percentage of vacancies in the lattice and so the frequency of atom/vacancy jumps increases exponentially according to Arrhenius' law. As explained in Chapter 8, the migration of vacancies to a dislocation can cause dislocation climb, one of the mechanisms postulated for creep in metals. It follows, then, that creep is a diffusion controlled process and steady-state creep rates vary with temperature according to an Arrhenius type expression.

In the atom/vacancy jump example illustrated in Figure 12.4, all atoms are identical. However, if the crystal structure contains a second component in solid solution, it will be evident that substitutional atoms can move through the lattice via the mechanism of atom/vacancy jumps. This migration of a second component through the lattice is diffusion. Atoms in interstitial solution can move by 'jumping' from one interstice to another. The diffusion of one substance into or through another is of importance in many aspects of materials. Examples are: diffusion of carbon in steels giving both carburising and decarburising effects, depending on condition, ageing and precipitation hardening effects in alloys, diffusion bonding, controlled diffusion of impurity elements into silicon to form p–n junctions.

12.4 Fick's laws of diffusion

Adolph Fick, in 1855, proposed a mathematical treatment for diffusion in one dimension at a constant temperature. For steady state diffusion, in which the concentration gradient is constant, he stated that such steady state diffusion could be described by the equation.

$$J = -D\frac{dC}{dx}$$

where J is the flux (kg m^{-2}s^{-1}) (the quantity of material crossing a unit area normal to the flux direction in unit time), D is the coefficient of diffusion (m^{2}s^{-1}) and dC/dx is the concentration gradient ($kg\ m^{-4}$). This is known as Fick's first law of diffusion. As stated earlier, the process of diffusion is a thermally activated process and, so, the diffusion coefficient, D, is not a constant but varies with temperature according to the Arrhenius equation.

A more common instance in practice is where the concentration of the diffusing substance changes with time. This means that both the flux and the concentration gradient also change with time. This dynamic condition is described by Fick's second law which can be written

$$\frac{dC_x}{dt} = D\frac{d^2 C_x}{dx^2}$$

where C_x is the concentration of the diffusing species at some distance x from a surface and t is the time (s).

In the case of solid state diffusion, where the concentration of the diffusing species at the surface of the material, C_s, remains constant as, for example, in the case carburising of steels, the solution to this differential equation may be written

$$\frac{C_s - C_x}{C_s - C_0} = \text{erf}\left\{\frac{x}{2(Dt)^{\frac{1}{2}}}\right\}$$

where C_0 is the concentration at the start of the process ($t = 0$) and erf is the Gaussian error function or normalised probability.

For values up to about 0.6, the Gaussian error function can be approximated as erf $y \simeq y$, so that

$$\frac{C_s - C_x}{C_s - C_0} \simeq \frac{x}{2(Dt)^{\frac{1}{2}}}$$

Table 12.1 *Gaussian error function*

y	erf y	y	erf y
0.65	0.6420	0.85	0.7707
0.70	0.6778	0.90	0.7970
0.75	0.7112	0.95	0.8209
0.80	0.7421	1.00	0.8427

12.5 Carburising and decarburising

Carbon can diffuse through iron and this phenomenon is utilised in the carburising processes which are used to increase the carbon content in the surface layers of low carbon steels to give surface, or case hardened components (see also Chapter 16). The reverse process of decarburisation can also occur if a steel is heated in an oxidising atmosphere. In these circumstances, carbon can diffuse out from the steel. Some surface decarburisation may occur during heat treatments but this effect is unwanted as it gives a softer surface layer than is generally required. In the preparation of carbon steel tools, the thin decarburised surface layer is removed during tool grinding.

During carburising operations, it can be assumed that a high carbon content is immediately established at the surface and that this surface carbon concentration remains constant during the process.

Examples

1. A low carbon (0.20% C) steel component is to be surface hardened by carburising at 950°C. Calculate the length of treatment needed to give a carbon content of 0.35% at a distance of 1.0 mm below the surface. Assume that the carbon content at the surface is constant at 0.95% throughout the process and the diffusion coefficient, D, at 950°C is $1.74 \times 10^{-11} \, m^2 s^{-1}$.

This is a case of dynamic diffusion and Fick's second law applies. As the surface concentration is constant the solution

$$\frac{C_s - C_x}{C_s - C_0} = erf\left\{\frac{x}{2(Dt)^{\frac{1}{2}}}\right\}$$

applies. Substituting in this expression we have

$$\frac{0.95 - 0.35}{0.95 - 0.2} = 0.8 = erf\left\{\frac{1.0 \times 10^{-3}}{2(1.74 \times 10^{-11}t)^{\frac{1}{2}}}\right\}$$

When erf $y = 0.8$, $y = 0.9$ so that

$$0.9 = \frac{1.0 \times 10^{-3}}{2(1.74 \times 10^{-11}t)^{\frac{1}{2}}}$$

from which $t = 17738 \, s \equiv 4$ hours 56 minutes.

2. A component, made from a high carbon steel containing 0.8% carbon, is exposed to an oxidising environment at 1000°C for a period of one hour. At what depth below the surface will the carbon content of the material be reduced to 0.4%? D, the diffusion coefficient for carbon in steel is $3.11 \times 10^{-11} \, m^2 s^{-1}$ at 1000°C.

The assumption is made that the concentration of carbon at the extreme surface is zero throughout the exposure period. The expression

$$\frac{C_s' - C_x}{C_s - C_0} \simeq \frac{x}{2(Dt)^{\frac{1}{2}}}$$

can be used as, for a value of 0.5, erf $y \simeq y$

$$\frac{0 - 0.4}{0 - 0.8} = 0.5 = \frac{x}{2(3.11 \times 10^{-11} \times 3600)^{\frac{1}{2}}}$$

from which $x = 0.33 \times 10^{-3} m \equiv 0.33$ mm.

12.6 Precipitation hardening

Precipitation hardening is the form of dispersion hardening (see also Section 8.9) which can be brought about in certain alloys by heat treatment. The types of alloys in which precipitation hardening effects may be induced are those in which there is partial solid solubility (see Section 11.7) and where there is a fairly large difference between the high temperature and low temperature limits of solubility. In alloy systems of this type the alloys within the narrow composition band shown in Figure 12.5 may be capable of being precipitation hardened.

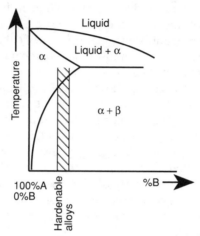

FIGURE 12.5   Portion of partial solubility phase diagram showing the range of alloy compositions which may respond to precipitation hardening

Consider an alloy with a composition within this range. When heated to a high temperature just below the solidus, the alloy is in the α phase field and the structure will consist of homogeneous α solid solution crystals. If the alloy is cooled very rapidly from this temperature by, for example, quenching in water, there may be insufficient time for the equilibrium structure of β precipitate within the α crystals to form and, instead, the high temperature structure will be retained at low temperature as a supersaturated α solid solution. This is a non-equilibrium state and, as such, is a metastable phase. In some alloy systems, such a supersaturated solid solution may remain indefinitely at room temperature without change while in other systems some diffusion and progression towards the stable equilibrium $\alpha + \beta$ structure may occur at room temperature. As diffusion is a thermally activated process, any increase in temperature will cause an increase in the rate of diffusion and change.

Diffusion of solute within the crystals is a necessary pre-requisite for precipitation as the supersaturated solution is, originally, homogeneous with a uniform distribution of solute atoms. There must be a diffusion of solute to build up the solute concentration in some areas of the lattice prior to precipitation of the second phase. This diffusion creates zones of high strain where there is a high solute concentration within the parent lattice. As the solution lattice is still fully coherent these zones are termed areas of coherent precipitate (see Section 8.9). These regions of high strain offer considerable resistance to the passage of dislocations and, hence, have a major strengthening effect.

If diffusion of solute proceeds fully and the coherent precipitate zones separate out into true precipitate particles which are non-coherent with the parent lattice then the strain in the solute lattice begins to be released and the strength of the material begins to reduce. When this phenomenon occurs it is termed *overageing*.

There are many alloys which respond to precipitation hardening including aluminium–copper, copper beryllium, magnesium alloys containing thorium, zirconium or zinc, some titanium alloys and maraging steels. The first precipitation hardening alloys to be developed and exploited were alloys of aluminium containing about 4% copper, one proprietary name for which is *duralumin*.

The heat treatments necessary for a precipitation hardening alloy involve firstly, a *solution heat treatment*. This involves heating the alloy to a temperature in the single phase region to create a homogeneous solid solution followed by rapid quenching to produce a supersaturated solution. The second stage of the heat treatment is heating at a relatively low temperature, the *precipitation treatment*, to permit diffusion and formation of coherent precipitate. Some alloys do not need this second heat treatment as diffusion and hardening will take place slowly at room temperatures. Such alloys are termed *age hardening* or *naturally ageing* alloys. However, the rate of hardening of these naturally ageing alloys can be accelerated by raising the temperature. Figure 12.6 shows the change of hardness with time at various temperatures for a naturally ageing alloy of aluminium and copper which has been solution heat treated. It will be seen that overageing will occur if this alloy is heated to a temperature of 200°C. If cooled to some temperature below room temperature the ageing of this alloy can be prevented. This phenomenon is put to good use in rivetted aluminium alloy structures. Aluminium–copper alloy rivets are placed in a refrigerator immediately after solution heat treatment to retard ageing and enable them to be kept for several days without hardening. They are then in the soft condition, permitting easy forming of the head during the rivetting operation, but subsequently age harden to full strength at room temperature.

Example

A heat-treatable aluminium alloy will fully age harden in 10 days at 20°C but will reach

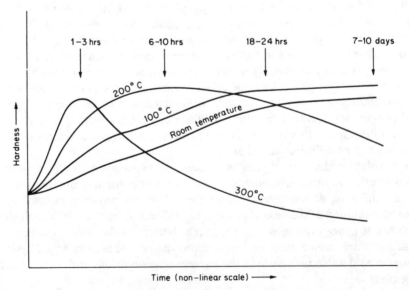

FIGURE 12.6   Hardness–time relationships for an aluminium–copper age hardening alloy

maximum hardness after 10 hours when precipitation treated at 180°C. How long will this alloy take to harden if maintained at -25°C after solution heat treatment?

The Arrhenius equation can be used for this example.
Hardening time at 20°C (293 K) is 240 hours and at 180°C (453 K) is 10 hours.

$$\frac{240}{10} = \exp\left\{\frac{B(453-293)}{453 \times 293}\right\} = 3.178 \text{ From which } B = 2636.3$$

Let x be the hardening time, in hours, at -25°C (248 K)

So

$$\frac{x}{10} = \exp\left\{\frac{2636.3(453-248)}{453 \times 248}\right\} = 122.8$$

From which $x = 1228$ hours $\simeq 51$ days.

12.7 Martensitic transformations

Martensite is the name of the metastable phase which is formed when a steel is rapidly cooled and the transformation into the equilibrium state, the eutectoid mixture, is prevented. Martensitic transformations can occur in certain other alloy systems also and are the result of rapid cooling through a eutectoid transformation.

Consider the eutectoid portion of the iron–carbon phase diagram (Figure 12.7). According to the phase diagram an alloy containing, say 0.6% C, would be a homogeneous solid solution of carbon in γ iron at a temperature of 900°C. During cooling the γ structure should begin to transform into α iron when the liquidoid temperature is reached (about 800°C). α iron is capable

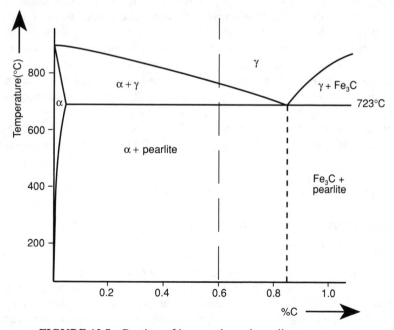

FIGURE 12.7 Portion of iron–carbon phase diagram

of holding very little carbon in solid solution and the concentration of carbon in the γ phase increases as cooling continues. On cooling through the eutectoid temperature, 723°C, all remaining γ phase should transform into the eutectoid mixture of α and the carbide Fe_3C. This eutectoid mixture in steels is termed *pearlite*. It will be evident that there must be a diffusion of dissolved carbon through the γ iron lattice to build up areas of carbon concentration prior to the formation of Fe_3C particles and eutectoid formation.

Transformations within the solid state are sluggish in comparison with liquid–solid transformations and in practical cases the structural changes will not occur at the equilibrium temperatures. There will be a certain amount of undercooling, the extent of the undercooling increasing as the rate of cooling is increased. If the rate of cooling from high temperature is very high, there will be insufficient time for any diffusion of carbon within the γ iron lattice to occur. While diffusion can be suppressed by a very rapid cooling from high temperature, the allotropic transformation from γ to α iron is not prevented. Consequently, when this change from the γ form to the α form occurs a large amount of carbon will be trapped in supersaturated interstitial solid solution creating extremely large strains. (The equilibrium solid solubility limit for carbon in α iron is only 0.006% at 20°C.) The body centred cubic α iron lattice is distorted into a body centred tetragonal structure and is referred to as the α' structure, or martensite. This highly strained structure is extremely hard and also extremely brittle.

The relationship between cooling rate and undercooling, and the conditions required for martensite formation can be shown on a time–temperature–transformation (T–T–T) diagram (Figure 12.8).

It will be noted that if the rate of cooling exceeds a particular value, the *critical cooling velocity*, the γ phase will transform into martensite rather than into pearlite, the transformation beginning at the M_s temperature, about 340°C for a 0.6%C steel.

As mentioned above, α' or martensite is a non-equilibrium phase which is metastable. It will not undergo any changes at room temperature as there is insufficient thermal energy to permit diffusion, but if the temperature is raised diffusion of carbon can occur leading to the precipitation of particles of Fe_3C from the martensite. This process can begin to occur if the temperature is raised to about 200°C or above. Any precipitation of carbon in the form of Fe_3C from the α' lattice will release some of the lattice strain with a consequent reduction in the hardness and brittleness. This phenomenon is utilised in the process of *tempering* of hardened steels. The higher the tempering temperature, the greater will be the amount of strain release and the greater will be the reduction in hardness and increase in ductility. It is important to note that during tempering the structure does not change to the equilibrium structure of α and pearlite, as formed during slow cooling, but rather to one containing very fine precipitated particles of Fe_3C.

Martensitic transformation has been described with reference to steels but it can occur in other systems containing eutectoids, a good example being the copper–aluminium system.

12.8 *Phase transformations in ceramics and glasses*

Solid state phase changes can occur in ceramic materials as well as in metals. Phase changes are generally accompanied by a volume change and in a non-ductile ceramic material such a volume change can lead to the formation of cracks within the material. For example, zirconia, the oxide of zirconium, ZrO_2, transforms on cooling from a tetragonal structure into a monoclinic structure with a volume change of about 3%. For this reason pure zirconia is rarely used as a ceramic material and stabilised zirconia products are manufactured containing additions of lime, CaO, magnesia, MgO or yttria, Y_2O_3. These form solid solutions with a cubic structure which is stable at room temperature.

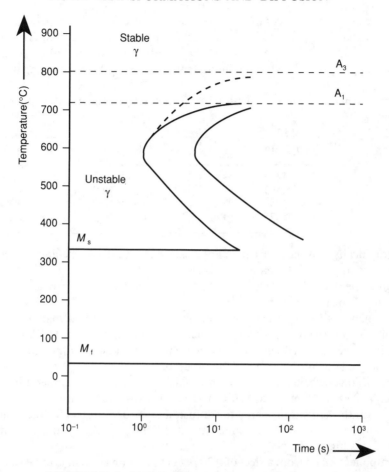

FIGURE 12.8   T–T–T diagram for a 0.6% carbon steel

The glass state is metastable and a glass may slowly transform into a crystalline structure. This process is termed *devitrification*, and is accompanied by a decrease in volume. This could cause cracks to form within the glass or crazing in glazed coatings. Commercial glasses are formulated so that devitrification will not occur at ordinary temperatures, but in some materials, the *glass ceramics*, the material is first formed as a glass and then converted into a fine grained crystalline material by controlled crystallisation heat treatments. Glass ceramics maintain their mechanical strength to much higher temperatures than glasses. LAS glass ceramics, based on $LiO–Al_2O_3–SiO_2$, have high strength coupled with very low thermal expansion coefficients.

12.9 Questions

12.1   For a thermally activated reaction with an activation energy of 100 MJ/kmol, calculate by what factor the rate of reaction will be increased if the temperature is raised from 15°C to 25°C.
12.2   The time required to recrystallise aluminium varies with temperature according to Arrhenius' law. Determine the activation energy of the recrystallisation process if recrystallisation is complete in 6 minutes at 330°C but required 100 hours at 260°C. (Assume $R_0 = 8314$ kJ/ kmol K).

12.3   The strength of a steel component designed for use at high temperatures is reduced to below its safe working limit after 50 per cent recrystallisation has occurred. This limit is reached after a time of 4250 s at a temperature of 580°C and after 3.6 s at 750°C. Assuming that the Arrhenius law applies, calculate the maximum temperature at which a working life of 10^4 hours is attainable.

12.4   The variation in thermionic emission from a tungsten filament with temperature is given in the Table below.

Temperature (K)	Current density (mA/mm²)
1470	7.63×10^{-7}
1543	4.84×10^{-6}
1761	4.62×10^{-4}
1897	4.31×10^{-3}

(a) Show graphically that the above data are consistent with the Arrhenius law and calculate the activation energy (work function) for tungsten.

(b) Determine the thermionic current density for a filament temperature of 2000 K. (Use Boltzmann's constant. $1 A = 1 Cs^{-1}$).

12.5   Joints made using an epoxy-based adhesive can be cured to maximum strength in 30 minutes at 150°C but requires 3 hours to reach maximum strength at 80°C. Assuming that the curing reactions obey the Arrhenius law, estimate the time required for complete curing at 25°C and determine the activation energy of the process.

12.6   The diffusion coefficient for carbon in steel is measured at two temperatures and the values found to be: $4.826 \times 10^{-12} m^2s^{-1}$ at 850°C and $1.805 \times 10^{-11} m^2s^{-1}$ at 950°C.

Calculate: (a) the activation energy for diffusion of carbon in steel for the temperature range 850–950°C, and (b) the value of diffusivity at 1000°C, assuming the activation energy is unchanged at this higher temperature.

12.7   A component made from a steel of 0.20 per cent carbon content is to be gas carburised at a temperature of 925°C. Calculate the time required to increase the carbon content to 0.40 per cent at a distance of 0.55 mm below the surface, assuming that the carbon concentration at the surface is maintained at 0.90 per cent throughout the process. (Take diffusivity of carbon in steel at 925°C as $1.28 \times 10^{-11} m^2s^{-1}$).

12.8   A steel component is to be nitrided and it is required, for optimum characteristics, that the concentration of nitrogen at 0.5 mm below the surface be 0.025 per cent. Calculate the time necessary for nitriding, assuming that the concentration of nitrogen at the surface of the steel is maintained at 0.125 per cent throughout and that the diffusion coefficient for nitrogen in steel is $3.48 \times 10^{-12} m^2s^{-1}$.

12.9   A steel containing 1.2 per cent carbon is heated at 1000°C in a strongly oxidising atmosphere. Estimate the depth below the surface at which the carbon concentration is reduced to 0.6 per cent after 1 hour's exposure. (Assume that the concentration of carbon at the surface is zero during the oxidation and D for carbon in steel at 1000°C is $3.1 \times 10^{-11} m^2s^{-1}$.)

12.10   There is a time delay, the incubation period, between water quenching and the commencement of hardening for a precipitation hardening aluminium alloy. Precipitation hardening tests show that, for a particular aluminium alloy, the incubation period is 30 s at a temperature of 150°C and 3 minutes at 100°C. To what temperature should the alloy be cooled, immediately after solution heat treatment, if it is desired to maintain the alloy for a period of three days without commencement of hardening?

13

Electrical and Magnetic Properties

13.1 Conduction

The conduction of electricity through a material is the transference of an electrical charge from one position to another. The charge may be transferred by movement of electrons, or by the migration of ions. Conduction in metals is due to electron migration, but it is the movement of ions that is responsible for the conductivity of electrolytes and for the very low conductivities observed in some insulating materials. In the class of materials known as *semiconductors* charge is carried by the motion of electrons and the movement in the opposite direction of positive 'holes'.

Electrical conduction in metals is due to the migration of electrons through the crystal lattice and metals are classed as good conductors because of the high mobilities of electrons. The resistivity of a metal increases with increasing temperature. As the temperature is raised the vibrational amplitude of the atoms is increased and there is a greater probability of interference and collision between moving electrons and atoms. However, an increase in pressure effectively increases the number of electrons per unit volume of material and so the resistivity is reduced. This property is useful and electrical strain gauges are based on this principle.

In the case of ionic conductivity an increase in temperature increases the mobility of ions and causes a rise in the conductivity. This is the reverse of the effect of heat on an electron conductor.

There is a complete spectrum of resistivities from the extremely low resistivity of silver, the best metallic conductor, to the very high resistivities of plastics and ceramic materials (Table 13.1). Resistivity is expressed in ohm metres (Ω m). The resistance of a piece of conducting material increases with the length of the conductor and decreases with an increase in the cross-sectional area (c.s.a.):

$$\text{resistance } (\Omega) = \text{resistivity } (\Omega \text{ m}) \times \frac{\text{length (m)}}{\text{c.s.a. (m}^2)}$$

13.2 Band structure

It is necessary to consider the band structure of atoms in order to understand electronic conduction in conductors and semiconductors. The electronic structure of atoms was discussed in Chapter 2 and it was stated (Section 2.8) that the energy levels of electrons become closer to one another as the distance from the nucleus increases. In the elementary description of metallic

Table 13.1   *Resistivities of some pure metals, alloys, and other materials*

Material	Resistivity Ω m @ 20°C	Material	Resistivity Ω m @ 20°C
Silver	1.6×10^{-8}	Silicon	8.5×10^{-4}
Copper	1.67×10^{-8}	Germanium	1.0×10^{-3}
Aluminium	2.66×10^{-8}	Urea-formaldehyde	
Magnesium	3.9×10^{-8}	(white Bakelite)	1×10^{6}
Sodium	4.3×10^{-8}	Fireclay	1.4×10^{8}
Zinc	5.92×10^{-8}	Alumina	1×10^{11}
70/30 brass	6.2×10^{-8}	Phenol-formaldehyde	
Nickel	6.84×10^{-8}	resins (Bakelite)	1×10^{12}
Iron	8.85×10^{-8}	Diamond	5×10^{12}
Tin	1.15×10^{-7}	Polyethylene	1×10^{13}
Mild steel	1.7×10^{-7}	Nylon	1×10^{14}
Lead	2.1×10^{-7}	Mica	9×10^{14}
50/50 cupronickel	5.5×10^{-7}	Quartz	9×10^{14}
18/8 stainless steel	7.0×10^{-7}	Pyrex glass	1×10^{16}
Mercury	9.6×10^{-7}	PTFE	2×10^{16}
Graphite	1.4×10^{-5}	Vitreous silica	1×10^{20}

bonding (Section 2.15) it was stated that when atoms are in close proximity the outer shell electrons tend to overlap. In an aggregate of atoms the overall effect of these two factors is to cause the discrete energy levels of the electron states to broaden into bands.

Figure 13.1 shows the broadening of energy levels for the element magnesium. The dotted line in the figure indicates the interatomic distance in crystalline magnesium. It will be seen that the 3*s* and 3*p* levels overlap. When considering a large aggregate of atoms, as would be the case in a metal crystal, the energy band is broad enough to contain very many energy levels, each one differing infinitesimally from the next. The Pauli exlusion principle still holds and no two electrons can possess exactly the same energy. In the case of the element sodium each atom

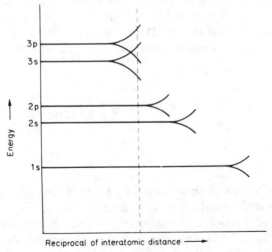

FIGURE 13.1   Broadening of the energy levels of the electron states of magnesium

contains one electron in the 3s state, although the 3s state could hold two electrons per atom, one with a spin quantum number $m_s = +\frac{1}{2}$ and the other with $m_s = -\frac{1}{2}$. In an aggregate of sodium atoms, if there are n atoms present there will be $2n$ available energy levels within the 3s energy band. With only one valence electron per atom, this means that the 3s band in sodium is only half filled. The valence electrons tend to occupy the lowest available energy levels. Energy must be given to an electron in order to move it from one position to another, that is, to conduct electricity through the material. In the case of sodium the energies of the 3s, or valence, electrons can be readily increased within the part-filled 3s band, and the metal is a good electrical conductor. In magnesium each atom possesses two 3s electrons and the 3s energy band is completely filled. But, in this element the 3s and 3p bands overlap and valence electrons may have their energies increased within the limits of the 3p band, and so magnesium conducts electricity readily. Adjacent permitted electron energy bands do not overlap in all materials. In diamond, for example, the four outer shell electrons per atom completely fill the valence band and there is a large energy gap between the valence and the next permitted energy range. The magnitude of this gap is a measure of the amount of energy an electron requires to break away from bonds and become a free electron. In diamond this energy is 8.33×10^{-19} J (5.2 eV). Consequently, diamond is an insulating material with a high resistivity. In other Group IV elements, silicon, germanium and tin, the energy gaps are smaller. In silicon the energy gap is 1.76×10^{-19} J (1.1 eV), in germanium it is 1.5×10^{-19} J (0.72 eV) and in grey tin it is only 0.13×10^{-19} J (0.08 eV) (Figure 13.2). The resistivities of these elements (see Table 13.1) reflect the magnitudes of the energy gaps. A small electrical field is sufficient to cause the valence electrons of tin to cross the energy gap and so tin is a reasonably good conductor of electricity. It is possible, though more difficult, for valence electrons in silicon and germanium to cross the energy gap; hence, these elements are semiconductors.

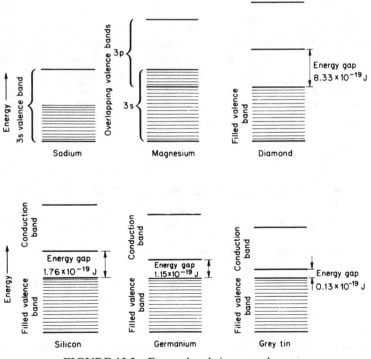

FIGURE 13.2   Energy bands in some elements

13.3 Conduction in metals

The good conductivity of a metal is due to the high mobility of the outer shell, or valence, electrons throughout the crystal lattice. Any factor that disturbs the regularity of a space lattice will, therefore, adversely affect the mobility of valence electrons and increase resistivity. The presence of lattice defects such as dislocations and interstitial or substitutional solute atoms will increase the probability of collisions between moving electrons and the cores of atoms and cause an increase in resistivity. Consequently, a solid solution of two elements will possess a higher resistivity than the pure metals. Refer to Section 11.10 and also compare the resistivities of copper, nickel, and a cupronickel alloy quoted in Table 13.1.

If a metal or alloy is cold worked the grain structure is fragmented and distorted and the number of dislocations in the material is increased. This also leads to an increase in resistivity.

An increase in temperature will cause an increase in the amplitude of vibration of atoms and, hence, increase the probability of electrons colliding with an atom core. This is the reason for the increase in resistivity observed when the temperature of a metallic conductor is raised.

13.4 Semiconductors

It has already been stated that silicon and germanium are semiconductors. Their conductivity is due solely to the distribution of electron energies within the pure material. Materials of this type are termed *intrinsic* semiconductors. Silicon and germanium possess the diamond type of crystal structure with each atom covalently bonded to four other atoms. The freeing of one valence electron to cross the energy gap into the conduction band will create a situation with one atom within the crystal lattice possessing only three bonds. The gap in the bonding is termed an *electron hole*. The bonding electrons are constantly in motion and may switch from one atom to another. Movement of an electron in one direction will cause an electron hole to move through the lattice in the opposite direction (Figure 13.3). As the motion of a hole is opposite in direction to the drift of bonding electrons the hole can be regarded as a positive charge carrier. In intrinsic semiconductors conduction is due to both the movement of freed valence electrons in the conduction band and the motion of positive holes. The energy necessary to free an electron and allow it to cross the energy gap can be provided by thermal activation. From this it can be seen that increasing the temperature of an intrinsic semiconductor will increase the conductivity. This is the opposite of the effect of increasing the temperature of a fully conducting metal.

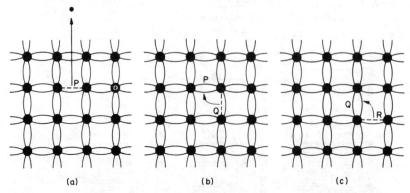

FIGURE 13.3   Electrical conduction by movement of 'holes'. (a) An electron is freed from a covalent bond creating an electron hole at P. (b) Electron transfer to P from adjacent site. Hole effectively moves to Q. (c) Electron transfer from R to Q. Hole moves to R

Another type of semiconductor is the impurity, or *extrinsic*, semiconductor. Silicon and germanium have four outer shell electrons per atom and a balanced covalently bonded crystal structure is formed. If some atoms of an element with only three outer shell electrons are substituted for some of the parent atoms there will be a shortage of electrons for full covalent bonding and some electron holes will be present in the crystal lattice. Conduction due to the migration of positive holes can then occur. This type of semiconductor is known as *p-type*. A small amount of aluminium dissolved in silicon or germanium is an example of this type. Alternatively, if an element such as phosphorus with five outer shell electrons is introduced into the covalent silicon lattice there will be additional electrons present. These additional electrons can be freed to enter the conduction energy band relatively easily. This type of semiconductor, in which transference of electrical charge is by the motion of freed electrons, is termed *n-type*.

By the introduction of small controlled amounts of various elements into silicon or germanium, semiconductors can be 'designed' to meet specific requirements.

Other semiconductor materials may be ceramic compounds made by combining a Group III element and a Group V element, or a Group II element with one from Group VI, or oxides with a variety of structures. The important compound semiconductors are gallium arsenide, GaAs, indium phosphide, InP, indium antimony, InSb, zinc oxide, ZnO, zinc sulphide, ZnS, and cadmium sulphide, CdS. Three-component semiconductor materials have also been developed and some important ones are InGaAs, GaAsP and CdHgTe. Elements used as doping additions in Group III–Group V compound semiconductors are zinc or cadmium to make *p*-type, and selenium and sulphur for *n*-type. Semiconductor materials are used in integrated circuits and electronic devices.

13.5 The p-n junction

A *p*-type semiconductor, although it contains positive holes due to the presence of atoms of a Group III element, is electrically neutral. Similarly, although an *n*-type crystal contains free electrons from the Group V donor atoms it is still electrically neutral. If a *p*-type semiconductor is joined to an *n*-type semiconductor* the result is a *p-n* junction and this is effectively a one-way valve. At the interface between the *p*-type and *n*-type materials there will be some interaction between positive holes and free electrons. The combination of a hole and an electron destroys both as mobile charge carriers. The small interface zone in which the charge carriers are lost is termed the depleted layer. This zone is very narrow, being only about 10^{-6} m in thickness. Within the depleted layer the *p*-type impurity atoms have each captured an electron and have become negatively charged ions, and the *n*-type impurity atoms have lost a free electron to become positively charged ions. These are termed acceptor and donor atoms respectively (Figure 13.4). The depleted layer is thus highly charged, carrying a negative charge on one side and a positive charge on the other. This charged zone is a potential barrier and effectively prevents charge carriers from passing freely across it.

Now let us consider what happens when a voltage is applied across the ends of a crystal containing a *p-n* junction. If the *n*-type end is made positive with respect to the *p*-type end (Figure 13.5(a)) the positive charge on the *n*-type side of the depleted layer will be increased. The height of the potential barrier at the depleted layer will thus be increased and it becomes extremely difficult for charge carriers to pass across the junction. Conversely, if an applied voltage is applied in the opposite sense (Figure 13.5(b)) the height of the potential barrier will be reduced and current will be able to flow fairly freely across the junction. Figure 13.5(c) shows the voltage-

---

* *p-n* junctions are normally made by introducing *p*-type and *n*-type impurities into opposite ends of a crystal of silicon or germanium, rather than by joining two separate crystals.

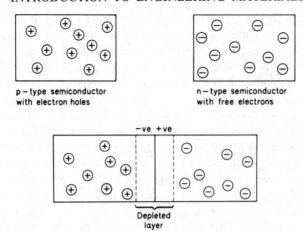

FIGURE 13.4   *p-n* junction showing depleted layer

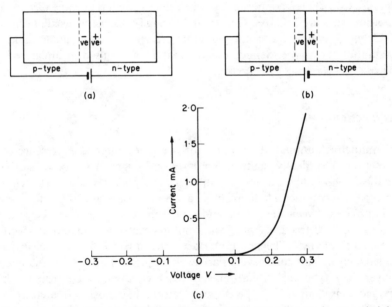

FIGURE 13.5   *p-n* junction characteristics. (a) Applied voltage tending to increase height of potential
barrier. (b) Applied voltage tending to decrease height of potential barrier, (c) Voltage–
current relationship for silicon *p-n* junction

current relationship for a silicon *p-n* junction. The reverse current with *V* negative, is extremely
small and is of the order of one microampere (1 µA).

13.6 Insulating materials

Insulating materials are those in which the bond structure is ionic, or covalent, or a mixture of
both types. In these bonds the electrons are tightly bound and are not 'free' as in metals. Electron
and ion mobility is very low and measured resistivities are 10^6 Ωm or greater (see Table 13.1).

The energy band diagram for an insulating material is similar to that for a semiconductor with a completely filled valence band and an energy gap between the valence band and the next permitted energy band. However, in the case of insulating materials the energy gap is so large that electrons are unable to cross it under normal conditions. An occasional electron may possess a sufficiently high energy to cross the gap, and it is this that accounts for the very high, but measurable resistivities of insulators. At high temperatures there will be a greater probability of an occasional electron possessing the necessary energy for conduction and so the resistivities of insulators show a decrease with increasing temperature. In the case of some materials containing ionic bonds there is also the possibility that small ions in a relatively open lattice could migrate, giving some small degree of conductivity. Again, at high temperatures ions become more mobile and can diffuse more readily, so reducing the measured resistivity.

Insulating materials are also known as *dielectrics*, and have the capacity to store an electrostatic charge. When a dielectric is placed in an electric field some polarisation occurs. If the material already contains polar molecules the effect is to cause the molecules to align themselves in the direction of the applied field. In non-polar materials the effect of the field is to create dipoles by causing small atomic movements leading to slight separation of positive and negative ions. When an applied electric field induces polarisation in a material there is a back effect which modifies the field. This effect is termed *permittivity*.

The relative permittivity, ε_r, or dielectric constant is a parameter of dielectric materials and is the ratio of the capacity of a capacitor (condenser) with the material filling the space between the plates, to the capacity with a vacuum between the condenser plates. The parameter is also known as relative permittivity or specific inductive capacity. The value of the dielectric constant, ε_r, is within the range of 2 to 10 for many polymers and ceramics, but the value tends to reduce as frequency increases. High permittivity ceramics based on barium titanates have been developed for the production of miniaturised capacitors. These possess values of ε_r, of the order of 1000.

Some energy is lost as heat when a dielectric material is polarised in an alternating electrical field. Energy losses are due to two factors, dipole friction and current leakage. The fraction of the energy lost during each reversal is termed the dielectric loss, W, and is given by $W = \omega C V^2 \delta$, where V is the root mean square of voltage, C is capacitance, δ is the power factor or loss angle and $\omega = 2\pi f$, where f is the frequency. A high power factor is an indication of greater power in an insulator being dissipated as heat. Low loss steatite ceramics were developed for high frequency insulation, for example, in radio communication equipment.

Insulating materials, or dielectrics, can break down at very high voltages. The applied electrical field may be sufficiently high to raise the energies of some electrons above the energy gap and thus free them. The possibility of this occurring is increased if the material contains some impurities or defects. The high energy electrons may then disrupt covalent bonds within the insulator liberating more electrons to take part in a chain reaction. This type of breakdown is referred to as *cascading* and is irreversible. The *dielectric strength* of a material is the electrical stress in volts per unit volume required to cause breakdown. During the testing of the strength of dielectrics, the electrical stress is maintained for one minute. Far more common than cascading is surface breakdown. The presence of moisture or an accumulation of dirt on the surface of an insulator may provide a surface-conduction path. To minimise the incidence of surface breakdown, ceramic insulator surfaces are glazed and also made with a shaped or corrugated surface to greatly increase the surface path length.

13.7 *Ferroelectric materials*

In many dielectric materials, the polarisation which occurs within an electrical field is directly proportional to the strength of the field and all polarisation is lost when the field decreases to zero. Some materials retain a net polarisation when the electrical field is removed and these are termed *ferroelectric*. In such a material the dipoles line up with the field when an electrical field is applied until all the dipoles are in alignment and a saturation polarisation is achieved. When the field is removed the degrees of dipole alignment largely remains and there is a remanent polarisation (compare with magnetic remanence—Section 13.9). In order to remove the polarisation completely, the electrical field must be reversed and increased in strength to a specific value (compare with magnetic coercivity—Section 13.9). In an alternating electric field, a ferroelectric hysteresis loop is described. The ferroelectric materials are ceramics, such as barium titanate, with a distorted perovskite structure (see Section 5.6) and are dielectric materials with a high dielectric constant and are suitable for use in capacitors. Also, the ability to retain polarisation gives them the capability for holding and storing information, thus making them suitable for use in computer circuitry. Ferroelectric materials may also be *piezo-electric* (see Section 13.8).

Ferroelectric behaviour is temperature dependent and is lost above a critical value of temperature, the Curie temperature. The value of the Curie temperature varies from one material to another and, for example, is 120°C for barium titanate and 570°C for lead niobate.

Some dielectric materials show a different phenomenon, known as anti-ferroelectric behaviour. In anti-ferroelectric materials, such as lead zirconate and sodium niobate, individual dipoles occur but they arrange themselves so as to be antiparallel to adjacent dipoles. As a result, the net spontaneous polarisation is zero.

13.8 Piezo-electric effect

Some crystalline dielectric materials polarise when strained mechanically and an electrical voltage or field is created. This is termed the *piezo-electric effect*. It is a reversible phenomenon and such a material, when subject to an electric field, will strain. Piezo-electric materials are crystals which do not possess a centre of symmetry. A good example is barium titanate which has a distorted perovskite structure (see Section 5.6). Natural quartz is a piezo-electric material and, for many years, was used in the manufacture of transducers. Today, quartz has been largely superseded by inorganic crystalline materials such as barium titanate, lead niobate, lead zirconium titanate and lithium sulphate. Transducer crystals made from these materials are polycrystalline with all the crystals orientated in the same direction. This is achieved by allowing the material to crystallise within a strong electrical field.

There are two piezo-electric coefficients, d and g. The electrical field, ζ(V/m), caused by a stress σ (Pa) is given by: $\zeta = g\sigma$ and the strain, ε, produced by a field is given by: $\varepsilon = d\zeta$. The two coefficients are related: $g = 1/dE$, where E is the modulus of elasticity.

Piezo-electric materials are used in transducers which convert electrical fields into mechanical vibrations, generating acoustic waves, and vice versa.

13.9 Magnetic behaviour

The motion of an electrical charge creates a magnetic field. Consequently, the motion of electrons within an atom produces a magnetic effect. In Chapter 2 it was stated that electrons are not just moving in orbitals around the nucleus but can also be regarded as spinning about their own axes. The fourth, or spin quantum number, m_s, for an electron is either $+\frac{1}{2}$ or $-\frac{1}{2}$ signifying one of two opposing spin directions. In most instances an element with an even number of electrons has as many electrons spinning in one sense as in the other, while an element with an odd number of electrons will just have one more electron spinning in one direction than in the other. Because of this, most elements possess very weak magnetic properties and are either *paramagnetic*, very weakly attracted to a magnet, or *diamagnetic*, very slightly repelled by a magnet. Four elements, however, possess structures in which several more electrons spin in one direction than in the opposing sense and these elements are strongly attracted by a magnet. This type of behaviour is termed *ferromagnetic* and the elements that behave in this manner are iron, cobalt, nickel and gadolinium. In these ferromagnetic elements each atom behaves as a small magnet as a result of the out-of-balance electron spins.

Within a ferromagnetic material there is a sub-structure known as a *domain* structure. A magnetic domain is a small section of a crystal in which the atomic magnets are aligned parallel to one another. The dimensions of an individual domain are very small and the volume of a domain is of the order of $1 \times 10^{-4} \, mm^3$. In the unmagnetised condition the domains possess random orientation, but when the material is placed in a magnetic field all the domains tend to realign themselves in the direction of the field. The material is now magnetised. When the external magnetic field is removed the domain alignment may remain and the material be permanently magnetised. When a material behaves in this way it is said to be magnetically *hard*. A *soft* magnetic material, on the other hand, will lose its magnetism when the external field is removed, with the domain structure reverting to a random orientation.

When a ferromagnetic material is magnetised the degree of magnetisation induced does not increase linearly with increasing magnetic field strength but follows the curved path OA in Figure 13.6 until magnetic saturation is reached at point A. If the intensity of the magnetising field is reduced the induced magnetisation falls off slowly, but still retains a positive value when the magnetising field has been completely removed. The extent of magnetisation remaining, B_r, is termed the *remanence*. To reduce the magnetisation, B, to zero the direction of the applied magnetic field must be reversed and its magnitude increased to à value H_c. This value, H_c, is

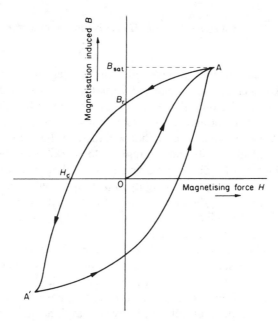

FIGURE 13.6 Magnetisation curve for ferromagnetic material

termed the *coercivity*. If the magnetising force is increased in a reverse sense beyond the value of H_c the material will be magnetised in the opposite direction until saturation is reached at A'. Reducing the magnetising field from A' and then increasing it in the positive direction will complete the magnetisation, or *B-H*, curve. For a material to be a good permanent magnet the remanence, B_r, should be close to the saturation value, and the coercivity, H_c, should be large. The area enclosed within the *B-H* loop is proportional to the energy used to realign the magnetic domains during each magnetisation cycle. This energy is then dissipated as heat. If a ferromagnetic material is used as the core of a coil through which an alternating electric current is passed it will be magnetised, first in one direction and then in the other, with every cycle of current. A magnetically soft material with a low remanence and low coercivity is necessary for this type of application. Figure 13.7 shows the difference in shape of *B-H* curves for two ferro-magnetic materials.

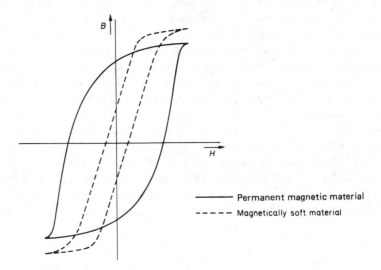

FIGURE 13.7   *B–H* curves for two materials

A ferromagnetic material that is physically soft is likely to be magnetically soft, and vice versa. In a material such as soft annealed iron the magnetic domains will have random orientation but will readily become aligned on magnetisation. When the external magnetic field is removed, the majority of the domains will return to a state of randomness and the material will have a low remanence value. Any factor that will disturb the regularity of the iron lattice and strengthen the material physically, for example cold working or the introduction of alloying elements, will impede the rotation of domains and movement of domain boundaries and so will increase the coercivity of the material. Highly alloyed materials containing iron, cobalt, nickel, and other elements are used for the manufacture of permanent magnets. These alloys, which can be precipitation hardened to give a very highly dispersed precipitate, are most suitable because domain movement is very greatly restricted.

If a magnetised material is annealed to soften it the domains can revert to a random orientation. The characteristics of some magnetic materials are given in Tables 13.2 and 13.3.

In addition to metals and alloys some complex oxides can also be strongly magnetic. Magnetite, the lodestone of the ancient navigators, is magnetic. Magnetite has the formula Fe_3O_4 but of the three iron atoms per molecule one is divalent and the other two are trivalent. Magnetite is crystalline and possesses the spinel type structure (refer to Section 5.6). These are ceramic mixed oxides of iron and another metal, M, such as copper, cobalt, manganese, nickel or zinc with the general formula $MO.Fe_2O_3$. These materials are *ferrimagnetic* and are generally referred to as *ferrites*. Unlike metals, these materials are insulators and, thus, become suitable materials for use as magnetic cores at ultra-high frequencies as eddy currents cannot be established within them.

The characteristics required in a material for a permanent magnet are high values of both remanence and coercivity, and high power. The power of a magnet is related to the size of the hysteresis loop and is the maximum product of B and H. Such a material is termed magnetically hard. Originally, fully hardened high carbon steels were used for permanent magnet manufacture. These were followed by the development of alloy steels containing cobalt, chromium and tungsten. In the 1930s aluminium–nickel steel was developed in Japan as a magnet material and since then many alloy developments have taken place producing a series of Fe–Al–Ni–Co alloys with additions of copper, niobium and titanium. The coercivities and permeabilities of these alloys can be improved further by *magnetic annealing* which is the slow cooling of a ferromagnetic material from the annealing temperature within a strong magnetic field. This produces a permanent alignment of magnetic domains in a specific crystal direction. Other magnetic alloys include Co–Pt, Cu–Co–Ni and Cu–Ni–Fe materials, while other permanent magnet materials are hard ferrites including barium iron oxide and strontium iron oxide materials made by powder compaction and sintering. The properties of some permanent magnet materials are given in Table 13.2.

Materials with low remanence, low coercivity and a very small hysteresis loop are required in components, such as transformer and electromagnet cores, which operate in a constantly alternating magnetic field. Soft iron is very useful for this purpose, as is silicon steel sheet containing 3–5 per cent silicon and virtually no carbon. This latter material, in the fully annealed condition, is used for laminated transformer cores. The silicon content gives the material a high electrical resistivity and thereby reduces eddy current losses. Another group of magnetically soft materials are those with low hysteresis but high permeability and used in communications equipment where a high permeability will utilise fully the small currents. These materials include Mumetal (a Ni–Cu–Mn–Fe alloy), Permalloy (78% Ni/22% Fe) and Supermalloy (79% Ni/16% Fe/5% Mo). Other magnetically soft materials include some ferrites which are suitable for use as cores in high frequency transformers because of their low hysteresis losses and the fact that eddy

Table 13.2 *Properties of some permanent magnet materials*

Material	B_r (T)	H_c (A/m)	$B-H_{max}$
1% C steel	1.0	4000	2000
Alni (25% Ni; 13% Al; 4% Cu; 58% Fe)	0.56	46000	9500
Alnico I (21% Ni; 12% Al; 5% Co; 62% Fe)	0.71	35000	14000
Alnico V (14% Ni; 8% Al; 24% Co; 3% Cu; 51% Fe)	1.31	53000	60000
Alnico XII (18% Ni; 6% Al; 35% Co; 8% Ti; 33% Fe)	0.58	76000	16000
Cunico (50% Cu; 21% Ni; 29% Co)	0.34	53000	8000
Cunife (60% Cu; 20% Ni; 20% Fe)	0.54	44000	15000
Co_2Sm	0.95	760000	
Co–Pt (23% Co; 77% Pt)	0.45	216000	40000
Ferrite $BaO.Fe_2O_3$	0.4	190000	
Ferrite $SrO.Fe_2O_3$	0.34	256000	

currents cannot be generated within them. Some soft ferrites have square hysteresis loops and are used for information storage in computers. The properties of some soft magnetic materials are given in Table 13.3.

Table 13.3 *Properties of some magnetically soft materials*

Material	B_r (T)	B_{mat} (T)	H_c (A/m)	Applications
Soft iron	1.3	2.16	7	electromagnets
Silicon steel (3% Si)	0.8	1.75	24	transformers
Silicon steel (4.25% Si)	0.7	1.90	12	transformers
Mumetal (74% Ni; 5% Cu; 1% Mn; 20% Fe)	0.6	0.8	4	magnetic shielding
Permalloy (55% Fe; 45% Ni)	1.1	1.6	24	magnetic shielding
Supermalloy (79% Ni; 16% Fe; 5% Mo)	0.5	0.8	160	magnetic shielding
Ferrite $NiO.Fe_2O_3$	0.11	0.27	950	h.f. transformers
Ferrite $MnO.Fe_2O_3/ZnO.Fe_2O_3$	0.14	0.36	50	h.f. transformers

13.10 The Hall effect

If a current is passed through a thin slab of a conducting material and the conductor is situated in a magnetic field, such that the direction of the field is normal to both the surface of the slab and the current direction, an electric field (hence a potential difference) will be developed across the width of the conductor. This is known as the Hall effect (see Figure 13.8).

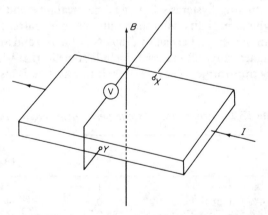

FIGURE 13.8   The Hall effect—a potential difference is developed between X and Y

The magnitude of the electric field, E (V/m), that is developed is given by

$$E = R_H I B$$

where I is the current density in the conductor (A/m^2), B is the magnetic field strength (T) and R_H is the Hall coefficient of the material. The Hall effect is observed in metals and in semiconductors. The polarity of the electric field produced in a semiconductor depends on whether the material is *p*-type or *n*-type, that is, whether the charge carriers are positive holes or electrons.

The magnitude of the Hall coefficient, R_H, is small in metals, but is very much greater in semiconductor materials and the Hall voltage developed across a crystal may be large enough to be put to practical use. The effect is used in a device for the measurement of magnetic-field strength. A Hall probe may be made small in size, for example 3 mm × 1 mm × 10 mm, and such an instrument may be used for the determination of magnetic-field strengths in confined spaces. One novel use of a Hall effect device is in the grading of ball- and roller-bearings. Bearings moving along an inspection line are magnetised before passing a scanning device consisting of a Hall probe, where the coercivity, H_c, is measured. There is a direct relationship between coercivity and the physical hardness of the bearing. The measuring device feeds information to an electronic sorting-gate so that all bearings with coercivities outside the acceptable limits for the product are separated from the rest. The satisfactory bearings are demagnetised before use. A schematic diagram of such an inspection line is shown in Figure 13.9.

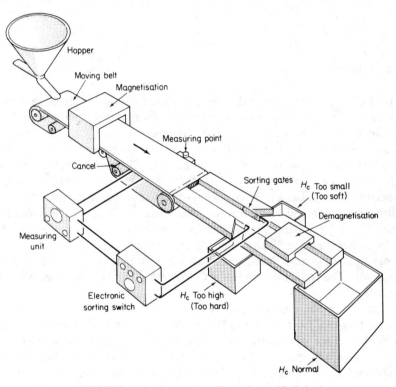

FIGURE 13.9   Inspection line using a Hall device

14

Optical, Thermal and Other Properties

14.1 Electron excitation

In Chapter 2 the build-up of electron shells was considered and it was stated that each electron shell and sub-shell within an atom represents a particular energy level (Section 2.6). In the normal state the electrons of an atom occupy the lowest permissible energy levels. This is called the 'ground state' for the atom. As mentioned Section 2.6 it is possible for the energy of an electron to be raised to a higher level. This is called the excited state. The excited state is an unstable state and excited electrons quickly revert to their ground state and in so doing cause energy to be emitted. This energy is emitted as a discrete packet or *quantum*. These quanta released by excited atoms are called *photons* and photons travel at the speed of light. The magnitude of energy quanta is of the order of 10^{-19} J and it is customary to use the electron volt as a unit of energy for these low values, rather than the SI unit of energy, the joule.

The electron volt is the change in energy caused when an electron moves through a potential difference of one volt. The charge of one electron is 1.602×10^{-19} C, therefore one electron volt (1 eV) is equal to 1.602×10^{-19} J.

The photons emitted by an excited atom will interact with other atoms in their paths and this could result in excitation of further atoms. The photon is a means of transporting energy from one atom to another. An energy quantum may be regarded sometimes as a photon particle, but it may also be considered as a packet of electromagnetic waves of a particular frequency. The relationship between quantum energy, E, and the frequency of electromagnetic waves is Planck's quantum relationship

$$E = h\nu$$

where ν is the frequency of the energy emission and h is Planck's constant (6.625×10^{-34} J s). The concept of energy then is a dual one and depending on the circumstances, the energy quanta may be thought of as photons or waves.

Atoms may be raised from the ground state to the excited state in several ways. If crystals of common salt are placed in a flame, the flame changes to a characteristic yellow. This is because the heat energy excites the atoms making up the salt. When the electrons within the sodium atoms present revert to the ground state photons possessing energies of about 2 eV are emitted. From Planck's equation this would be equivalent to electromagnetic waves with a frequency of about 5×10^{14} Hz. This is a frequency in the yellow portion of the visible spectrum. Atoms of other elements, when heated in this way, will emit photons of different energies corresponding to waves of different frequencies. For example, strontium will give a red colouration and copper will give a green coloured flame.

Atoms may be excited in other ways than by heating in a flame. When an electric arc is struck between two electrodes the atoms in the electrode materials are excited into photon emission. Similarly when an electrical discharge is made through a vapour, as in a sodium or mercury vapour lamp, photon emission will occur.

When the atoms of an element are excited there are different levels of excitation and consequently photons of several energies are emitted (see Figure 14.1). The photons of any one energy value will give rise to a radiation emission of a particular frequency but with photon emission at many discrete energy values there will be many characteristic frequencies of radiation. An excited element will therefore yield a line spectrum containing many lines characteristic of that element. This fact is used as the basis of an analytical tool and spectroscopy may be used for both qualitative and quantitative chemical analysis.

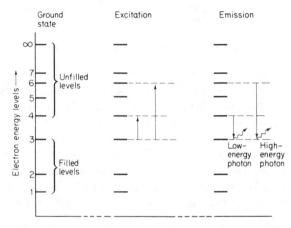

FIGURE 14.1   Levels of electron excitation

When a material is bombarded by a stream of high-energy electrons the degree of excitation within the atoms of the material is very great resulting in the emission of photons of very high energies. Such photons, with energies of the order of 10^2 to 10^5 eV, are X-rays with frequencies in the range 10^{17} to 10^{20} Hz. This principle is employed for the generation of X-rays.

If the energy input to an atom is large enough it is possible to excite one or more electrons sufficiently to allow them to be freed from the atom. If this occurs and an electron leaves the atom completely it leaves behind a positively charged ion. The minimum quantity of energy required to ionise an atom is termed the ionisation energy or ionisation potential of the element.

14.2 The energy spectrum

Energy may be transmitted in the form of a wave-like radiation termed electromagnetic radiation. Electromagnetic radiations may possess an extremely wide range of frequencies ranging from low radio-frequencies, through infra-red, visible light and ultra-violet into the X-ray and γ-ray range. Electromagnetic waves travel through a vacuum at a velocity of

$c = 3 \times 10^8$ m/s (c = velocity of light). The frequency, ν, and wavelength, λ, of any radiation are related, since their product is a constant

$$\nu\lambda = c$$

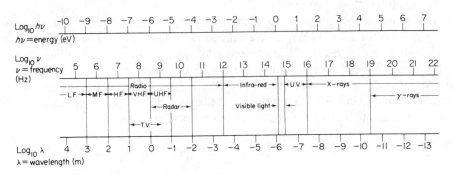

FIGURE 14.2 Electromagnetic spectrum

As stated in Section 14.1 electromagnetic waves can be regarded as a series of energy quanta, or photons. Planck's quantum relationship $E = h\nu$ relates the energies of photons to the frequency of the radiation; h is Planck's constant (6.625×10^{-34} J s).

The energy of a photon for a low-frequency radio wave of frequency 200 kHz is very small and has a value of about 13×10^{-29} J (8×10^{-10} eV). The energy of a photon of light in the visible sector of the spectrum, frequencies of almost 10^{15} Hz, is about 2 eV while for X- and γ-radiation with frequencies in excess of 10^{16} Hz, photon energies will range between 100 and 10^6 eV. High-energy photons may cause changes in a material when the material is subject to the radiation (see Section 28.7).

14.3 Absorption and Transmission

In metals the outer-shell, or valency, electrons are in partially filled conduction bands (see Section 13.2). When photons of comparatively low energy are incident upon a metal the photon energy can be transferred to valency electrons, raising the energies of these electrons within the conduction band. When this occurs the incident energy is absorbed by the material and not transmitted through it. Metals are opaque to radio waves, infra-red, visible light and ultra-violet but the high-energy photons of X- and γ-rays can penetrate through the metals. There is some absorption of this high-frequency radiation and the total amount of X- or γ-radiation that is transmitted through the metal is inversely proportional to the thickness. The absorptive power of a metal for X-radiation is also in proportion to the density of the metal. X- and γ-radiation are used for the radiography of metals and greater penetration is achieved by the harder (higher-frequency) radiation. There is some reflection of X-rays from atomic nuclei and because of this fact X-rays can be used for investigation of crystal structures.

Polished metal surfaces will largely reflect, rather than absorb incident low-energy photons. This reflection is partially selective. For example, copper absorbs visible light at the short wavelength end of the visible spectrum, hence its reddish colour.

Dielectric materials possess filled electron-energy bands and forbidden energy gaps. There cannot be transference of energy from photons to electrons in these materials in the same way as in metals and so these materials generally do not absorb electromagnetic radiation but transmit it.† Good radio reception is obtained within a building since the brick and glass are largely transparent to radio waves, but radio reception within a metal-bodied car is difficult without an external aerial.

The explanation of absorption and transmission given above is a very simple one and does not cover the situation fully. Non-metallic materials are not fully transparent to all types of electromagnetic radiation and many materials show absorption peaks at particular frequencies. For example, window glass is opaque to the infra-red section of the spectrum. Also certain materials show definite colours because they strongly absorb certain frequencies within the visible part of the energy spectrum.

These colour effects are caused by the presence of certain types of crystal defects, termed *colour centres*. Impurity ions in certain materials may create donor or acceptor energy levels which cause selective absorption of photons of particular wavelengths. It is this selective absorption which is responsible for the varied colour of materials. Many types of defect (colour centre) may occur, the best known example being the F-centre which is an electron trapped on an anion vacancy and this has a series of energy levels available to it. The energy required to transfer from one level to another falls within the visible part of the electromagnetic spectrum giving the characteristic colour, which depends on the host crystal. F-centres in sodium chloride, NaCl, give the characteristic greenish-yellow colour.

Semiconductor materials possess filled valence bands but with forbidden energy gaps of a much smaller size than those in insulators. Low-frequency radiation is not absorbed by semiconductors but if the frequency of the incident radiation is such that the photon energy is equal to the magnitude of the energy gap, energy may be transferred from photons to electrons. In other words, the incident energy is absorbed and electrons are excited into the conduction band. This gives rise to the effect termed *photo-conduction* since the conductivity of the material increases in a similar manner to that caused by raising the temperature of the semiconductor.

14.4 Refraction and polarisation

A transparent material possesses a refractive index, μ, and this is defined as being the ratio of the velocity of light in a vacuum, c, to the velocity of light in the material, v.

$$\mu = \frac{c}{v}$$

There is a relationship between the refractive index of a material and the dielectric constant, K, for the material

$$\mu^2 = K \quad \text{or} \quad \mu = \sqrt{K}$$

The velocity of light and the refractive index of a material are functions of frequency. An example of this is the separation of white light into the colour spectrum by a glass prism; this effect is called *dispersion*.

There are very many different types of glass available for optical purposes and they possess refractive indices in the range 1.5 to 1.7. A material with a refractive index of this order of magnitude is ideal for use in lenses and similar devices. Refractive index is not the only important

† Most non-metallic materials are transparent to light in thin section. Their opaqueness in thick section is largely due to repeated reflection and refraction at internal surfaces.

property, and an optical material must also possess a high transmission value. The transmission ability of optical glass is about 98 per cent (that is, 98 per cent of incident light would be transmitted through a section of 10 mm thickness). A number of plastics materials are also suitable for optical applications, possessing similar refractive indices to glass and with transmission abilities of 90 per cent or more.

A glass is isotropic, but crystalline materials are not. If crystallisation takes place in a glass or glassy polymer the crystals will possess a different refractive index from the glass. Reflections at internal interfaces between glass and crystal regions will lead to the material becoming translucent.

Due to the anistropy of crystals light will have a different velocity in the directions of the different crystal axes (except in cubic crystals where the velocity of light is the same for each major axis.) Crystals in which the velocity, and hence the refractive index, differs from one crystal axis to another are termed *birefringent*. Birefringent crystals are doubly refracting. A beam of light incident upon the crystal is split into two beams. There is another interesting effect in that the two beams are polarised at right angles to each other. Some crystals have the effect of absorbing one polarised component of the split beam to a very much greater extent than the other resulting in the emergent beam being of polarised light. This is the effect when light passes through Polaroid.

The phenomenon of birefringence is used in the micro-examination of rocks and minerals under polarised light and it can also be used to study fibre orientation in fibre reinforcd composites.

An isotropic material, such as glassy polystyrene, becomes anisotropic to some extent when it is physically strained. When physically strained the material becomes birefringent. This stress-induced birefringence is termed the *photoelastic* effect. In photoelastic stress-analysis a model of a component, say a chain hook, is made from clear plastic. The model is loaded and, due to the birefringence, interference fringes can be observed within the model. The fringe pattern will indicate the stress distribution within the model and the number of fringes observed is in direct proportion to the magnitude of the stress induced.

14.5 Diffraction

Electromagnetic radiation can suffer diffraction. Consider a beam of monochromatic light passing through a narrow slit of width w (Figure 14.3(a)). The wavefront diverges as it emerges from the slit such that the light intensity on a plane surface set at some distance from the slit varies in the manner shown in Figure 14.3(b). The first point of minimum intensity is at an angle θ from the incident beam where $\theta = \sin^{-1} \lambda/w$, λ is the wavelength. If, instead of passing through a single slit, the light beam is passed through a finely ruled grating on glass then diffraction at each of the narrow slits in the grating produces regions where emergent coherent beams overlap producing a series of fringes, narrow bright lines separated by dark zones. The angles, θ_n, at which these fringes occur is given by $n\lambda = d \sin \theta_n$ where n is an integer and d is the separation distance of the lines on the grating. The principle of optical diffraction fringe formation can be used as the basis of an extremely sensitive system of linear measurement. The regularly spaced planes of atoms in crystals act with respect to X-radiation or an electron beam in a similar manner to a diffraction grating and light. X-ray and electron beam diffraction patterns are used in the identification and measurement of crystal structures (see Section 5.8).

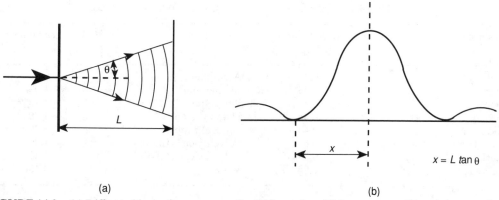

(a)   (b)

FIGURE 14.3   (a) Diffracted beam from narrow slit. (b) Intensity of light on screen, first minimum at half-cone angle θ, $\sin\theta = \lambda/w$

14.6 Luminescence

Luminescence is the general term covering the phenomenon of light emission from a material as a result of excitation of some type. During excitation, some of the valency electrons may be excited into a higher energy band, as mentioned in Section 14.1. When the excited electrons drop back to the ground state, or into some intermediate energy level, energy will be emitted from the material. An emission which occurs only for the duration of the excitation is termed *fluorescence*. A delayed emission which persists for a period of time after excitation has ceased is termed *phosphorescence*. The intensity of a phosphorescent emission at some time t after excitation has ceased, I_t, decays with time according to the expression: $I_t = I_0 \exp(-t/\tau)$ where I_0 is the initial intensity and τ is a constant for the material termed the *relaxation time*. Phosphorescent materials, or *phosphors*, are used to coat the surface of television and radar tubes. For a television tube, the relaxation time of the phosphors must not be too long or the images will begin to overlap but for the display tubes of search radar a phosphor with a longer relaxation time is selected to give an afterglow period of several seconds. Excitation of a luminescent material may be caused in one of several ways, by incident photons, termed *photoluminescence*, by an incident electron beam, termed *cathodoluminescence*, by the effects of an alternating electric field, termed *electroluminescence*, or by certain chemical reactions, termed *chemiluminescence*.

14.7 Heat capacity

The thermal properties of a material are influenced by atomic vibration. At any temperature above absolute zero, the atoms that make up a material are in a state of continual movement in the form of a vibration. The amplitude and frequency of the vibration varies with temperature. The vibration of atoms produces elastic waves, termed *phonons*, and the energy of a phonon, E, can be related to temperature as $E = kT$, where k is Boltzmann's constant (1.38×10^{-23} J/K) and T is temperature (K). The material gains or loses heat by gaining or losing phonons. The *heat capacity* of a material is the amount of energy required to raise the temperature of one kilomole of a substance through one degree K. For most practical purposes the quantity used is the *specific heat capacity*, or *specific heat*, which is the amount of heat energy required to raise the temperature of unit mass of substance through one degree K. The specific heat of a substance is not constant and increases as the temperature rises. The values of specific heat for a number of substances is given in Table 14.1. There are two specific heats quoted for gases, the specific heat at

Table 14.1 *Specific heats (at 15°C) for some substances*

Substance	Specific heat (kJ/kg K)	Substance	Specific heat (kJ/kg K)
Lead	0.13	Aluminium	0.88
Silver	0.23	Magnesium	1.03
Copper	0.38	Polystyrene	1.20
Nickel	0.44	Nylon (PA66)	1.70
Mild steel	0.44	Natural rubber	1.90
Plate glass	0.67	Polyethylene (LDPE)	1.90
Quartz	0.73	Polyethylene (HDPE)	2.31
Silica glass	0.84	Water	4.186

constant pressure, C_p, and that at constant volume, C_v, C_p being the larger of the two. The ratio of the specific heats, C_p/C_v, for gases is denoted by the symbol γ.

14.8 Thermal expansion

When a material is heated the vibrations of its constituent atoms increase. This is accompanied by an increase in the physical dimensions of the material, that is, expansion. The change in length, δl, that occurs when a material of length l_0 is heated from a temperature T_0 to T_1 is given by:

$$\delta l = l_0 \alpha (T_1 - T_0)$$

where α is the linear expansion coefficient for the material. The value of α, the expansion coefficient, is small for many ceramic materials and approximately half of that for many metals. Polymer materials tend to have very high thermal expansion coefficients. The thermal expansion coefficients for some materials are given in Table 14.2.

If the material is constrained during heating or cooling so that dimensional change is restricted, thermal stresses can be established (see Section 7.4) and this could lead to distortion or cracking. The risk of cracking is pronounced during the rapid heating or cooling of non-ductile materials such as ceramics and glasses. The magnitude of the stress developed in a non-ductile

Table 14.2 *Thermal expansion coefficients for some materials*

Material	$\alpha\ (K^{-1})$	Material	$\alpha\ (K^{-1})$
Pyroceram	0.4×10^{-6}	Nickel	12.8×10^{-6}
Silica glass	0.54×10^{-6}	Copper	16.2×10^{-6}
Invar	0.9×10^{-6}	Silver	18.8×10^{-6}
Diamond	1.2×10^{-6}	Aluminium	21.6×10^{-6}
Graphite	5.4×10^{-6}	Magnesium	25.2×10^{-6}
Quartz (parallel to axis)	7.5×10^{-6}	Polystyrene	63×10^{-6}
Quartz (normal to axis)	13.7×10^{-6}	Phenol-formaldehyde	72×10^{-6}
Plate glass	9.0×10^{-6}	Natural rubber	80×10^{-6}
Building brick	9.0×10^{-6}	Nylon (PA66)	100×10^{-6}
Mild steel	11.0×10^{-6}	Polyethylene (HDPE)	120×10^{-6}
Concrete	12.6×10^{-6}	Polyethylene (LDPE)	200×10^{-6}

material during rapid heating or cooling is proportional to α and, so, the susceptibility to crack formation through thermal shock is much less in materials with very low expansion coefficients. LAS ($LiO–Al_2O_3–SiO_2$) glass ceramics with extremely low expansion coefficients have been developed for applications in which a high resistance to thermal shock is required. The proprietary name for one such glass ceramic is Pyroceram. Some alloys with very low expansion coefficients have been developed, the best known being 'Invar', a steel containing 36% Ni and 0.2% C.

When a material melts, generally there is a marked increase in volume accompanying the transition from solid to liquid. In the case of metals the expansion at the melting point is of the order of 6 per cent. There are certain exceptions to this, a notable one being antimony, which contracts on melting. Use is made of this peculiarity in the manufacture of type metals. These contain some antimony to prevent solidification shrinkage thus ensuring that a good clean type-face is obtained. Ice is another solid that contracts on melting, with the reverse process, expansion during freezing, giving rise to burst water-pipes.

14.9 Thermal conductivity

If one end of a bar of material is heated the atoms in that portion will be stimulated into increased vibration. Energy will then be transmitted along the bar from the hot to the cool end in an attempt to reduce the energy at the heated end. For an electrically insulating material the energy transference along the bar will be a result of atomic vibrations only. The transmission of heat energy along the bar will be a function of the separation distance between atoms, the density of the material and the specific heat. The thermal conductivity of a glass or liquid will be less than the conductivity of the crystalline version of the same material, since the atoms in a glass or liquid are not in a fully ordered array and interatomic separation distances are variable.

In metals, with valence electrons existing as a free cloud or gas, energy may be transmitted along a bar through the medium of electron movement as well as by means of atomic vibration. Metals are, in general, good conductors of both electricity and heat. The class of materials known as electrical semiconductors are also reasonably good thermal conductors because heating the material can cause some electrons to be excited and cross the energy gap into the conduction band. The excited electrons may then take part in the transmission of heat energy from one part of the material to another.

The thermal conductivities of some materials are listed in Table 14.3. The units of thermal conductivity are watts per metre per degree Kelvin (W/m K).

It will be noticed that the thermal conductivities of gases, apart from hydrogen, are very much lower than the conductivities of other non-metallic materials. Hence the reason for most thermal insulating materials being porous. Cavity-wall construction and double glazing in buildings are other examples in which use is made of the insulating properties of gases. There is no thermal conduction at all in a vacuum, since there are no atoms present to vibrate and transmit energy. This principle is used in the vacuum flask in which there is an evacuated space between two layers of glass.

A quantity related to conductivity is *thermal diffusivity*. The thermal diffusivity, D_T, is given by: $D_T = k/(\rho C_p)$ where k is thermal conductivity, ρ is density and C_p is the specific heat capacity. The units of diffusivity are $m^2 s^{-1}$. Thermal diffusivity can be regarded as the diffusion coefficient for thermal energy. The resistance of refractory ceramic materials to spalling tends to increase as the diffusivity increases.

The rate of heat transmission through a material is given by the expression $kA(T_1 - T_2)/t$ where

Table 14.3 *Thermal conductivities of some materials at 20°C*

Material	Conductivity (W/m K)	Material	Conductivity (W/m K)
Silver	420	Building brick	0.63
Copper	395	Polyethylene	0.33
Aluminium	240	Nylon (PA66)	0.25
Magnesium	168	Phenol-formaldehyde	0.25
Graphite	125	Natural rubber	0.13
Nickel	88	Asbestos	0.12
Mild steel	51	Polystyrene	0.08
Lead	33.5	Cork	0.046
Quartz	5.1	Sand	0.042
Silica glass	1.25	Hydrogen*	0.17
Concrete	1.05	Oxygen*	0.025
Mica	0.84	Nitrogen*	0.024
Plate glass	0.75	Perfect vacuum	zero

* At atmospheric pressure

k is the thermal conductivity, A is the surface area of the material, T_1 and T_2 are the temperatures of each surface of the material and t is the thickness.

Example

Determine the heat loss through a brick wall 4 m by 3 m of 0.25 m thickness if the inner surface is maintained at 20°C and the outer surface temperature is 5°C. The thermal conductivity of brick is 0.5 W/m K.

$$\text{Heat loss} = \frac{kA(T_1-T_2)}{t}$$

$$= \frac{0.5 \times 4 \times 3 \times 15}{0.25}$$

$$= 360 \text{ W}$$

14.10 Sound absorption and damping

Sound energy is transmitted through a medium both as a compression wave and as a shear wave. Sound cannot travel through a vacuum because there are no atoms available to vibrate and transmit energy. The velocity of sound within a material is related to both the density and the elastic moduli. The velocity, v_1, of a longitudinal compression wave within a material is given by

$$v_1 = \sqrt{\frac{K + (\frac{4}{3})G}{\rho}}$$

where K is the bulk modulus of elasticity, G is the modulus of rigidity, and ρ is the density of the material. The shear wave velocity, v_s, is given by

$$v_s = \sqrt{\frac{G}{\rho}}$$

The elastic constants of a material can be determined from a knowledge of the velocity of sound through it. All the elastic constants are related to one another (see Section 7.2) and so values for the modulus of elasticity, E, and Poisson's ratio, v, can also be determined in this way.

When sound travels through a medium some of the energy is absorbed by the material. This absorption, or damping of vibrations, is due to the internal friction of the material. The internal friction of a material is a function of the degree of crystallinity and the nature of the interatomic bonds within the substance, but it is also greatly affected by the presence of discontinuities within the material, such as impurities, and cracks. These serve to increase the internal friction. The internal friction of metals, and a number of non-metallic crystalline solids, is generally low. The internal friction of brick and concrete is about 100 times greater than that of many metals, while that of timber is about 1000 times greater than that of many metals. Most thermoplastic materials have internal friction values similar to or greater than those for timber.

For sound-insulation purposes the most effective type of material is one of low density and high internal friction. Foamed plastics are an excellent example of this. When sound energy reaches an interface between two different media there will be some reflection and this reflection may approach 100 per cent at an air/dense-solid interface. It is for this reason that double glazing is such an effective sound insulator since there are four air/glass interfaces between the external noise source and the ear, with a low density medium, air, between the two glass layers.

A material of high internal friction will have great ability to attenuate vibrations, and vice versa. This ability is termed the damping capacity of the material. The damping capacity, ζ, is given by

$$\zeta = \frac{1}{n}$$

where n is the number of vibrations in free attenuation from an amplitude A to an amplitude A/e, that is, $0.368\,A$. For the manufacture of a bell a material of very low damping capacity is necessary, hence the use of the copper alloys, brass and bronze. Cast iron is a metal with a very high damping capacity because of the presence of graphite flakes and small flaws in its structure. It is therefore a most suitable material for components where vibrations should be absent, as in a lathe bed. For antivibration mountings, very high internal friction materials such as rubber and cork are used.

14.11 Interface effects and surface tension

The properties at a surface are different from those within the heart of a material. An atom or molecule in the centre of a liquid is completely surrounded by other molecules and it is attracted equally by all its neighbours. A molecule at the surface, however, will be attracted inwards by the molecules within the bulk material. The net result of this inward pull is that the surface of the liquid tends to contract to the smallest possible area. It is for this reason that liquid droplets and gas bubbles within liquids tend to be spherical. The surface behaves as if it is in a state of tension, and this phenomenon is termed *surface tension*. The term signifies a separation between two substances, for example liquid-gas, and the value of surface tension is dependent on both substances present. The recorded values of surface tensions for liquids are generally for a liquid-air interface. The surface tension value may well be different for the interface between the liquid and some other gas. The interface surface tension, γ, has the units of newtons per metre (N/m) and the values of γ for some liquids are given in Table 14.4.

Table 14.4 *Surface tensions of some liquids in air (at 20°C)*

Liquid	γ (N/m)	Liquid	γ (N/m)
Ethyl alcohol	0.022	Benzene	0.029
Acetone	0.024	Water	0.073
Carbon tetrachloride	0.027	Mercury	0.485

It will be noticed that the surface tension of water is high compared with organic liquids, even though it is very low compared with mercury.

Whether a liquid wets the surface of a solid or not depends on the values of interface surface tensions for the three interfaces involved, namely liquid-air (γ_{la}), liquid-solid (γ_{ls}) and air-solid (γ_{as}). Consider a liquid resting on the surface of a solid. At equilibrium there will be an angle of contact, θ. The angle θ is measured within the liquid (see Figure 14.4). At equilibrium, the interface forces will be in balance, and this can be represented by the statement

$$\gamma_{as} = \gamma_{ls} + \gamma_{la} \cos \theta$$

If $\gamma_{as} > \gamma_{ls}$ then $\cos \theta$ is positive and $\theta < 90°$; if $\gamma_{as} < \gamma_{ls}$ then $\cos \theta$ is negative and $\theta > 90°$. When the contact angle, θ, is less than 90° the liquid is said to wet the solid surface. This is the case with water on glass. If the contact angle, θ, is greater than 90°, as with mercury on glass, the liquid does not wet the solid surface.

The air-liquid interface for liquid in a tube will be curved. When the liquid wets the tube the surface will be concave upwards but with non-wetting liquids the surface will be convex upwards. The curved liquid surface is termed *meniscus* and readers are probably familiar with meniscus curves for both water and mercury in glass tubes.

The pressure of liquid close to a curved surface is different from that close to a plane surface by an amount, δP, given by

$$\delta P = \frac{2\gamma}{r}$$

where r is the radius of curvature. When the liquid surface is convex δP is positive and when the surface is concave δP is negative. When an open-ended fine-bore, or capillary, tube is partially immersed in water the meniscus within the tube will be concave with a small radius r, compared with the flat surface of water outside the tube. The pressure just below the meniscus in the tube

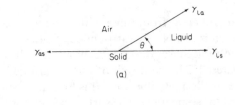

(a)

(b)   (c)

FIGURE 14.4   (a) Liquid-solid contact angle. (b) Liquid wetting solid. (c) Non-wetting liquid

will be less than the pressure just below the plane water-surface in the main vessel by an amount $2\gamma/r$ and so the water level within the tube will rise through some height h, where h is the height of a column of water having a pressure of $2\gamma/r$ at its base. (The level of a liquid with an angle of contact, θ, greater than 90°, for example mercury, will be depressed). This phenomenon is termed *capillary attraction* or *capillarity*. In general, liquids of low surface tension will have low contact angles and show high capillarity.

This phenomenon is utilised in *liquid phase sintering* of ceramics. In this form of sintering of ceramic powder compacts one of the constituents melts and wets the surface of the solid particles. The resulting capillary pressure can cause movement of some particles to give a better densification and elimination of porosity in the sintered product.

The surface of a solid-gas interface is in a state of tension in a similar manner to a liquid-gas surface. In a strongly bonded crystalline solid the atoms in the surface layers have their bonds only partially satisfied and they can react or join with atoms from the other medium, providing an adsorbed film on the surface of the solid. In some cases the forces of attraction between atoms on the solid and the adsorbed layer are of the van der Waals type but in other cases there may be ionic or covalent type bonding between the solid surface and atoms from the surrounding medium. This type of effect occurs between clean metal surfaces and oxygen and is an important first stage in the surface oxidation of metals. When there are ionic or covalent bonds formed at the surface the phenomenon is called chemisorption.

14.12 Friction and lubrication

The interface between two solids is of considerable importance in engineering. The surface of a solid is normally far from smooth when examined on a microscale. A lathe-turned or flat-ground surface will possess innumerable 'hills and valleys'. The height difference between hills and valleys for a fine-ground surface is of the order of 1–3 μm. When two metal surfaces are put together the true contact area, because of surface irregularities, will be very much less than the apparent area of contact. In consequence the stresses involved at points of contact may be sufficiently high to cause localised deformation and possibly welding. If a shearing force is applied in such a way as to cause sliding between the contacting surfaces, sliding motion will be resisted. The resistance to sliding is termed friction and is due to both the roughness of the surfaces and localised welding that may have occurred at points of contact.

When dry solid surfaces are in sliding contact the work done in overcoming friction is converted to heat energy and there will be considerable abrasion between them. This leads to severe wear of the surfaces. The surface temperature build-up could be sufficient to cause the two surfaces to weld over an extensive area. This is termed *seizure*. The extent of the friction between surfaces is considerably reduced by the presence of a lubricant, that is, a fluid layer interposed between the two solid surfaces.

In normal fluid lubrication, also termed hydrodynamic lubrication, the two solid surfaces are kept apart by means of a fluid layer and provided that the separation distance between the surfaces does not fall below about 0.1 μm almost any fluid that does not react with the solids may be used. Most lubricating fluids are mineral oils but high-pressure air, as a fluid, is used in low friction air bearings. Most lubricating oils contain boundary-layer additives. Such an additive is composed of polarised molecules which become strongly adsorbed on a metal surface. When conditions are such that hydrodynamic lubrication may break down, the adsorbed layers will still keep the solid surfaces apart. Organic fatty acids are used as boundary-layer additives in oils, as is also molybdenum disulphide.

14.13 Questions

14.1 Upon excitation an element emits radiation at a wavelength of 532 nm. Determine the energy transition involved.

14.2 The energy change when an excited electron within an element falls back to 'ground state' is 18 eV. What type of radiation is emitted and at what frequency?

14.3 Thermionic electrons emitted from the filament of an X-ray tube are accelerated to the target by a tube voltage of 50 kV. Calculate the maximum frequency of the X-radiation produced when high energy electrons strike the target material.

14.4 A beam of monochromatic light of wavelength 560 nm passes normally through a narrow slit of width 0.08 mm to a screen set 1.5 m from the slit. At what distance from the axis of the beam is the first point of minimum light intensity?

14.5 Red monochromatic light of wavelength 600 nm falls on a diffraction grating with 1×10^5 lines per metre. Calculate the angular separation between fringes.

14.6 (a) Estimate the percentage reduction in dimensions which occurs when a magnesium casting cools from its melting temperature (649°C) to 15°C. (b) What size should the mould cavity be for casting a 0.55 m diameter wheel in magnesium? (α for magnesium $= 25.2 \times 10^{-6}$).

14.7 A room has two external walls measuring 3.3 m × 2.4 m and 5.4 m × 2.4 m. The walls are of brick of 0.3 m thickness and each contains a window of 3 mm thick glass. The window dimensions are 1.4 m × 2.0 m and 2.0 m × 3.0 m respectively. Calculate the amount of heat energy lost through these walls per day if the room is maintained at 21°C and the outer surface temperature is 7°C. (Assume: thermal conductivity of brick is 0.6 W/m K and that of window glass is 0.75 W/m K.)

PART III
THE MATERIALS OF
ENGINEERING

15

Non-ferrous Metals and Alloys

15. Introduction

Although there are very many metallic elements, it is customary to divide metals and alloys into two major categories, ferrous and non-ferrous. The former category covers the element iron and its alloys, while all the other metallic elements (some 70 in number) and their alloys are classified as non-ferrous. The division is not quite as unbalanced as might at first appear, because iron occupies a very special position among metallic materials, owing to its availability, its comparatively low cost, and the very useful ranges of alloys that are formed when iron is alloyed with carbon and other elements. Some 94 per cent of the total world consumption of metallic materials is in the form of steels and cast irons. On the other hand, out of all the non-ferrous metals, only a few, aluminium, copper, lead, magnesium, nickel, tin, titanium and zinc, are produced in moderately large quanties. Many of the other metallic elements play an important part in engineering, both as alloying elements and as metals in their own right. The role of some of these, including beryllium, cobalt, molybdenum and tungsten, as alloying elements in some steels and non-ferrous alloys is mentioned in this and the following chapters but a detailed coverage of the metallurgy of these metals is outside the scope of this volume.

15.2 Aluminium

Aluminium is one of the most abundant elements in the earth's crust, but, owing to its high affinity for oxygen, it cannot be reduced to the metallic state by reduction with carbon or carbon monoxide, as is the case with many metallic oxides. Aluminium was first produced in the early part of the nineteenth century by reduction with potassium. This was a very expensive process, and until 1850 the price of aluminium was about £250 ($437.5) per kilogramme and, because of its great cost at that time, some European royal houses used aluminium cutlery at state banquets! In 1886, Charles Hall devised a new method for the relatively cheap production of aluminium. The Hall method involved the electrolysis of a molten solution of alumina in the mineral cryolite (Na_3AlF_6) at a temperature of about 950°C.

The economic ore of aluminium is bauxite, a hydrated alumina, and this mineral normally has iron oxides and other minerals associated with it. The bauxite has to be purified before electrolytic reduction can take place. Pure alumina is obtained by the Bayer process, in which bauxite is digested with caustic soda. Aluminium hydroxide is then precipitated from solution and calcined to give a pure alumina, about 99.6 per cent Al_2O_3, which can be electrolysed by the Hall process. The aluminium produced by the Hall process is of a purity ranging from 99.5 to 99.8 per cent, with the major impurities being iron, silicon, and manganese. Aluminium of purities up to 99.9 per cent can be produced by refining molten aluminium from the reduction cell in another electrolytic cell, the Hoopes process, with a molten electrolyte.

The principal properties of pure aluminium (99.9 per cent) are given below.

Melting point	660°C
Crystal structure	face centred cubic
Density	$2.70 \times 10^3 \, kg/m^3$
Young's modulus, E	70.5 GPa
Tensile strength	45 MPa
Electrical resistivity	$2.66 \times 10^{-8} \, \Omega m$ at 20°C
Corrosion resistance	very good

Aluminium possesses a number of properties that make it an extremely useful engineering material. Its good corrosion resistance and low density make it particularly suitable for applications in the field of transportation—land, sea, and air. The pure metal has a very low strength but, by alloying with other elements, the strength may be increased considerably to give alloys with very good strength to weight ratios. Young's modulus for aluminium is only one-third of the value for steels, however, and so the specific modulus (E/density) is almost identical for both materials. Like all metals with a face centred cubic crystal structure, aluminium is highly ductile and can be shaped easily by a wide variety of methods. The good electrical conductivity of the metal makes it suitable for many applications in the electrical industry.

Aluminium has an extremely high affinity for oxygen and any fresh metal surface will oxidise rapidly. The surface oxide layer that forms is only a few atoms in thickness, but it is impermeable to oxygen and protects the surface from further attack. The corrosion resistance of aluminium is due to the presence of this oxide coating. The corrosion resistance of aluminium may be improved further by anodising. In an anodising operation the aluminium is made the anode of an electrolytic cell. Oxygen is liberated at the anode and the thickness of the protective aluminium oxide film is increased. The anodic coating is cellular in structure and immersion of the anodised material in boiling water will seal this cellular structure and give a hard, smooth surface. An anodised aluminium part with a cellular oxide layer may be dipped in a vat of dye, prior to the sealing operation. Dye will be absorbed into the pores to produce the type of colour-anodised finishes widely used for decorative purposes.

High-purity aluminium (purities of 99.5 per cent aluminium, and above) is too weak to be used for many purposes and its main applications are for use as corrosion-resistant linings for vats and other vessels in the food and chemical industries. The material commonly termed pure aluminium is, in reality, an aluminium–iron alloy. Commercial-purity aluminium alloy is made by adding up to 0.5 per cent of iron to the metal produced in the Hall process. This small iron addition gives a considerable increase in strength (see Table 15.1), although there is some reduction in ductility and corrosion resistance. Commercial-purity aluminium is used extensively, and accounts for about nine-tenths of aluminium product sales. Its principal applications are: thin foils for packaging, thermal insulation, and capacitor manufacture; cold-rolled sheet for panelling, and the production of kitchenware, rod and wire for electrical transmission cables, and armature windings.

15.3 Aluminium alloys

Aluminium may be alloyed with a number of elements to produce a series of useful engineering materials. The alloys of aluminium may be sub-divided into those that are not heat-treatable, and those that are heat-treatable. Alloys in the first group may be strengthened by cold working operations, and the only type of heat treatment that may be given is an annealing treatment to soften a work hardened sample. The heat-treatable alloys, on the other hand, may be strengthened by giving them a special type of heat treatment.

The non-heat-treatable alloys are those of the Al–Mn, Al–Mg, and Al–Si systems. Manganese enters solid solution in aluminium to a limited extent, and the only commercial alloy in this system is that containing 1.25 per cent of manganese. This alloy is some 15 per cent stronger, and has a slightly better corrosion resistance, than commercial purity aluminium. Its principal uses are for the production of domestic utensils, including pressure cookers and, in the form of corrugated sheet, for the cladding of buildings.

Magnesium enters solid solution in aluminium to a limited extent and gives a considerable solid solution strengthening effect. Magnesium also improves the resistance to corrosion, particularly to corrosion in a marine environment. A range of aluminium-magnesium alloys is marketed, with magnesium contents of up to 10 per cent. The various alloys in this range are used for the production of pressed and deep-drawn articles, where high strength is required, for example, boxes, crates and milk churns, and as a constructional material for ship building. The majority of aluminium-magnesium alloys are manufactured using commercial purity aluminium as a base, and so contain a range of impurity elements. Some alloys are made from very high purity ingredients and these can be polished to give a bright, attractive surface with a very high reflectivity. These highly pure alloys are used for decorative trim applications as a substitute for chromium-plated mild steel.

Silicon forms a eutectic with aluminium, and it also increases the fluidity of the molten aluminium. Alloys containing silicon are very suitable for the manufacture of both sand and die castings. The most widely used casting alloys are those containing 10–13 per cent of silicon, namely those compositions close to the eutectic composition. Normally, the eutectic mixture would be of fairly coarse structure, rendering the alloy brittle. The melt is treated immediately before casting by making a very small addition (about 0.01 per cent) of sodium. This is termed modification of the alloy and it produces a fine-grained eutectic structure. The modification treatment also causes a shift in the position of the eutectic point (Figure 15.1).

The principal heat-treatable alloys are those of the following systems: Al–Cu, Al–Cu–Ni, Al–Mg–Si, Al–Zn–Cu, and Al–Li. These various alloy types respond to age hardening, or precipitation hardening, treatments (see also Sections 8.9 and 12.6). This type of hardening is possible in alloy systems that show partial solid solubility, with a wide difference between the low-temperature and high-temperature solubility limits. The aluminium-rich end of the aluminium–copper phase diagram is of this type (Figure 15.2). Aluminium can hold 0.2 per cent, by weight, of copper in solid solution at room temperature, but it can hold up to 5.7 per cent, by weight, of copper in solution at 548°C. If an alloy containing, say, 5 per cent, by weight, of copper (2 atomic per cent of copper) is slowly cooled from high temperature, the microstructure will be as shown in Figure 15.2(b), with the compound $CuAl_2$ precipitated at the aluminium grain boundaries. If, alternatively, the alloy is heated to a high temperature and rapidly cooled, the structure at room temperature will consist of a supersaturated solid solution of copper in aluminium (Figure 15.2(c)). Such a solution is unstable, or metastable, and there will be a tendency for copper atoms to diffuse through the aluminium space lattice prior to precipitation of $CuAl_2$ particles. The concentration of copper in the compound $CuAl_2$ is 53.5 per cent, by weight, or 33.3 atomic per cent, so that considerable diffusion of copper must take place in order to produce copper-rich areas within the aluminium space lattice, as a prerequisite of the precipitation of $CuAl_2$ particles. As this pre-precipitation diffusion occurs there will be a considerable increase in the amount of lattice strain within the aluminium lattice, and, consequently, there will be an increase of hardness and strength with time following a solution

Table 15.1 *Properties of some aluminium alloys*

Alloy number British	Alloy number American	Approximate composition	Condition*	Tensile strength (MPa)	Type of product	Uses
1080	1060	99.99% Al 99.8% Al	O O	45 75	Sheet, strip Sheet, strip	Linings for vessels in food and chemical plant
1200	1200	99.0% Al	O H14 H18	90 120 150	Sheet, strip, wire, extruded sections	Lightly stressed and decorative panelling, wire and bus bars, foil for packaging, kitchen and other hollow-ware
3103	3103	Al + 1.75% Mn	O H14 H18	110 160 210	Sheet, strip, extruded sections	Hollow-ware, roofing, panelling, scaffolding tubing
5251	5052	Al + 2% Mg	O H24	180 250	Sheet, plate, tubes and extrusions	Stronger deep-drawn articles; ship and small boat construction, and other marine applications
5154A	5454	Al + 3.5% Mg	O H24	240 300		
5056A	5056A	Al + 5% Mg	O H24	280 335		

		Composition	Condition		Product form	Application
LM6	S12C	Al + 12% Si	M	180†	Sand and die castings	Excellent casting alloy
			M	210‡		
6082	6082	Al + 0.9% Mg; 1% Si; 0.7% Mn	T4	220	Sheet, forgings, extrusions	Structural components for road and rail transport vehicles
			T6	320		
2014	2014	Al + 4.5% Cu; 0.5% Mg; 0.8% Mn	T4	440	Sheet, forgings, extrusions, tubing	Highly stressed parts in aircraft construction and general engineering
			T6	480		
2024	2024	Al + 4.5%Cu; 1.5% Mg; 0.6% Mn	T3	480		
2L95 L160		Al + 5.6% Zn; 1.6% Cu; 2.5% Mg	T6	500	Plate	Aircraft construction
7075	7075	Al + 7% Zn; 1.75% Cu; 2%Mg	T6	620	Rod and bar Sheet and extrusions	
2090	2090	Al + 2.2% Li; 2.7% Cu; 0.12% Zr	T6	580	Sheet, plate,	Aircraft construction
8090	8090	Al + 2.5% Li; 1.3% Cu; 0.7% Mg	T6	495	Sheet, plate,	

Key: * The symbols for condition are: O = annealed; M = as cast; H14 = partly work hardened; H18 = fully work hardened; H24 = work hardened and partially annealed; T3 = solution treated, cold worked and aged; T4 = solution treated and naturally aged; T6 = solution treated and precipitation hardened. † Sand cast. ‡ Chill or die cast.

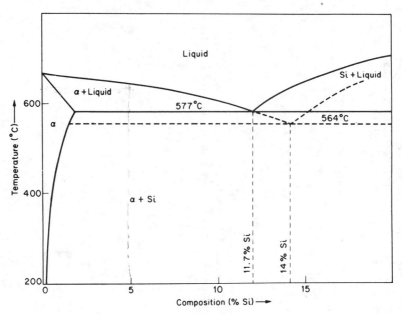

FIGURE 15.1   Aluminium–silicon phase diagram. The addition of 0.01 per cent sodium modifies the diagram to position shown by broken lines

heat treatment. With aluminium–copper alloys, this diffusion process and consequent hardening, will occur slowly at room temperature but in some alloy systems hardening does not take place spontaneously at room temperature. In these latter cases, the alloy has to be heated to some comparatively low temperature before the diffusion process will occur. Owing to there being insufficient thermal energy available at room temperature, the diffusion of copper in aluminium–copper alloys does not occur to the extent necessary for precipitation of $CuAl_2$ unless the temperature is raised. An increase in temperature will bring about an increase in copper diffusion rates, and if the temperature is high enough true precipitation of $CuAl_2$ particles will begin to take place. Once true precipitation occurs, the strain in the aluminium lattice will be relieved and softening of the alloy will take place. This softening is termed *overageing*. Figure 12.6 shows the relationship for hardness with ageing temperature and time for an alloy of aluminium and copper. When an alloy slowly hardens with time on being kept at ordinary temperatures after receiving a solution heat treatment, the alloy is termed an age-hardening alloy. If, after solution heat treatment, it is necessary to heat an alloy to some temperature to cause diffusion and hardening to occur, this second heat treatment is termed precipitation heat treatment, even though the treatment is stopped before true precipitation and overageing can occur.

The age-hardening alloys of aluminium and copper are used for structural applications, particularly in aircraft construction. The presence of copper reduces the corrosion resistance of the material, and sheet and plate forms of the alloy are often clad with a thin layer of high-purity aluminium to improve the corrosion resistance. This thin layer is applied at an early stage in the production sequence by sandwiching the cast ingot between two plates of high-purity aluminium prior to hot rolling. Pressure welding of the layers occurs during hot rolling.

In the ternary aluminium–copper–nickel alloys, both copper and nickel contribute to the age-hardening process. The presence of nickel also increases the temperatures at which overageing occurs. Consequently, these alloys may be used for service at elevated temperatures. The

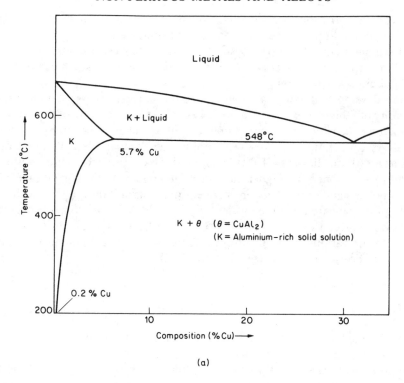

(a)

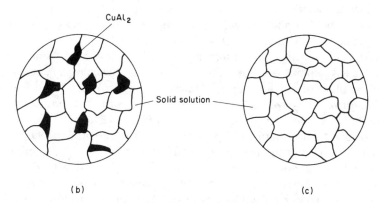

(b) (c)

FIGURE 15.2 (a) Aluminium-rich end of Al–Cu phase diagram. (b) 5% Cu alloy slow-cooled from 550°C showing CuAl$_2$ precipitate. (c) 5% Cu alloy rapidly cooled from 550°C—supersaturated solid solution

structure of the supersonic aircraft, Concorde, is made of this type of alloy. These alloys are also used for the manufacture of cylinder heads, cylinder blocks, and pistons for internal combustion engines.

Magnesium and silicon form an intermetallic compound, Mg$_2$Si, and alloys of aluminium that contain magnesium and silicon in the correct proportions are effectively binary alloys of aluminium and the compound Mg$_2$Si. The aluminium-rich end of the Al–Mg$_2$Si phase diagram is similar in form to the Al–Cu diagram and heat-treatable alloys result. These alloys are relatively strong, and have good corrosion resistance. They are used in quantity for structural applications, and for panelling, in road and rail vehicles, and transit containers.

Alloys of the aluminium–zinc–copper system possess the highest strengths of any aluminium alloys. These alloys are difficult to work, and they are also subject to stress corrosion. They are mainly used in aircraft construction.

Considerable development work has been carried out in recent years on aluminium–lithium alloys. The density of lithium is extremely low (534 kg/m^3) and an addition of lithium will provide an alloy with a lower density than other aluminium alloys. The tensile strengths obtainable in heat-treated alloys containing 2–2.5 per cent lithium are comparable with those of the high strength Al–Cu and Al–Zn alloys (see Table 15.1) and this is coupled with a modulus of elasticity, E, value which is about 10 per cent higher than other aluminium alloys. Weight saving in aircraft construction can offer significant reductions in operating costs and these alloys, with densities of around 2600 kg/m^3 compared with densities of about 3000 kg/m^3 for other high-strength aluminium alloys, show considerable potential for use in the aerospace industries.

In addition to the various alloys mentioned above, alloys of aluminium with tin form an important group of bearing materials (see Section 15.12).

15.4 Copper

Copper is one of the oldest metals known to man, and, together with tin in the alloy bronze, has been worked for over 5000 years. Copper and bronze were used for both utilitarian and decorative purposes in the early civilisations. Today, there are very many extremely useful copper alloys but, owing to the present high price of the metal, copper and its alloys are being replaced by cheaper materials, such as aluminium and plastics, in many applications.

The principal ores of copper are generally complex mixtures of copper and iron sulphides, and copper production from the ore may either be carried out by a smelting process, or by electrolytic extraction. Smelting operations are carried out in a blast furnace, or reverberatory furnace, and roasted sulphide ore, coke, and flux, make up the charge. A *matte*, which is a high-density double sulphide of iron and copper, collects at the base of the furnace and the gangue, or waste material, forms a lower-density slag, which separates easily from the matte. The matte is reduced to copper in a converter, similar to a Bessemer converter. The copper produced in this way is highly oxidised, and is termed blister copper. It is of the order of 99 per cent pure and it may be refined to higher purities by either electrolytic, or fire refining, techniques.

For electrolytic extraction, the finely divided ore is digested with dilute sulphuric acid. Copper compounds are dissolved, but most other minerals in the ore are not attacked. The solution obtained is then purified and electrolysed. The copper produced in this way is termed cathode copper, and may be of purity ranging from 99.2 to 99.7 per cent copper. Cathode copper may be refined either electrolytically, or in a reverberatory furnace.

The fire refining of blister or cathode copper is an interesting process. The copper is melted under oxidising conditions in a reverberatory furnace. Most impurities are oxidised and form a slag which floats on the surface of the copper. The bath of highly oxidised copper is then subjected to reducing conditions to reduce the oxygen content, the resulting product being termed *tough pitch* copper with a residual oxygen content of about 0.03 per cent. Formerly the necessary reducing conditions were obtained by *poling* the liquid bath. Poles, or trunks, of green wood were immersed in the bath of copper and the stream of reducing gases evolved removed the bulk of the oxygen from the copper. Poling is stopped when the oxygen content of the bath is about 0.03 per cent. Over-poling would cause the copper to become contaminated with excess hydrogen. The purity of tough pitch copper is not less than 99.85 per cent. Tough pitch copper is unsuitable for welding owing to the oxygen content of the material. It is also unsuitable for applications requiring high electrical conductivity. Copper is commonly deoxidised by adding phosphorus, in the form of a copper–phosphorus alloy, but this is not a suitable method for the production of electrical grades of copper. For high-conductivity purposes tough pitch, or electrolytically refined copper is remelted under controlled conditions and deoxidised, if necessary, with lithium. Some grades of copper contain up to 0.5 per cent of arsenic. Arsenical copper, which has improved tensile properties and an improved resistance to oxidation at high temperatures, is made by adding up to 0.5 per cent of arsenic to tough pitch or deoxidised copper.

The principal properties of pure copper are given below.

Melting point	1083°C
Crystal structure	face centred cubic
Density	$8.93 \times 10^3 \, kg/m^3$
Young's modulus, E	122.5 GPa
Tensile strength	220 MPa
Electrical resistivity	$1.67 \times 10^{-8} \, \Omega m$ at 20°C
Corrosion resistance	very good

Some applications of the various grades of pure copper are: wire, for electrical windings and wiring; sheet, for architectural cladding, and for shaping into articles such as domestic water tanks and vessels used in the food and chemical industries; tubing for heat exchangers and domestic installations.

Copper may be alloyed with a number of elements to provide a range of useful alloys. The important alloy systems are:

copper–zinc (brasses)
copper–tin (zinc) (bronzes and gun-metals)
copper–aluminium (aluminium bronzes)
copper–nickel (cupronickels)

Small additions of beryllium or chromium to copper give high strength alloys, a small addition of cadmium gives a significant increase in strength with little loss of electrical conductivity, while an addition of tellurium to copper gives an alloy with very good machine-ability.

15.5 Copper alloys

The phase diagrams for the binary systems copper–zinc, copper–tin, and copper–aluminium are complex and a number of intermediate phases are formed in each system (Figure 15.3). In each case, the α (copper-rich) phase possesses a face centred cubic structure, and all the α alloys are ductile and suitable for cold working. The β phase in each system is body centred cubic in structure, and alloys containing this phase cannot be readily cold worked. They are used in the cast form, and for making hot worked products. (In the Cu–Sn and Cu–Al systems it will be noticed that β phase exists only at elevated temperatures.) The γ phase (and δ phase in bronzes) possesses a complex cubic crystal structure, and is a very hard and brittle phase.

Brasses. The range of alloys known as the brasses are, primarily, alloys of copper and zinc. The addition of zinc to copper brings about an increase in strength because the zinc enters into solid solution in the copper. An unusual feature is that the ductility of the alloys also increases

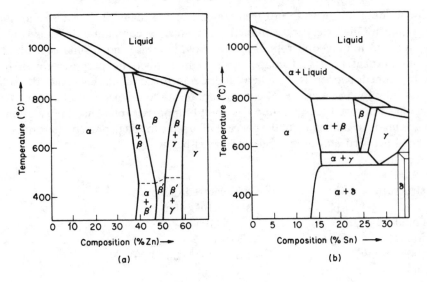

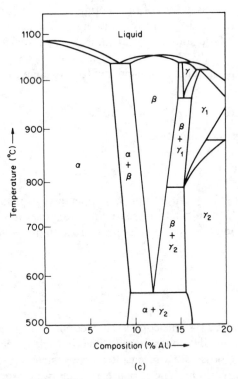

FIGURE 15.3   Binary phase diagrams for copper alloys: (a) copper–zinc; (b) copper–tin; (c) copper–aluminium

with dissolved zinc content, reaching a maximum value at a zinc content of 30 per cent. The ductility of the material then decreases with further zinc content until at a composition of 37 per cent of zinc, the limit of the α phase, the ductility of the alloy is of similar value to that of pure copper. The α brasses are widely used for deep drawing, spinning, and tube manufacture, because of their high ductilities. The alloy containing 30 per cent of zinc, 70/30 brass, is known as *cartridge brass* and is used for the manufacture of cartridge and shell cases because of its high ductility. For general purpose cold working applications, an α brass containing 35 per cent of zinc is normally used. The increased zinc content causes a reduction in ductility, but the alloy is somewhat cheaper than 70/30 brass. The properties of α brasses may be modified by the addition of small amounts of other elements. Small additions of tin or aluminium improve resistance to marine corrosion, while an addition of 1 or 2 per cent of lead improves the machineability of the alloy. Cold worked α brasses are prone to a form of stress corrosion termed *season cracking*, unless they are properly stress relieved before being put into service.

The (α + β) alloys are generally formed into shapes by hot working or by direct casting. The most common composition for an (α + β) brass is 60/40, that is, 40 per cent of zinc, and this alloy is termed *Muntz metal* (refer to Table 15.2 for properties of brasses). Small additions of other elements may be made to (α + β) brasses in order to improve properties. A small addition of lead will impart good free-machining characteristics to the alloy, while the addition of elements such as manganese, tin, aluminium, iron, or nickel will considerably increase the strength of the material. The alloy containing an addition of manganese was formerly known as *manganese bronze*, but the improved strength alloys are more properly termed *high-tensile brasses*. The structures of high-tensile brasses are generally (α + β), but may sometimes be wholly β. They are used in the form of castings and hot worked products. A typical application is for the manufacture of ships' screws and marine fittings. β brass is used as the basis for some brazing alloys. No commercial alloys are made containing more than 50 per cent of zinc, as above this value the presence of γ phase in the structure would embrittle the alloy. (Photo-micrographs of some brass structures are shown in Figures 15.4, 15.5 and 15.6.)

Molten brasses do not require deoxidation before casting. This is because zinc possesses a very low boiling point (907°C) and the volatilisation of some zinc from the melt expels oxygen, and also the heavy zinc vapour forms a protective blanket over the liquid metal surface.

Bronzes. The bronzes are high-strength alloys with a good resistance to corrosion. Although, according to the phase diagram, copper can hold up to 14 per cent of tin in solid solution, cast alloys containing more than 5 per cent of tin will show some δ phase in the microstructure. This is because pronounced dendritic coring occurs owing to the wide freezing-temperature range of these alloys. Coring may be removed by a lengthy annealing treatment. The α phase cold working alloys do not normally contain more than 7 per cent of tin. These α alloys are ductile and are suitable for cold working, although they work harden rapidly. The uses of these alloys include the manufacture of non-ferrous springs, coinage, and sheet metal for ornamental work. Alloys containing the brittle δ phase are unsuitable for working, and are used in the 'as cast' condition. Bronzes need to be deoxidised before casting and phosphorus is used for this purpose. Sufficient phosphorus is normally added to give a certain residual phosphorus content in the deoxidised alloy. The materials are then termed *phosphor bronzes*. The effect of the residual phosphorus is to strengthen the alloy. α bronzes contain about 0.1 per cent of phosphorus, and the residual phosphorus content in an (α + δ) bronze may be as high as 1.0 per cent.

A major application for (α + δ) bronze is as a bearing material (see also Section 15.12). The duplex structure of the material is most suitable for this purpose as the α phase matrix will be highly resistant to shock, and the very hard δ phase particles will sustain some of the load and reduce wear. Some of the residual phosphorus in (α + δ) bronze will form copper phosphide, Cu_3P, and this extremely hard constituent is associated with δ phase in the microstructure, resulting in an improvement in the bearing characteristics (see Figure 15.8).

Zinc and lead may be added to bronzes. Zinc can be used to cheapen the alloy, and it also improves the casting performance of the material. Alloys containing zinc are termed *gun-metals*, as they were used in former days for the casting of heavy ordnance. Small lead additions improve

Table 15.2　Properties of copper and some copper alloys

Alloy	Approximate composition	Condition*	Tensile strength (MPa)	Type of product	Uses
Pure copper	99.95% Cu	O	220	Sheet, strip, wire	High conductivity electrical applications
		H	350		
	99.85% Cu	O	220	All wrought forms	Chemical plant, deep drawn and spun articles
		H	360		
Arsenical copper	99.25% Cu; 0.5% As	O	220	All wrought forms	Retains strength at elevated temperatures. Heat exchangers, steam pipes
		H	360		
Brasses					
Gilding metal	90% Cu; 10% Zn	O	280	Sheet, strip, wire	Imitation jewellery and decorative work
		H	510		
Cartridge brass	70% Cu; 30% Zn	O	325	Sheet, strip	High-ductility brass for deep drawing
		H	700		
General cold working brass	65% Cu; 35% Zn	O	340	Sheet, strip, extrusions	General purpose cold working alloy
		H	700		
Muntz metal	60% Cu; 40% Zn	M	375	Hot rolled plate and extrusions	Condenser and heat exchanger plates
High tensile brass	35% Zn; 2% Mn; 2% Al; 2% Fe; balance Cu	M	600	Cast and hot worked forms	Ships' screws, rudders and high-tensile applications
Bronzes	95.5% Cu; 3% Sn; 1.5% Zn	O	325	Strip	British 'copper' coinage
		H	725		
	5.5% Sn; 0.1% P; balance Cu	O	360	Sheet, strip, wire	Springs and steam turbine blades
		H	700		
	10% Sn; 0.5% P; balance Cu	M	280	Castings	General purpose castings and bearings

Name	Composition	Key	Form		Uses
Gunmetal	10% Sn; 2% Zn; balance Cu	M	Castings	300	Pressure-tight castings, pump and valve bodies
Aluminium bronze	95% Cu; 5% Al	O	Strip, tubing	400	Imitation jewellery and condenser tubes
		H		770	
	10% Al; 2.5% Fe; 2–5% Ni; bal. Cu	M	Hot worked and cast products	700	High strength castings and forgings
Cupronickel	75% Cu; 25% Ni	O	Strip	360	British 'silver' coinage
		H		600	
	70% Cu; 30% Ni	O	Sheet, tubing	375	Condenser tubing, excellent corrosion resistance
		H		650	
Monel	29% Cu; 68% Ni; 1.25% Fe; 1.25% Mn	O	All forms	550	Excellent corrosion resistance, used in chemical plant
		H		725	
Beryllium–copper	$1\frac{3}{4}$–$2\frac{1}{2}$% Be; $\frac{1}{2}$% Co; balance Cu	WP	Sheet, strip,	1300	Springs, non-spark tools
Cadmium–copper	99% Cu; 1% Cd	O	Wire, rod	285	Overhead electrical wire, spot-welding electrodes
		H		500	
Chromium–copper	0.4–0.8% Cr; bal. Cu	WP	Wrought forms and castings	450	Welding electrodes, commutator segments
Tellurium–copper	0.3–0.7% Te; bal. Cu	O	Wrought forms	225	Free-machining properties
		H		300	

* Key: O = annealed; H = work hardened; M = as manufactured (cast or hot worked); WP = solution heat treated and precipitation hardened

FIGURE 15.4   Photomicrograph of cold worked and annealed deoxidised copper (etched) × 200. Some of the equiaxial crystals show pronounced annealing twins. α brass and other α-phase copper alloys have a similar microstructure. (Courtesy of The Copper Development Association)

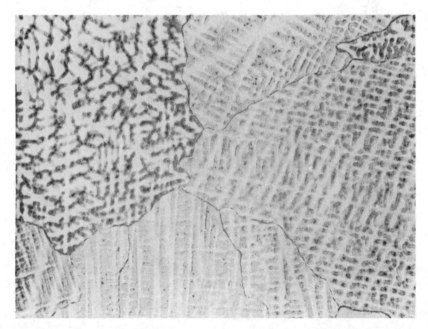

FIGURE 15.5   Photomicrograph of 70/30 brass, sand cast (etched) × 100 showing cored α crystals. The dendritic growth pattern is visible. (Courtesy of The Copper Development Association)

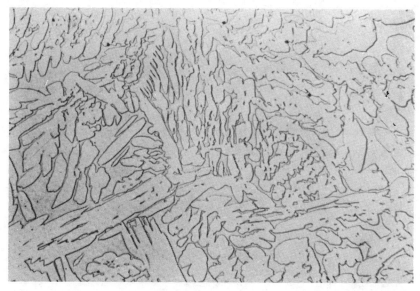

FIGURE 15.6 Photomicrograph of 60/40 brass, sand cast (etched) × 100. A two-phase structure with α precipitate in a β matrix. (Courtesy of The Copper Development Association)

FIGURE 15.7 Photomicrograph of 95/5 bronze, cold worked (etched) × 200. The α crystals are distorted and show strain bands. (Courtesy of The Copper Development Association)

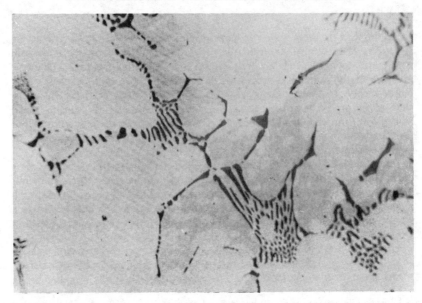

FIGURE 15.8   90/10 phosphor bronze, sand cast × 500, showing the very hard Cu_3P phase (dark in colour) and the $\alpha + \delta$ eutectoid. The light background is α phase. (Courtesy of The Copper Development Association)

the machining characteristics of bronzes and gun-metals, while large additions, up to 25 per cent of lead, are added to some bearing metals (see Section 15.12).

Aluminium bronzes. The aluminium bronzes possess similar strengths to bronzes. Again, the α alloys are cold working materials, while the $(\alpha + \gamma)$ alloys are used for the manufacture of castings and hot worked products. The main characteristic of aluminium bronze is its excellent corrosion resistance, particularly in marine environments. The α alloys are used in the form of sheet and tubing for applications including condensers and heat exchangers. With aluminium contents of greater than 9.5 per cent, the alloy possesses an $(\alpha + \gamma_2)$ structure at ordinary temperatures, although the γ_2 phase will transform into β phase on heating to temperatures above 565°C. On cooling through 565°C, the β phase undergoes a eutectoid transformation into α and γ_2 phases. The $(\alpha + \gamma_2)$ eutectoid consists of alternate layers of ductile α and brittle γ_2 (compare with pearlite eutectoid in steels, see Chapter 16). Slow cooling through the eutectoid temperature will give a coarse lamellar structure, which will tend to be brittle, while more rapid cooling will give a fine structure which will be tough. Consequently, sand castings should be heat treated followed by cooling in air, or by quenching, in order to produce a fine-grained structure. If the β phase is rapidly quenched, the formation of γ_2 phase may be suppressed, and a structure similar to martensite in steels produced. This martensite structure may be tempered by heating the quenched alloy at about 500°C to produce a very fine-grained $(\alpha + \gamma_2)$ microstructure. Aluminium bronzes containing about 10 per cent of aluminium are used for ships' screws, valves and pumps for marine use.

Cupronickels. Nickel is soluble in solid copper in all proportions (see phase diagram Figure 11.8(a)), and the cupronickel alloys are strong, ductile alloys with a face centred cubic crystal structure. Cupronickels possess an excellent resistance to corrosion. British 'silver' coinage is a cupronickel containing 25 per cent of nickel. Other applications include use for condenser tubing and heat exchangers.

Nickel silvers. The addition of zinc to copper and nickel produces the alloys termed *nickel silvers.* These are also face centred cubic in structure, and ductile. A major use for nickel silver alloys is the manufacture of cutlery and decorative articles. Generally these articles are given a thin coating of electrodeposited silver to enhance their appearance. The term E.P.N.S. (electroplated nickel silver) characterises products made in this way.

Beryllium–copper. The copper-beryllium materials, containing up to 2.7 per cent of beryllium, are the highest strength copper alloys available. The alloys may be strengthened by solution heat treatment and precipitation hardening (compare with Al–Cu) and tensile strengths of up to $1400 \, MN/m^2$ can be obtained. Normal beryllium–copper contains 2 per cent of beryllium but other alloys with beryllium contents ranging from 0.4–2.7 per cent beryllium and with small additions of cobalt or nickel have been developed. The alloys are used for the manufacture of springs, pressure diaphragms and capsules, and non-sparking tools.

Cadmium–copper. The addition of 1 per cent cadmium to copper gives an alloy with a tensile strength approximately 50 per cent higher than that of high conductivity copper but with little loss of electrical conductivity. In the annealed condition the conductivity is only about 7 per cent less than that of annealed high conductivity copper while the fully work hardened material has a conductivity which is about 85 per cent of that of annealed pure copper. The main applications of this material are for overhead contact wires for electric railways, telephone wires and electrodes and electrode holders for electric resistance welding equipment.

Chromium–copper. Chromium–copper is a precipitation hardening alloy containing, typically, 0.5 per cent of chromium, which has a high strength coupled with high electrical and thermal conductivities. A tensile strength of about 400 MPa is achieved by precipitation hardening and, in this condition, the alloy has a conductivity of about 80 per cent of that of annealed high conductivity copper. Chromium–copper is used mainly for the manufacture of spot and seam welding electrodes.

Tellurium–copper. This material contains 0.3–0.7 per cent tellurium and is a free-machining alloy of high electrical conductivity. Tellurium is virtually insoluble in copper and appears in the microstructure as small particles of copper telluride. These second phase particles act as internal chip-breakers giving discontinuous chip formation during machining.

15.6 Lead

The principal properties of lead are:

Melting point	327°C
Crystal structure	face centred cubic
Density	$11.34 \times 10^3 \, kg/m^3$
Young's modulus, E	16.5 GPa
Electrical resistivity	$2.1 \times 10^{-7} \, \Omega m$ at 20°C
Corrosion resistance	very good

Lead is a metal that has been used by man for over 4000 years. It is soft and malleable, and possesses an excellent resistance to corrosion. Because of these properties it has been used for water pipework, and weather flashing on buildings. Because of its high density it is very suitable as a radiation-shielding material. Lead has largely been replaced by other material nowadays for water pipework and waste disposal, but the metal is widely used as the basis of a number of alloys. The principal alloys of lead are those with tin and antimony. Some of the important alloys of lead are listed in Table 15.3.

Table 15.3 *Some lead alloys*

Alloy	Composition (per cent)					Applications
	Pb	Sb	Sn	Bi	Hg	
Antimonial lead	99	1	–	–	–	Cable sheathing
Hard lead	94	6	–	–	–	Lead-acid batteries, lead shot
Plumber's solder	60	2.5	37.5	–	–	⎫
Common solder	50	–	50	–	–	⎬ Soft solders
Tinman's solder	38	–	62	–	–	⎭
Type metal	62	24	14	–	–	⎫ For casting into printing type
Linotype metal	81	14	5	–	–	⎭
Wood's alloy	24	–	14	50	–	(+ 12% Cd) Alloy with M.Pt. of 71°C used for fusible plugs in sprinkler systems
Rose's alloy	28	–	22	50	–	Low melting point alloy with M.Pt. of 100°C
Dental alloy	17.5	–	19	53	10.5	Dental cavity filling (M.Pt. 60°C)

A major application for lead is in the manufacture of lead–acid storage batteries. The battery plates are grids of a hard lead–antimony alloy, into which a paste of lead compounds is pressed. Storage batteries account for almost 30 per cent of the annual world consumption of lead.

The soft solders are basically lead–tin alloys. The lead–tin phase diagram is shown in Figure 15.9. The low melting temperatures of these alloys make them very suitable for use as solders. The soft solders often contain a little antimony for added strength. The eutectic composition alloy, known as tinman's solder, freezes at 183°C and is used for electrical jointing and the sealing of tinplate cans. Plumber's solder contains a much higher lead content, 60 per cent of lead, and solidifies over a range of temperature. This enables it to be used for making wiped joints in lead pipework.

Alloys of lead, tin and antimony are used for casting printer's type. Tin, usually in amounts of up to 25 per cent, increases the fluidity of the melt, and increases the toughness of the alloy. Antimony, unlike most metals, expands as it solidifies from liquid. An alloy containing between

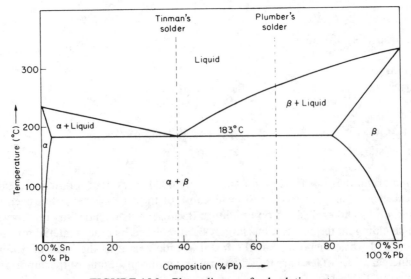

FIGURE 15.9   Phase diagram for lead–tin system

20 and 30 per cent of antimony shows a negligible volume change on solidification, and this ensures the production of a good, clean type face.

Alloys with very low melting points have been developed. These are based on lead, bismuth and tin and are used for the manufacture of fusible plugs for use in automatic fire alarm water-sprinkler systems and also for making patterns for the production of precision castings.

Lead is also a major constituent of many bearing alloys, and these will be discussed in the Section 15.12.

15.7 Tin

The principal properties of tin are as follows:

Melting point	232°C
Crystal structure	body centred tetragonal (β or white tin)
Density	7.29×10^3 kg/m^3
Young's modulus, E	40.8 GPa
Electrical resistivity	1.15×10^{-7} Ωm at 20°C
Corrosion resistance	very good

Tin can exist in two allotropic forms, α (grey tin) possessing a diamond-type crystal structure and β (white tin) with a body centred tetragonal structure. The equilibrium phase transition temperature is 18°C but, in practice, white tin does not transform into grey tin until it is cooled to sub-zero temperatures. When the change begins areas of grey powder begin to appear on the bright white tin surface. This was referred to as *tin disease*.

The attractive appearance of tin, its low melting point, good corrosion resistance and its lack of toxicity made tin an ideal metal for coating mild steel to make tin-plate for the manufacture of cans. When first produced, tin-plate was made by dipping sheets of mild steel into baths containing molten tin. A small amount of tin-plate is still made in this way, but the vast majority is made by the continuous electrodeposition of a small thickness of tin of the surface of mild steel strip as the strip passes through a bath of electrolyte at speeds of 12–30 m/s. The electro-deposited layer of tin is a matt grey in appearance and after the strip has passed through a drying zone this tin layer is flash-melted to give the bright finish associated with tin-plate.

Another major use of tin is in the manufacture of white metal bearings (see Section 15.12), and this probably accounts for about 40 per cent of the total world tin production. The bulk of the remaining tin production is used for alloying purposes with other elements.

Tin can be rolled into very thin foils, and tin foil was used for packaging before World War II. The high price of tin has led to its replacement by aluminium in this field.

Pewter is an alloy of tin that is widely used for making decorative articles. Earlier forms of pewter contained lead but in recent years non-toxic lead-free pewters have been developed. Pewter or Britannia metal, is tin alloyed with 6 per cent of antimony and 2 per cent of copper. Antimony greatly reduces the $\beta \to \alpha$ transition temperature and tin disease does not occur with this alloy.

15.8 Magnesium

The principal properties of magnesium are given below:

Melting point	649°C
Crystal structure	close packed hexagonal

Density	1.74×10^3 kg/m^3
Young's modulus, E	44 GPa
Electrical resistivity	3.9×10^{-8} Ωm at 20°C
Corrosion resistance	very good

Perhaps the most important property of magnesium is its very low density of 1.74×10^3 kg/m^3, compared with 2.71×10^3 kg/m^3 for aluminium. Pure magnesium is comparatively weak and is not used unalloyed. Magnesium can be strengthened considerably by suitable alloying. The principal alloying element is aluminium, in amounts up to 10 per cent. Additions of aluminium, thorium, zirconium or zinc will all give alloys than can be strengthened by precipitation hardening heat treatments (see Section 12.6). Small amounts of other elements may also be added to magnesium alloys. Manganese will improve the corrosion resistance while both thorium and cerium give increased strength and improve creep resistance at elevated temperatures.

Because of its hexagonal crystal structure magnesium is not as ductile as aluminium and is not as easy to cold work as that metal. Magnesium alloys are readily hot worked and can be easily cast into shapes by both sand and die casting methods.

The corrosion resistance of magnesium is good and this can be further improved by anodising.

Table 15.4 *Some magnesium alloys*

Composition (per cent)						Condition	Tensile strength (MPa)
Al	Mn	Zn	Zr	Th	Ce		
3	0.3	1	–	–	–	Annealed	250
						Cold rolled	280
6	0.3	1	–	–	–	Forged	280
8	0.3	0.7	–	–	–	Cast	140
						Heat treated	210
10	0.3	0.7	–	–	–	Cast	200
–	–	3	0.7	–	–	Rolled	260
–	–	4	0.7	–	1	Cast	170
						Heat treated	220
–	–	–	0.7	3	–	Heat treated	210
–	1	–	–	–	3	Rolled	280

Some of the major uses of magnesium alloys are in aircraft construction, in the form of castings, forgings and sheet material for such components as landing wheels, airscrew blades, fuel tanks and engine components. The alloys are also used for cast car and motorcycle wheels, petrol engine crankcases and as a canning material for nuclear fuel elements.

The composition and properties of some magnesium alloys are given in Table 15.4.

Pure magnesium is used as sacrificial anode material for the cathodic protection of steel structures and ship's hulls against corrosion. The magnesium anodes are corroded and have to be renewed periodically while the steelwork is protected against corrosion. Zinc may also be used for this application.

15.9 Nickel

The principal properties of nickel are summarised below:

Melting point	1455°C
Crystal structure	face centred cubic
Density	$8.88 \times 10^3 \, \text{kg/m}^3$
Young's modulus, E	210 GPa
Electrical resistivity	$6.84 \times 10^{-8} \, \Omega\text{m}$ at 20°C
Corrosion resistance	very good

Nickel is one of the most important metals in this technological age. It is used to a large extent as an alloying addition in many steels and cast irons, and in a number of non-ferrous alloys. Nickel is also used as a pure metal in its own right, and as the base for a number of useful engineering materials. The use of nickel as an alloying element in steels and irons will be dealt with in Chapter 16.

Pure nickel possesses an excellent resistance to corrosion by alkalis and many acids and, consequently, is used in chemical engineering plant, and in the food industry. For cheapness, nickel is frequently used as a cladding of thin sheet on a mild steel base. Nickel may also be electroplated on to a number of materials, and an intermediate layer of electrodeposited nickel is essential in the production of chromium-plated mild steel.

The principal nickel base alloys used industrially are *Monel*, *Inconel*, and the *Nimonic* series of alloys. Monel is an alloy containing 68 per cent of nickel, 30 per cent of copper, and 2 per cent of

Table 15.5 *Composition and uses of some nickel alloys*

Name	Composition (per cent)										Uses
	Ni	Cu	Cr	Fe	Mo	W	Ti	Al	Co	C	
Monel	68	30	–	2	–	–	–	–	–	–	Chemical engineering plant. Steam turbine blades
Inconel	80	–	14	6	–	–	–	–	–	–	Chemical engineering plant. Electric cooker heating elements. Exhaust manifolds
Brightray	80	–	20	–	–	–	–	–	–	–	Heating elements for kettles, toasters, electric furnaces
Hastelloy C	55	–	15	5	17	5	–	–	–	–	Chemical engineering plant.
Hastelloy X	47	–	22	18	9	1	–	–	–	–	Furnace and jet engine components
Nimonic 75	77	–	20	2.5	–	–	0.4	–	–	0.1	Thermocouple sheaths, furnace components, nitriding boxes
Nimonic 90	56.5	–	20	1.5	–	–	2.4	1.4	18	0.06	Gas turbine discs and blades
Nimonic 115	56.5	–	15	0.5	4	–	4	5	15	0.1	Gas turbine discs and blades

iron. It is a single-phase alloy with an exceptionally good corrosion resistance, and it is used for steam turbine blades, and in chemical engineering plant. Inconel contains 80 per cent of nickel, 14 per cent of chromium, and 6 per cent of iron. Inconel combines good corrosion resistance with good properties at elevated temperatures, and it is used in chemical engineering plant, and as heater sheaths for electric cooker elements.

The Nimonic series of alloys are basically nickel–chromium alloys containing between 55 and 80 per cent of nickel, and around 20 per cent of chromium. In addition, the various alloys in the range contain small amounts of titanium, cobalt, iron, and aluminium, together with a very small amount of carbon. These alloys are high-temperature creep resistant, and have been very fully developed to meet the requirements of gas turbine designers. The alloys are used for the manufacture of gas turbine discs and blades, and for flame tubes. They owe their high-temperature strengths to finely dispersed precipitates of intermetallic compounds of nickel with titanium and iron, and also some metallic carbides. Other nickel–chromium and nickel–chromium–iron alloys that have been developed include the proprietary alloys Brightray and Hastelloy (see Table 15.5). These materials, together with the Nimonic alloys, all developed to meet the very stringent demands of service at elevated temperatures are sometimes referred to as *superalloys*. (See also Section 15.13.)

15.10 Titanium

The principal properties of titanium are summarised below:

Melting point	1660°C
Crystal structure	α close packed hexagonal up to 880°C
	β body centred cubic above 880°C
Density	$4.54 \times 10^3 \, \text{kg/m}^3$
Young's modulus, E	106 GPa
Tensile strength	30 MPa
Corrosion resistance	Excellent

The useful properties of titanium are its relatively high strength coupled with a low density, and its excellent corrosion resistance with respect to most acids, alkalis and chlorides. It does possess some characteristics, though, that make processing both difficult and costly. At very high temperatures, titanium will readily dissolve oxygen, nitrogen and carbon, all of which will cause embrittlement. Consequently, titanium is produced by reducing titanium chloride with magnesium in a closed reaction vessel followed by vacuum distillation to remove any unreacted magnesium. The pure titanium sponge cannot be melted and cast in the same manner as other metals. Instead, the sponge is made into an electrode and this is arc-melted into a water-cooled copper crucible either under an inert gas atmosphere or in vacuo. The liquid titanium is then poured into a graphite-faced ingot mould, again either in vacuo or in an inert atmosphere. Ingots of titanium alloys may be made by compacting the required alloying additions with the sponge when making the electrode.

Titanium oxidises rapidly in air at all temperatures in excess of 700°C and oxygen will dissolve in the metal. Small amounts of oxygen will cause strengthening, with loss of ductility, while larger amounts will cause almost total brittleness. The α phase (hexagonal) is difficult to plastically deform and the amount of cold working that can be accomplished is limited. Hot forging and hot rolling of titanium alloys can be carried out at temperatures between 700°C and 1100°C. After hot working operations, surface contamination is removed by grinding and machining followed by pickling in hydrofluoric acid and nitric acid solutions. Castings in titanium alloys can be made by melting and casting in vacuo into graphite-lined moulds.

The principal alloying element in titanium alloys is aluminium and up to 6 per cent will enter solid solution in α titanium giving a strengthening effect. Some alloying elements including iron, chromium and molybdenum tend to stabilise the β phase and give alloys with $\alpha + \beta$ structures. Other alloying elements, including copper, will produce alloy structures that respond to precipitation hardening. Tensile strengths of up to 1500 MPa can be achieved with some of the complex alloys.

The main use of titanium alloys is in aircraft construction, as structural forgings, sheet for bulkheads and panelling, and for discs and blades for the compressor stages of turbofan engines. The materials also find use in chemical engineering plant where an excellent corrosion resistance is required.

15.11 Zinc

The principal properties of zinc are summarised below:

Melting point	419°C
Boiling point	907°C
Crystal structure	close packed hexagonal
Density	$7.14 \times 10^3 \, kg/m^3$
Young's modulus, E	90 GPa
Electrical resistivity	$5.92 \times 10^{-8} \, \Omega m$ at 20°C
Corrosion resistance	good

Pure zinc, having a melting point only slightly in excess of 400°C, makes an excellent material for die casting purposes. The die-casting alloys contain 4 per cent of aluminium and, in some cases, 1 or 2 per cent of copper. These alloying additions increase the tensile strength of the zinc, in the die cast state, to about 250 MPa. The zinc used for die casting purposes must be of extremely high purity, as traces of cadmium, tin, or lead, in the material would lead to the incidence of intercrystalline corrosion failure in service. Zinc alloy die castings are used extensively, and their applications include automotive components, such as carburettor bodies, fuel pump bodies, and door handles, and the production of accurately scaled model toys.

Metals possessing hexagonal crystal structures are normally difficult to work cold. Zinc is quite brittle at ordinary temperatures, but it can be plastically deformed with ease at temperatures within the range of 100°C to 150°C. Sheet zinc is produced by rolling the metal at these temperatures. Sheet zinc is used in the manufacture of battery cases for dry cells, the zinc being the negative electrode of the battery.

There is a eutectoid transformation in the zinc–aluminium alloy system at a composition of 22 per cent aluminium. The eutectoid alloy can be produced with an extremely fine grain size and is superplastic at the eutectoid temperature of 275°C (see Section 8.12). This material is used to manufacture deep-drawn products such as instrument cases.

In addition to the above, a major use for zinc is for galvanising, that is the coating with zinc of steel sheet, structural sections, and nails, as a protection against corrosion. Like magnesium, zinc may also be used as a sacrificial anode to give cathodic protection to steel structures.

15.12 Bearing materials

There are several requirements which should be possessed by a successful bearing material. It should possess sufficient hardness and wear resistance so that it does not wear away during service but its hardness should be low relative to the shaft or journal in order to avoid wear or damage to the shaft, particularly during a start-up when, because of low oil pressure, there may be metal-to-metal contact. The strength of the bearing material should be sufficient to sustain the load without deformation and yet possess considerable toughness to resist shock loading. Many of these apparently conflicting requirements can be met by metallic alloys possessing a duplex structure with hard constituents, to sustain the load, embedded in a softer and tough matrix.

There are very many bearing materials, each suited to particular types of loading, running speed and service conditions. The main types are: white metals, copper-base alloys, aluminium-base alloys, cast iron, plastics materials and ceramics.

White metals. These bearing metals are either tin-base or lead-base, of which the former are the more widely used. The tin-base bearing metal alloys are of better quality than the cheaper lead-base variety and are suitable for use in many low- and medium-duty bearing applications, particularly in the automotive industry. The lead-base alloys tend to be used for low pressure/low speed bearing applications. Both types contain about 10 per cent of antimony and usually 1.5–4.0 per cent of copper. The tin-base bearing metals are known as Babbit metals after the original patentee, Isaac Babbit. Tin and antimony combine to form a very hard intermetallic compound, with the formula SbSn. The compound SbSn appears in the microstructure as small hard cubic crystals, termed cuboids. Within the tin–lead–antimony alloys these hard SbSn cuboids are dispersed within a soft matrix of ternary tin–lead–antimony eutectic. During the solidification of bearing metals, the hard SbSn cuboids solidify from the liquid first and, being of low density, they tend to rise to the surface of the liquid metal. This segregation of the hard constituent may be overcome by introducing about 3 per cent of copper to the alloy composition. In the alloys containing some copper, it is a hard intermetallic compound of copper and tin, Cu_6Sn_5, that

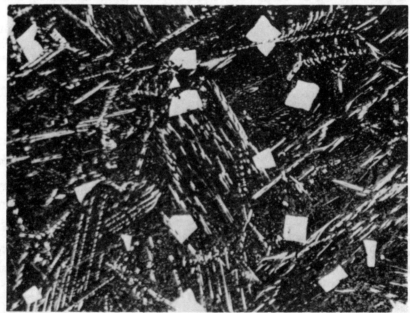

FIGURE 15.10   Photomicrograph of white bearing metal (80% Sn, 11% Sb, 6% Pb, 3% Cu) (etched) × 60, showing cuboids of SbSn and network of Cu_6Sn_5 needles

solidifies first. This compound forms as a series of needle-chaped crystals, creating a network throughout the melt. This network largely prevents the SbSn cuboids from segregating to the surface of the melt when they begin to solidify (see Figure 15.10).

Bearing bronze. There are several types of bronze used in bearing applications; straight Cu–Sn alloys, phosphor bronzes or leaded bronzes. Generally, a bearing bronze contains 10–15 per cent tin and has an $\alpha + \delta$ structure. The hard δ phase, present as part of an $(\alpha + \delta)$ eutectoid within a softer α matrix, is capable of sustaining heavy loads. In a phosphor bronze, the very hard phase Cu_3P is associated with the $\alpha + \delta$ eutectoid and provides improved load bearing capacity. A typical composition for a leaded bronze is 2% Sn, 24% Pb, 74% Cu. Lead is insoluble in copper and the presence of lead globules in the microstructure reduces the dry coefficient of friction of the material. There is a tendency for major segregation of lead during solidification and the addition of a small amount of nickel to the alloy reduces this tendency. The leaded alloys are less strong than other bearing bronzes but will sustain higher loads at higher speeds than white metals. They find application in aero, diesel and automobile engines.

Some bronze bearings are made by compacting and sintering a powder. The powder particle size and compaction pressures are controlled to produce a porous bearing which can be saturated with lubricant before installation. This type is used widely in applications where regular lubrication is difficult, as for example in automotive starter motors and alternators.

Aluminium bearings. Aluminium-base alloys containing tin, copper and nickel are used widely as bearing materials and have replaced the more expensive tin-base bearing metals in many applications, particularly in the automotive industry. Typical compositions lie within the ranges 5.5–7.0 per cent tin, 0.7–1.3 per cent copper and 0.7–1.3 per cent nickel. The basis alloys tend to be of low strength and frequently are bonded to a steel backing shell.

Some polymeric materials, notably PTFE and the polyamides (nylons), have very low coefficients of friction and are used for low load applications and where the presence of a lubricating oil is undesirable.

Ceramic materials have been used as bearings in small precision instruments for many years, for example jewel bearings in watch movements. Modern industrial ceramics such as alumina are used as bearings in high speed precision movements.

15.13 Superalloys

The development of the aero gas turbine, with its requirements for materials capable of operating at high levels of stress at high temperatures for extended periods, has been the stimulus for much research into the development of creep resistant alloys. Superalloy is a general term used to describe the nickel-base and cobalt-base alloys which have been developed for use at elevated temperatures. These alloys tend to contain many alloying elements and are complex in structure. An alloy that will resist high-temperature creep successfully must possess structural elements that will greatly impede the movement of dislocations. Some alloying elements will enter solid solution and strengthen the matrix while others, together with other elements, will form intermetallic compounds and carbides. When carbides and other precipitate particles are present in a highly dispersed state in high density it will be extremely difficult for dislocation movement to occur within the material, even at very high temperatures. Examples of some of the stable precipitate particles that are used to good effect in superalloys are Ni_3Al, Cr_7C_3, TiC and Ni_3Ti. An alloying addition that will promote a large grain size or that will cause grain boundary strengthening will also be beneficial in resisting creep because the viscous flow of grain boundaries contributes to the creep process at temperatures close to the melting point. The

Table 15.6 *Functions of elements added to superalloys*

Function	Al	B	C	Cr	Co	Hf	Mo	Nb	Ta	Ti	W	Zr
To strengthen the matrix				x	x		x				x	
To strengthen the grain boundaries		x	x				x					x
To form carbides				x			x	x	x	x	x	
To create protective oxide scale	x			x								

functions of the various alloying additions that are made to nickel in the creation of superalloys is given in Table 15.6.

Development of processing technology has taken place alongside alloy composition developments. The introduction many years ago of vacuum melting and casting gave clean gas-free materials and resulted in improved alloy performance. Some ten years ago, DS casting (directional solidification) in vacuo was introduced for the production of turbine blades. In a DS casting the crystal grains are aligned in the direction of maximum direct stress and there are very few grain boundaries normal to this direction. A logical, but more recent, development was the production of single crystal turbine blades by a directional solidification casting process.

16

Iron and Steel

16.1 Introduction

The principal ores of iron are oxides and carbonates, and iron ores are widely distributed throughout the world. Until the eighteenth century, oxide ores of iron were reduced to the metal using charcoal as a reducing agent, and steel, which is basically an alloy of iron and carbon, was produced by the *cementation* process, in which wrought iron bars were packed, together with charcoal, in cast iron boxes and heated to about 1000°C. At the high temperature carbon would slowly diffuse into the iron. The quality of steel produced in this way tended to be erratic, and in the mid-eighteenth century the *crucible* process was invented. Bars of steel, produced by cementation, were melted in a crucible and the homogeneous melt obtained was cast into bars. The major ferrous constructional materials until the latter part of the nineteenth century were cast iron and wrought iron, and many road and railway bridges built in these materials are still in service. The pattern of working was revolutionised in 1856, when Sir Henry Bessemer introduced his *converter* process for steel making. The *open-hearth* process followed a decade later, and these two processes held sway for almost a century until the introduction, in the 1950s, of oxygen steel-making processes.

16.2 Production of iron

Iron ores are smelted to iron in large shaft-type furnaces, termed *blast furnaces*. The ore, together with coke as a fuel, and limestone to make a fusible slag with the *gangue* or waste material, is charged to the top of the furnace and a high-pressure blast of hot air is blown into the furnace through a series of jets, or *tuyeres*, situated just above the hearth. The coke is partially burnt and the hot carbon monoxide produced reduces the oxides of iron to the metal. Liquid iron collects in the hearth and the waste material collects as a liquid slag which floats on top of the iron. Both slag and iron are tapped separately from the furnace at regular intervals. The iron that is produced is impure. The principal impurities that it contains are carbon (about 4 per cent) derived from the coke, silicon (about 2 per cent) and manganese (about 1 per cent) reduced from the ore, and sulphur. Many ore deposits have phosphorus associated with them, and iron smelted from phosphatic ores will also contain up to 2 per cent of phosphorus.

The iron from the blast furnace may be cast into moulds to give small ingots, or *pigs*, or transferred in liquid form to the steel-making furnaces as *hot metal*. The blast furnace slag can, in many cases, be used as hard core or concrete aggregate. Some solidified slags have a high porosity, resemble pumice, and are used as lightweight aggregate for low-density concretes. Also, fibre pulling from liquid slag is used to make slag wool for insulation purposes.

It is appropriate, at this stage to list some of the terms encountered in iron and steel making.

Pig iron is the impure product of the blast furnace when cast into small ingots, or pigs.

Cast iron is selected grades of pig iron remelted in a cupola and cast into shapes in sand moulds.

Acid iron is a pig iron of low phosphorus content, which may be refined to steel by an acid process.

Basic iron is a pig iron containing up to 2 per cent of phosphorus, and this may only be refined into steel using a basic process.

Wrought iron is refined iron containing a certain amount of slag. Wrought iron was made by the *puddling* process, in which molten iron from the blast furnace was refined by oxidation, using a slag rich in iron oxide as an oxidising agent. As the impurities are removed, so the melting temperature of the iron is raised. In the eighteenth and nineteenth centuries furnaces were not capable of giving temperatures high enough to keep pure iron molten. Consequently, as the iron became refined in the puddling furnace, it began to solidify, and had to be removed from the furnace as a pasty mass containing entrapped slag. The iron was then worked by forging and rolling. Some slag is exuded from the material during working, but much remains, giving wrought iron a fibrous texture. Today, wrought iron is not made by the puddling process, but by adding slag to refined liquid iron.

Ingot iron, or *Swedish iron* is the purest commercial form of iron. It is made by giving iron from the blast furnace the maximum possible refining in a steel-making furnace.

Steel. Steels are, essentially, alloys of iron and carbon, containing up to 1.5 per cent of carbon. Steel is made by oxidising away the impurities that are present in the iron produced in the blast furnace.

16.3 Steel production

The steel-making process is an oxidation process. The impurities present in pig iron are oxidised, and the oxidation products, apart from carbon monoxide, are collected in a slag layer. Initially, many iron ores yielded acid iron, that is an iron of low phosphorus, and this could be refined using a refining slag based on silica, an acidic oxide. The furnace or converter for working with siliceous or acid slags was lined with an acidic refractory, namely silica brick. The process was termed the acid process and the product termed *acid steel*. Most ores worked today are phosphatic and the blast furnace iron formed contains up to 2 per cent phosphorus. Phosphorus cannot be removed from the iron during acid steel-making and can only be removed effectively if the refining slag is rich in lime, a basic oxide. The use of lime-rich, or basic slags, requires that the furnace or converter be lined with a basic refractory. *Dolomite*, the mixed carbonate of magnesium and calcium, is used to produce basic refractory materials. The steel produced using a basic refining slag is termed *basic steel*. By the end of the refining process the basic slag will have become rich in phosphates. After it has solidified, basic slag can be ground into powder and sold as an agricultural fertiliser.

Sulphur cannot be removed efficiently from iron under oxidising conditions and the molten iron from the blast furnace is treated with soda ash before it is charged to a converter or an open-hearth furnace for refining into steel.

$$FeS + Na_2CO_3 = FeO + Na_2S + CO_2$$
$$\text{(soda ash)} \qquad \text{(enters slag)}$$

In the original Bessemer process a blast of hot air was blown through the molten pig iron, from the base of the converter, in order to oxidise impurities. Bessemer steels tended to possess lower ductilities than other steels, owing to the steel absorbing some nitrogen. Some ten years after the introduction of the Bessemer process open-hearth steel making was introduced. The open-hearth furnace is a large reverberatory-type furnace, and oxidation of impurities from the charge is carried out by means of a slag rich in iron oxide. The other major constituent of the slag is either silica or lime, as outlined above. The slag, which is less dense than iron, floats as an upper liquid layer on the molten iron, and the refining reactions occur at the slag/metal interface. The impurities enter the slag, except for the carbonaceous gases, which escape to atmosphere.

$$Mn + FeO = Fe + MnO$$
$$\text{(enters slag)}$$
$$Si + 2FeO = 2Fe + SiO_2$$
$$\text{(enters slag)}$$
$$C + FeO = Fe + CO$$
$$2P + 5FeO + 3CaO = 5Fe + Ca_3(PO_4)_2 \text{ (basic process only)}$$
$$\text{(enters slag)}$$

For very many years both processes existed alongside one another as viable production methods. In the Bessemer process a charge of about 20 tonnes of iron could be converted into steel in about 20 minutes, but the presence of nitrogen rendered the steel unsuitable for some applications involving deep drawing or some other cold forming processes. The open-hearth process gave a better-quality product, but the length of the refining process was much greater, increasing the cost of the steel. This was partly offset by building large furnaces with capacities of 200 and 300 tonnes. With a charge of 50 per cent of hot metal from the blast furnace and 50 per cent of steel scrap, the refining time for a 200 tonne charge would be about 12 hours. A major revolution in steel-making practice has taken place during the last forty years with the availability of liquid oxygen in large quantity. Pure oxygen cannot be used for Bessemer blowing as the rate of heat evolution within the converter would be too great for full control to be exercised over the process (all the chemical reactions involved are exothermic) but, for some time, Bessemer converters were used with an oxygen/dry steam blast. This resulted in Bessemer steel of very low nitrogen content. In the early 1950s important European developments in steel making were the introduction of the *Kaldo* and *L–D* (Linz–Donawitz) processes. In the L–D process the molten iron is placed in a converter, a large cylindrical vessel, and pure oxygen is blown into the top of the converter. The Kaldo process also utilises top blown oxygen for refining, but the Kaldo converter is continuously rotated to ensure good mixing of the charge and efficient refining. Most modern steelworks use one or both of these processes with converters of about 100 tonnes capacity. With pure oxygen blowing the full refining cycle for a 100 tonne charge takes about one hour.

By the end of the refining process, whether it has been in a converter or in an open-hearth furnace, the iron will contain only small amounts of carbon, silicon, manganese, sulphur and phosphorus (less than 0.05 per cent of each), but it will be in a highly oxidised state. For the production of all steels, except those of rimming quality, the liquid metal must be deoxidised. Silicon and manganese, in the form of ferro-alloys, are generally used as deoxidisers and sufficient is added to give some residual silicon and manganese in the steel. After deoxidation is complete, anthracite is added to bring the carbon content up to the desired level. The molten steel is then ready for casting into ingot moulds.

As stated in Section 16.2, a steel is essentially an alloy of iron and carbon, but it will contain some silicon and manganese, traces of sulphur and phosphorus, and it may also contain other elements, such as nickel and chromium, present as alloying additions. Two further definitions are now given.

A *plain carbon steel* is a steel containing up to 1.5 per cent of carbon, together with not more than 0.5 per cent of silicon and not more than 1.5 per cent of manganese, and with only traces of other elements.

An *alloy steel* is one that contains either silicon or manganese in amounts in excess of those quoted above, or that contains any other element, or elements, as the result of deliberately made alloying additions.

Many alloy steels are made in electric furnaces. The direct arc type of furnace is generally used for the production of low-alloy steels containing nickel and chromium, while high-alloy and special steels are made by crucible methods, usually using high-frequency induction furnaces.

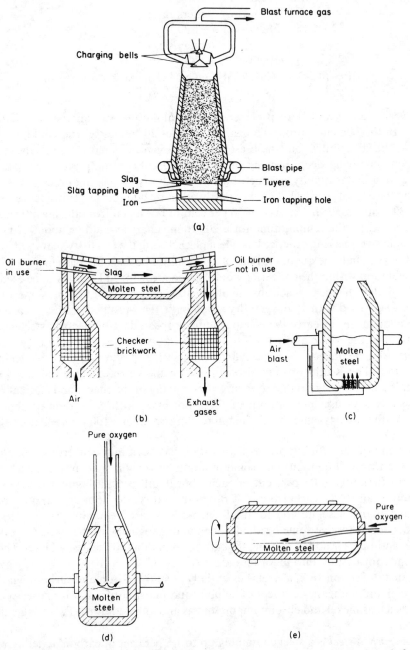

FIGURE 16.1   Iron-making and steel-making methods: (a) blast furnace; (b) open-hearth furnace; (c) Bessemer converter; (d) L–D converter; (e) Kaldo rotary converter

16.4 Constituents in steels

There are a number of special names that are used to denote the various phases and types of structure that occur in steels, and it is appropriate to list, and explain, these terms at this stage. These terms will be used extensively in the remainder of the chapter.

Ferrite. Ferrite is the name given to the body centred cubic allotropes of iron, α and δ iron, and to body centred cubic solid solutions.

Austenite. Austenite is the name given to the face centred cubic, or γ, variety of iron, and to face centred cubic solid solutions.

Cementite. Cementite is the name given to the carbide of iron; formula Fe_3C. This is an extremely hard and brittle constituent.

Pearlite. Pearlite is the eutectoid mixture of ferrite and cementite, and is formed when austenite decomposes during cooling. It consists of alternate thin layers, or lamellae, of ferrite and cementite.

Martensite. This is the name given to the very hard and brittle constituent that is formed when a steel is very rapidly cooled from the austenitic state. It is a ferrite, highly supersaturated with dissolved carbon.

Sorbite and troostite. These are names given to the structures produced when martensite or bainite is tempered, that is, heated to some temperature not exceeding 700°C for the purpose of reducing brittleness and hardness.

Bainite. This is the term that is given to the decomposition product that is formed when austenite decomposes by either isothermal transformation, or at a cooling rate intermediate between the very rapid cooling necessary for martensite formation and the slower rate of cooling at which pearlite is formed.

16.5 The iron–carbon phase diagram

As mentioned in Section 11.11, iron is allotropic. The crystal structure of pure iron at ordinary temperatures is body centred cubic. This form is known as α iron. α iron is stable at all temperatures up to 908°C. On heating through 908°C the crystal structure of iron changes to face centred cubic. This form is known as γ iron. γ iron changes into δ iron when heated through 1388°C, and this high-temperature form possesses a body centred cubic structure. δ iron is the stable form from 1388°C up to the melting temperature of pure iron at 1535°C.

Iron loses its ferromagnetic characteristics when it is heated to temperatures above 768°C, but the ferromagnetism returns when the metal is cooled to below this temperature. This temperature is termed the *Curie temperature*. Originally, iron at temperatures between 768°C and 908°C was termed β iron. The use of the term β iron was discontinued when it was discovered that no change in crystal structure took place at the Curie temperature.

The phase diagram for the iron-carbon system is shown in Figure 16.2. This diagram may appear at first sight, to be extremely complex, but it can be divided into sections that, in themselves, are straightforward. For the consideration of steels it is convenient to consider only that portion of the diagram up to a carbon content of 1.5 per cent, and up to a temperature of 1000°C (see Figure 16.3).

It will be seen that ferrite cannot hold carbon in solid solution to any great extent, the limits being 0.04 per cent of carbon at 723°C and 0.006 per cent of carbon at 200°C. Austenite, however, can hold a considerable amount of carbon in solid solution, ranging from 0.87 per cent at 723°C to 1.7 per cent at 1130°C. Carbon is held in interstitial solid solution in both ferrite and austenite. The eutectoid point occurs at a temperature of 723°C and at a carbon content of 0.87 per cent.

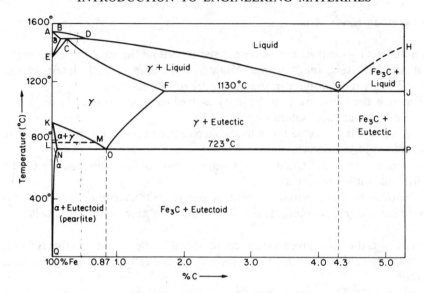

FIGURE 16.2   Fe–Fe₃C phase diagram

The terms *hypoeutectoid* and *hypereutectoid* are used to denote steels that contain less carbon than, and more carbon than, the eutectoid composition, respectively.

The presence of carbon depresses the α-γ transformation temperature of iron. Line KMO in the figures denotes this transformation temperature and its dependence on composition. Lines OF and QN are solvus lines and denote the maximum solubility limits of carbon in γ and α iron respectively. Point O is the eutectoid, or pearlite, point. The line LMOP indicates the Curie temperature.

If a sample of a steel is heated or cooled, and accurate measurements are taken, thermal arrest points will be noted corresponding to the phase transformation lines (and Curie temperature) on the phase diagram. The phase line NOP is known as the A_1 transformation, the Curie temperature, LM as the A_2 transformation, the phase line KMO as the A_3 transformation, and the line OF as the A_{cm} transformation. These arrest points, or transformation temperatures, are also known as the *critical points*, or *critical temperatures*, for the steel. The eutectoid temperature, A_1, is known as the *lower critical temperature*, and the α to γ transformation, line KMO, is known as the *upper critical temperature*. If a steel is heated or cooled very slowly, so that equilibrium conditions are approached, the measured arrest temperatures will agree with the values shown on the iron-carbon phase diagram. With more rapid heating or cooling rates, the measured arrest points will differ from equilibrium values. They will be higher than the equilibrium values when determined during heating, and lower when determined during cooling. Values measured during heating are written as Ac_1, Ac_2, and Ac_3, while values determined during cooling are written as Ar_1, Ar_2, and Ar_3.*

16.6 Structures of plain carbon steels

Let us now consider the changes that occur during the cooling of steels of various compositions. Refer to Figure 16.3 and consider first the cooling of a hypoeutectoid steel of composition (1). At

---

* From the French: *chauffage*—heating; *refroidissement*—cooling.

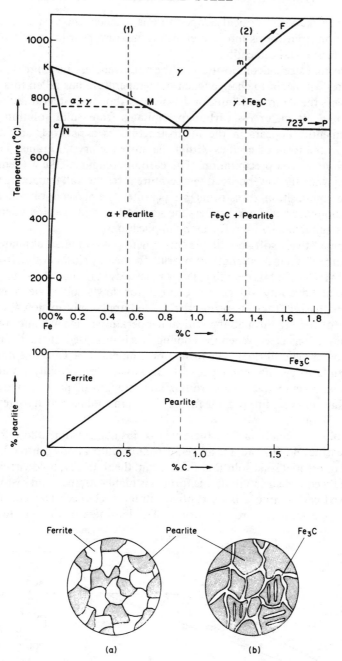

FIGURE 16.3   Steel portion of the Fe–Fe₃C diagram: (a) microstructure of hypoeutectoid steel (1); (b) microstructure of hypereutectoid steel (2)

a high temperature the steel structure will be composed of homogeneous crystals of austenite solid solution. On cooling to the upper critical temperature, point 1 on the diagram, austenite will begin to transform into ferrite. The ferrite can hold very little carbon in solid solution and so the remaining austenite becomes enriched in dissolved carbon. As the temperature falls, more ferrite is formed, and the composition of the remaining austenite increases in carbon content, following the line KMO. When the lower critical temperature is reached, the austenite, which is now of eutectoid composition, transforms into the eutectoid mixture pearlite, a mixture composed of alternate layers of ferrite and cementite.

For a hypereutectoid steel of composition (2) the homogeneous austenite structure that exists at high temperatures will begin to change when the temperature has fallen to a point m, on the solvus line OF. This is the saturation limit for dissolved carbon in austenite and on cooling below the temperature of point m, excess carbon precipitates from solid solution in the form of cementite. The cementite appears in the microstructure as a network around the austenite crystals, and also in the form of needles within the austenite crystal grains. This latter type is termed *Widman–Statten* type precipitation. The carbon content of the austenite reduces with further cooling and when the lower critical temperature is reached all remaining austenite, which is now of eutectoid composition, transforms into pearlite. The presence of cementite in the form of needles, or as a boundary network, renders the steel brittle, and heat treatment is necessary to put the steel into a suitable condition for many applications.

Ferrite is a comparatively soft and ductile constituent possessing an ultimate tensile strength of about 280 MN/m². The tensile strength of pearlite formed by slow cooling from the austenitic range is about 700 MN/m², but its ductility is very much less than that of ferrite. The variation of strength, hardness, and ductility with carbon content, for slowly cooled steels, is given in Figure 16.4. It must be emphasised that the properties quoted apply only to slowly cooled steels. An increase in the rate of cooling through the critical temperature range will alter the structure, and hence the properties, of any steel. When the cooling rate is increased, there is some undercooling of austenite to below the equilibrium transformation temperatures. Once the phase change from undercooled austenite to pearlite commences it takes place very rapidly resulting in very fine lamellae of ferrite and cementite. The hardness and strength of pearlite is dependent on the interlamellar spacing, and very fine pearlite formed by rapid cooling may have tensile strengths of the order of 1300 MN/m².

If a steel is cooled extremely rapidly there will be insufficient time allowed for austenite to decompose into pearlite and, instead, the austenite changes into a body centred lattice with all the carbon trapped in interstitial solid solution. Ferrite can, theoretically, hold virtually no carbon in solution. The rapidly cooled structure that is formed is highly strained and distorted by the large amount of dissolved carbon into a body centred tetragonal lattice. This constituent is termed martensite, and it is extremely hard and brittle. The hardness of martensite depends on the

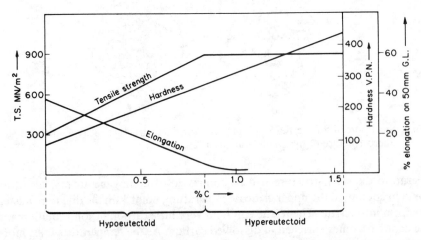

FIGURE 16.4 Properties of normalised plain carbon steels, and their variation with carbon content

carbon content, and is greatest in high carbon steels, that is, the greatest degree of lattice strain. Martensite, under the microscope, appears as a series of fine needle-like (*acicular*) crystals. A martensite structure can be formed by rapidly quenching a heated steel, from the austenitic state, into water or oil. This is the treatment termed *hardening*.

16.7 T–T–T diagrams

Martensite and bainite are non-equilibrium phases and do not appear in the iron–carbon phase diagram. In order to show the influence of cooling rates (that is, time) on the transformation of austenite, another type of diagram is necessary. This is the time–temperature–transformation, or T–T–T diagram. T–T–T diagrams are sometimes known as 'S curves' because of their general shape. (See also Section 12.7).

A typical T–T–T diagram for plain carbon steel is shown in Figure 16.5. It will be seen that a slow cooling rate will lead to the formation of coarse pearlite, with little undercooling of austenite, while a faster cooling rate will give a greater amount of undercooling and the formation of fine pearlite. If the critical cooling velocity is exceeded, the non-equilibrium phase, martensite, will be formed. Bainite, which is a finely divided dispersion of carbide particles in ferrite, may be formed by the isothermal transformation of undercooled austenite, namely, by the rapid quenching of the steel to a temperature below the nose of the T–T–T curve and then maintaining constant temperature until transformation is complete.

The position of the nose of the curve, and hence the value of the critical cooling velocity, is not constant for all steels. An increase in the carbon content of the steel, or an increase in the content of other alloying elements, will reduce the value of the critical cooling velocity, that is, move the T–T–T curve toward the right. The rate of cooling possible by quenching a steel in water is about equal to the critical cooling velocity of a plain carbon steel containing 0.3 per cent

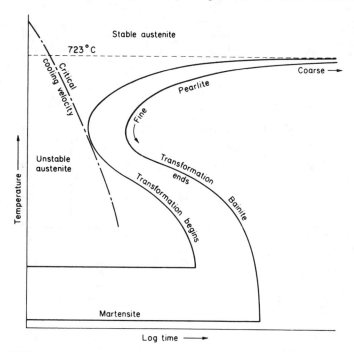

FIGURE 16.5   T–T–T diagram (S curve) for a plain carbon steel

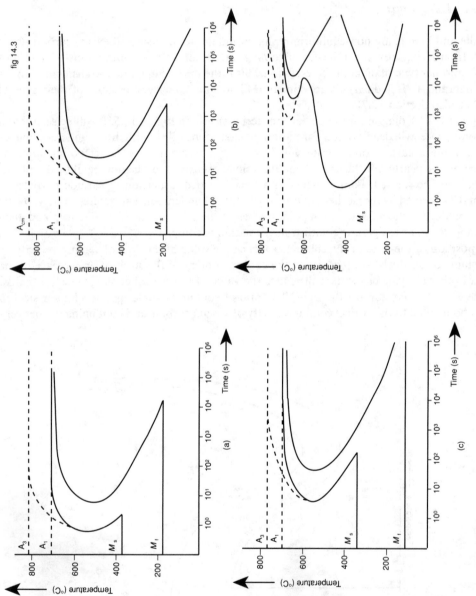

FIGURE 16.6 T–T–T diagrams for some steels. (a) carbon steel 0.4% C; 0.6% Mn. (b) carbon steel 1.2% C; 0.7% Mn. (c) low alloy steel 0.4% C; 1.25% Ni; 0.6% Cr. (d) low alloy steel 0.3% C; 1.8% Ni; 0.8 Cr; 0.25% Mo

of carbon. Consequently, it is impracticable to harden, by quenching, plain carbon steels with a lesser carbon content than this.

T–T–T diagrams for a few different steels are given in Figure 16.6. The T–T–T diagrams for some alloy steels show a double nose and there are two critical cooling velocities, a high rate for martensite formation and a lower rate for bainite formation.

16.8 Hardenability

As stated earlier, the hardness of a martensite is related to the carbon content of a steel. The term hardenability, however, refers not to the maximum hardness obtainable, but to the depth of hardening that may be obtained. When a heated bar of a steel is quenched in water from a high temperature, the surface is cooled extremely rapidly, but the interior of the bar is cooled at a slower rate. This means that, while the surface layers may be converted into martensite, the rate of cooling at some little distance below the surface of the bar may be less than the critical cooling velocity of the steel, and pearlite is formed. The depth of hardening obtained in a plain carbon steel may be only 2 or 3 mm. Alloy steels, because of their lower critical cooling velocities, will possess much greater hardenabilities than plain carbon steels and, in some cases, may show martensite at distances of 50 mm, or more, below the surface of the material. Alloy steels of high hardenability may be air hardening, that is, martensite is formed at the surface of the steel even when it has been cooled in air from a high temperature. A test that is frequently used for assessing the hardenability of a steel is the *Jominy* test. In this test a bar of steel is heated within the austenite range and it is then quenched by directing water on to one end only (see Figure 16.7). The rate of cooling at points along the length of the bar will approximate to the rate of cooling at points below the surface of a large section that has been quenched in water. When the test bar has fully cooled, surface flats are carefully ground, and a series of hardness measurements are then made along the length of the bar, working from the quenched end. Standard conditions are laid down for the dimensions of the test-piece, and for the quenching procedure, in the Jominy test (see BS 4437 (1969) and ASTM A 255-89).

16.9 Tempering

Martensite, although very hard, is also extremely brittle and a hardened steel requires a further heat treatment, known as *tempering*, before it can be put into service. When a martensite structure is heated it becomes possible for the carbon trapped in supersaturated solid solution to diffuse through the lattice and precipitate from solution in the form of iron carbide particles. This precipitation will relieve the strain within the lattice and cause the hardness and brittleness of the material to be reduced. This diffusion process can commence at temperatures of about 200°C, but the rate of diffusion is extremely slow at this temperature. An increase in the temperature will cause an increase in diffusion and precipitation rates and, therefore, increase the extent of the softening. At temperatures up to 450°C the carbide precipitate particles are much too fine to be resolved under the optical microscope, although their presence may be detected by using more sophisticated techniques. At higher temperatures the carbide particles increase in size and, at 700°C, the cementite coalesces into a series of fairly large, and roughly spheroidal particles (700°C is just below the lower critical temperature). This gives rise to a soft, but incredibly tough, material. Originally, the microstructures observed in tempered steels were given the names *troostite* and *sorbite*, but these names have largely gone out of use today, and structures are generally described as *tempered martensite*.

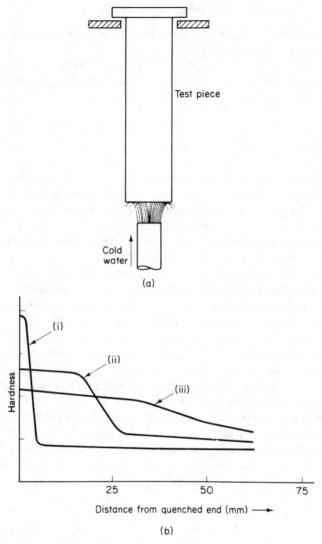

FIGURE 16.7   (a) Jominy test. (b) Jominy end quench curves for (i) steel containing 1 per cent C, 0.6 per cent Mn, (ii) steel containing 0.6 per cent C, 0.7 per cent Cr, 0.6 per cent Mn, (iii) steel containing 0.4 per cent C, 1.5 per cent Ni, 1.1 per cent Cr, 0.3 per cent Mo, 0.6 per cent Mn

Table 16.1   *Tempering temperatures*

Tempering temperature (°C)	Suitable for
220	Hacksaw blades
230	Planing and slotting tools, hammers
240	Milling cutters, drills, reamers
250	Taps, shear blades, punches, dies
260	Stone-cutting tools, drills for wood
270	Axes, press tools
280	Cold chisels, wood chisels, plane blades
290	Screwdrivers
300	Wood saws, springs
450–700	To produce great toughness, but at the expense of hardness

Tempering temperatures and times have to be fairly accurately controlled in order to produce the desired properties in the material. In small workshops this is accomplished by observing the colour that develops when a clean steel surface is heated. The colour developed is an interference colour, owing to the thickness of the oxide film formed on the metal surface, and this colour ranges from a pale-yellow, at 220°C, through brown, at 250°C, to dark-blue at 300°C.

16.10 Heat treatments for steels

Some of the effects of heat treatments on steels have been mentioned in earlier paragraphs, but it is now convenient to give fuller definitions of the various steel heat treatments.

Full annealing is the treatment given to produce the softest possible condition in a hypoeutectoid steel. It involves heating the steel to a temperature within the range 30–50°C above the upper critical temperature and then allowing the steel to cool slowly within the furnace. This produces a structure containing coarse pearlite.

Full annealing is an expensive treatment and when it is not absolutely essential for the steel to be in a very soft condition, but a reasonably soft and ductile material is required, the process known as *normalising* is used instead. Normalising involves heating the steel to 30–50°C above the upper critical temperature, but followed by cooling in still air. The structure of the pearlite produced is finer than in the fully annealed state, giving a slightly harder and stronger material.

A different process has to be used for softening hypereutectoid steels. If a hypereutectoid steel is slowly cooled from above the A_{cm} temperature, a brittle grain boundary network of cementite will be formed in the structure. Hypereutectoid steels are softened by giving them a *spheroidising* treatment, sometimes known as *spheroidising annealing*. This involves heating the steel to a

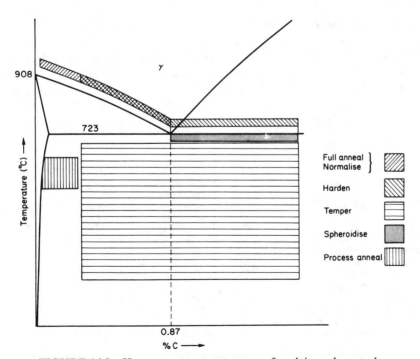

FIGURE 16.8   Heat treatment temperatures for plain carbon steels

temperature just below the lower critical temperature. During this treatment cementite forms as spheroidal particles in a ferrite matrix, putting the steel into a soft, but very tough, condition.

Process annealing, or *sub-critical annealing*, is often used for softening cold worked low carbon, or *mild*, steels. To fully anneal such a steel would involve heating to a temperature of more than 900°C, with subsequent high cost. In a mild steel ferrite makes up about 90 per cent of the structure, and the recrystallisation temperature of cold worked ferrite is only about 500°C. Annealing a cold worked mild steel in the temperature range 550–600°C will result in complete recrystallisation of ferrite, although the cold worked pearlite will be largely unaffected.

Hardening. This involves rapidly quenching the steel, from a high temperature, into oil or water, in order to bring about the formation of martensite. Hypoeutectoid steels are heated to 30–50°C above the upper critical temperature prior to quenching. If hypereutectoid steels are quenched from above the A_{cm} temperature, it is possible that some cementite would be precipitated at austenite grain boundaries. This would have an embrittling effect, which would still be evident after subsequent tempering. Consequently, hypereutectoid steels are hardened by quenching from 30–50°C above the lower critical temperature. At this temperature the structure becomes one of spheroidal cementite particles in an austenite matrix. Quenching transforms the austenite into martensite. (Note that it is impracticable to harden, by quenching, plain carbon steels containing less than 0.3 per cent of carbon.)

Tempering. Hardened steels may be tempered by heating them within the temperature range 200–700°C. This treatment will remove internal stresses set up during quenching, remove some, or all, of the hardness, and increase the toughness of the material (see also Section 16.9). For most steels, cooling from the tempering temperature may be either cooling in air, or quenching in oil or water. Some alloy steels, however, may become embrittled if slowly cooled from the tempering temperature, and these steels have to be quenched.

The various heat treatments quoted above apply to both plain carbon and low-alloy steels.

16.11 Types of steels and their uses

As has already been stated, the hardness and strength of steels vary very considerably with both carbon content and type of heat treatment. Certain names, which relate to the carbon content, are used in connection with steels. *Mild* or *low carbon steels* are those containing up to 0.3 per

Table 16.2 *Compositions and typical applications of steels*

C (per cent)	Name	Applications
0.05	Dead mild steel	Sheet and strip for presswork, car bodies, tin-plate; wire, rod, and tubing
0.08-0.15	Mild steel	Sheet and strip for presswork; wire and rod for nails, screws, concrete reinforcement bar
0.15	Mild steel	Case carburising quality
0.1-0.3	Mild steel	Steel plate and sections, for structural work
0.25-0.4	Medium carbon steel	Bright drawn bar
0.3-0.45	Medium carbon steel	Shafts and high-tensile tubing
0.4-0.5	Medium carbon steel	Shafts, gears, railway tyres
0.55-0.65	High carbon steel	Forging dies, railway rails, springs
0.65-0.75	High carbon steel	Hammers, saws, cylinder linings
0.75-0.85	High carbon steel	Cold chisels, forging die blocks
0.85-0.95	High carbon steel	Punches, shear blades, high-tensile wire
0.95-1.1	High carbon steel	Knives, axes, picks, screwing dies and taps, milling cutters
1.1-1.4	High carbon steel	Ball bearings, drills, wood-cutting and metal-cutting tools, razors

cent of carbon. Steels containing between 0.3 and 0.6 per cent of carbon are termed *medium carbon steels*, and these may be hardened and tempered. Steels containing more than 0.6 per cent of carbon are always used in the hardened and tempered condition, and these are known as *high carbon steels*, or *tool steels*.

Table 16.2 gives some typical uses of steels of various carbon contents. Some photomicrographs are shown in Figure 16.9.

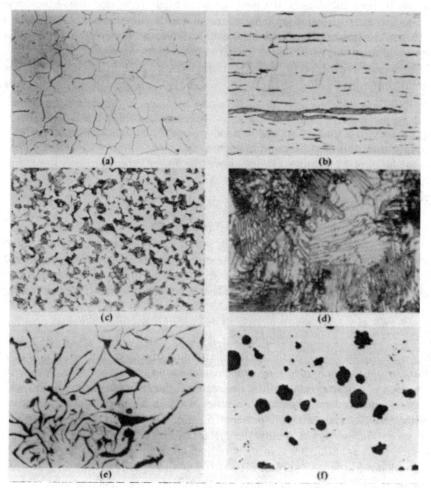

FIGURE 16.9   (a) Ingot iron (etched) × 600 showing crystals of ferrite. (b) Wrought iron (etched) × 150 showing crystals of ferrite and elongated slag particles. (c) 0.2 per cent steel (etched) × 300 showing ferrite and pearlite. The pearlite has etched dark. (d) Pearlite at high magnification (× 3000) showing lamellar structure. (e) Grey cast iron (unetched) × 200 showing irregular flakes of graphite. (f) Spheroidal graphite cast iron (etched) × 100 showing graphite spheroids and ferrite crystals

16.12 Surface hardening

Generally, the toughness of a material decreases as the hardness and strength increase. There are very many service conditions where the requirement is for a tough material of very high surface hardness. This is the case with shafts and gears, in particular. An ideal solution to the material problem would be to create a very hard surface layer on a comparatively soft and tough material. The surface hardening of steels is readily accomplished by a variety of methods.

Case carburising is one of the methods used for producing a hard surface on a ductile steel. It involves the introduction of additional carbon into the surface of a mild steel, effectively producing a composite material consisting of a low carbon steel with a thin case, about 0.5– 0.7 mm in thickness, of high carbon steel. The principal methods used are *pack carburising* and *gas carburising*. In pack carburising the parts to be treated are heated above the upper critical temperature in contact with wood or bone charcoal and barium carbonate, within a cast iron container. In gas carburising the parts are heated above the upper critical temperature in a furnace with an atmosphere of methane, or mixed hydrocarbon gases.

Considerable grain growth occurs in the material during a carburising treatment, and a three-stage heat treatment must be given to the carburised parts to produce the desired final properties. This heat treatment involves:

(a) refining the core structure by heating to a value above the upper critical temperature for the core composition, followed by rapid quenching;
(b) hardening the case by heating to just above the lower critical temperature, followed by quenching;
(c) tempering the case.

Another commonly used surface hardening process is *cyaniding*. In this process the steel components are heated in a bath of molten sodium cyanide and sodium carbonate at a temperature of about 950°C. During this treatment both carbon and nitrogen diffuse into the surface of the steel. The formation of hard iron nitrides contributes to the surface hardening of the material. After cyaniding, the parts require the same type of three-stage heat treatment as is given to case-carburised articles.

Carburised and cyanided articles have to be ground to final dimensions after all heat treatments have been given, as some dimensional changes occur during the processes.

Nitriding is another process for the surface hardening of a steel. Nitriding is not suitable for the hardening of plain carbon steels, as iron nitrides could be formed to a considerable depth below the surface of the steel, embrittling the material. Steels for nitriding are low-alloy steels containing chromium and molybdenum, together with, in some cases, nickel and vanadium. Some nitriding steels contain about 1 per cent of aluminium. Aluminium nitride is extremely hard, and does not diffuse very far below the surface (about 0.8 mm). Nitriding is carried out by heating the steel parts at about 500°C in a gas-tight chamber, in an atmosphere of ammonia. The ammonia dissociates at the steel surface into nitrogen and hydrogen and some of the former is absorbed by the steel. The major advantages of the nitriding process are:

(a) an extremely hard surface is formed,
(b) the treatment is conducted at comparatively low temperatures, minimising cracking and distortion,
(c) no subsequent heat treatment is necessary.

Parts for nitriding are first heat treated, to produce the best core properties. Machining to final dimensions is then carried out, while the material is in a soft condition, allowing for the small growth of 0.02 mm that occurs during nitriding.

Flame hardening and *induction hardening* are very rapid processes for the production of hard surfaces on steels with carbon contents in excess of 0.35 per cent. The steel surface is rapidly raised to the critical temperature range by means of a flame, or by being enclosed within a high-frequency induction coil. The surface is then rapidly quenched to promote martensite formation. After quenching, the outflow of heat from the core to the surface is sufficient to temper the surface layers. These processes are widely used, and they are particularly suitable for hardening the journals of shafts.

The above-mentioned and other methods for surface hardening are summarised in Table 16.3.

16.13 The effects of alloying elements in steels

It is possible to improve the properties of steels greatly by alloying with other elements. The general effects of alloying elements in steels are as follows.

(a) An alloying element entering into solid solution will increase the strength of the steel. An increase in carbon content will increase the strength of the steel, but with a considerable loss of ductility. Many of the commonly used alloying elements increase the strength of steels with little or no reduction in ductility.
(b) Almost without exception, alloying elements reduce critical cooling velocities and, so, increase the hardenability of a steel.
(c) The alloying element will have an effect on the α to γ transformation temperature of iron, and may either increase or decrease this (Figure 16.10). Chromium, tungsten, and silicon are some of the elements that raise the transformation temperature, while nickel and manganese are elements that lower it. There will be similar alterations in the critical temperatures of steels that contain alloying elements in addition to carbon. It will be noted that if a large amount of alloying element is present, a transformation could be eliminated and the steel become either wholly ferritic, or wholly austenitic.
(d) The alloying element could form stable carbides. Manganese, chromium, and tungsten are some of the elements that have a strong carbide-forming tendency, and these carbides are extremely hard.
(e) The alloying element could cause the breakdown of cementite, resulting in the appearance of

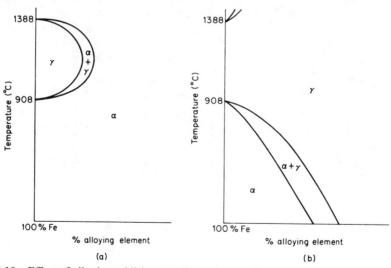

FIGURE 16.10   Effect of alloying addition on allotropic transformation temperatures of iron. (a) Ferrite-stabilising (γ loop) element—(Cr, W, Si). (b) Austenite-stabilising element—(Ni, Mn)

Table 16.3 *Surface hardening methods*

	Carburising	Nitriding	Cyaniding	Carbonitriding
Effect	A high carbon surface is produced on a low carbon steel and is hardened by quenching	A very hard nitride-containing surface is produced on the surface of a strong tough steel	A carbon and nitride-containing surface is produced on a low carbon steel and is hardened by quenching	Carbon and nitrogen are added to the surface of a low carbon steel and permit hardening by an oil quench
Suitability	Suitable for plain carbon or alloy steels containing about 0.15 per cent C	Nitralloy steels containing aluminium; a typical nitriding steel contains 0.3 per cent C, 1.6 per cent Cr, 0.2 per cent Mo, 1.1 per cent Al. This steel is hardened by oil quenching from 900°C and tempered at 600–700°C before being nitrided	Suitable for plain carbon or alloy steels containing about 0.15 per cent C	Suitable for plain carbon steels containing about 0.15 per cent C
Method	Low carbon steel is heated at 850–930°C in contact with gaseous, liquid, or solid carbon-containing substances for several hours. The high carbon steel surface produced is then hardened by quenching	The steel is heated at 500–540°C in an atmosphere of ammonia gas for 50–100 hours. No further heat treatment is necessary	Low carbon steel is heated at 870°C in a molten 30 per cent sodium cyanide bath for about one hour. Quenching in oil or water from this bath hardens the surface of the steel	Low carbon steel is heated at 700–870°C for several hours in a gaseous ammonia and hydro-carbon atmosphere. Nitrogen in the surface layer increases hardenability and permits hardening by an oil quench
Result	Case depth is about 1.25 mm. Hardness after heat treatment is H_{RC} 65 (H_D 870). Negligible dimension change caused by carburising. Distortion may occur during heat treatment	Case depth is about 0.38 mm. Extreme hardness (H_D 1100). Growth of 0.025–0.05 mm occurs during nitriding. Case is not softened by heating for long times up to 420°C. Case has improved corrosion resistance	Case depth is about 0.25 mm. Hardness is about H_{RC} 65. Negligible dimension change caused by cyaniding. Distortion may occur during heat treatment	Case depth is about 0.5 mm. Hardness after heat treatment is H_{RC} 65 (H_D 870). Negligible dimension change occurs. Distortion is less than in carburising or cyaniding
Applications	Typical uses are for gears, camshafts and bearings	Typical uses are for valve guides and seatings, and for gears	Typical uses are for small gears, chain links, nuts, bolts and screws	Typical uses are for gears, nuts and bolts

Surface hardening methods

	Flame hardening	Induction hardening	Siliconising (Ihrigising)	Hard chromium plating
Effect	The surface of a hardenable steel or iron is heated by a gas torch and quenched	The surface of a hardenable steel or iron is heated by a high-frequency electro-magnetic field and quenched	A moderately hard corrosion-resistant surface containing 14 per cent silicon is produced on low carbon steels	A hard chromium plate is applied directly to the metal surface
Suitability	Steel containing 0.4–0.5 per cent carbon or cast iron containing 0.4–0.8 per cent combined carbon may be hardened by this method	Steel containing 0.4–0.5 per cent carbon or cast iron containing 0.4–0.8 per cent combined carbon may be hardened by this method	Suitable for plain carbon steels containing 0.1–0.2 per cent carbon	Generally used on steels, low or high carbon, soft or hardened
Method	A gas flame quickly heats the surface layer of the steel and a water spray or other type of quench hardens the surface	The section of steel to be hardened is placed inside an induction coil. A heavy induced current heats the steel surface in a few seconds. A water spray or other type of quench hardens the surface	The steel parts are heated at 930–1000°C in contact with silicon carbide and chlorine gas for two hours. No further heat treatment is required	The steel parts are plated in the usual plating bath but without the usual undercoat of nickel. The plating is a thousand times thicker than decorative chromium plating
Result	The hardened layer is about 3 mm thick. Hardness is H_{RC} 50–60 (H_D 500–700). Distortion can often be minimised	The hardened layer is about 3 mm thick. Hardness is H_{RC} 50–60 (H_D 500–700). Distortion can often be minimised. Surface remains clean	Case depth is about 0.63 mm. Hardness is about H_D 200. Case has good corrosion resistance. Growth of 0.025–0.05 mm occurs during siliconising	Plating thickness is about 0.125 mm. Extreme hardness H_D 900. Plating has good corrosion resistance and a low coefficient of friction
Applications	Used for gear teeth, sliding ways, bearing surfaces, axles and shafts	Used for gear teeth, sliding ways, bearing surfaces, axles and shafts	Typical uses are for valves, tubing and shafts	Typical uses are for dies, gauges, tools and cylinder bores

graphite in the structure of the steel. Nickel and silicon have this effect and, for this reason, are not added to high carbon steels.

(f) The alloying element may confer some property, characteristic of itself, on the steel. For example, chromium, which is corrosion resistant, will make a steel corrosion resistant, or stainless, if it is present in the steel in amounts in excess of 12 per cent.

16.14 Alloy steels

Alloy steels are generally classified into two major categories, *low-alloy* steels and *high-alloy* steels. The main effects conferred by specific alloying elements are given in Table 16.4.

Low-alloy steels. A low-alloy steel generally contains up to 3 or 4 per cent of one or more alloying elements and is characterised by possessing similar microstructures to, and requiring similar heat treatments to, plain carbon steels. Frequently they are referred to as pearlitic alloy steels as the normalised structure contains the eutectoid pearlite. The presence of alloying elements provides enhanced properties such as increased strength without loss of toughness and increased hardenability. The applications of low-alloy steels are similar to those quoted in Table 16.2 for plain carbon steels of equivalent carbon content, for example, 0.4 per cent carbon is quoted as typical for the manufacture of shafts—an alloy steel containing 1 per cent nickel and 0.4 per cent carbon is an ideal shaft material with improved strength and toughness over a plain carbon steel and also an increase in the resistance to fatigue conferred by the presence of nickel.

A major category of low-alloy steels is *high strength low-alloy steel* (HSLA). This is a group of low-alloy steels produced with a very fine grain size. This is achieved by the addition of small controlled amounts of niobium, titanium or vanadium and these elements form carbonitride precipitates which inhibit austenite grain growth during hot working operations. This effect, coupled with close control of hot working temperatures and a controlled cooling rate, gives an extremely fine grain structure which, together with the effect of dispersed precipitate, results in a range of low-alloy steels for structural applications with tensile yield strengths between 350 and 560 MPa.

High-alloy steels. High-alloy steels are those that possess structures, and require heat treatments, that differ considerably from those of plain carbon steels. Their room temperature structures after normalising may be austenitic, martensitic or contain precipitated carbides. Generally, they contain more than 5 per cent of alloying element. A few examples of some high-alloy steels are given below.

(a) *High-speed tool steels.* Tungsten and chromium form very hard and stable carbides. Both elements also raise the critical temperatures and, also, cause an increase in softening temperatures. High carbon steels rich in these elements provide hard wearing metal-cutting tools, which retain their high hardness at temperatures up to 600°C. A widely used high-speed tool steel composition is 18/4/1 steel, containing 18 per cent of tungsten, 4 per cent of chromium, 1 per cent of vanadium and 0.8 per cent of carbon. In a steel of such a composition very hard and stable chromium carbides, vanadium carbide and complex iron-tungsten carbides are formed instead of the normal cementite. The annealed material, produced by very slow cooling from a 'soaking' temperature of about 900°C will possess a microstructure of ferrite and spheroidal carbide particles, with most of the alloy element content combined as carbide. The heat treatment that is needed to put the material into a hard and tough condition is heating at a very high temperature (1200–1300°C), to take a proportion of the alloying elements and carbides into solution, followed by oil quenching. The oil-quenched structure will comprise carbides, martensite and a high proportion of

Table 16.4 *Effects of alloying elements in steels*

Alloying element	General effects	Typical steels
Manganese	Increases the strength and hardness and forms a carbide; increases hardenability; lowers the critical temperature range, and when in sufficient quantity, produces an austenitic steel; always present in a steel to some extent because it is used as a deoxidiser	Pearlitic steels (up to 2 per cent Mn) with high hardenability used for shafts, gears, and connecting rods; 13 per cent Mn in Hadfield's steel, a tough austenitic steel
Silicon	Strengthens ferrite, raises the critical temperatures; has a strong graphitising tendency; always present to some extent because it is used, with manganese, as a deoxidiser	Silicon steel (0.07 per cent C; 4 per cent Si) used for transformer cores; used with chromium (3.5 per cent Si; 8 per cent Cr) for its high-temperature oxidation resistance in internal combustion engine valves
Chromium	Increases strength and hardness, forms hard and stable carbides; raises the critical temperatures; increases hardenability; amounts in excess of 12 per cent render steel stainless	1.0–1.5 per cent Cr in medium and high carbon steels for gears, axles, shafts, and springs, ball bearings and metal-working rolls; 12–30 per cent Cr in martensitic and ferritic stainless steels; also used in conjunction with nickel (see below)
Nickel	Marked strengthening effect, lowers the critical temperature range; increases hardenability; improves resistance to fatigue; strong graphite-forming tendency; stabilises austenite when in sufficient quantity	0.3–0.4 per cent C with up to 5 per cent Ni used for crankshafts and axles, and other parts subject to fatigue
Nickel and chromium	Frequently used together in the ratio Ni/Cr = 3/1 in pearlitic steels; good effects of each element are additive, each element counteracts disadvantages of the other; also used together for austenitic stainless steels	0.15 per cent C with Ni and Cr used for case carburising; 0.3 per cent C with Ni and Cr used for gears, shafts, axles and connecting rods; 18 per cent, or more, of chromium and 8 per cent, or more, of nickel give austenitic stainless steels
Tungsten	Forms hard and stable carbides; raises the critical temperature range, and tempering temperatures; hardened tungsten steels resist tempering up to 600°C	Major constituent in high-speed tool steels; also used in some permanent magnet steels
Molybdenum	Strong carbide-forming element, and also improves high-temperature creep resistance; reduces temper-brittleness in Ni–Cr steels	Not normally used alone; a constituent of high-speed tool steels, creep-resistant steels and up to 0.5 per cent Mo often added to pearlitic Ni–Cr steels to reduce temper-brittleness
Vanadium	Strong carbide-forming element; has a scavenging action and produces clean, inclusion-free steels	Not used on its own, but is added to high-speed steels, and to some pearlitic chromium steels
Titanium	Strong carbide-forming element	Not used on its own, but added as a carbide stabiliser to some austenitic stainless steels
Aluminium	Soluble in ferrite, also forms nitrides	Added to nitriding steels to restrict nitride formation to surface layers
Cobalt	Strengthens, but decreases hardenability	Used in 'Stellite' type alloys, magnet steels, and as a binder in cemented carbides
Niobium	Strong carbide former, increases creep resistance	Added for improved creep resistance and as a stabiliser in some austenitic stainless steels
Copper	Increases strength and corrosion resistance. >0.7% Cu permits precipitation hardening	Added to cast steels to improve fluidity, castability and strength. Used in corrosion resistant architectural steels
Lead	Insoluble in iron	Added to low-carbon steels to give free-machining properties

retained austenite. Quenching is followed by a double tempering process. The first-stage tempering at 300–400°C will allow most of the retained austenite to transform into martensite while the second tempering at 550–600°C relieves internal stresses and forms a structure of carbide particles in a toughened martensite matrix. This high-alloy content martensite does not soften appreciably until it is heated at temperatures in excess of 600°C making them usable as cutting tools at high cutting speeds.

(b) *Stainless steels.* When chromium is present in amounts in excess of 12 per cent, the steel becomes highly resistant to corrosion, owing to the protective film of chromium oxide that forms on the metal surface. Chromium also raises the α to γ transformation temperature of iron, and tends to stabilise ferrite in the structure. There are several types of stainless steel, and these are summarised below.

(i) Ferritic stainless steels contain between 12 and 25 per cent of chromium and less than 0.1 per cent of carbon. These steels are ferritic in structure at all temperatures up to their melting points. As austenite cannot be formed, it is impossible to form hard martensitic structures by quenching the steels from high temperature. This type of stainless steel cannot be heat treated, but may be strengthened by work hardening.

(ii) Martensitic stainless steels contain between 12 and 18 per cent of chromium, together with carbon contents ranging from 0.1 to 1.5 per cent. The presence of carbon restores the α to γ transition, and these steels can be hardened by quenching from the austenite range.

(iii) Austenitic steels contain both chromium and nickel. When nickel is present, the tendency of nickel to lower the critical temperatures over-rides the opposite effect of chromium, and the structure may become wholly austenitic. The austenitic stainless steels, which are also non-magnetic, are widely used in chemical engineering plant, for cutlery, and for architectural work. A notable alloy is 18/8 stainless steel, which contains 18 per cent of chromium, and 8 per cent of nickel. Carbon contents are kept below 0.15 per cent in the austenitic steels in order to minimise the formation of chromium carbides in the structure, as this would cause a reduction in corrosion resistance. Carbides may form in these steels if they are allowed to cool slowly from high temperature, or if they are reheated in the range 500–700°C. As this latter condition may apply in the heat-affected zones adjacent to welds, the type of corrosion failure that can occur, owing to the presence of carbide particles, is known as *weld decay*. Austenitic stainless steels that are required for welding contain small stabilising additions of titanium, or niobium, and these stabilisers prevent the intercrystalline corrosion, weld decay.

(c) *Maraging steels.* These are very high-strength materials that can be hardened to give tensile strengths of up to 1900 MPa. They contain 18 per cent nickel, 7 per cent cobalt and small amounts of other elements such as titanium. The carbon content is low, generally less than 0.05 per cent.

The heat treatments that are necessary to develop maximum properties are a solution treatment at 800–850°C, in order to produce a uniform austenite structure, followed by rapid quench. A martensitic structure is formed after quenching but this is relatively soft as the carbon content of the steel is very low. This is followed by precipitation treatment at about 450°C. A very finely dispersed precipitate of complex intermetallic phases occurs which causes great strengthening.

A major advantage of maraging steels is that after the solution treatment they are soft enough to be worked and machined with comparative ease. The precipitation treatment is at a fairly low temperature when distortion of machined parts is negligible. Although the basic material cost of a maraging steel is very high the final cost of a complex component made from a maraging steel may be less than if made from a more conventional and cheaper high-strength steel because of the much lower machining costs. One major use of maraging steels is for the manufacture of aircraft undercarriage components.

(d) *Hadfields manganese steel*. This is a high-alloy steel that contains 12–14 per cent of manganese, and 1 per cent of carbon. The high manganese content renders this steel composition austenitic at all temperatures. It is also, therefore, non-magnetic. This type of steel is unique in that any attempt to deform, or abrade, the material greatly increases the surface hardness. Because of the exceptionally high resistance of this type of steel to abrasion, coupled with the fact that the core of the material remains comparatively soft and tough, it is used for pneumatic drill bits, rock crusher jaws, excavator bucket teeth, and railway points and switches.

16.15 Cast irons

Cast irons are selected grades of pig iron, which are remelted and cast in sand moulds. The carbon content of cast irons is generally between 2 and 4 per cent. Although normal grades of cast irons are not particularly strong, and are quite brittle, they are widely used as engineering materials, because of their cheapness, ease of melting and casting, very good machineability and a high damping capacity. According to the phase diagram (Figure 16.2) the phases that should exist at ordinary temperatures are ferrite, cementite, and pearlite. Cementite, however, is only metastable, and most cast irons contain graphite in their structures. Cast irons may be classified as either *white* or *grey*. These terms arise from the appearance of a freshly fractured surface. An iron that contains graphite in its structure shows a fracture surface that is grey and dull in colour. If, however, the whole of the carbon content of an iron is in the combined form, as cementite, a fresh fracture surface appears bright and silvery, and this is termed a white iron structure. White irons are exceptionally hard, but brittle. White irons are not often used, except as the basis for the production of malleable irons, although portions of many sand castings are made to solidify white, by the insertion of chills in the sand mould, in order to produce hard and wear resistant surfaces in parts of the castings.

According to the phase diagram (Figure 16.2), the eutectic occurs at 4.3 per cent carbon, but the presence of other elements, notably silicon and phosphorus, will affect the position of the eutectic. The effects of these elements can be assessed by assigning carbon replacement values to them. The carbon equivalent value, C.E., is given by:

$$C.E. = \%C + 0.33\,(\%Si + \%P)$$

For example, a cast iron with a composition of 3.0% C, 2.8% Si and 1.1% P would have a carbon equivalent value of: C.E. = 3.0 + 0.33 (2.8 + 1.1) = 4.3%, the eutectic composition.

The structure of cast irons is affected by the following factors.

(a) The rate of solidification. Slow rates of solidification allow for graphite formation and castings made in sand moulds tend to solidify grey. More rapid solidification will tend to give white iron structures. Metal chills are sometimes inserted into parts of sand moulds in those areas where a high surface hardness is required.

(b) Carbon content. The higher the carbon content of the iron, the greater will be the tendency for it to solidify grey.

(c) The presence of other elements. Some elements promote the formation of graphite in an iron structure. Silicon and nickel have strong graphitising tendencies. Figure 16.11 illustrates the effect of silicon on the structure of a cast iron.

(d) The effect of heat treatment. The prolonged heating of a white iron will cause graphitisation to occur. This phenomenon is used as the basis for the production of malleable irons (refer to Section 16.16). Graphite is less dense than cementite and, if cementite decomposes into ferrite and graphite during service, this change will be accompanied by a reduction in the density of

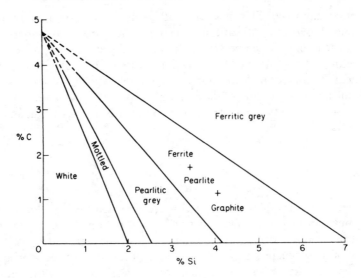

FIGURE 16.11   Effect of silicon and carbon content on the structure of sand-cast irons

the iron, and a corresponding increase in dimensions. This phenomenon is termed the *growth* of cast irons, and irons for high-temperature service must be in a fully graphitised state before being put into use.

When an iron solidifies from the molten state, the crystals of graphite usually form as irregularly shaped flakes, many of which may be quite large (see Figure 16.9(e)). The presence of graphite in this form gives the material a low value of fracture toughness. Cast iron containing flake graphite of this type is perfectly suitable for many low- and medium-duty applications but it is possible, by using special melting and/or melt treatment procedures, to produce high strength irons. These treatments are given below.

Lanz–Perlit iron. This is a high strength iron with a microstructure of fine graphite flakes and pearlite produced by casting an iron with controlled low carbon and silicon contents into a mould which has been preheated to about 400°C. This composition iron, if cast in the normal way, would solidify to give either a white or mottled structure but, with the slower solidification rate within a heated mould, it solidifies giving a structure of fine graphite flakes and pearlite, even in thin sections. The growth of graphite in thicker sections of the casting is restricted because of the low silicon content of the iron. This means that castings of variable section will have fairly uniform micro-structures. This type of cast iron is used for high quality castings, including automotive cylinder blocks.

Meehanite. This is the proprietary name for a high strength cast iron. It is an iron composition which, ordinarily, would solidify white but is treated by superheating the melt to about 1500°C and inoculating it with calcium silicide immediately before casting. This produces a grey iron structure with very small graphite flakes, resulting in improved mechanical properties.

S.G. iron. S.G. (spheroidal graphite) or nodular cast irons possess structures in which graphite is present in the form of roughly spherical-shaped nodules rather than as flakes (see Figure 16.9(f)). This type of structure is achieved by inoculating the melt, prior to casting, with small additions of magnesium and/or cerium. The amounts required are about 0.05% Mg or

0.01% Ce. The presence of spheroidal, as opposed to flake, graphite improves the strength and fracture toughness of the material. S.G. irons may be made in a wide range of compositions and the properties may be improved further by means of alloying additions.

Table 16.5 *Composition and properties of some cast irons*

Approximate composition	Tensile strength (MN/m^2)	Type and uses
3.2% C, 1.9% Si	250	Pearlite and graphite. Motor brake drums
3.25% C, 2.25% Si	220	Pearlite and graphite. Engine cylinder blocks
3.25% C, 2.25% Si, 0.35% P	185	Ferrite, pearlite and graphite. Light machine castings
3.25% C, 1.75% Si, 0.35% P	200	Ferrite, pearlite and graphite. Medium machine castings
3.25% C, 1.25% Si, 0.35% P	250	Pearlite and graphite. Heavy machine castings
3.6% C, 2.8% Si, 0.5% P	370	Wear resistant. Piston rings
3.6% C, 1.7% Si	540	Pearlitic S.G.
3.6% C, 2.2% Si	415	Ferritic S.G.
2.8% C, 0.9% Si	310	Blackheart malleable
3.3% C, 0.6% Si	340	Whiteheart malleable
2.9% C, 2.1% Si, 1.75% Ni, 0.8% Mo	450	Shock resistant. Crankshafts for petrol and diesel engines
2.9% C, 2.1% Si, 15% Ni, 2% Cr, 6% Cu	220	Ni-resist. Corrosion-resistant austenitic iron. Used in chemical plant
2.5% C, 5% Si	170	Silal. A growth-resistant iron for high-temperature service

16.16 Malleable irons

Malleable irons are produced by the heat treatment of certain white irons. There are two types of process used giving rise to *blackheart* and *whiteheart* irons, respectively. These names arise from the appearance of the fracture surface of a treated iron.

In the blackheart process, white iron castings are heated at a temperature of about 900°C for between 2 and 3 days, and then cooled very slowly, at about 3°C per hour. A neutral atmosphere is maintained in the furnace during the treatment. Cementite in the white iron structure breaks down into ferrite and roughly spherical aggregates of graphite. If the iron is cooled more rapidly from the treatment temperature, the resulting structure will consist of graphite in a matrix of ferrite and pearlite.

The whiteheart process involves packing the white iron castings into boxes with haematite ore, and heating to a temperature of 900°C for a period of 2–5 days. Because the casting is in contact with haematite, carbon is oxidised away from the surface, and the resulting structure is composed of ferrite at the edge of the casting and ferrite, pearlite, and some graphite nodules at the centre.

16.17 Alloy cast irons

Alloying elements may be added to cast irons in order to improve their properties, and alloy cast irons have been developed to give high-strength materials, hard and abrasion-resistant materials, corrosion resistant irons, and irons for high-temperature service. For the production of high-strength irons, additions of chromium, nickel, and molybdenum, are commonly made. The addition of about 5 per cent of nickel causes the formation of a martensitic structure at the surface of a sand casting. This very hard constituent makes this type of iron very suitable for the manufacture of metal working rolls.

The best irons for corrosion resistance are those containing 15–25 per cent of nickel. These irons are austenitic in structure, and non-magnetic. They usually contain some chromium and copper, in addition to the nickel content, to give some further improvement in corrosion resistance.

For prolonged service at elevated temperature it is important that the iron does not grow. In order to ensure that the structure of the iron is completely graphitic when cast, the carbon content is kept down to about 2 per cent, and the alloy contains about 5 per cent of silicon, or silicon and nickel. The presence of silicon also reduces the rate of oxide scale formation at high temperatures.

16.18 Questions

16.1 Microexamination of two plain carbon steels shows that steel 1 contains about 60 per cent ferrite and 40 per cent pearlite and steel 2 contains about 95 per cent pearlite and 5 per cent free cementite. Estimate the percentage carbon in each steel.

16.2 Refer to the Fe–C phase diagram, Figure 16.12. For each of two plain carbon steels, one containing 0.4 per cent C, the other containing 1.2 per cent C, estimate:

(a) the value of upper critical temperature,
(b) the proportion of austenite present at 724°C,
(c) the proportion of pearlite present after slow cooling to 15°C.

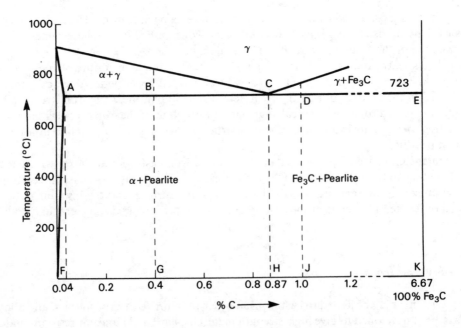

FIGURE 16.12

16.3 The graphs showing dimensional change with temperature for a hypoeutectoid steel during heating and cooling, both at 15°C per minute, are shown in Figure 16.13.

(a) Determine the equilibrium values of the critical temperatures for this steel and explain any differences from values expected from consideration of the iron–carbon phase diagram.
(b) Estimate the percentage carbon in the steel.

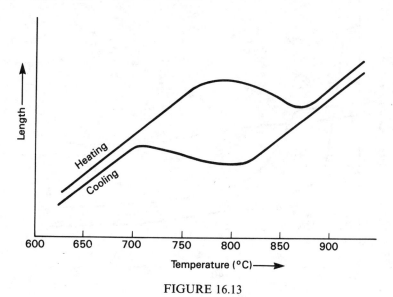

FIGURE 16.13

16.4 The T–T–T diagram for a steel is shown in Figure 16.14. Use the diagram to estimate:

(a) the critical cooling velocity,
(b) a cooling rate to give a coarse pearlitic structure,
(c) a cooling rate to give very fine pearlite with no bainite or martensite,
(d) the minimum holding time, after rapid quenching to 425°C, to give a fully bainitic structure.

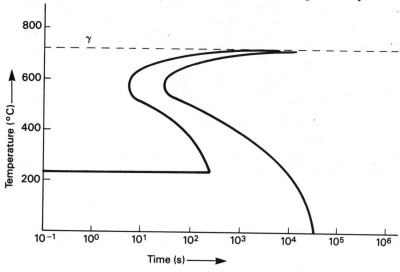

FIGURE 16.14

16.5 (a) Distinguish between hardness and hardenability. (b) Jominy hardenability curves for three steels are shown in Figure 16.15. (i) Which curve relates to the steel of greatest hardenability? (ii) Which curve relates to the steel of highest carbon content? (iii) Estimate the ruling section of steel B.

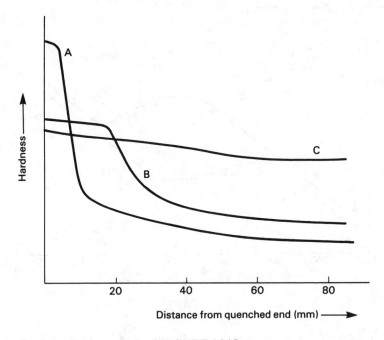

FIGURE 16.15

16.6 The compositions of several cast irons are given in the Table below.

	C %	Si %	Mn %	P %
1	3.5	1.2	0.8	0.1
2	3.3	2.2	0.5	0.3
3	3.6	1.8	0.5	0.8
4	3.0	5.0	0.3	0.1
5	3.3	0.6	0.5	0.1

(a) Determine the carbon equivalents of these irons.
(b) State, giving reasons, which iron would:
    (i) be most likely to solidify white when sand cast,
    (ii) be most suitable for prolonged service at high temperatures.

17

Thermoplastics

17.1 Introduction

The group of materials known as plastics can be sub-divided into *thermoplastic materials*, *elastomers* and *thermosetting materials*. Elastomers and thermosets are dealt with in Chapters 18 and 19 respectively. The plastics industry is based almost entirely on petrochemicals derived from petroleum and natural gas. The volume of plastics materials used is continually increasing with a high rate of growth. This has been of the order of 7 per cent per annum for a long period and looks set to continue for some time to come. Also, in recent decades the industry has continued to develop and introduce new *engineering plastic materials*.

In many instances, plastics materials have replaced metals, but it should not be thought that plastics materials will eventually replace all metals. There are many areas of application in which the selection of a metal or alloy is the only feasible solution to a design requirement. However, in many instances a plastic material can provide a better solution than a metallic material. A simple example is the domestic washing up bowl. A bowl made from polyethylene is far more suitable for its task than its predecessor, the bowl formed from low-carbon steel sheet with a vitreous enamel finish.

The major increase in the use of plastics materials has been due to their many useful properties, coupled with comparatively low cost, which more than offset their disadvantages. Some of the useful properties are their low densities, high resistance to chemical attack, good thermal and electrical insulation properties, and their ease of fabrication into a wide variety of both simple and complex shapes. The main disadvantages of plastics are the low strength and elastic modulus values, as compared with metals, the comparatively low softening and thermal degradation temperatures and, also, their comparatively high thermal expansion coefficients. However, although the innate strength and moduli values are low they can often be improved to some extent by using fillers as reinforcement.

The plastics materials as supplied to a component manufacturer in the form of, say a granular moulding compound, is not a pure polymer but contains, in addition, quantities of fillers and reinforcing materials, dyes and pigments, stabilisers, antistatic agents, flame retardants, plasticisers, and so on. The exact nature of and the proportions of additives in, say, a polypropylene moulding compound may vary considerably from one supplier to another. The apparently wide variation in some of the properties quoted for plastics materials in this chapter is a reflection of the compositional variations that can exist in polymer moulding compounds.

Details of the properties and applications of some of the more important thermoplastic materials are given in the subsequent sections of this chapter, while those of elastomers and thermosets are given in Chapters 18 and 19. Composite materials based on plastics materials are discussed in Chapter 21.

17.2 Polyethylene (PE)

Polyethylene, or as it is more commonly known, polythene, is made by the polymerisation of ethene (ethylene) $(CH_2 = CH_2)$. The chemical notation for the polymerisation of ethylene indicates a simple reaction:

$$n(CH_2 = CH_2) \rightarrow (-CH_2-CH_2-)_n$$

However, the properties and applications of this polymer vary over wide ranges depending on the molecular weight, the method of manufacture, and differences in structure and density. There are several varieties of polyethylene made, these being *high density* (HDPE), *low density* (LDPE), *linear low density* (LLDPE), and *ultra-high molecular weight* (UHMWPE).

The raw polymer is compounded with various additives to produce a material suitable for moulding. The principal additions that may be made are: (1) up to 2 per cent of carbon black to improve the stability of the material when exposed to sunlight; (2) up to 10 per cent butyl rubber to prevent in-service cracking in blow moulded containers for oils and detergents; (3) pigments to produce the desired colour.

The various types of polyethylene have excellent resistance to most solvents and chemicals and are tough and flexible over a wide range of temperature. With the exception of UHMWPE, they have low softening temperatures and can be readily moulded and formed into shapes. They are also non-toxic and possess good electrical insulation properties.

High density polyethylene (HDPE). HDPE is made by a low pressure polymerisation process carried out in solution. The result is a fully linear polymer with a high degree of crystallinity and a density of 960–970 kg/m³. The material has a good impact resistance at low temperatures, excellent chemical resistance but is susceptible to stress cracking and has a low resistance to ultraviolet light. There is also a high mould shrinkage. HDPE is used for the manufacture of bowls, buckets, bottles, toys, refrigerator parts, pipes and pipe fittings for water, and packaging film.

Low density polyethylene (LDPE). LDPE is made by polymerising in the vapour phase at high temperatures and pressures. The polymer molecules are not completely linear but show some branching. This reduces the crystallinity and gives a material with a density of 915–930 kg/m³. This is a low cost material with good low temperature impact properties, good chemical resistance and excellent electrical properties but it has a very low strength and stiffness and is susceptible to stress cracking. Its uses include packaging film, bowls and other kitchen ware, squeeze bottles, toys, seals and gaskets, and wire and cable insulation.

Linear low density polyethylene (LLDPE). LLDPE is made by a low pressure vapour phase process and has a regular structure with a number of short branch chains. It possesses higher tensile and impact strengths, and a greater ductility than LDPE and has a higher puncture resistance. It is possible to blow films of thinner gauge than with LDPE but the temperature range at which LLDPE film can be heat sealed is narrower than for LDPE. The applications for the material are similar to those for LDPE.

Ultra-high molecular weight polyethylene (UHMWPE). UHMWPE contains extremely long unbranched chains—the degree of polymerisation is of the order of 2×10^5. This material, which is not melt processable, has an outstanding toughness, excellent chemical resistance and good resistance to cutting and abrasive wear. Some of the applications for the material are as chopping boards, gears, bearings and other machine parts.

Crosslinked polyethylene (XLPE). Polyethylene can be crosslinked to give a material with higher strength and stiffness at elevated temperatures. The crosslinking may be brought about either by irradiation (see Section 4.5) or by chemical reaction with peroxides. The former method requires the use of expensive equipment while the latter method may cause a deterioration in

Table 17.1 *Properties of polyethylene*

Property	HDPE	LDPE	HDPE (30% glass filled)
Density (kg/m^3)	960–970	915–930	1170
Tensile strength (MPa)	22–38	1–16	70
E (GPa)	0.42–1.4	0.12–0.3	7
% elongation	20–1300	90–650	2–3
Impact strength (J/m^2)	80–100	no break	5.5
	An easily moulded thermoplastic that does not soften in boiling water	An easily moulded thermoplastic that softens in boiling water. It is flexible down to $-75°C$	

electrical properties. Applications of XLPE include piping for domestic water and heating systems, water tanks and film for shink-wrapping.

Polyethylene foam. A thermoformable polyethylene foam with excellent barrier properties and chemical resistance is available. It possesses better resilience than polystyrene foams but is more expensive. It finds use as a packaging material for delicate equipment and as acoustic insulation.

Polyethylene can also be filled with glass fibre to give materials of high stiffness and increased strength.

17.3 Ethylene copolymers

A number of useful copolymers based on ethylene have been developed. Ethylene and propylene can be copolymerised in various proportions to produce a range of different products. One of these, *polyallomer*, is discussed in Section 17.4. Other ethylene–propylene copolymers are ethylene–propylene thermoplastic elastomer, the elastomers EPM and EPDM, this last being a terpolymer between ethylene, propylene and a diene (see Section 18.4).

Ethylene–vinyl acetate copolymers (EVA) are block copolymers which have a better low temperature flexibility and tear resistance than LDPE but their chemical resistance is poorer than that material.

They are alternative materials to PPVC and do not have to be compounded with additional plasticising agents, as is the case with PVC. Major applications for EVA copolymers are for the manufacture of ice-cube trays, flexible toys, disposable gloves, vacuum cleaner hose and flexible pipe connectors. EVA can be produced as film and this, which has a high clarity, is used for shrink-wrapping.

Ethylene may be copolymerised with some fluorocompounds (see Section 17.6).

Ionomer. Ionomer is the name given to a copolymer of ethylene with an organic acid, for example methacrylic acid, together with metallic ions such as Na^+, K^+, Mg^{2+} or Zn^{2+}. Ionomer is a soft flexible transparent material with a better resistance to stress-cracking than LDPE. Ionic cross-links exist between polymer molecules at ordinary temperatures and this makes the material more rigid than LDPE. These ionic links break when the material is heated, thus allowing the polymer to be moulded with the same degree of ease as LDPE. The ionic cross-links

Table 17.2 *Properties of some ethylene copolymers*

Property	EVA (7.5% vinyl acetate)	EVA (15% vinyl acetate)	Ionomer
Density (kg/m³)	920	970	930–960
Tensile strength (MPa)	16	17	20–30
E (GPa)	0.16	0.045	0.07–0.38
% elongation	700	700	320–520

reform as the material cools. Ionomer is used for making blow moulded bottles to hold vegetable oils and cosmetics, for tubing to convey liquid foods and beer, and to make disposable hypodermic syringes.

17.4 Polypropylene (PP)

The monomer propene (propylene), $CH_2 = CH-CH_3$, is related to ethene and, like ethene, possesses a double covalent bond that may split to allow addition polymerisation to occur.

$$n(CH_2\!\!=\!\!CH\!-\!CH_3) \rightarrow -CH_2-\underset{CH_3}{CH}-CH_2-\underset{CH_3}{CH}-CH_2-\underset{CH_3}{CH}-$$

Polyproplyene is a highly crystalline olefinic polymer and the presence of CH_3 groups attached to the linear chain molecules gives polypropylene a greater strength and stiffness than HDPE although its density is similar to that of LDPE. It has a higher softening point than HDPE and can be used to make bottles that may be boiled or steam sterilised without softening. It has a good resistance to most forms of chemical attack and excellent fatigue resistance. One very useful property of polypropylene is its excellent fatigue resistance and its ability to be bent repeatedly without fear of cracking. An example of how this property has been utilised by designers is the integrally moulded hinge. Many small boxes and components such as car accelerator pedals are produced as one-piece mouldings in polypropylene with an integrally moulded hinge.

Polypropylene is brittle at low temperatures and, to overcome this, high impact grades are available. High impact polypropylene may be made by mechanically blending the polymer with small amounts of EPM or EPDM elastomers. A toughened grade of polypropylene, which is a block copolymer of propene with between 2 and 10 per cent of ethene, is also available. This block copolymer is also known as *pollyallomer*. The brittleness temperature of polyallomer is reduced to about −40°C.

There is a high mould shrinkage with this material and it has poor resistance to ultraviolet light. The main applications are for components requiring a reasonable strength and rigidity and include buckets, bowls, battery cases and bottle crates, toys, bottles and bottle caps, automotive parts and components for consumer durables, including vacuum cleaner bodies, and washing machine drums. Pollyallomer is used for the manufacture of containers with integrally moulded hinges, in particular for the holding of frozen foods. In fibre form, polypropylene is used for carpeting and ropes, and in film form for packaging. Other grades available include mineral filled and fibre reinforced grades, fire retardant grades and structural foams.

Table 17.3 *Properties of polypropylene*

Property	PP	High impact PP	Polyallomer	PP (30% glass filled)
Density (kg/m³)	900–910	900–910	900–910	1120–1130
Tensile strength (MPa)	27–40	19–35	21–28	47–103
E (GPa)	0.5–1.9	0.8–1.3	0.5–0.8	4–7
% elongation	30–200 +	30–200 +	400–500	2–4
Impact strength (J/m²)	2–12	8–64	30–180	3–16

17.5 Polyvinyl chloride (PVC)

The monomer vinyl chloride, $CH_2 = CHCl$, is a derivative of ethylene and can be readily polymerised by an addition process to give the linear polymer PVC, $(-CH_2-CHCl-)_n$ a relatively cheap thermoplastic material with a good resistance to acids and alkalis. The presence of chlorine atoms causes the molecular chains to be polarised and there are, in consequence, polar attractive forces between adjacent molecules. This gives a material which is quite hard and rigid at ordinary temperatures.

About one-quarter of the PVC produced is used in the unplasticised form (UPVC), while the remainder is produced as plasticized PVC (PPVC).

UPVC. UPVC is a strong, rigid and tough material with good resistance to ultraviolet light. The main applications for the material include many products for the building industry, pipes and pipe fittings, guttering, cladding panels, and window frames, and it is used also for the manufacture of bottles and gramophone records.

PPVC. The raw polymer is compounded with a semisolvent or an oil as a plasticising agent to make PPVC, a soft, flexible, rubbery material. The amount of plasticiser may range from 5–50% of the final weight of the moulding material, the amount and type determining the final properties. Also, the properties of the material may vary with time due to migration of plasticiser. The chemical resistance of PPVC is less than that for UPVC. The main uses for PPVC include seals, gaskets, simulated leathercloth for clothing, furniture and luggage, floor and wall coverings, electrical wire and cable insulation, hoses, bottles and sachets. Inert fillers may be added to PPVC, both to cheapen and also to improve the properties for some applications, for example in flexible floor tiles and floor coverings.

PVC has a tendency to decompose with the liberation of hydrochloric acid gas when it is heated or exposed to sunlight. To prevent this occurring a small amount of a lead salt, lead silicate or lead stearate, is blended into the polymer at the compounding stage. The lead salt acts as a *stabiliser*. Other additives that may be compounded with PVC are pigments to produce the desired colour, plasticisers and fillers. PPVC may suffer from microbial attack unless fungicides and bactericides are included in the formulation.

Copolymers. Vinyl chloride may be copolymerised with other monomers, including vinyl acetate and acrylonitrile. In copolymers with either of these two monomers, no external plasticising agent need be compounded with the material as both vinyl acetate and acrylonitrile act as *internal plasticisers.*

There are two other polymers that are related to PVC. These are CPVC (chlorinated PVC), a polymer of dichloroethylene, $(-CHCl-CHCl-)_n$ and PVDC (polyvinylidene chloride), with the formula $(-CCl_2-CH_2-)_n$.

CPVC has a softening point of some 10–15°C higher than PVC and is used for the manufacture of piping to carry hot water and hot aqueous solutions.

PVDC is a polymer with extremely low water vapour permeability and a good resistance to organic solvents. It is a pliable transparent polymer of low strength, which tends to some loss of mechanical properties on heat ageing. The principal commercial uses are for pipes, seals and gaskets in chemical engineering plant and as coatings for other polymers used in the packaging industry.

The properties of PVC and related compounds are given in Table 17.4.

17.6 Polytetrafluoroethylene (PTFE)

PTFE (polytetrafluoroethylene), formula $(-CF_2-CF_2-)_n$ is the most widely used fluorocarbon polymer. It is a highly crystalline material that does not soften appreciably on heating. It loses its crystallinity above 327°C but does not flow significantly under stress even at these high temperatures. Generally, components in PTFE are made by hot pressing and sintering the polymer powder, in a manner similar to that used for ceramic and powder metal parts.

The carbon–fluorine covalent bond is extremely strong and the presence of these bonds gives a material of extreme stability, PTFE is almost completely chemically inert, will not stick to most materials and possesses an extremely low coefficient of friction. Its main uses are as chemically resistant coatings, seals and gaskets in chemical plant, non-stick coatings and low friction bearing bushes. Certain properties can be improved by using a filler, for example, chopped glass fibre filled PTFE has greatly increased creep resistance, while the use of a bronze powder or bronze-graphite mix as a filler gives material suitable for bearing applications.

The main properties of PTFE are:

Density	$2.1 - 2.25 \times 10^3 \text{ kg/m}^3$
Tensile strength	17–28 MPa
E	350–620 MPa

There are other fluorocarbon polymers, these being PVDF, PVF and PFA. PVDF (polyvinylidene fluoride) $(-CH_2-CF_2-)_n$ has the highest tensile strength of all the melt processable fluorocarbon materials, together with good resistance to abrasion and chemical attack. Its main applications are for pipes, pipe fittings, bearings and cable insulation for the chemical processing industry, and for chemical laboratory apparatus. PVF (polyvinyl fluoride) $(-CH_2-CHF-)_n$ is a crystalline polymer with an excellent resistance to weathering, staining and chemical attack. It is very slow burning and has low permeability to vapours. Its low thermal stability normally precludes injection moulding and the material is usually only available as film. It is used mainly as a coating material for metals, glazing panels and as a film covering in aircraft interiors to reduce flammability. PFA (perfluoroalkoxyethylene) is a melt processable thermoplastic material with the highest temperature resistance of any of the melt processable fluoroplastics and a resistance to chemical attack similar to that of PTFE.

$$(-CF_2-CF-)_n$$
$$|$$
$$O$$
$$|$$
$$C_x F_{2x+1}$$

Table 17.4 *Properties of polyvinyl chloride and related materials*

Property	UPVC	PPVC (low plasticiser content)	PPVC (high plasticiser content)	CPVC	PVDC	Glass-filled (20 per cent) PVC
Density (kg/m³)	1400–1540	1300–1350	1200–1550	1380–1580	1650–1750	1580
Tensile strength (MPa)	24–62	28–42	7–56	35–62	21–34	96
E (GPa)	2.4–4.1	0.35–0.4	0.003–0.02	2.5–3.2	0.3–0.5	7.6
Percentage elongation	2–40	200–450	200–450	4.5–65	up to 250	3
Impact strength (J/m²)	2–100	—	—	5–30	2–5	8
	Tough and horny	Tough but pliable	Soft, flexible and rubbery	Higher softening point than PVC. Can be used up to 100°C	More resistant to organic solvents than PVC	

It is a material with a high toughness which possesses a similar stiffness and strength to PTFE at room temperatures and retains its stiffness and strength at elevated temperatures. It is, however, more expensive than PTFE. Its main applications include sheathing for heater cables and chemically resistant linings for pumps and pipes that also require a high temperature resistance.

Tetrafluoroethylene can be copolymerised with ethylene to produce the copolymer ETFE. This material possesses a very high impact strength and, for a fluorocarbon polymer, a high stiffness. It has a poorer chemical resistance than PTFE and is very expensive. The main applications for it include chemical apparatus, components for pumps and valves and as resistant linings. ECTFE is another fluoro-copolymer, being a copolymer of ethylene and chlorotrifluoro-ethylene, $(CFCl = CF_2)$. The material is very expensive and its main applications include chemically resistant linings, pump and valve components, and tubing for optical fibres.

17.7 Polystyrene (PS)

Polystyrene is a low cost, brittle, glassy ($T_g = 100°C$), transparent polymer with the formula:

$$(-CH_2-CH-)_n$$
$$|$$
$$C_6H_5$$

It is easy to process and has good dimensional stability. It has a poor chemical resistance and is susceptible to degradation if exposed to ultraviolet light. Unmodified polystyrene is used for toys, food boxes, light diffusers and components for domestic appliances.

High impact polystyrene is a hybrid material composed of small particles of an elastomer (BR or SBR) dispersed throughout a polystyrene matrix with some graft polymerization having taken place at the boundaries between the two phases. This is a much tougher material than unmodified polystyrene and again is used for toys, mouldings for domestic and office appliances, boxes and cases.

Much polystyrene is used in expanded form as a solid foam. The foam is available in a rigid massive form, which can be cut to size and shape or as 'sandwich' sheets for decorative shapes and mouldings. Expandable beads and shapes are available also, and these may be expanded in place or pre-expanded and moulded. Styrene polymer compounded with 6% *n*-pentane, a gas-releasing chemical, is one of the most common expandable foam polymer grades for in-place expansion. The polymer beads are heated to about 120°C in a jacketed mould, at which temperature the polymer softens and the expanding agent decomposes. Major applications of moulded, expanded foam is for food trays, cups and food packaging. In pre-expansion the beads are heated in a steam bath. This causes the beads to expand to about 30 times their previous size. These beads are broken apart and can be used for insulation and packaging.

Styrene is also used as the basis of some addition copolymers. The copolymer of styrene and butadiene (SBR) is an elastomer and is discussed in Chapter 18. Other copolymers are ABS, a terpolymer of acrylonitrile, butadiene and styrene, and SAN (styrene–acrylonitrile).

ABS has good mechanical properties, outstanding impact strength, and is dimensionally stable over a variety of conditions. It is also resistant to many acids, alkalis and many petroleum-based solvents. It is readily mouldable and can be calendered into sheet for vacuum forming. It has many applications for products such as refrigerator linings, vacuum cleaner bodies, food-mixers, mouldings for telephones, typewriters and business machines, automotive parts, safety helmets, bathroom fittings, including shower pedestals. It is also used in furniture manufacture and some small boat hulls are made by vacuum forming large sheets.

SAN, which generally contains some 20–30 per cent of acrylonitrile, possesses a good combination of strength, toughness, rigidity and transparency, with a better chemical resistance than polystyrene. Applications for the material include cups, trays, containers, toothbrush handles, battery cases, dials, knobs, switches and lenses.

All the styrene based materials may be glass-filled for added strength and rigidity. Table 17.5 gives the properties of some styrene based materials.

17.8 Acrylic materials

Acrylic materials are thermoplastic polymers based on the polymerisation of esters of acrylic acid and/or methacrylic acid. The most widely used acrylic polymers are PMMA (polymethyl methacrylate) and PAN (polyacrylonitrile). Other polymers in the group are based on ethyl, propyl, or butyl acrylate or methacrylate.

PMMA is produced by the addition polymerisation of methyl methacrylate according to the following reaction and is commonly known as *perspex* or *plexiglass*.

$$n \left(\begin{array}{c} CH_3 \\ | \\ CH_2{=}C \\ | \\ C{-}O{-}CH_3 \\ \| \\ O \end{array} \right) \xrightarrow[\text{addition to}]{} \left(\begin{array}{c} CH_3 \\ | \\ {-}{-}{-}{-}{-}CH_2{-}C{-}{-}{-}{-}{-}{-} \\ | \\ C{-}O{-}CH_3 \\ \| \\ O \end{array} \right)_n$$

The polymer is in the amorphous state at ordinary temperatures and is well below its T_g of 110°C. PMMA is a hard, rigid and high impact strength thermoplastic that is highly transparent to light and can be readily formed by most of the forming techniques used for thermoplastics. It is resistant to most household chemicals but is attacked by petrol, acetone and cleaning fluids. It is one of the most widely employed optical materials, because, in cast form, it has a higher transparency than many types of optical glass. The material can be readily coloured and its colour and decorative properties are excellent. Moulded components are often decorated by hot stamping, spray painting and vacuum metallisation. Although the material is classed as slow-burning, acrylic smoke is a hazard. The applications are numerous and include aircraft glazing, outdoor signs, baths and shower units, automotive lenses, light covers and instrument and architectural panels. The principal properties of PMMA are:

Density	1180 kg/m³
Tensile strength	50–75 MPa
E	2.7–3.5 GPa
% elongation at break	5–8

Polyacrylonitrile (PAN) is an acrylic polymer made by the addition polymerisation of acrylonitrile: $n(CH_2 = CHCN) \rightarrow (-CH_2-CHCN-)_n$. Polyacrylonitrile is remarkably stable and is resistant to acids and greases. It is used mainly for synthetic fibre manufacture. As its density is considerably lower than that of cotton or wool, it is used as a fibre in the manufacture of attractive, but lightweight, garments. Some proprietary names for acrylonitrile fibre are *Acrilan*, *Courtelle* and *Orlon*. It is also a precursor for carbon fibre manufacture, which gives a higher yield than rayon. Filaments of the polymer are highly orientated and decompose without melting when heated to the carbonisation temperature.

Some polyacrylates are elastomers (see Section 18.4).

Table 17.5   *Properties of polystyrene, ABS and SAN*

Property	Polystyrene	Toughened polystyrene	Glass-filled (25 per cent) polystyrene	ABS	Glass-filled (30 per cent) ABS	SAN	Glass-filled (25 per cent)
Density (kg/m³)	1040–1100	980–1100	1200–1330	1040–1700	1230–1360	1070–1100	1200–1460
Tensile strength (MPa)	35–84	12–62	63–105	17–62	60–133	63–84	80–140
E (GPa)	2.8–3.5	0.9–3.5	5.9–9	0.9–2.8	4.2–7	2.8–3.8	2.8–9.8
Percentage elongation	1–4.5	7–60	0.75–1.3	10–140	2.5–3.0	1.5–3.7	1–4
Impact strength (J/m²)	1.3–3.4	2.7–40	2–22	17.5	5–12	1.7–2.5	2–20
	An easily moulded thermoplastic with good resistance to most foods and household acids but attacked by petrol and cleaning fluids			Mouldable thermoplastic resistant to acids and alkalis and to many petroleum solvents			

17.9 Polyamides (nylons) (PA)

The polyamides, or *nylons*, are the products of condensation reactions between an amine and an organic acid (or the self condensation of an amino acid) and they possess amide groups as an integral part of a linear chain. This family of polymers was developed as a synthetic replacement for natural silk in the late 1930s. The polyamides resemble linear polyesters in that they are long carbon chains with recurring functional groups in the chain:

$$(-N-R-C-N-R'-C-)_n$$

The original reactants were hexamethylene diamine and adipic acid. As both of these reactants contain six carbon atoms, the product is termed PA 6.6.

$$\underset{\text{Hexamethylene diamine}}{N-(CH_2)_6-N} \quad + \quad HO \mid OC-(CH_2)_4-COOH$$

Adipic acid

The nomenclature of the polyamides is based on the numbers of carbon atoms in the monomer molecules. In the example above, both monomers have six carbon atoms. PA 6 is made from *caprolactam*, or its parent compound *ω–amino caproic acid*, both compounds with 6 carbon atoms.

$$\underset{\text{(caprolactam)}}{N-(CH_2)_5-CO} OH \quad + \quad N-(CH_2)_5-COOH$$

$$\rightarrow \quad H_2O \; + \; \ldots(CH_2)_5-CO-NH-(CH_2)_5\ldots$$

PA 6.10 is made from hexamethylene diamine (6 carbon atoms) and sebacic acid (10 carbon atoms), PA 11 from 11 amino-undecanoic acid (11 carbon atoms), and so on.

The polyamides, as a group, are very strong and tough and possess good abrasion resistance. They are flexible and have high impact strengths. They are thermoplastic but possess high softening temperatures, in excess of 200°C and so moulding is difficult. Much PA material is produced as fine fibre for conversion into fine fabrics and rope but they are also good moulding compounds. In general they possess a good resistance to most solvents and chemicals but are affected by phenols. They tend to absorb water, with a consequent reduction in strength, but the amount of absorption is less with some types than others. Fillers and property modifiers may be used and very many grades are available with mineral fillers, chopped carbon or glass fibre reinforcement, with solid phase lubricants such as molybdenum disulphide or PTFE, and many others. Some of the characteristics of the main PA polymers are given below.

PA 6. This polymer has the highest rate of water absorption and the highest equilibrium water content of all the polyamides, but it is easier to process than PA 6.6 and is also castable and may be reaction injection moulded (RIM). The material has a good fatigue resistance but a lower strength and stiffness than PA 6.6. The stiffness may be improved by addition of glass fibre. The range of applications for PA 6 is generally similar to that for PA 6.6, namely as gears, cams, bearings, nuts and bolts, electrical connectors, coil formers, car fuel tanks and kitchen utensils. The choice between PA 6 and PA 6.6 is often made on the basis of availability and cost or a familiarity with the product rather than on analysis of all the properties.

PA 6.6. PA 6.6 has a good abrasion resistance, better than PA 6, and is the strongest and stiffest aliphatic polyamide. It possesses a good resistance to fatigue and has a better low temperature toughness than either PA 6, PBT or acetal. It is relatively difficult to process as it has an exceptionally low melt viscosity. It also shows high mould and post-moulding shrinkage. It has a high water absorption (8% saturated). The main applications are as gears, cams, bearings, nuts and bolts, wheels, power tool casings and many automotive components, including rotationally moulded fuel tanks, radiator tops, timing chain covers and fan blades.

PA 6.10. This material has a lower brittle temperature and a lower water absorption (4% saturated) than either PA 6 or PA 6.6 and is stronger than PA 11, PA 12 or PA 6.12. This material shows a high mould shrinkage and is more expensive than PA 6.6. Its main uses include zip fasteners, electrical insulators and filaments for brushes.

PA 6.12. PA 6.12 has a similar water absorption (4% saturated) and brittle temperature to PA 6.10 but has a better heat resistance than that material. It possesses a lower mould shrinkage than PA 6, PA 6.6 or PA 6.10. It is more expensive than PA 6 or PA 6.6 and its uses include zip fasteners, mechanical components, filaments and abrasion resistant coatings.

PA 11. This material has a low water absorption (2.5% saturated) and a better resistance to ultraviolet light than other polyamides. It is expensive and its applications include compressed air, hydraulic and petrol hosing, gears, cams and tool handles.

PA 12. This material is very expensive but has the lowest water absorption of any polyamide (2% saturated). Its applications include precision engineering components and components requiring good low temperature toughness. It is used as a sports shoe sole material also.

PA (RIM). RIM (reaction injection moulded) polyamide, a proprietary name for which is Nyrim, shows a better stiffness than RIM polyurethanes. Although more difficult to mould because of the higher temperatures required and having some dimensional instability due to moisture absorption, this material is used for large structural parts, such as automotive panels and housings for business machines, because of its combined properties of stiffness and toughness.

The properties of some polyamides are given in Table 17.6.

17.10 Polycarbonate (PC)

Polycarbonate is a linear heterochain polymer made from the condensation of bisphenol A and carbonic acid having the structural formula:

Polycarbonate is a tough, transparent material with a high impact resistance at temperatures down to $-40°C$ and can also be used continuously up to $115°C$. It does, however, have low fatigue and wear resistance, is attacked by some organic solvents and is susceptible to stress cracking. It is, though, the second engineering thermoplastic, after polyamides, in terms of

Table 17.6 *Properties of polyamides*

Property	PA 6.6 dry	PA 6.6 50% RH	PA 6.6 30% glass filled dry	PA 6 dry	PA 6 50% RH	PA 6.10 dry	PA 6.10 50% RH	PA 6.12 dry	PA 6.12 50% RH	PA11 dry	PA12 dry
Tensile strength (MPa)	70	58	117–200	65–86	24–53	50–70	48	60	51	55	49–65
E (GPa)	3.2	2.6	9.6	2.6–3.0	1.4	1.9	1.1	2.0	1.2	1.3	1.1–1.4
% elongation at break	60	300–540	3	20	180–300	85	300	150	340	100–300	120–350
Impact strength (J/m²)	5	10	4.5–11	5.3	16	3.2	8.5	5.3	7.5	7–11	6.4–27.4
Density (kg/m³)	1130–1150		1340–1420	1120–1140		1070–1090		1060–1080		1040	1010

tonnage produced, being used for such applications as safety helmets, shields and goggles, lenses, glazing, lighting fittings, instrument casings and machine housings, and in electrical switchgear. It is used also for sterilisable medical components, kitchen ware and, in the form of laminated sheet, for bullet-proof glazing. Other grades of polycarbonate available include glass and carbon fibre reinforced materials, PTFE lubricated polymer, structural foams and ultraviolet stabilised and fire retardant materials. Polycarbonate is also a component in hybrid polymers, the main examples being PC/ABS and PC/PBT.

The properties of unfilled polycarbonate are:

Tensile strength	59–70 MPa
E	2.2–2.4 GPa
% elongation at break	80–120
Impact strength	65–85 J/m^2

17.11 Acetal (POM)

Acetal, also known as *polyacetal* and *polyoxymethylene* (POM), is based on the polymerisation of formaldehyde and has the structural formula: $(-CH_2-O-)_n$. Acetals have good strength and toughness, coupled with a low coefficient of friction. They also retain their properties at temperatures up to 120°C. Known also by the proprietary name *Delrin*, the materials' main applications are as a substitute for metals in the manufacture of gears, cams, bearing bushes, levers, shafts, fans, pump parts and water taps. The properties are:

Density	1370–1430 kg/m^3
Tensile strength	62–70 MPa
E	3.5 GPa
% elongation at break	25–60
Impact strength	7–12 J/m^2

17.12 Saturated polyesters

An *ester* is the product of the condensation reaction between an organic acid and an alcohol. PET (polyethylene teraphthalate), also known as PETP, is a linear polyester made by the condensation polymerization of ethylene glycol, and teraphthalic acid.

PET is a high strength, high stiffness thermoplastic produced as a fibre, as transparent film and as a moulding material. Moulding grades may be reinforced with glass fibre. Fibre is extruded from the liquid state in a similar manner to polyamide fibre and is drawn at a temperature above its T_g to give maximum molecular orientation. One proprietary name for PET fibre is *terylene*. PET suffers from a high mould shrinkage and a limited resistance to hydrolysis but possesses good electrical and thermal properties and can operate continuously at temperatures of up to 120°C. It is used for electrical parts such as transformer coil bobbins and components for tuners and fire alarms, and the manufacture of blow-moulded bottles for carbonated and alcoholic drinks, perfumes and drugs. In film form it is used for packaging, photographic and radiographic film, audio/video tapes, capacitor film and drawing office transparencies.

PBT (polybutylene teraphthalate), with a structural formula:

is a thermoplastic with a good combination of stiffness and toughness. It can withstand continuous service at 120°C and possesses good arcing and tracking resistance. However, its high mould shrinkage and poor resistance to hydrolysis are problems. The main applications for the material are for electrical connectors, switches, bobbins, integrated circuit carriers, light fittings and reflectors, parts for domestic goods, such as domestic iron handles, power-tool casings, gears, and for under-bonnet and exterior automotive components. Glass fibre reinforced and fire retardant grades of PBT are also available.

PCDT (poly (1.4) cyclohexylene dimethylene teraphthalate), with the formula:

is more expensive than PET but possesses higher temperature stability and it has a superior resistance to strong alkalis. It is used for similar applications to PET.

The properties of PET and PBT are given in Table 17.7.

The only thermoplastic polyesters are of the saturated type. Unsaturated polyesters form the basis of some useful thermosetting materials and are dealt with in Chapter 19.

Table 17.7 *Properties of PET and PBT*

Property	PET	PBT
Density (kg/m³)	1320–1340	1300
Tensile strength (MPa)	55	52
E (GPa)	2.2–2.5	2.6–2.7
% elongation	300	200–250
Impact strength (J/m²)	2	2.5

17.13 Cellulosics

Cellulose is a naturally occurring high polymer, formed by the photosynthesis of glucose, and it is the chief constituent of most forms of plant life, being one of the most widely distributed chemical materials available. It forms the framework of cell walls in trees and most green plants. The composition of wood comprises about 50% cellulose although the exact amount varies from species to species. The natural fibre cotton is about 98% cellulose. The most important sources for chemical processing are cotton, linen and wood pulp. The structure of cellulose is shown in Section 21.2.

Cellulose, generally in the form of cotton linters, can be treated to produce a range of materials, cellulose nitrate (CN), cellulose acetate (CA), cellulose acetate butyrate (CAB), cellulose propionate (CP), ethyl cellulose (EC), viscose and viscose rayon.

The first synthetic resin derivative of cellulose was *celluloid* or cellulose nitrate (CN). Cellulose fibres, in the form of cotton linters or purified wood cellulose, is nitrated in an aqueous solution of 55% nitric and 25% sulphuric acids to give cellulose nitrate, often referred to as nitro-cellulose. After separation and washing, the cellulose nitrate is mixed with camphor, as a plasticiser, and stabilisers and compacted into a block. This block can be sliced into celluloid sheets in thicknesses ranging from 0.125 to 25 mm. A major disadvantage of celluloid is its very high flammability. Celluloid is used for brush backs, knife handles, and table tennis balls.

Cellulose acetate (CA) is made by reacting cellulose with acetic anhydride and acetic acid, using sulphuric acid as a catalyst. Another process is the reaction between cellulose, acetic anhydride and methylene chloride with perchloric acid as a catalyst. It is mainly the diacetate which is formed but by modifying the process cellulose triacetate can be produced. Cellulose acetate is a transparent, glossy thermoplastic which can be injection moulded and formed into film and sheet. Acetate fibre can be formed by spinning from a solution of cellulose acetate in acetone, the fibre hardening as the solvent evaporates. The diacetate suffers from high moisture absorption, with associated dimensional instability, and is flammable, though much less so than celluloid. The triacetate has a lower moisture resistance than the diacetate and also has a lower flammability. Applications for CA include brush backs, cosmetic containers, spectacle frames, tool handles and toys, as sheet and film for packaging and, the triacetate, as the base for photographic film, and as fibres. Acetate fibre is weaker than rayon but is more elastic and resistant to wrinkling.

Cellulose acetate butyrate (CAB) is made by reacting cellulose with a mixture of acetic and butyric anhydrides. It is a transparent, glossy polymer which is tougher and has a lower moisture absorption than cellulose acetate and has a better resistance to weathering than either cellulose acetate or cellulose propionate. Applications include bathroom fittings, decorative automobile trim, gun stocks, pens, set squares, stencils and other drawing aids, tool handles and for packaging, especially blister packaging.

Cellulose propionate (CP) is made by reacting cellulose with propionic anhydride. It is a transparent, glossy polymer which is stiffer, has a better low temperature impact strength, but poorer resistance to weathering than both cellulose acetate and cellulose acetate butyrate. It is more expensive than both CA and CAB but is used for a similar range of applications.

Ethyl cellulose (EC) is a polymer with good toughness, thermoplasticity, low flammability, and low water absorption. The unplasticised grades remain flexible down to temperatures as low as about $-40°C$. However, the material is readily attacked by organic liquids. An important use for ethyl cellulose is in the production of protective hot dips by mixing with oils, platicisers and stabilisers. Metal components such as tools, gauges and valves are dipped in the hot molten mix to give a tough, transparent, strippable coating. The material is readily injection moulded and extruded and applications include helmets, furniture trim, luggage and tubing. Ethyl cellulose is also used as an additive in laquers, varnishes and adhesives to improve toughness.

Cellulose reacts with sodium hydroxide and carbon disulphide to form a compound called cellulose xanthate. Viscose is a solution of cellulose xanthate in dilute sodium hydroxide, from which regenerated cellulose can be produced. The degree of polymerisation of regenerated cellulose is much lower than that of natural cellulose. Viscose is made into two main products, a fibre termed viscose rayon and a sheet material, *cellophane*. Cellophane is a transparent and tough material used widely in packaging.

17.14 PEEK

PEEK (polyether etherketone) is a linear crystalline heterochain polymer developed in the 1970s for service at high temperatures. It has a melt temperature of about 330°C and, though melt processable, is difficult to process, but can be used continuously at temperatures up to 250°C. The chain structure of the polymer is:

$$\left[\underbrace{\bigcirc}_{} - \overset{\overset{\displaystyle O}{\|}}{C} - \underbrace{\bigcirc}_{} - O - \underbrace{\bigcirc}_{} - O - \right]_n$$

PEEK has good fatigue and chemical resistance and is an inherent fire retardant material. Glass fibre and carbon fibre reinforced grades are available. The main uses are in aerospace applications and in radiation environments as injection moulded components, wire covering and as a resin in fibre prepregs. The properties of PEEK are:

Density	$1300 \, kg/m^3$
Tensile strength	$240 \, MPa$
E	$3.8 \, GPa$
% elongation at break	4

17.15 Polyphenylenes

PPO (polyphenylene oxide), also known by the proprietary name *Noryl*, is an amorphous linear heterochain polymer developed in the 1960s with good mechanical properties and possessing a high impact strength over a wide range of temperature ($-40°$ to $+150°C$), and suitable for continuous use at temperatures up to 80°C. Its structural formula is:

$$\left[\underbrace{\bigcirc}_{\substack{CH_3 \\ CH_3}} - O - \right]_n$$

PPO has a low mould shrinkage and shows very good dimensional stability over a wide range of temperature. It has a good resistance to hydrolysis but not to many solvents. PPO is compatible with polystyrene and the two are miscible in all proportions. Modified PPO is a single phase hybrid polymer of PPO and polystyrene and is a more easily processable material. It is relatively expensive but is widely used for such applications as electrical fittings, components for dishwashers and washing machines, television components, VDU cabinets and for car fascia panels. It has found widespread use as a replacement for zinc-based diecastings in business machines, pumps and impellers and is particularly suitable for pumping fluids at high temperatures.

PPS (polyphenylene sulphide) is a related material and is also a linear heterochain polymer which was developed in the 1970s. It has a good temperature stability and is suitable for continuous use at temperatures up to 190°C. Its structural formula is:

$$\left[-\hexagon-S- \right]_n$$

PPS has good resistance to chemicals and solvents, and good flame resistance, but is difficult to pigment and, hence, is only available in dark colours. PPS, although a thermoplastic, has some of the characteristics of a thermoset and can be cured by heating at about 200°C. It is thought that chain length extensions and some crosslinking are brought about by heat treatment. The main uses of the polymer are for electrical connectors and coil bobbins, components for chemical pumps, pipes and gaskets and as chemical and abrasion resistant coatings.

Table 17.8 *Properties of PPO and PPS*

Property	PPO		PPS	
	Unfilled	Glass-filled (30%)	Unfilled	Glass-filled (30%)
Density (kg/m³)	1060	1270	1340	1640
Tensile strength (MPa)	61	118	75	150
E (GPa)	2.5	8.4	3.3	7.8
% elongation	60	5	1.6	1.3
Impact strength (J/m²)	21	9		

17.16 Polysulphones and polyarylates

Polysulphones and polyarylates are tough, glassy polymers with a value of T_g in the range 200–250°C. They are melt processable at temperatures of the order of 350°C and can be extruded and injection moulded. They possess a very good resistance to oxidation and creep and can be used at elevated temperatures. They are both heterochain aromatic polymers. The main members of these groups are PSU (polysulphone), PES (polyethersulphone), PPSU (polyphenylenesulphone), PAS (polyarylsulphone) and PAE (polyarylether). Another heterochain aromatic polymer with good high temperature stability is PPS (polyphenylene sulphide).

PSU (polysulphone) is a heterochain linear thermoplastic polymer containing $-SO_2$ groups in the chain. PSU, which has been available since the mid 1960s, has a good high temperature resistance, up to 170°C, is non toxic, possesses an excellent resistance to acids, alkalis and oils, and has good electrical properties. Its structural formula is:

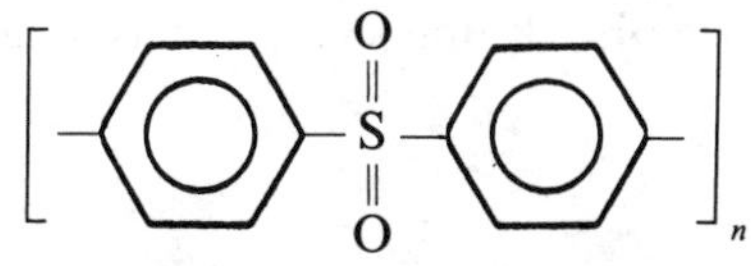

PSU is melt processable but processing temperatures of about 350°C are required. The main applications are for integrated circuit boards, coil bobbins, high-temperature and flame resistant cable insulation, parts for milking machines and for food dishes and trays.

PES (polyethersulphone) is a linear heterochain thermoplastic developed in the 1970s and suitable for continuous service at temperatures up to 180°C. Its formula is:

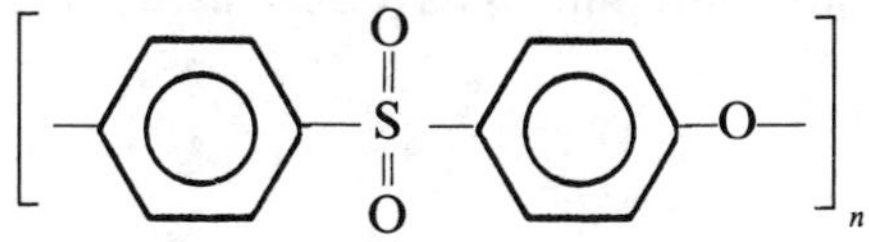

PES is transparent and yellow and is expensive and difficult to process, requiring processing temperatures of the order of 350°C. Its uses include electrical components to operate at elevated temperatures, such as coil bobbins, circuit boards and lamp holders, and in medical, agricultural and food industry components which require repeated sterilisation in service. Glass and carbon fibre reinforced grades of this material are also available.

PPSU (polyphenylene sulphone) is one of the group of polysulphone thermoplastics. Like other polysulphones, PPSU has good electrical properties, excellent chemical and thermal resistance and can be used for continuous service at temperatures up to 180°C. PPSU is not used on a wide scale but it does find application for electrical parts, such as connectors, where high temperatures are involved, and in the construction of fuel cells.

Table 17.9 *Properties of some polysulphones and polyarylates*

Property	PSU	PES	PPS	PAS	PAE
Density (kg/m³)	1245	1370	1290	1360	1140
Tensile strength (MPa)	60–74	84	72	90	55
E (GPa)	2.5	2.44	2.14	2.55	2.21
% elongation	50–100	40–80	60	13	25–90

Polyarylates are completely aromatic thermoplastics developed in Japan in the 1970s and they possess a linear heterochain structure of the general type shown below:

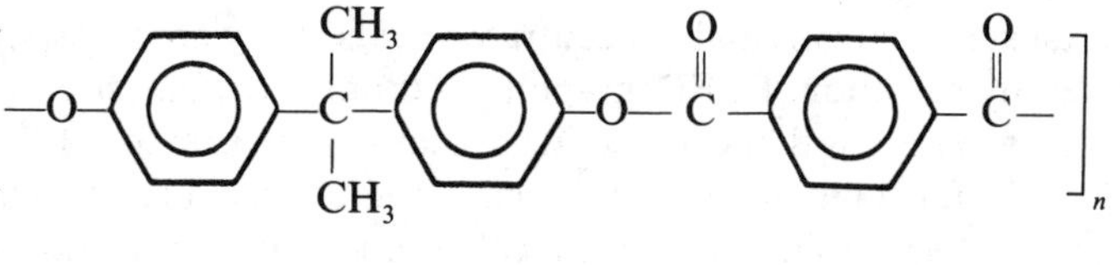

This structure gives a tough, glassy, transparent material with a high T_g value of about 200°C. The materials have a similar impact resistance to that of ABS and are resistant to ultraviolet radiation. However, the polymers are susceptible to stress cracking in contact with hydrocarbon solvents and require high processing temperatures, of the order of 350°C. Applications for the materials include electrical connectors, glazing, components in hot drinks vending machines and housings for appliances.

17.17 Polyimides (PI)

The polyimides are a group of linear aromatic polymers, some of which are thermoplastic and some are thermosets. They possess good mechanical properties with a low coefficient of friction and they possess excellent thermal resistance up to about 250°C and with inherent fire retardance. They are produced by a condensation reaction between pyromellitic dianhydride and a diamine. Several diamines, mainly aromatic amines have been used for the production of polyimides. The reaction is shown below:

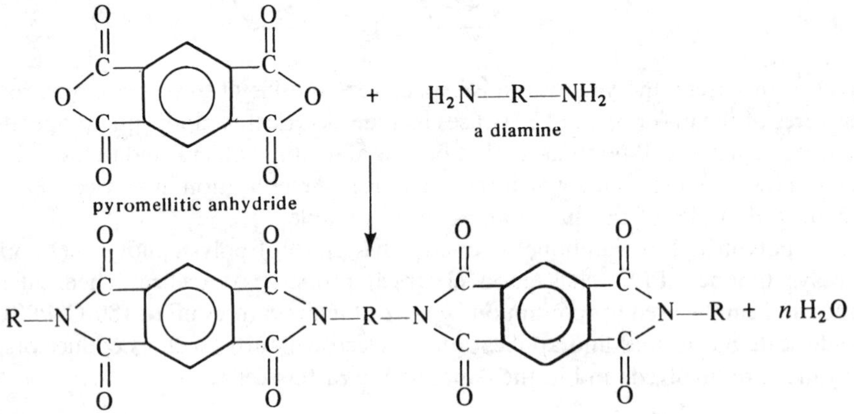

pyromellitic anhydride

A typical polyimide structure is:

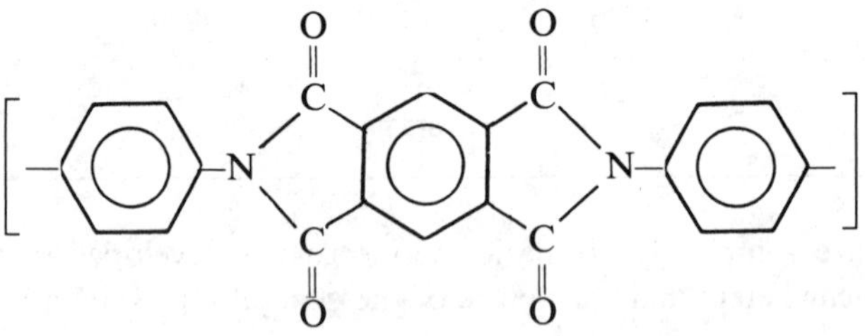

These materials, though expensive and difficult to process, have good resistance to organic solvents but are attacked by alkalis and concentrated acids. They are transparent to microwaves and are not affected by radiation. PTFE, graphite and molybdenum disulphide can be used as fillers to create bearing materials that can operate at temperatures up to 250°C. The main uses are as bearings and seals for high load/high speed applications, printed circuit boards, high temperature electrical cable insulation, mechanical components exposed to irradiation and wet or dry spun textile fibres. The resins may also be used as composite matrices. They have similar properties to epoxies at room temperature, but are stable up to higher temperatures than them.

Polyetherimide (PEI). This is a thermoplastic polymer related to polymide but with a greater chain flexibility and, hence, a better processing performance than polyimide. The structural formula is given below.

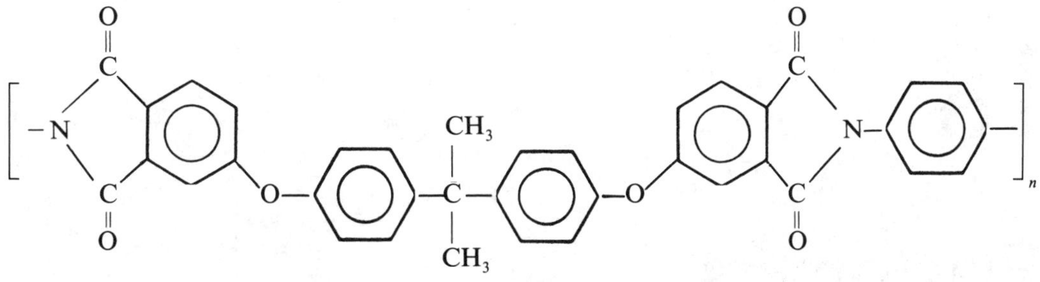

This polymer can be used continuously at temperatures up to about 170°C (compares with 180°C for PES). The material possesses a higher strength than PES at ambient temperature and is less expensive than either PES or PEEK. It does have a low value of notched impact strength and is susceptible to stress cracking in contact with chlorinated solvents. The main applications include electrical switchgear and connectors for high temperature service, cookware for microwave ovens and automotive components.

Polyamideimide (PAI). Related to polyetherimide and the polyimides, this is a linear aromatic polymer for use at high temperatures. The chain repeat unit contains an aromatic amide group:

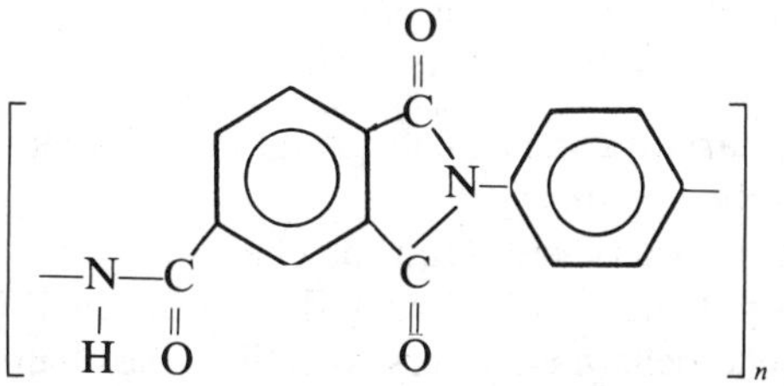

This material has properties approaching those of polyimide and can be used at temperatures up to about 230°C, yet it is easily melt processable. Unfortunately, the material is expensive and is attacked by alkalis. Applications include valves, bearings, electrical connectors, printed circuit boards and components for gas turbine and spark-ignition engines. Grades with glass or graphite reinforcement are available.

Table 17.10 *Properties of polyimides*

Property	Injection moulded polyimide	Graphite filled (40 per cent) polyimide	15 per cent graphite, 10 per cent molybdenum disulphide filled polyimide
Density (kg/m³)	1430	1650	1550
Tensile strength (MPa)	120	46	42
Compressive strength (MPa)	210	125	127
E (GPa)	1.33	—	—

18

Elastomers

18.1 Introduction

Elastomers are materials that have low elastic moduli which show great extensibility and flexibility when stressed but which return to their original dimensions, or almost so, when the deforming stress is removed. In essence, the molecules act like a series of small helical springs. There is some cross-linking between molecules (see Section 4.6), the flexible cross-links causing the material to return to its original shape. An increase in the number of cross-links formed will increase the hardness and stiffness of the material.

There are several classes of elastomers, these being:

R class — unsaturated carbon chains (This group includes natural rubber)
M class — saturated carbon chains
O class — heterochain polymers with O in the chain
U class — heterochain polymers with C, O and N in the chain
Q class — silicones—heterochain polymers with Si in the chain

18.2 Natural rubber (NR)

Natural rubber is *cis*-polyisoprene and this is an unsaturated linear molecule, that is, there is double covalent bonding between some atoms in the molecular chain.

$$\left[-CH_2-\underset{\underset{}{|}}{\overset{\overset{CH_3}{|}}{C}}=\underset{\underset{}{|}}{\overset{\overset{H}{|}}{C}}-CH_2-\right]_n$$

The source of natural rubber is the milky liquid known as *latex*, which is obtained from the cultivated tree *Hevea Brasiliensis*. The major rubber plantations are in Malaysia. To obtain the latex, a cut is made in the bark of the tree and the milky liquid slowly flows out into a collecting cup. Latex is an emulsion containing tiny particles of rubber. The latex is treated with an organic acid, such as acetic acid, to coagulate the rubber particles and precipitate them from the emulsion. The separated rubber is then dried thoroughly. It is necessary to mill the raw rubber between heavy rolls in order to make it plastic and suitable for forming. During this milling operation, the mechanical shearing effect breaks up some of the polymer chain molecules, reducing the average molecular weight of the rubber. The material is now in a plastic state and can be compounded with fillers and other additives, including plasticisers and antioxidants. In order to restore rubber-like elasticity to this plastic mass, an addition of a controlled amount of sulphur is made to allow for some cross-linking, or vulcanisation, to take place. The plastic rubber mass is then moulded or extruded to the desired shape and, finally, the article is *cured* by heating at some temperature in the range 100–200°C to vulcanise. During vulcanisation, some of the double covalent bonds in the rubber molecules split allowing sulphur atoms to form cross-links between adjacent molecules (see Section 4.6). For ordinary soft rubber, about 4 per cent of sulphur is added to the mix and this will cause cross-linking between approximately 10 per cent of the double bonds in the polymer. If the sulphur content is increased, the amount of cross-linking is increased giving a harder and less flexible product. Full vulcanisation, requiring about 45 per cent of sulphur, will give rise to a fully rigid network structure. Fully vulcanised natural rubber is the hard elastic solid termed *ebonite*.

Oxygen or ozone will react with the double bonds in natural rubber molecules forming cross-links in a similar manner to the vulcanisation reaction. Oxygen cross-links are not flexible and they embrittle the material. The rate of oxidation, or *perishing*, of the rubber can be retarded by compounding antioxidants and antiozonants into the material.

The role of fillers in rubber is two-fold. The use of a cheap inert filler will cheapen the material, but a filler may also be used to improve certain properties. Carbon black and silica are two fillers commonly used in rubbers and both of these give reinforcement and increase the resistance to abrasion and tearing.

Cis-polyisoprene can be synthesised and the nomenclature for synthetic polyisoprene rubber is IR. IR is less expensive to produce than natural rubber and it is often blended with natural rubber for reasons of economy.

Natural rubber accounts for some 30–35 per cent of the elastomeric materials used and synthetic isoprene makes up about 10 per cent of world output of synthetic elastomers.

18.3 Synthetic R class elastomers

BR. There are a number of synthetic elastomers with unsaturated carbon chains which permit vulcanisation. An important member of this group is polybutadiene (BR), made by the addition polymerisation of butadiene.

$$n(CH_2 = CH-CH = CH_2) \rightarrow (-CH_2-CH = CH-CH_2-)_n$$

BR is a cheap and resilient elastomer with a good abrasion and tear resistance. However, it is difficult to process and has a poor resistance to ozone and oxidizing chemicals. Its sensitivity to heat and light results in poor ageing characteristics and the poor stability of the carbon bonds often results in low pigmentation stability. BR is mainly used for making physical blends with NR and SBR, for example a blend of 25% BR with 75% SBR is the most common material for rubber tyres. This is used also for footwear, seals, domestic floor tiles and sponges, a considerable quantity of BR production is used as the dispersed rubber phase in high impact polystyrene.

SBR. SBR is a random copolymer of styrene and butadiene and accounts for over half of the world production of synthetic rubbers. This material is cheaper than natural rubber and is used, sometimes alone, and sometimes in physical blends with NR or BR.

NR, SBR and BR absorb organic solvents, petrol and oils and swell considerably. Two types of synthetic rubber which may be used in contact with oils and other solvents are the types CR and NBR.

CR. CR is polychloroprene, the structural formula being: $(-CH_2-CHCl = CH-CH_2-)_n$. It is also known as *chloroprene* and *neoprene*. This polymer can be vulcanised by heat alone but its properties are enhanced by compounding with zinc oxide and magnesium oxide, both of which act as vulcanising agents. It has an outstanding resistance to oils and many other chemicals.

NBR. NBR is a copolymer of butadiene with between 10 and 40 per cent of acrylonitrile. It is also known as *nitrile rubber*. The nitrile rubbers possess excellent resistance to fuels and oils, even at temperatures up to 160°C.

IIR. This rubber, also known as *butyl*, is a copolymer of iso-butylene with a small amount of isoprene. Isobutylene has the structural formula:

$$n\left(\begin{array}{c} CH_3 \\ | \\ C{=}CH_2 \\ | \\ CH_3 \end{array}\right) \rightarrow \left(\begin{array}{c} CH_3 \\ | \\ -C-CH_2- \\ | \\ CH_3 \end{array}\right)_n$$

IIR is fully saturated when polymerised. The presence of a small amount of isoprene provides some double bonds, and hence cross-linking sites, within the molecules. The rubber is fully saturated after vulcanisation and this makes the material more resistant to oxidation than NR or SBR. IIR has an extremely low permeability to air and other gases and its main applications are for hoses, tubing and inner tubes. It is also possible to react the material with halogens and substitute chlorine or bromine for some of the hydrogen atoms in the molecular chains. These derivatives, chlorobutyl and bromobutyl rubbers have even greater resistance to oxidation and improved properties at elevated temperatures.

18.4 M class elastomers

The M class of elastomers are saturated carbon chain elastomers. EPM and EPDM are, respectively, a copolymer of ethylene and propylene and a terpolymer of ethylene, propylene and diene. EPM is fully saturated but cross-links (*vulcanisation*) can be brought about by reaction with peroxides. In EPDM, the presence of a small amount of diene introduces a small number of double covalent bonds into the structure, thus permitting vulcanisation by sulphur. Both of these elastomers are highly resistant to oxidative degradation.

Polyethylene can be chlorinated by direct exposure to gaseous chlorine or to a solution of chlorine. There is a simple substitution of chlorine for some hydrogen atoms in the linear chain molecules. Similarly, polyethylene can be reacted with both chlorine and sulphur dioxide to give chlorosulphonated polyethylene. Both of these derivatives can be cross-linked to produce elastomers. They are identified by the nomenclatures CM and CSM respectively. Both of these elastomers have excellent resistance to oxygen, ozone and many chemicals and good high temperature characteristics. FKM is the term for a fluorinated hydrocarbon elastomer. This type of material is suitable for use at temperatures up to 230°C and possesses an excellent resistance to oils and chemicals.

Some polyacrylates are elastomers (see Chapter 18). These are a group of synthetic elastomers based on ethyl acrylate copolymerized with a small amount of an active monomer which will permit some subsequent cross-linking. Acrylate elastomers possess excellent resistance to oxidative degradation. ACM is polyethylacrylate and ANM is a copolymer of ethyl acrylate with acrylonitrile.

18.5 O class elastomers

A polyether is a heterochain polymer containing the –C–O–C– ether element in the chain. Epichlorhidrin can be polymerised to give a polyether with elastomeric properties, with the nomenclature CO or ECO. This is an elastomer produced in relatively small quantities for special purposes. It possesses an excellent resistance to petrols, oils and greases but with superior thermal and oxidation resistance to NBR.

Polysulphide rubbers. Polysulphides are heterochain polymers which are sulphur analogues to the polyethers with some carbon–sulphur–sulphur–carbon links within the molecular chain. Polyethylene polysulphide is an elastomer which has an extremely high resistance to swelling in the presence of organic solvents. It is, however, weak but finds application as an adhesive and sealant.

18.6 U class elastomers

The U class of elastomer is characterised by having a carbon–oxygen–nitrogen heterochain structure. These are the materials termed *polyurethanes* and there are two types, AU and EU, these being polyester- and polyether-urethanes respectively. The term polyurethane is given to the series of polymers produced by the reaction between aromatic di-isocyanates and low molecular weight polymer molecules which may be polyesters, polyethers or a mixture of a polyester and a polyamide. They can be tailored chemically to give a range of products in several forms, some of the polyurethanes being elastomers. These materials have excellent abrasion and tear resistance but have a major disadvantage in that isocyanates are highly toxic and there is a major evolution of poisonous smoke when these materials degrade and burn.

18.7 Q class elastomers

The Q class are heterochain polymers with silicon in the chain, in other words, *silicones* or *siloxanes*. The backbone chain in a silicone is not composed of silicon alone but are repeating units of silicon and oxygen, –Si–O–. As silicon is tetravalent and oxygen is divalent, the –Si–O– unit has four valency bonds. Two of these are used up in linking with other –Si–O– units and the other two bonds link with hydrogen or organic groups, such as the methyl group, –CH_3, or the phenyl group, –C_6H_5. The nomenclature code for these materials lists the types of organic groups attached to the siloxane chain, for example, MQ is methyl silicone, PMQ is phenyl-methyl silicone, and VMQ is vinyl-methyl silicone.

The major advantage of silicone rubbers is that they possess rubber-like elasticity over a very wide temperature range and some of them retain their flexibility and elasticity down to temperatures as low as $-80°C$. They are, however, very expensive. Applications of silicone rubbers include cable insulation, seals for hydraulic systems, and door and canopy seals for aircraft.

18.8 Thermoplastic elastomers

The large group of materials known as thermoplastic elastomers are materials which have elastomeric properties at ordinary temperatures but can be processed in the same way as thermoplastics. The thermoplastic elastomers do not have the type of cross-linking possessed by other elastomers, for example the sulphur links in a vulcanised rubber. A typical thermoplastic elastomer structure is one in which the very large polymer molecules consist of a block copolymer with three segments, a large central segment of a 'soft' flexible polymer with two somewhat smaller blocks of a harder more rigid polymer at either end. In styrene-butadiene thermoplastic elastomers, for example, the central flexible polybutadiene block may show a degree of polymerisation of between 600 and 2000, while the outer polystyrene blocks may be about one-tenth of that size. A similar structure occurs in ethylene–propylene thermoplastic elastomers, in this case with the polyethylene forming the central flexible segment. The hard end segments in this type of copolymer tend to come together into small groups giving, in effect, a series of small hard blocks, or domains, within a softer elastomeric matrix. These domains act as cross-links.

Table 18.1 *Properties of some elastomers*

Type	Temperature range (°C)	Maximum % elongation	Hardness (Shore A)
NR	−50 to +100	700	30–90
BR	−60 to +100	600	40–80
SBR	−60 to +70	600	40–90
CR	−20 to +130	600	30–90
NBR	−20 to +120	600	40–90
IIR	−50 to +125	800	40–80
EPM	−40 to +150	600	30–90
ACM	−40 to +130	400	40–85
CSM	−40 to +120	500	50–90
FKM	−40 to +230	300	60–95
AU	−50 to +100	700	40–100
EU	−50 to +100	700	40–100
MQ	−60 to +230	700	30–85
PMQ	−60 to +230	700	30–85

Table 18.2 *Applications of some elastomers*

Type	Applications
NR	Low cost, good properties. Tyres, engine mountings, flexible hoses
BR	General purpose rubber with better abrasion resistance than NR. Used for hose, shoe soles, tyre treads
SBR	Low cost, similar properties to NR. Tyres, transmission belts
CR	Good heat resistance and resistance to oils. Car radiator hoses, gaskets and seals, conveyor belts
NBR	Best resistance to oils and heat. Seals, hoses and linings
IIR	Good resistance to ozone, low permeability. Inner tubes
EPM	Good mechanical properties. Resistant to hot water and superheated steam. Steam hoses, conveyor belts
ACM	Good heat resistance. Excellent resistance to hot oils. High temperature gaskets and seals
CSM	Excellent resistance to chemicals. Acid hoses
FKM	High temperature resistant, no swell in oils. 'O'-rings and gaskets in automotive applications
MQ/PMQ	High cost. Mainly used for seals and gaskets. Aircraft applications

When the temperature is increased and the hard block domains 'melt' the material becomes truly thermoplastic and can be processed using any of the usual techniques for thermoplastics.

Thermoplastic elastomers can also be made by blending an elastomer with a thermoplastic material to make an 'alloy'. Examples of this are blends between natural rubber, SBR or NBR, with polypropylene.

19

Thermosetting Materials

19.1 Introduction

Unlike a thermoplastic material which will soften when heated, a thermosetting material undergoes chemical changes when first heated and is converted from a plastic mass into a hard and rigid material. During the heating, or *curing*, process further polymerisation occurs and a full network molecular structure is formed. Figure 19.1 is a schematic representation of a three-dimensional network structure. There are also a number of materials that will set hard and rigid at ordinary temperatures. These *cold-setting* materials are of the same general type as the thermosetting plastics since the curing process is fundamentally the same in both cases.

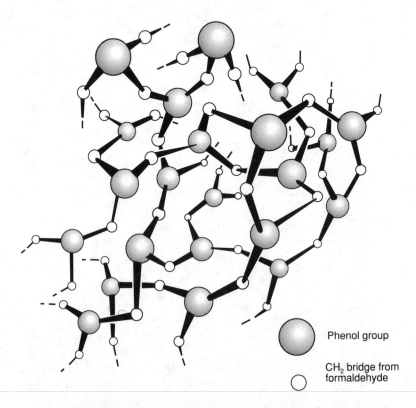

Phenol group

CH$_2$ bridge from formaldehyde

FIGURE 19.1 Schematic representation of a network structure

19.2 Phenolic materials

The first thermosetting materials to be commercially exploited were polymers based on phenol and formaldehyde. These two compounds polymerise by a series of condensation reactions. During the reaction the phenol molecules become linked by $-CH_2-$ groups. The $-CH_2-$ linkages may be formed at three positions in each phenol molecule, namely at the 2, 4, and 6 sites, giving a full network polymer.

Within a cyclic carbon compound the sites may be numbered thus:

The resultant pure polymer is opaque and milky-white in colour, but it darkens with time. Consequently, the phenolic moulding powders are always mixed with dark pigments to give constant dark-coloured materials. One trade name for phenolic materials is 'Bakelite'.

In the production of moulding powder polymerisation is allowed to proceed only to a limited extent to give low molecular weight resin that can be readily melted. This material is then ground into powder and compounded with suitable fillers and other ingredients. When the part-polymerised powder is next heated it melts and can flow under pressure into a mould. Curing is accompanied by a volume contraction of the resin and an inert filler must be mixed in with the resin powder in order to reduce the effect of this shrinkage. The use of a filler will also tend to reduce the cost of the moulding material and many fillers also give an improvement in some properties. Other additives used in the compounding of thermosetting moulding powders are catalysts to accelerate the curing process, pigments to colour the resin, and lubricants to prevent the material from sticking to the moulds. A variety of materials may be used as fillers and the filler generally accounts for between 50 and 80 per cent of the total weight of moulding powder. Coconut shell flour and wood flour, a fine soft-wood sawdust, are two widely used fillers. Wood flour increases the impact strength of the material. Coconut shell flour, a very cheap filler, would, if used alone, give rise to mouldings with comparatively poor properties, and so it is generally used in conjunction with wood flour.

Cotton flock and chopped fabric are often added as fillers where a high resistance to impact is required. Paper pulp and nylon fibre may also be used for this purpose. Other fillers that may be used to improve specific properties are asbestos, for improved resistance to heat, mica, for improved electrical resistance, and graphite, to reduce the coefficient of friction and give good sliding properties.

The compounded moulding powder may then be processed into a wide range of shapes by compression or transfer moulding. The properties of some phenolic resin moulding materials are given in Table 19.1.

Table 19.1   *Properties of phenolic moulding materials*

	PF unfilled	PF wood flour filled	PF asbestos filled	PF glass filled
Density (kg/m^3 × 10^{-3})	1.25–1.3	1.32–1.45	1.6–1.85	1.5–2.0
Tensile strength (MPa)	35–55	45–60	28–55	40–70
Compressive strength (MPa)	165	180	140–250	260–280
E (MPa)	5200–7000	5500–8000	9000–11 500	7000–14 000
Percentage elongation at break	1–1.5	0.4–0.8	0.1–0.2	0.5
General properties	Hard, rigid, thermosetting materials; stable up to about 150°C; low thermal conductivity; good electrical insulation characteristics; good resistance to oils, greases, and most common solvents			

The principal uses for phenol-formaldehyde (PF) mouldings are as electrical equipment, plugs, sockets, and switches, ash trays, door knobs and handles. Many laminates are made from phenolic resins. In these laminated materials the reinforcement may be paper or a fabric. Nylon, cotton, linen, and woven glass fabrics may be used. The major applications of these laminates are for electrical insulation, the manufacture of small gear wheels, and as water-lubricated bearings. These laminates are generally made in the form of sheet and rod material and they may be shaped easily by conventional machining methods.

Some phenolic resins are cold-setting. These have a syrupy consistency and the addition of a catalyst causes curing to occur at ordinary temperatures. These frequently form the basis for glues and adhesives, and may be used for the bonding of plywood and in the manufacture of hardboard. Powdered resin may be used as a binder in the production of shell moulds and cores for metal casting (refer to Section 22.6).

19.3 Amino-formaldehyde materials

Compounds containing an amino group, $-NH_2$, can condense with aldehydes to form rigid polymer. Of these, the most important are urea and melamine. The condensation reaction between urea and formaldehyde is shown below.

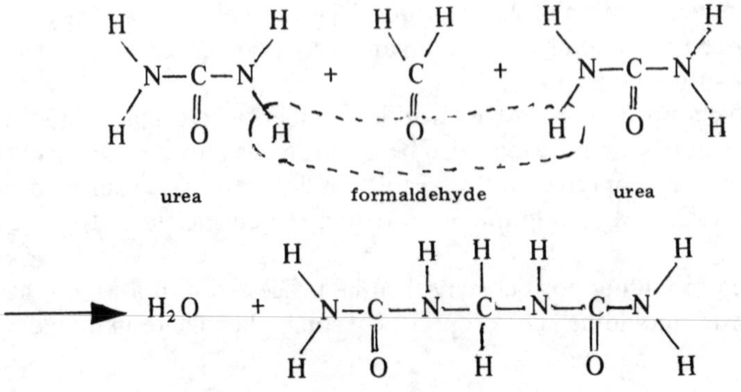

It is possible for both of the hydrogen atoms in each —NH_2 group to take part in a condensation reaction with an aldehyde molecule to give a highly rigid network polymer structure. In the same way as in the preparation of phenolic moulding powders, urea and formaldehyde are first part-polymerised and the low molecular weight polymer that is produced is then ground into powder and compounded with fillers, pigments, and a catalyst to make the finished moulding powder. The amino-formaldehyde resins are light in colour and pigments can be added to give a wide range of coloured moulding materials. Urea–formaldehyde (UF) mouldings are used for the manufacture of electrical fittings and also for making bottle caps and cups, saucers, and plates, as UF materials do not impart a taste to foodstuffs.

Melamine, a cyclic amino compound, can also be condensed with formaldehyde to give a strong and rigid material. The general properties of melamine–formaldehyde (MF) materials are similar to those of UF materials, but the MF resins have a better resistance to heat and can be used at temperatures of up to 95°C. UF mouldings have good stability at temperatures up to 80°C but they tend to darken and decompose at higher temperatures than this.

melamine

In addition to their use as moulding powders, both UF and MF materials, like the PF resins, are used in the manufacture of laminates, as adhesives, and as binding agents in shell moulding sands.

Table 19.2 *Properties of amino-formaldehyde moulding materials*

	UF cellulose filled	UF wood flour filled	MF cellulose filled	MF glass filled
Density (kg/m³ × 10⁻³)	1.5–1.6	1.5–1.6	1.5–1.6	2.0
Tensile strength (MPa)	40–90	40–55	41–70	35–138
Compressive strength (MPa)	175–315	140–190	280–330	140–245
E (MPa)	7–10	7–10	7–11	12–17
Percentage elongation at break	0.5–1.0	0.5–1.0	0.5–1.0	0.5
General properties	Hard, rigid, thermosetting materials; good electrical insulation properties; good resistance most chemicals			

19.4 Polyester materials

As mentioned in Chapter 17 (Section 17.12), an ester is formed when an acid condenses with an alcohol. If an unsaturated polyester is made, that is, the molecules contain reactive double covalent bonds, it becomes possible to cause cross-links to form between molecules creating a rigid network structure. Unsaturated polyesters are principally used as the basis of fibre reinforced plastics, and glass fibre is the most usual reinforcing material.

There are many polyester formulations possible but the ester produced by the condensation

of maleic acid,
$$\begin{array}{c} CH—COOH \\ \parallel \\ CH—COOH \end{array}$$
, and propylene glycol,
$$\begin{array}{c} CH_2OH \\ | \\ CH.OH \\ | \\ CH_3 \end{array}$$
, is typical of the type.

The unsaturated polyester is blended with an unsaturated monomer, such as styrene. Subsequently, on mixing in an *initiator*, an addition polymerisation will occur between the styrene and the polyester, giving a hard and rigid molecular network. Benzoyl peroxide is generally used as an initiator and curing will occur rapidly at elevated temperatures, up to 100°C. When certain amino compounds are present with benzoyl peroxide the curing process will take place at ordinary temperatures and the application of heat is not necessary. It is customary to mix the initiator thoroughly into the polyester resin immediately before use, and then make up a resin/fibre laminate over a former, as outlined in Section 23.11. The final properties of the laminate will depend, not only on the nature of the resin, but also on the type and form of reinforcing fibre used.

Polyester laminates are produced to meet a number of specific requirements and separate grades are available for general purposes, heat resistance, chemical resistance, water resistance and for good impact resistance. Some typical applications of the various grades are

General purpose   — boats, baths and bathroom fittings, trays
Heat resistant   — aircraft components and electronic parts
Chemical resistant — tanks, pipes and ducts
Water resistant   — structural panels
High impact   — safety helmets, machine guards

19.5 Epoxide materials

These are a group of materials that, although two or three times as expensive as polyesters, are finding a growing use as engineering materials. The uncured resin consists of polymer molecules with a highly reactive epoxy group at both ends of each molecular chain.

$$\begin{array}{ccc} O & & O \\ / \backslash & & / \backslash \\ CH_2—CH———————CH—CH_2 \end{array}$$

Most of the polymers used are based on the polymerisation of bisphenol A and epichlorhydrin. By controlling the degree of polymerisation, the physical state of the resin produced can vary from solid to fully liquid. In order to convert the epoxy resin into a fully rigid material it is necessary to mix a hardener into it. There are several types of hardener that may be used but some of the more commonly used types are reagents possessing amino groups, $-NH_2$. The amine hardener will form cross-links between the epoxy polymer molecules. It is necessary that the hardener possesses two or more amino groups. The equation showing cross-linking is given below.

$$NH_2—R—NH_2 + \overset{\displaystyle O}{\overset{\displaystyle \triangle}{CH_2—CH—}}\text{ -- -- --} \rightarrow NH_2—R—\overset{\displaystyle H}{\overset{|}{N}}—CH_2—\overset{\displaystyle OH}{\overset{|}{CH—}}\text{ -- -- -- --}$$

Both hydrogen atoms in every amino group can react with an epoxy group in this manner. Consequently, considerable cross-linking can occur and an extremely hard and rigid molecular structure can result. It should be noted that this is not a condensation reaction, and that there is no by-product evolved. There is virtually no shrinkage associated with this cross-linking reaction.

Epoxide materials are generally used with reinforcing fibres, such as carbon and glass fibre, to produce hard and strong structural materials. Carbon reinforced epoxy resins are extremely strong materials with a very high value for Young's modulus, compared with other plastics materials, and they are finding applications in the aerospace field as compressor blades in gas turbine engines, wing leading edges and flaps, and rocket motor cases.

Table 19.3 *Properties of polyester and epoxide laminates*

	Polyester reinforced with chopped glass mat	Polyester reinforced with plainweave glass fabric	Polyester reinforced with carbon fibre	Epoxide reinforced with plainweave glass fabric	Epoxide reinforced with carbon fibre
Reinforced content (per cent)	25–35	60–65	40	60–65	60
Density (kg/m³ × 10⁻³)	1.5–1.6	1.8	1.54	1.8	1.5
Tensile strength (MPa)	70–105	140–300	720	200–420	1100–1350
E (GPa)	5.5–12.5	7–20	150	21–25	125–150

Another property of the epoxide materials is that they will adhere to virtually all materials. Consequently, they find a use as high-strength adhesives, particularly in the aerospace industries. Other important applications of epoxides are for the encapsulation of miniature electronic components and for road surfacing. The extreme resistance to wear of these materials makes them suitable for use at busy road junctions. The high cost of the material is offset by the greatly reduced need for repair operations and the consequent disruptive effect on traffic.

19.6 Polyurethanes

Polyurethane is a family name given to a series of polymers that are produced by the reaction between aromatic di-isocyanates and low molecular weight polymer molecules, which can be polyesters, polyethers or a mixture of a polyester and a polyamide. The reactions are of the type:

(1) OCN—R—NCO + HO—R′— — — —

di-isocyanate

$$OCN—R—\overset{\displaystyle H}{\overset{|}{N}}—\overset{\displaystyle O}{\overset{\|}{C}}—O—R'\text{— — — —}$$

or (2)   OCN—R—NCO + H₂N—R′— — —

$$OCN-R-\underset{\underset{H}{|}}{N}-\underset{\underset{}{||}}{\overset{O}{C}}-\underset{\underset{H}{|}}{N}-R'- - -$$

or (3)   OCN—R—NCO + HOOC—R′— — —

$$OCN-R-\underset{\underset{H}{|}}{N}-\overset{O}{\underset{||}{C}}-R'- - - + CO_2$$

There are a number of different polyurethanes with differing characteristics. They all possess a very good wear resistance and are resistant to oils, greases and petrol but, depending on the exact formulation, they may be thermosets, thermoplastics, which can be processed by injection moulding or extrusion, elastomeric materials or, as is the case with materials formed by reactions of type (3) above, foams, and such foams may be either flexible or rigid.

The principal uses for polyurethane moulded components are for hoses, car bumpers, shoe heel tips, hammer heads, gears, including gears for quiet running in film cameras and projectors.

19.7 Allyl resins

Allyl phthalates can be polymerised in the presence of an alkaline catalyst to give a thermoset product. Diallyl phthalate (DAP) is a thermoset resin used in the manufacture of laminates, as a coating resin and as a moulding compound. It has the structural formula:

$$CH_2{=}CH{-}CH_2{-}\overset{O}{\underset{||}{C}}{-}O{-}\hexagon{-}O{-}\overset{O}{\underset{||}{C}}{-}CH_2{-}CH{=}CH_2$$

diallyl phthalate

DAP possesses a higher strength than polyester or epoxy materials and has good electrical properties, and a better resistance to tracking than phenolic resins. The electrical properties are maintained at high temperatures (up to 160°C) and high humidities. The resin is non-toxic and, unlike epoxies, does not cause dermatitic reactions. It is used as a laminating resin for the manufacture of radomes, trunking, printed circuit boards and high temperature applications. Typical moulded components include electrical connectors, insulators, tool handles and pump impellers. As a coating resin it is used as a wood sealant. Glass fibre reinforced and mineral filled grades are available.

Diallyl isophthalate (DIAP) is a material similar to DAP but has a poorer resistance to alkaline chemicals. It has a better temperature resistance than DAP (up to 180°C) and a higher strength. It is used for applications similar to DAP but where a better temperature capability is required.

20

Ceramics and Glasses

20.1 Introduction

The word *ceramic* conjures up different images with different people. For many, the word means porcelain and pottery, while the engineer usually thinks of the new industrial ceramics, namely sintered oxides, carbides, and nitrides. In this chapter we will give brief consideration to a wide range of crystalline inorganic materials, including not only the materials normally termed ceramics but also stone, cement, concrete, and bricks. Building stone and bricks are materials that have been used since earliest times for constructional purposes and, like ceramics generally, are usually hard and brittle materials that are comparatively strong in compression, but weak in tension. *Glasses* are amorphous, brittle materials composed of metal oxides and silicates. There is also a group of new materials, termed *glass ceramics*. These are fine-grained polycrystalline materials produced by a controlled crystallization of glasses. Table 20.1 gives a summary of some of the main groupings of ceramic and glass materials.

Table 20.1 *Types of ceramic*

Class	Material types	Applications
Whitewares	porcelain, china, stoneware	tableware, wall tiles, sanitary ware, insulators.
Heavy clayware	clay-based material	bricks, roof tiles, pipes.
Refractories	acid, intermediate, basic	furnace linings.
Stone	granite, sandstone, limestone	concrete aggregate, building stone
Glass	various glasses, glass-ceramics	containers, flat glass.
Abrasives	diamond, carborundum, corundum	grinding wheels, abrasive cloth & paper
Cements	Portland cement, alumina	concrete, structural products.
Engineering ceramics	oxides, carbides, nitrides, cermets	dies, seals, bearings, spark-plugs, diesel and turbine engine components
Electrical and magnetic ceramics	many types	various, including capacitors, thermistors, varistors, magnets
Nuclear ceramics	oxides and carbides of uranium	nuclear fuel elements

20.2 Building stone

The rocks that make up the earth's crust vary considerably, both in composition and structure. Rocks may be classified as *igneous, sedimentary,* or *metamorphic* according to the method of their formation. Igneous rocks are those that have formed by solidification from hot liquid rock material—*magma*. The structure of igneous material is one of interlocking crystal grains. The various constituent minerals crystallise by a process of nucleation and growth from liquid to give a compact mass of irregular shaped crystals. Igneous rocks, which include *granite* and *dolerite*, are hard, weather-resistant materials and possess compressive strengths in the range 110–180 MN/m^2. Quartz (silica), which is a very hard mineral, is a major constituent in granites.

Because of the interlocked grain structure, igneous rocks fracture in an irregular manner. There are not normally major lines or planes of weakness in an igneous rock mass, and quarrying and dressing this type of stone into regular shapes for building purposes is both difficult and costly. These days comparatively little stone is used for building, but granites and other igneous material are widely used in the form of crushed chips for road surfacing, as railway track ballast, and as aggregate for concrete. The irregular intercrystalline fracture surfaces allow for extremely good bonding with cement.

Much of the earth's crust is made up of sedimentary rocks. These are derived from sediments laid down hundreds of millions of years ago, which have become consolidated into rock with the passage of time. Sedimentary rocks are often well stratified and show well-defined bedding planes. Frequently, they are also well jointed, joints being normal to the bedding planes and formed by lateral contraction of the sediment during consolidation. Many sedimentary rocks can be split and cleaved easily, both in the bedding direction and normal to the bedding planes, to make building blocks of fairly regular shape. The properties of sedimentary rocks vary considerably, depending on the nature of the sediment and the type of bonding that exists between adjacent sediment grains. *Sandstones*, formed by the consolidation of sand deposits, consist principally of small rounded grains of silica. In some sandstones the silica grains are bonded together with a clay or other soft mineral. Sandstones of this type will be comparatively soft and will weather rapidly. In other types of sandstones the silica grains may be bonded with hard minerals. In the *millstone grits* of the carboniferous period a very hard type of rock is found composed of silica grains bonded with a silica cement.

Other types of sedimentary rocks are *limestones* and *dolomites*, formed by the precipitation of calcium carbonate, or calcium and magnesium carbonates, from solution. Some limestones and dolomites are very hard, with compressive strengths of the order of 150 MN/m^2, while others, such as chalk, are very soft. Portland stone is a well-bedded and jointed limestone which can be easily dressed into regular shapes for building purposes.

Some sedimentary deposits have been caused to recrystallise, during geological time, by the action of heat or pressure to give new structures, termed *metamorphic* rocks. *Slate* is a metamorphic material formed by the recrystallisation of a clay or shale. In the recrystallisation to form slate, very small plate-like crystals are formed, all with a similar orientation. The hard, weather-resistant slate that results may be readily cleaved into thin sheets.

20.3 *Cement and concrete*

Portland cement powder is made by firing a mixture of limestone with shale or clay in a rotating kiln. The maximum temperature in the cement kiln is about 950°C, and at this temperature the lime and clay partially fuse together as a hard clinker. Cement clinker is ground into powder and mixed with a small amount of *gypsum* (calcium sulphate) to produce dry cement powder. The function of the gypsum is to control the setting characteristics of the cement.

When cement powder is mixed with water a series of complex chemical reactions occurs, forming hydrated silicates and aluminates of calcium. It is these chemical reactions that cause the wet cement to harden and set as a rigid material. Some of the hydration reactions take place very slowly and, although the cement will set fairly rapidly, great strength and hardness will not be developed for several days, or weeks, depending on the composition of the cement, the amount of water added, and the temperature. The amount of water added to dry cement powder is important. If insufficient water is added, full hydration of the cement particles will not occur. For full hydration and the development of maximum strength, a water/cement ratio of about 0.4/1 is necessary. If the water/cement ratio is much in excess of this value the strength of the hardened cement will be reduced.

Concrete consists of a mixture of cement, sand, and an aggregate of small stones. When water is added to a dry concrete mix, the cement paste formed should fully coat all sand and aggregate particles, and fill in the void spaces between aggregate particles. The cement paste hardens, owing to the hydration reactions, and bonds the inert sand and aggregate together.

The final properties of a concrete will be dependent on a number of factors, including the relative proportions of water, cement, sand and aggregate in the material, the average size of aggregate particles, the type of aggregate stone used, and the surface texture of the aggregate. The compressive strengths of plain concrete may be up to $65 \, MN/m^2$, but the strength in tension is only about one-tenth of the value of compressive strength. (This compares with a compressive strength of about $100 \, MN/m^2$ for hardened cement).

When a concrete is subjected to stress, failure probably commences at the interface between aggregate and cement. Aggregate particles with rough surfaces will give concretes of higher strengths than will smooth-surfaced aggregate. The tensile strength of concrete is low (up to $5 \, MN/m^2$) and, to overcome this disadvantage, concrete fabrications are very often reinforced with steel. In plain reinforced concrete, a network of steel rods or bars is assembled and the concrete is allowed to set around this framework. The steel reinforcement is positioned in the portion of the concrete member that will be subjected to tensile stresses. For example, in a simply supported beam (Figure 20.1(a)) the steel lies along the lower portion of the beam. There is a purely mechanical bonding between concrete and steel, and the reinforcement bars are often twisted, or possess surface projections (these may be formed by rolling the bars through patterned rolls) in order to increase the adhesion between steel and concrete. Another form of reinforced concrete is known as *pre-stressed* concrete. The concrete is put into a state of

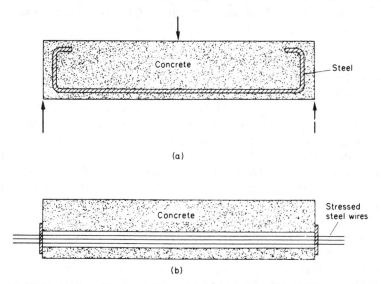

(a)

(b)

FIGURE 20.1 (a) Plain reinforced concrete beam. The upper layers of the beam are in compression and the lower layers are in tension. Steel bars help sustain the tensile stress. (b) Pre-stressed concrete beam. Steel wires in tension exert a compressive force on the concrete

compression by means of highly stressed steel wires (Figure 20.1(b)). When a pre-stressed concrete beam is in service, the initial compressive stresses must be overcome before tensile stresses can be developed within the material. Concrete may be pre-stressed by pre-tensioning, or by post-tensioning. In the former method steel wires are placed in tension before being surrounded by concrete. The externally acting stress on the steel is removed when the concrete has set. In post-tensioning, the concrete is allowed to set and harden around a tube, or tubes. Steel wires are then put through the tubes and these wires are stretched and anchored to the concrete.

20.4 Clay and clay products

Muds and *clays* have been used as the basis for artifacts, including structural products, throughout human history. Clay is a general name for minerals containing small plate-like crystals which have been formed from the decomposition of felspar in igneous rocks, such as granite, by the action of air and water over long periods of time. A major product of this decomposition is the clay mineral *kaolinite*, a hydrated aluminium silicate. Clay formed from the parent rock may be deposited at its place of origin, and this is known as residual or primary clay, or it may be transported by water and redeposited at some distance and this is known as sedimentary or secondary clay. The latter is usually of much finer particle size and contains more impurities than residual clay deposits. Many types of clay occur in nature and they are mixtures of the clay minerals with non-clay minerals such as mica and quartz. The kaolinitic clays include *china clay*, *ball clay* and *fireclay*. Ball clay is a fine textured clay with a high plasticity which consists of kaolinite, quartz and, in some cases, mica. China clay is almost pure kaolinite. It is coarse and has a low plasticity but fires to give a pure white product. The various types of clay products can be broadly classified into *whiteware*, which includes china, porcelain and stoneware, heavy clayware, which includes brick, roofing tiles, earthenware pipes and insulators, and clay refractories. This last group is dealt with in Section 20.5.

Wet clays are plastic and can be moulded or extruded to form clay products such as bricks, tiles and earthenware articles. After the plastic clay has been formed into the required shapes, the products are allowed to dry out. This drying process is followed by firing the clay in a kiln within the temperature range 800–1500°C. During the firing process, water of hydration is driven off with resulting crystal structural changes within the material. A partial melting of some phases occurs and, on cooling, these molten areas vitrify and the glass phase cements the fired clay particles together. The amount of glass phase formed depends on both the clay composition and the firing temperature, a higher firing temperature giving a higher glass content. After firing the clay product is both porous and brittle. Many clay products are glazed after the first firing to produce a non-porous surface layer that will render the product impermeable. The glaze, which is a slurry of mixed oxides—mainly silica, soda and lime—is sprayed or painted onto the surface and the article kiln-fired a second time. During this second, or *glost*, firing the surface coating is converted into a hard glass.

Whiteware and *porcelain* are general terms for fine textured clay ceramics including china, stoneware and fine earthenware. They are mixtures of ball and china clays compounded with silica as a filler and a flux, usually felspar, to form a glass phase. Firing temperatures, and the flux content, for porcelains are higher than for clay products such as bricks and tiles so as to give a high glass content with low levels of porosity. Both china and stoneware are highly vitrified, which gives high strength. The compressive strength of a low-porosity porcelain may be as high as 600 MPa.

20.5 Refractories

Refractories are mineral products that are stable at high temperatures and that are used in the construction of furnaces and allied equipment. Refractory materials must also resist the action of molten metals, slags, glasses, abrasive particles and hot gases. Refractory materials are classified as acidic, basic and neutral. Acidic refractories are rich in silica, SiO_2, an acidic oxide, while basic refractories are composed mainly of magnesite, dolomite and lime. There are many neutral refractories and the group includes alumina, fireclays, chrome, carbon and other pure metal oxides such as zirconia.

Acid refractories. Silica, SiO_2, is an excellent refractory as its melting point is in excess of 1700°C, and materials with a high silica content (over 93 per cent) are used in metallurgical furnaces to resist high temperatures and attack by acid slags, namely those with high silica and a low basic oxide content. Silica refractories are also used as regenerator bricks and for the roofs of furnaces in the steel and glass industries. The raw material for silica refractories is *quartzite* and *ganister*. These are both naturally occuring siliceous rocks containing 96–98 per cent of silica. The rock is crushed and broken down into a fine state of sub-division. Pressings made from dry silica grains would possess virtually no strength, so a small amount of an organic binder is added to the powder to aid the production of moulded silica shapes. A small amount of lime is also added to the silica before moulding. The moulded shapes are then dried and fired in a kiln. The lime content forms calcium silicate during firing and this acts as a bonding agent for the silica grains. Silica refractories can be used successfully at temperatures up to 1700°C.

Basic refractories. These are refractory materials consisting mainly of basic oxides such as magnesia, MgO, and lime, CaO, used in metallurgical and other furnaces to resist high temperatures and attack by basic oxides and slags. Naturally occurring magnesite, $MgCO_3$, calcined to the oxide has been used as a refractory material but most magnesite deposits contain some silica and this adversely affects the quality of the refractory. Much of the magnesia, MgO, used for refractories is derived from sea-water. The most widely used raw material for basic refractories is *dolomite*, a natural rock which is a mixed carbonate of magnesium and calcium. When calcined at 800–900°C, dissociation occurs to give calcium oxide, magnesium oxide and carbon dioxide. The mixture of the two oxides is susceptible to hydration and is referred to as *caustic dolomite*. In refractory bricks made from dolomite, the calcined grains are usually chemically bonded. A bonding agent, frequently tar, is used to coat the grains which are then pressed into bricks at pressures up to 70 MPa and fired or 'burnt' at 1500–1700°C. The resulting product is more resistant to hydration due to partial sintering and is referred to as *dead-burnt dolomite* and used as a refractory in the steelmaking and glass industries. Lime, calcium oxide, CaO, produced from limestone has been used as a basic refractory. It needs to be stabilized by tar impregnation or surface treatment to prevent hydration, but has been used for lining rotary furnaces and converters in steelmaking.

Neutral refractories. A major neutral refractory is *chrome magnesite*. The term covers a range of refractory materials made from chrome ore and calcined magnesite. When chrome predominates the resulting refractories are *neutral* whereas when magnesite predominates, the materials are slightly *basic* and the term *magnesite-chrome* is used. Chrome-magnesite refractories are more resistant to thermal shock than magnesite. Courses of chrome-magnesite brick are used as a separating layer between basic and acid refractories used in the linings of some types of high temperature furnace. Chrome-magnesite is used also as a structural material in furnaces for handling lead, copper and aluminium. Magnesite-chrome refractories are neutral or slightly basic refractories and they have a higher degree of refractoriness under load than the chrome-rich materials. They are used as structural materials in furnaces for copper, aluminium, and other metals.

Fireclays are naturally occurring sedimentary clays, often found underlying coal seams, consisting predominantly of kaolinite ($Al_2O_3.2SiO_2.2H_2O$) and quartz. They are of varied composition, depending on location, and may contain between 25 and 40 per cent of alumina, between 50 and 70 per cent of silica and low in Fe_2O_3, MgO and alkalis. They are refractory clays with high softening temperatures, depending on composition, in the range from 1400 to more than 1600°C. Fireclays are used in the manufacture of refractories for use as linings for blast-furnaces, lime and cement kilns, pottery kilns, furnaces for aluminium melting, and for gas heaters and domestic fire-backs. Some fireclays possess a high silica content and these siliceous clays, containing 75 per cent or more of silica, are used for the manufacture of semi-silica brick, which may be used at higher temperatures than normal fireclays.

Another range of successful refractories is the high-alumina group of materials. These range from aluminous fireclays containing > 45 per cent alumina up to almost pure alumina. They are made by a variety of processes from bauxite, sillimanite, mullite, fused and calcined aluminas and fireclays. They are suitable for use at temperatures up to 1600°C and find application in steel making furnaces, ceramic kilns, cement kilns, glass tanks and as crucibles for melting metals. Castable refractories can be made from a mixture of high alumina cement and a refractory grog blended with water. They are used for repairing furnace linings or building up monolithic structures.

Carbon is used successfully as a refractory material in applications where reducing conditions exist. A good example is the use of carbon as a blast furnace refractory. Carbon refractories made from coke or anthracite, bonded with pitch or tar are used for blast-furnace hearth linings. Clay-bonded graphite is also used as a refractory for some purposes.

20.6 Industrial ceramics

Considerable attention has been focused in recent years on the oxides, carbides, borides, and nitrides of metals. These compounds are generally of high hardness, and possess high elastic moduli. They are also of high stability and possess high melting points. The principal

Table 20.2 *Properties of some pure oxides, carbides, and nitrides*

Compound	Melting point (°C)	Density (kg/m³ × 10⁻³)	Hardness (Mohs' scale)
Al_2O_3	2015	3.97	9
CaO	2600	3.32	4.5
Cr_2O_3	2265	5.21	
MgO	2800	3.~8	6
SiO_2	1723	2.32*	6–7
ThO_2	3300	6.5	6.5
TiO_2	1840	4.24†	5.5–6
Cr_3C_2	1890	6.68	7+
NbC	3560	7.82	9
SiC	2830	3.21	9
TaC	3870	14.48	8+
TiC	3250	4.25	8
WC	2630	15.5	9
ZrC	3530	6.7	9
AlN	2230	3.05	7+
Si_3N_4	1950	3.44	9
TiN	2930	5.43	8+
ZrN	2980	7.32	8+

* Value for quartz.
† Value for rutile.

disadvantage of all these materials is their brittle nature. Their use is, in general, only considered when the conditions to be met in service preclude the use of other materials. Industrial ceramics can be considered for use when the requirements are one or more of the following:

(a) high resistance to abrasion and wear,
(b) high strength at high temperature,
(c) good chemical stability,
(d) good electrical insulation characteristics.

Some of the areas in which ceramic materials are used is as abrasives, cutting tools, bearings, insulators, magnetic materials, piezo-electric materials, superconductors, crucibles for metals of high melting point, and pumping equipment for liquid metals.

The properties of a number of oxides, carbides and nitrides are given in Table 20.2.

20.7 Alumina

High alumina ceramics, containing 75–99.8 per cent Al_2O_3, have been extensively developed. Most properties are improved as the alumina content increases (see Table 20.3). Precise control of crystal size, degree of calcination and levels of impurities is required in the manufacture of commercial aluminas and a variety of ceramics based on alumina are available including fusion-cast alumina and various grades of sintered alumina. High density, gas impermeable ceramics can be obtained by sintering alumina with small quantities, about 0.25% of magnesium oxide, MgO, to prevent exaggerated grain growth. High alumina ceramics are used for the manufacture of spark-plug insulators, ceramic/metal assemblies in vacuum tubes, substrates for the deposition of electronic microcircuits, and metal-cutting tool tips. Also, because of the low neutron absorption cross-section of alumina, the material finds application in nuclear equipment.

Table 20.3 *Properties of alumina ceramics*

	Alumina (per cent)			
	75	86–94	94–98	> 98
Density (kg/m³ × 10⁻³)	3.2	3.3	3.5	3.7
Hardness (Mohs' scale)	8.5	9	9	9
Compressive strength (MN/m²)	1250	1750	1750	
Flexural strength (MN/m²)	270	290	350	380
Maximum working temperature (°C)	800	1100	1500	

20.8 Silicon nitride

Silicon nitride, Si_3N_4, is a very useful engineering ceramic which is fully resistant to most strong acids and to molten aluminium and other low melting point metals. It exists in two hexagonal crystalline structural modifications, α-$Si_{12}N_{16}$ and β-Si_6N_8 and silicon nitride powder, consisting mainly of α, is made by a number of routes including the nitriding of silicon, the carbothermal reduction of silica in nitrogen, and the reaction of silicon tetrachloride with ammonia. There are several different types of silicon nitride ceramics:

reaction bonded silicon nitride (RBSN),
hot pressed silicon nitride (HPSN),
sintered silicon nitride (SSN),

and variants of those including hot isostatically pressed or pressure sintered silicon nitride, and sintered RBSN. Silicon nitride is used for a range of engineering applications at high temperatures and as wear parts.

Reaction bonded silicon nitride (RBSN). Silicon nitride ceramic made by reaction bonding in which a silicon powder compact, normally isostatically pressed, is converted into silicon nitride by heating in a nitrogen atmosphere at temperatures in the range 1100–1450°C according to the reaction: $3Si + 2N_2 \rightarrow Si_3N_4$ which occurs by a gas/solid, solid/liquid/gas or gas vapour reaction depending on temperature and the levels of impurity in the silicon and nitrogen. The resulting microstructure consists of sub-micron sized grains of silicon nitride with 15–30% porosity, including some large pores of up to 50 μm size, as a result of the melting out of silicon/iron impurity particles during nitriding, and small interconnected channels which may be <0.1 μm in diameter. No shrinkage occurs during the process as the silicon nitride which is formed fills up the pore space of the original silicon compact. Thus, after partial nitriding to give the compact sufficient strength, the process may be stopped and the compact machined, using conventional tools, to its final shape and dimensions. This material is used as pouring tubes for aluminium, and for high temperature engineering components.

Hot pressed silicon nitride (HPSN). This is a dense silicon nitride engineering ceramic produced by uniaxially pressing α-silicon nitride containing sintering additives such as MgO and Y_2O_3 in an induction heated graphite die at temperatures in the range 1700–1850°C, during which an oxynitride liquid is formed which allows densification and the transformation of α to β silicon nitride. A residual glassy or crystalline grain-boundary phase remains on cooling. The process is restricted to fairly simple shapes.

Sintered silicon nitride (SSN). This is a dense silicon nitride engineering ceramic produced by mixing α-silicon nitride, with its associated surface silica, and one or more metal oxides, such as MgO or Y_2O_3, as sintering additives. The powder mixture is formed into a shaped body and fired under a nitrogen atmosphere, protected by a powder bed of boron nitride. During the process an oxynitride liquid is formed which aids densification. On cooling, the oxynitride remains as a grain boundary phase, either amorphous or crystalline, depending on the type and amount of oxide used and this controls the high temperature properties of the ceramic.

Pressureless sintered silicon nitride (PSSN). This is a silicon nitride ceramic fabricated by the addition of an oxide or combination of two oxides to α-silicon nitride powder, consolidation to the required shape and firing in a nitrogen atmosphere at a pressure of one atmosphere at 1700–1800°C. During the firing, the oxide or oxides react with surface silica, SiO_2, and some nitride to form a M–Si–O–N liquid phase which aids densification and converts the α to β-silicon nitride. The M–Si–O–N phase, on cooling, is retained as a secondary phase at the grain boundaries which controls the high temperature properties of the ceramic. This material is under development for engine components.

The properties of reaction-bonded and hot-pressed silicon nitride are given in Table 20.4.

Table 20.4 *Properties of silicon nitride ceramics*

	RBSN	HPSN
Density (kg/m³ × 10⁻³)	2.3–2.6	3.12–3.18
Open porosity (per cent)	18–28	0.1
Hardness (Mohs' scale)	9	9
Young's modulus (GN/m²)	160	290
Flexural strength at 20°C (MN/m²)	110–175	550–680
Flexural strength at 1200°C (MN/m²)	210	350–480

20.9 Sialons

The sialons are derivatives of silicon nitride and the name sialon is an acronym derived from the elements involved, namely Si–Al–O–N. Much research and development work is being done on these materials and, at present, two major groups of sialon ceramics are manufactured commercially. These are known as low-substitution and high-substitution sialons. The former, and principal, material, is a solid solution of β-silicon nitride, Si_6N_8, in which Si is partially substituted by Al and nitrogen is partially substituted by oxygen, to give an expanded structure designated β'-sialon with a range of composition given by $Si_{(6-z)}Al_zO_zN_{(8-z)}$. β'-sialons sinter by a liquid-phase mechanism which aids densification and, on cooling, a glassy phase exists at grain boundaries. This glass phase, which softens at high temperatures, limits the operating temperature for the material to about 1000°C. In the highly substituted type of sialon, an yttrium aluminium garnet (YAG) crystalline phase forms at grain boundaries and these materials can operate at temperatures up to 1400°C. The sialons possess higher strength and hardness than silicon nitride and are used as cutting tool materials and for engine components and bearings.

20.10 Silicon carbide

Silicon carbide has been used as an abrasive for many years but is now also available as an engineering ceramic material.

Silicon carbide, SiC, is a hard, semiconducting ceramic existing in two crystalline structural modifications, α-SiC (hexagonal), made by the reduction of silica sand with carbon in an arc furnace, and β-SiC (cubic), produced by vapour phase reactions. The α form converts to the β form at temperatures above 2000°C. There are a number of different types of silicon carbide ceramics available, these being:

reaction bonded silicon carbide,
clay-bonded silicon carbide containing up to 50% fire clay; this material is coarse grained and used as a refractory for thermal shock applications,
hot-pressed silicon carbide containing 2% alumina as a densification aid and to give a fine grained product with zero porosity,
sintered silicon carbide containing boron and carbon as dopants to aid densification,
recrystallized silicon carbide, sintered to a less dense, fairly coarse grained product for use as furnace elements, refractory worktubes etc., and
nitride bonded silicon carbide produced by the reaction bonding of a material containing coarse α-SiC and silicon powders in a nitrogen atmosphere at 1350–1450°C used for refractory tubes and containers.

Reaction bonded silicon carbide (RBSC) is made by reaction bonding in which a porous compact, consisting of silicon carbide, carbon (graphite) with a binder, is first formed and the binder is carbonized in the pores. The body is then heated in contact with molten silicon, during which process the silicon moves through the material under the influence of capillary forces, reacting with the graphite and converting it to silicon carbide which deposits epitaxially on the surfaces of the original silicon carbide particles, thus bonding them together. All the carbon reacts and excess silicon remains. The final microstructure consists of the original silicon carbide grains overcoated with the silicon carbide formed *in situ*, and interconnected silicon-filled pores. It was originally developed as a cladding material for high temperature nuclear-reactor fuel elements and is now used for mechanical seals, bearings and engine components. It is also known as reaction sintered silicon carbide or self bonded silicon carbide.

Silicon carbide is also formed as fibres and whiskers for use as reinforcement in composite materials.

20.11 Boron nitride and other ceramics

Boron nitride is a synthetic nitride of boron, BN, which is isoelectronic with carbon and exists in two structural modifications similar to graphite (hexagonal) and diamond (cubic). The hexagonal form, with a density of 2270 kg/m^3, is soft and platelike and is manufactured as a ceramic by hot-pressing. This produces a material exhibiting anistropy because the platelets line up in planes perpendicular to the direction of pressing. Thin walled shapes, such as crucibles for special glasses, may be produced by a chemical vapour deposition process. The cubic form, with a density of 3480 kg/m^3, is the hardest substance known, next to diamond. It is produced from the hexagonal form at high temperature and pressure. The conditions for the formation of the cubic form are about 1500°C at a pressure of about 60 GPa. It is used as grinding material in grit form, or as a solid ceramic for the manufacture of metal cutting tools, and is referred to as CBN (cubic boron nitride). It is possible to sinter together small crystals of CBN to produce large polycrystalline masses. The resulting material, PCBN (polycrystalline cubic boron nitride) has a structure of randomly orientated crystals and is isotropic.

Aluminium nitride, AlN, is a ceramic manufactured by the method of *tape casting* (see Chapter 24) with suitable densification additives, such as CaO or Y_2O_3, mixed with the nitride. Suitable shapes for forming substrates are blanked out from the cast product and sintered at 1700–1800°C. Aluminium nitride is used for *chip carriers* where its high thermal conductivity allows good heat dissipation. It is also used as a raw material for sialons.

Aluminium titanate is a mixed oxide of alumina and titania with the formula Al_2TiO_5. The ceramic is formed at high temperature but it slowly dissociates below 1150°C into alumina and titania and this gives a microstructure with a network of fine microcracks. This structure gives the material an effective thermal expansion coefficient of zero up to 700°C. It is used, therefore, in applications where a high resistance to thermal shock is required.

Titania is the ceramic dioxide of titanium, TiO_2, occurring naturally as the mineral rutile and crystallizing as a body-centred tetragonal structure. Titania was formerly used as a dielectric material (dielectric constant = 100) to replace mica and electric porcelains (dielectric constants, 5–10) in applications requiring high permittivity but it has now been replaced for these applications by barium titanate and its derivatives (dielectric constants, 10^4–10^5). Titania is still used as a silicate-bonded ceramic for mechanical applications such as thread guides, the material being manufactured from titania with a small amount of clay and a calcium based flux. It is fired to 1350–1400°C in a reducing atmosphere to give a black non-stoichiometric ceramic which is semiconducting, thus allowing the discharge of static electricity from synthetic textile fibres.

Titanium carbide, TiC, is a hard ceramic material with a density of 4920 kg/m^3, usually used in combination with alumina, Al_2O_3, to give added strength and a higher elastic modulus. Compositions typically contain 20–40% TiC and are hot-pressed to give a fine-grained two phase material which is used as tool inserts for metal cutting.

Zirconia. The oxide of zirconium, ZrO_2, existing in three different crystalline modifications. The high temperature form is cubic and, during cooling, this transforms first into a tetragonal form and then, at a lower temperature, into a monoclinic phase. Transformation on cooling causes cracking because of a 3% volume change and so it is difficult to fabricate pure ZrO_2 ceramics. For this reason, 'stabilizing' oxides, for example, MgO, CaO, or Y_2O_3, are added and these form solid solutions of a cubic structure at room temperature.

The different types of alloyed zirconia ceramics are as follows:

fully stabilized ZrO_2 (FSZ), containing CaO, and having a cubic structure.
partially stabilized ZrO_2 (PSZ), containing MgO, having a duplex structure of cubic crystals with precipitated tetragonal phase.

tetragonal zirconia polycrystals (TZP), containing Y_2O_3 and having a fully tetragonal microstructure.

zirconia toughened alumina (ZTA) which is a composite polycrystalline alumina ceramic containing up to 20 volume percent ZrO_2 as a dispersed phase.

20.12 Glass ceramics

Glass ceramics are complex silicates that can be readily fabricated by the conventional forming techniques for glasses but which are then converted into fine grained polycrystalline materials by controlled crystallisation brought about by heat treatment. The resulting materials have considerably higher strengths than most glasses, and retain their strengths to much higher temperatures than glasses, being capable of use in stressed conditions at temperatures up to 900°C. In addition, ceramic glasses possess excellent resistance to thermal shock.

The controlled crystallisation or devitrification is brought about by a two-stage heat treatment. The first stage is to promote the formation of a large number of nuclei for crystal growth. The glass compositions generally contain small additions of oxides such as TiO_2 and P_2O_5 to act as nucleating agents. The material is then heated at a higher temperature to give controlled grain growth until the desired grain size structure is achieved. There is very little dimensional change during the crystallisation process. The degree of crystallinity achieved is in the range 70–100 per cent and the final structure is dense with no porosity. A range of glass-ceramic compositions exists in both the Li_2O–Al_2O_3–SiO_2 and Li_2O–MgO–ZnO–SiO_2 systems. The former type, LAS glass ceramics have very low thermal expansion coefficients together with a very high thermal shock resistance, while the latter type possess very high thermal expansion coefficients.

The applications for glass ceramics include missile nose cones, domestic ovenware, bearings and ceramic-to-glass seals. The proprietary name for one commercial LAS glass ceramic is *Pyroceram*.

20.13 Glasses

Glass is an amorphous, brittle material obtained by melting together a mixture of various compounds and supercooling the liquid to a temperature where its viscosity exceeds 10^{14} poise (10^{13} Pa s). There is no discontinuity in the cooling curve and the viscosity gradually increases throughout cooling. At low temperatures the material retains the internal structure of the liquid but with a viscosity so high that, for all practical purposes, it behaves as an elastic solid. There are several points or temperatures of importance with respect to glasses. The *glass transition temperature*, T_g, or *fictive temperature*, is the temperature at which viscous flow ceases. This is taken as the temperature at which the viscosity is 10^{17} poise (10^{16} Pa s) (see also Section 6.2). The temperature at which the viscosity is 10^{14} poise (10^{13} Pa s) is known as the *strain point*. At this level of viscosity, there is little atom movement and a glass can be rapidly cooled from this temperature without causing the development of internal stresses. The *annealing point* for a glass is that temperature at which the viscosity is 10^{13} poise (10^{12} Pa s) and at this temperature any internal stresses are rapidly relieved. Glasses are usually worked and formed into shapes at temperatures where the viscosity is in the range 10^3–10^6 Pa s.

Silica is the principal glass forming oxide but it has a high melting point ($>1700°C$). Other oxides, particularly alkalis, lower the softening point so that lower melting temperatures can be employed.

The major commercial glasses are:

Silica glass ($>99.5\%$ SiO_2),
96% silica glass,
Soda–lime–silica glass,
Lead glasses,
Borosilicate glasses,
Alumino-silicate glasses.

Silica glass, also referred to as vitreous silica or fused silica, is a glass consisting entirely of silica, SiO_2, with good chemical stability, a very low thermal expansion coefficient ($0.5 \times 10^{-6}K^{-1}$), and usable up to 900°C. It is made in two forms, (1) translucent, requiring temperatures >2000°C for 'melting' but still containing gas bubbles, and (2) transparent, produced according to the following high temperature reaction:

$$SiCl_4 + 2H_2O \rightarrow SiO_2 + 4HCl$$

the silica condensing as an optical disc. There is also a silica glass containing 96 per cent SiO_2 and 4 per cent B_2O_3. This has good thermal properties and is used for space and missile applications.

Soda–lime–silica glass is the most common type of glass with an approximate composition of 15% Na_2O (soda), 10% CaO (lime) and 75% SiO_2 (silica). Potash, K_2O, may be used as partial replacement of soda and magnesia, MgO, as partial replacement for lime, particularly in flat glass, to avoid premature devitrification during drawing. This type of glass is used also in containers and light bulbs. The electrical properties of the glass varying with composition.

Lead glasses have compositions within the range 30–70% SiO_2, 18–65% PbO, 5–20% Na_2O or K_2O. The original name for lead crystal glass was *flint glass*. It was developed for optical purposes using powdered flint as the source of high purity silica needed to maintain a colourless product. Now the term is used to refer to any colourless glass other than flat glass. Lead-crystal glass refers to a range of lead–alkali–silicate glasses with 18–32% PbO, 5–10% each of K_2O and Na_2O, 50–70% SiO_2, used for high quality tableware. It is also called crystal ware because of its especially lustrous quality as a result of its high refractive index. It is softer than other glass, thus making it easy to cut and grind the material to achieve brilliant decorative effects.

Borosilicate glasses are glasses in which the major constituents are silica (60–80%) and boric oxide (10–25%). They also contain 1–4% Al_2O_3 and 1–4% Na_2O. Borosilicate glasses have low thermal expansion, high chemical resistance, high dielectric strength and a higher softening point than soda glass. They are used for heat-resistant ovenware, laboratory ware, industrial piping and in electrical applications.

Alumino-silicate glasses contain 10–60% SiO_2, 20–40% Al_2O_3, 5–50% CaO, 0–18% B_2O_3 and possess a high chemical resistance and low thermal expansion.

Halide glasses. These are single halide ion special glasses, the most stable of which are based on fluorides of zirconium and barium (50–60% ZrF_4, 30–40% BaF_2 with LaF_3 and AlF_3 to improve stability). They possess a broad transparency range from near ultraviolet (300 nm) to the mid–far infrared (6–9 μm) and are under development for fibre-optic communications.

21

Composite Materials

21.1 Introduction

There are very many situations in engineering where no single material will be suitable to meet a particular design requirement. However, two materials in combination may possess the desired properties, and provide a feasible solution to the materials-selection problem. In this chapter some of the composites in current use, and under development, will be mentioned.

The term composite can refer to any multi-phase material. However, it is usually restricted to 'tailor made' materials in which two or more phases have been combined to yield properties not provided by the constituents alone. The continuous phase in a composite is referred to as the matrix, other phases provide reinforcement. The term includes fibre, whisker and platelet reinforced materials, particulate composites such as dispersion strengthened alloys and cermets, laminates and sandwich materials.

The principle of composite materials is not new. The use of straw in the manufacture of dried mud bricks, and the use of hair and other fibres to strengthen plasters, dates back to ancient civilisations. Also, many of the materials that have been discussed in earlier chapters could be regarded as composite in character, for example, multi-phased metals, case hardened metals, filled plastics, and concrete.

The first material that will be discussed, timber, may seem out of place in this chapter, but woods are complex in structure, and are, effectively, composite materials.

21.2 Timber and plywood

Wood has been used as a structural material for the whole of man's history. Timbers may be classified into two broad types, *soft-wood* and *hard-wood*. Soft-wood is obtained from trees of the coniferous type. This group includes pine, fir, cedar, spruce, and redwood. The term hard-wood is used for wood obtained from deciduous trees, and this includes oak, ash, elm, walnut, beech, hickory, and mahogany.

Wood contains the natural polymer cellulose. This polymer is also the basis of other natural materials such as cotton, hemp, and flax.

The cross-section of a tree shows a characteristic ring structure. In growth, a ring is added each year. In a mature tree, the inner rings, which were the first to form, have ceased to carry sap. These inner rings are termed *heart wood*. The outer rings still carry sap and are, thus, known as *sap wood*. The widths of individual rings vary, not only from one type of tree to another but also within the same species, as the rate of growth of a tree is governed by climate and soil condition. The rate of growth of a tree is greatest in spring, and is almost zero during late autumn and winter.

In soft-woods the cellulose molecules form fibrous tubular cells, termed *tracheids*, and these lie parallel to the trunk and limbs of the tree. In addition, there are also some cells set at right angles to the fibrous tracheids, and lying in a radial direction. These radial cells are termed *rays* and it is through these that plant food passes across the section of a tree. The spaces between cells are mainly composed of *lignin*, a high molecular weight polymer of carbon, oxygen, and hydrogen. The structure of hard-woods is more complex, and one major difference is that there is a higher proportion of radial cells than in soft-wood. The majority of the cells in wood are tracheids, and these give the wood its characteristic *grain*.

Green timber contains a high moisture content and this moisture content is reduced during seasoning. During seasoning, the timber is exposed to air for a period of several months. During this period the moisture content of the timber is reduced from about 33 per cent to about 12 per cent. After it has been seasoned the timber is kiln dried within the temperature range 70–85°C, and this process reduces the moisture content to about 3 per cent. As the moisture content of the wood decreases so the strength is increased.

Because of the fibrous nature of wood, it is highly anisotropic and the tensile and compressive strengths are very much greater along the grain than in a direction at right angles to the grain. The shearing strength along the grain is, however, considerably lower than the shearing strength across the grain. The low shear strength along the grain makes it difficult to use wood as tension members of a structure, but it is a good material for compression members.

One way in which the anisotropy of wood may be overcome is by bonding thin layers together as *plywood*. Plywood is built up of thin layers of wood bonded with a water-resistant glue or a thermosetting resin, with the grain of successive layers at right angles to each other. Plywoods contain an odd number of layers, in order that shrinkage stresses shall be symmetrical about the centre with a consequent minimum of warping.

21.3 *Fibre reinforced materials*

Very hard and strong materials are generally brittle, while the soft, but very tough materials, tend to have low yield strengths. In a fibre-reinforced material, high-strength fibres are encased within a tough matrix. The functions of the matrix are to bond the fibres together, to protect them from damage, and to transmit the load from one fibre to another. The greatest reinforcing effect is obtained when fibres are continuous and parallel to one another, and maximum strength is obtained when the composite is stressed in tension in a direction parallel to the line of fibres. When such a composite is stressed within the elastic range, the strain developed in both fibres and matrix will be the same and so the stresses induced in fibres and matrix will be proportional to the values of Young's modulus for each component. (Refer to Section 7.3.)

For a composite with a ductile metal matrix, the matrix will begin to yield in a plastic manner when the stress in the matrix exceeds its yield stress. Beyond this point, the matrix will deform plastically but the fibres will continue to strain in an elastic manner. At this stage E_c will be approximately equal to $E_f V_f$. If fibres are discontinuous, their strengthening effect will be less than that of continuous fibres, and short discontinuous fibres will be considerably less effective than long fibres.

A material with all fibres aligned parallel will be highly anisotropic. In order to obtain some degree of isotropy in, say, flat sheet material, fibres would have to be arranged in the form of a two-dimensional mat. In this case the maximum strength in any direction would be about one-third of the maximum strength of a material possessing only one-directional fibres. In a composite containing a random three-dimensional fibre arrangement, the maximum strength would only be about one-sixth of that of a fully aligned material.

If very few fibres are added to a matrix the material is weakened rather than strengthened and there will be a critical value of volume fraction of fibres, V_f, below which strengthening will not occur. In the case of carbon fibre filled epoxides the critical value of V_f is 0.02 when all the fibres are fully aligned but V_{fcrit} is 0.052 for random distribution of fibres in one plane. Both of these values are considerably exceeded in the manufacture of carbon fibre filled epoxides. However, in the case of fibres distributed randomly in three dimensions the value of V_{fcrit} is 0.1. Because of the difficulty of packing a large proportion of fibre with a wholly random orientation into a matrix it is not always possible to produce strengthening because it may be impracticable to exceed the value of V_{fcrit}.

Many different matrix/fibre combinations have been developed over the years. Glass, carbon, polymer and ceramic fibres and very fine metal wire filaments have been used in conjunction with thermoplastic and thermosetting resin, glass, ceramic and metal matrices. Not all of these combinations have yielded viable commercial materials but much research and development work is being conducted into new composite materials.

A major stimulus for this effort is the requirement for materials of high specific strength and stiffness for the aerospace industry. The saving of even 1 kg of structural weight can have a significant effect on the operating costs of a civil aircraft or the performance limits of a military aircraft. Many modern aircraft incorporate comparatively large amounts of composites in their structures, for example, carbon/epoxide composites account for more than 25 per cent of the structural weight of the latest versions of the Harrier V/STOL aircraft, and considerable weight saving has been achieved in the Airbus 320 and Boeing 757 through the use of composites.

21.4 Fibres

A fibre is a thread-like form with length to diameter ratios of the order of 10^3 or greater. The term embraces thin metal wires as well as natural and synthetic fibres such as wool, cotton, polyacrylonitrile and nylon. Many materials exhibiting low or moderate strengths and elastic moduli in bulk form, demonstrate greatly enhanced properties when in the form of small diameter fibres. Strong, stiff fibres are produced from cold-drawn metals (normally referred to as wires or filaments), ceramics, and highly orientated polymers. They may be incorporated in polymer, metal, or ceramic matrices to form fibre-reinforced composites. Glass and carbon fibres are the principal materials used in fibre reinforced composites although other polymeric fibres are being used increasingly.

Glass in fibre form with diameters in the range 6–10 μm exhibits very high tensile strength when compared with the bulk material. This results from the much reduced size and number of flaws as the diameter of the fibre is less than the critical Griffith crack length. Two main types of glass are used for the production of glass fibres. These are known as e-glass, a non-alkaline borosilicate, and s-glass, a high strength magnesium alumino-silicate. Although it has a lower strength and stiffness, e-glass is more widely used than s-glass. The availability and relatively low price of glass fibre compared with carbon and other high modulus fibres makes glass an attractive reinforcement where high stiffness composites are not required. Typical properties of glass fibres are given in Table 21.1.

Carbon fibres, with diameters in the range 6–10 μm, possess high elastic moduli and strengths and are used as a reinforcing material in epoxy and polyester resins for the manufacture of high strength and stiffness composites. Carbon fibre reinforcement is used also in some thermoplastics, metals, and in carbon itself. There are several grades of carbon fibre, the main types being Type I (high modulus) and Type II (high strength). Two other grades are Type A and fibre made from mesophase pitch. Commercial production is by the carbonizing of polyacrylonitrile fibre at 900°C followed by a heat treatment, with the fibres under tension, to give the final structure which is a graphitic crystalline structure with the basal planes orientated approximately parallel to the fibre axes. Heat treatment at 2000–2500°C will give Type I and heat treatment at 1300–1500°C gives Type II fibre. Type A, with intermediate properties, is treated at an intermediate temperature. Typical properties for the various grades of carbon fibre are given in Table 21.1

An aramid is an aromatic polyamide and a polymer of this group which has been used successfully for the production of high strength, high modulus fibres is poly p-phenylene diamine teraphthalate. This is marketed under the proprietary name *Kevlar*. The high strength and stiffness result from the rigid nature of the benzene rings and rod-like molecules of the polymer. Aramid fibres are used in high performance composites. The best known examples of aramid fibres are Kevlar 49, a competitor with carbon fibre, and Kevlar 29, a similar fibre but with a lower elastic modulus. The properties of Kevlar 49 are given in Table 21.1.

Thermoplastic polymers, including polyethylene, polypropylene and polyamides can be produced as high strength, high modulus fibres. The development of high strength and stiffness is conditional upon creating highly aligned and orientated molecular structures. Polyethylene, usually regarded as a low strength and stiffness material, can be produced in fibre form with a value of E up to 100 GPa and a tensile strength as high as 3–4 GPa. Polyethylene and polypropylene fibres are only suitable for low temperature use as the highly orientated structure is lost when the material is heated.

Boron fibres, made by the pyrolytic deposition of boron on a core fibre of carbon or tungsten from the reaction between boron chloride and hydrogen at 1200°C, have been used in conjunction with epoxy and aluminium matrices to give high strength fibre reinforced compos-

Table 21.1   *Properties of some fibres and wires*

Material	Young's modulus (E) (GPa)	Tensile strength (GPa)	Density (kg/m$^3 \times 10^{-3}$)
Natural cellulose	48–100	0.4–1.0	1.5
Terylene	1.2	0.6	1.4
Polyamide (nylon)	2.9	0.8	1.1
Aramid (Kevlar 49)	130	3.6	1.4
e-glass	70	2.0	2.5
s-glass	84	4.6	2.6
Boron	410	2.9	2.5
Carbon (Type I)	400–410	2.0–2.8	2.0
Carbon (Type II)	200–240	3.0–3.5	1.7
Carbon (Type A)	220–300	2.4–3.6	1.85
Silicon carbide	400	6.0–8.0	3.2
Alumina	470	2.0	4.0
Beryllium	300	1.5	1.8
Tungsten	350	3.5	19.0
Steel	210	2.8	7.8

ites. Boron fibres made in this way are much thicker than carbon and glass fibres and have diameters of between 0.1 and 0.2 mm. E for the fibres is 400 GPa and the tensile strength is in excess of 3 MPa. There has been some use of boron fibre/epoxy composites in the US space programme.

Silicon carbide can be produced as light, stiff and strong fibres with a typical diameter of 0.1 mm by chemical vapour deposition onto a tungsten substrate, using chlorinated silanes. Reaction with the tungsten substrate makes the fibres unsuitable for use above 900°C. Finer fibres (0.001 mm diameter) may be produced from carbon/silicon polymers in a process similar to that used for the manufacture of carbon fibres. Considerable research into the use of silicon carbide as a reinforcing fibre in aluminium and titanium matrices is being carried out.

21.5 Matrix materials

The main requirements of a good matrix material are that it can fully infiltrate between fibres and form a good interfacial bond, that there should be no chemical interaction between fibre and matrix, other than that involved in bonding, and the matrix should not cause physical damage to the reinforcing fibres.

Thermosetting resins

Unsaturated polyester resins (see Section 19.4) were the first type of matrix material to be used in conjunction with glass fibre reinforcement and polyester GRP materials have been used as constructional materials for more than 40 years. They are relatively cheap, are easy to work and cure at room temperature, although cure rate can be accelerated by heating. Polyester resins are the most widely used matrix materials, the reinforcing fibres used with them being, mainly, glass and carbon. Many components, both large and small, are fabricated using polyester/glass composites, including machine housings, body assemblies for car and other motor vehicles, canoe and boat hulls. Some of the largest GRP mouldings have been the hulls for naval minesweepers.

Epoxide resins. Epoxides (see Section 19.5) possess several significant advantages over polyester resins: they are stronger and have greater stiffness, show very little cure shrinkage, have much better adherence to reinforcing fibres, and can be used at higher temperatures than polyesters. Epoxides are the main matrix materials for use with carbon and aramid fibres. Other thermoset matrix materials have been investigated and polyimides offer the possibility of fibre composites with a higher maximum service temperature than is possible with epoxides.

Thermoplastics

The use of thermoplastic matrices for fibre reinforcement can give materials with a greater fracture toughness than is possible with thermosets. Virtually all thermoplastics can be strengthened and stiffened by incorporating glass fibre but in many cases these are in the form of glass-filled polymers, that is polymer with chopped glass fibre as a filler, rather than as FRP's with continuous fibre reinforcement. One disadvantage of thermoplastics as matrix materials for continuous fibre reinforcement is that infiltration around all fibres is difficult owing to the high viscosities of polymer melts. Many materials have been tried as matrices, including PA 6.6, PAI, PBT, PET, PES, PPS, PSU and PEEK. Considerable effort is being placed into development of thermoplastic matrices, particularly PEEK, with carbon and aramid reinforcement, for high performance composites. The fracture toughness and fatigue strength of a PEEK based composite is considerably higher than for epoxide-based materials and these offer potential for use in aerospace applications.

Metal matrices

The use of a metal matrix offers the possibility of obtaining composites of high stiffness and strength and with the ability to operate at higher temperatures than polymer-based composites (see also Section 21.6). There are, however, difficulties in fabrication which result in these becoming high cost materials. Focus has been mainly centred on the use of aluminium, titanium and magnesium, together with their alloys, as possible matrix materials with silicon carbide and alumina as the main reinforcing fibres. The manufacturing processes used to achieve these composites include powder metallurgy techniques, vapour deposition, diffusion bonding and infiltration of liquid metal into fibre bundles under pressure.

21.6 Metal matrix composites (MMC)

The term metal matrix composite (MMC) covers a range of materials and not merely composites with continuous fibre reinforcement. Early materials, which today could be regarded as MMCs, were tungsten filament wire strengthened with a dispersion of thoria and sintered aluminium powder (SAP), effectively a fine dispersion of alumina in aluminium. SAP containing 10 per cent of dispersed alumina has a room temperature tensile strength of 390 MPa, compared with a strength of 90 MPa for annealed aluminium of commercial purity. At 400°C the tensile stength of SAP is as high as 180 MPa although at this temperature the conventional high strength aluminium alloys possess little, if any, strength. Other materials, strengthened in similar manner, developed later were TD (thoria dispersion) nickel alloys for high temperature service. Other examples have been the development of DS (directionally solidified) eutectic nickel based superalloys for high temperature service. These can be regarded as *in situ* composites in which rod or plate shaped precipitate particles, produced during a controlled solidification process, provide the reinforcement.

Much interest has been given in recent years to the development of MMCs, mainly based on aluminium and titanium alloy matrices, with either continuous fibre reinforcement or discontinuous particle reinforcement, the main reinforcers being silicon carbide and alumina. Discontinuous particle reinforcement, using short fibres, whiskers, platelets or particles, appear to offer somewhat greater potential than the use of continuous reinforcement. Whiskers are very small filament-type single crystals produced in such a way that they contain few, if any, defects. Single crystal whiskers of ceramic materials are typically a few millimetres in length and about a micrometre in diameter. They are almost perfect crystals and have very smooth surfaces, hence strengths close to the theoretical values are obtainable. Production costs are high and their incorporation in matrices is technically difficult. A number of compounds such as mica and silicon carbide can be made to grow in laminar form as platelets or microplates. When produced with optimum thickness and smooth edges, these materials can be used as a reinforcing medium in composites. The properties of some whiskers and platelets are given in Table 21.2.

Table 21.2 *Properties of some whiskers and platelets*

Material	Young's modulus (E) (GPa)	Tensile strength (GPa)	Density (kg/m³ × 10⁻³)
Whiskers			
Alumina	410–480	8.0–9.6	4.0
Silicon carbide	840*	21.0	3.2
Beryllium oxide	720*	6.9	3.0
Graphite	675	21.0	2.2
Platelets			
Mica	226 (max)	3.1	2.8
Silicon carbide	480	10.0	3.2
Aluminium diboride	500	6.0	2.7

* Value in the most favourable orientation

21.7 Cermets

The dispersion-hardened materials normally contain less than 15 per cent of dispersed ceramic. The term *cermet*, however, refers to ceramic-metal composites containing between 80 and 90 per cent of ceramic.

The earliest cermet, also known as *hard metal*, was cemented tungsten carbide. This material was first developed in the early part of the twentieth century for use in wire drawing dies, and for metal-cutting tool tips. Originally, tungsten carbide powder was hot pressed with cobalt and iron, but subsequent developments have resulted in hard metal composed of a mixture of several carbides, such as tungsten carbide, titanium carbide, and tantalum carbide, in a matrix of cobalt. Cobalt is an exceptionally good binder for the carbides as it forms a thin continuous film around all the carbide particles. With cobalt contents of up to 15 per cent, the cermet is a wholly elastic and brittle material. If the cobalt content is increased beyond 15 per cent, the brittleness of the cermet is reduced, but there is an accompanying loss of hardness.

The possibility of other cermets being useful as engineering materials, particularly as potential gas-turbine engine materials, led to considerable investigation into oxide-based and silicon carbide-based composites with nickel, cobalt, and stainless steels. The major difficulties that were encountered in the development of cermets as possible turbine-blade materials were their very low impact strengths and low resistance to thermal shock. At present, research into cermets appears to have lost momentum.

21.8 Sandwich structures

Sandwich construction composites are being used increasingly in situations where high stiffness coupled with low structural weight is a major requirement. One of the first successful applications of sandwich construction for highly stressed structural components was the use of plywood skins with light balsawood as the sandwich filling to manufacture the main airframe structure of the Mosquito aircraft of World War II.

Sandwich structures generally are composed of two skins of high strength with a light weight core. They can be considered as miniature I-beams and possess a very high stiffness/weight ratio.

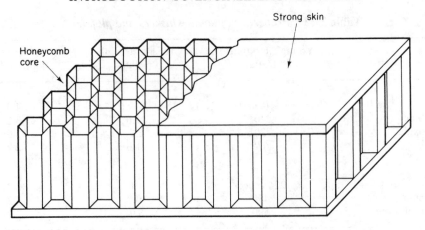

FIGURE 21.1   Honeycomb panel construction

Multiple layer combinations are also used widely in engineering. This arrangement provides a material with low density and high specific stiffness, where the maximum tensile and compressive stresses are carried by the skins. Examples of skin materials are aluminium and titanium alloys and fibre reinforced polymers and examples of light weight core materials are expanded polymers, such as polyurethanes, and honeycombs fabricated from metals or reinforced polymers, which resist shear stresses in the sandwich. The latter are more expensive, but provide superior properties. In addition to providing light weight structural materials for transport vehicles, particularly aircraft, sandwich materials offer useful thermal and sound insulation.

Honeycomb and structural foam-filled sandwich structures are being used increasingly as major structural elements in high-performance military and civil aircraft. This type of construction is expensive to fabricate but offers an excellent solution to design requirements in the aerospace industry.

There are very many other sandwich structures produced and used in a wide variety of applications. One of the simplest and one that has been used for very many years is the sandwich made with paper sheets bonded over a core of corrugated paper and used for the manufacture of cardboard cartons.

Many sandwich materials are used in building construction because in addition to a high stiffness/weight ratio, they may also possess good thermal and sound-insulation characteristics. A variety of combinations are used. Skin materials include sheet metal, plywood, plastics (including glass-reinforced plastics), concrete and plasterboard, with cores that may be metal or paper honeycomb structures, rigid plastic foams, chip board or low-density porous masses of glass fibre or rock wool bonded with a plastic resin.

PART IV
FORMING AND FABRICATION OF MATERIALS

22

Forming Processes for Metals

22.1 Introduction

Materials have to be processed into a great variety of shapes in order to make component parts of every type. The shapes required vary enormously, both in size and complexity, ranging from micro-electronic components to large castings and forgings of, perhaps, several hundred tonnes mass (1 tonne = 1000 kg = 1 Mg). The engineer must have some awareness of the range of manufacturing processes available, and of the advantages and limitations of the various processes. The properties of the material in the finished component are also influenced to a considerable extent by the type of shaping process employed, and by the conditions existing during processing. The whole range of shaping processes can be very broadly classified into four categories:

(a) casting, namely the pouring of liquid into prepared moulds,
(b) manipulative processes, involving plastic deformation of the material,
(c) powder techniques, in which a shape is produced by compacting a powder,
(d) cutting and grinding operations.

Most metallic materials are initially produced in the liquid phase and are then cast into shapes, either to give a casting which may require only machining operations to be carried out on it before sale, or into ingots which can be further processed by manipulative techniques such as rolling or forging. Conditions during melting and casting are very important in determining the subsequent quality of a metal or alloy and so a brief survey of some types of melting furnace is given before the discussion of metal-casting processes.

22.2 Melting furnaces

There are several types of furnaces used for the melting of metals, and these furnaces may be heated by solid, liquid, or gaseous fuels, or by electrical power. We will first consider furnaces fired by chemical fuels. In order to melt a metal efficiently it is necessary to have effective transference of the heat output from the burning fuel to the metal charge. Good thermal efficiency could be obtained with the metal to be melted in direct contact with the burning fuel, but this would lead to severe contamination of the metal. However, furnaces of this type are used in practice. Their application for metal melting is severely limited, owing to the contamination problem, and is confined to the melting of cast iron, a comparatively impure material (refer to Chapter 16). In this type of furnace, known as a *cupola*, the fuel used is coke and there will be some pick-up of carbon by the metal.

The risk of contamination will be lessened and thermal efficiency will still be reasonably good if only the hot products of fuel combustion are allowed to come into contact with the metal. Furnaces based on this principle are termed reverberatory furnaces (Figure 22.1(b)). In this type of furnace the hearth is normally rectangular with a relatively large surface area, but shallow in depth. Reverberatory furnaces may be fired by solid, liquid, or gaseous fuels. Heat transfer is both by direct contact between the hot gases and the metal charge, and also by radiation from the hot arched furnace roof. This type of furnace is used for the melting of very many metals and alloys, but furnace design differs in detail from one industry to another. For example, because aluminium tends to oxidise rapidly at high temperatures and is also susceptible to hydrogen contamination, the roof of a reverberatory furnace for aluminium melting would be somewhat higher than is customary in furnaces for the melting of other metals. Also, fuel burners would be angled away from the hearth and along the underside of the furnace roof so that fairly quiescent conditions would exist at the surface of the metal. Heat transfer to the metal charge would be mainly by radiation from the furnace roof.

Reverberatory furnaces may vary in size from liquid metal capacities of about one tonne to capacities of several hundred tonnes. In general, a high standard of cleanliness may be maintained with this type of furnace, and the type is used for the large-scale melting of very many metals and alloys of commercial purity. Many fuels, particularly fuel oils, possess a high sulphur content and the metal could pick up sulphur from the combustion products. Another possible source of contamination is the pick-up of hydrogen and/or oxygen from the dissociation of water vapour on the hot liquid metal surface. In order to reduce oxidation and contamination many metals are melted under the protective covering of a molten flux, or *slag*.

Where a higher quality is required than can be obtained from reverberatory furnace melting, crucible melting may be used. In this type the metal to be melted is placed within a crucible, or pot, and the external walls of the crucible are heated. In this way the metal is isolated from the major sources of contamination. Crucible capacities range from a few kilogrammes to about one tonne.

Electrical furnaces may be divided into crucible type and arc type. Electrical crucible furnaces operate on the induction principle, and this system of melting is widely used for the production of high-quality metals and alloys. The high-frequency induction furnace is suitable for the melting of metals of low electrical conductivity. The crucible is placed within a water-cooled coil. When the coil is energised, secondary eddy currents are induced within the metal charge. The heating effect of the induced currents is sufficient to raise the temperature of the charge to the melting range. When the charge is molten the circular paths of the eddy currents cause swirling of the metal and this ensures a thorough mixing. It is possible to operate this type of furnace *in vacuo*, and vacuum melting is sometimes used for the preparation of some high-purity metals and for some special high-performance metals and alloys.

When the induction principle is used for the melting of high-conductivity metals such as aluminium, copper, and their alloys, the low-frequency type of furnace is chosen. This is illustrated in Figure 22.1(d). A primary coil, with a very large number of turns, is positioned around the loop in the lower portion of the furnace. The furnace is primed with some molten metal to completely fill the loop and the primary coil is then energised. The loop metal effectively becomes the secondary winding, possessing one turn only, of a step-down transformer. A very large secondary current is induced within the liquid metal. The heating effect is proportional to the square of the current, and so there is a considerable heating of the metal in one limb of the loop. A convecting flow of liquid metal around the loop is established. Solid metal may then be charged to the furnace crucible and it is then melted cleanly and efficiently.

Direct arc furnaces, as illustrated in Figure 22.1(e), are used extensively in the production of alloy steels. The arc is struck between the carbon electrodes and the metal. There may be some

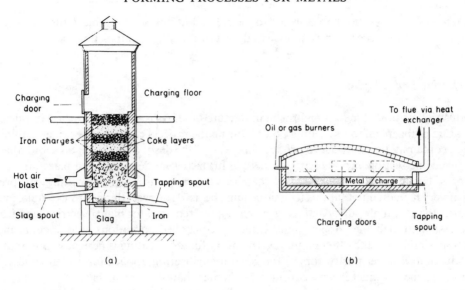

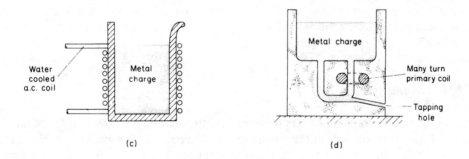

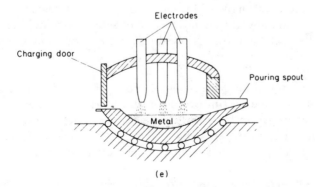

FIGURE 22.1   Types of furnace: (a) cupola; (b) reverberatory; (c) high-frequency induction; (d) low-frequency induction; (e) direct arc

transference of carbon from the electrodes to the metal, but this is acceptable in the production of many grades of steel, as carbon is an essential part of the alloy composition.

22.3 Melting and alloying

The industrial melting of metals and the manufacture of alloys is not simply a case of raising the temperature of the metal to its melting point. During the process care must be taken to minimise, or avoid, contamination in order to produce material of the desired quality at an acceptable cost. In the preceding section various types of melting furnace were described and placed in order of cleanliness, but it must not be assumed that because vacuum-induction melting can produce metals of very high quality that this technique must be used for everything. For example, in order to produce the very high-purity and oxygen-free copper required for many electrical applications it is necessary to melt the copper under carefully controlled conditions, and vacuum melting would be desirable. On the other hand, reverberatory furnace melting is perfectly satisfactory for the production of lower-purity copper for general engineering applications. Consequently, the cost of the former product is very considerably higher than that of the latter.

In some instances it is possible to make an alloying addition simply by adding ingots of the alloying element to a bath of molten parent metal. Magnesium can be added to aluminium in this way. Generally, however, once the parent metal has been melted additions are made in the form of hardener alloys, that is, pre-prepared alloys rich in alloying element. For example, to make an addition of up to 5 per cent copper to aluminium, the copper is added in the form of a hardener alloy containing 50 per cent, by weight, of copper. The use of hardener alloys greatly aids the solution of alloying elements in the basic metal.

22.4 Melt treatment

Once melting and alloying have been completed it is often necessary to treat the liquid metal prior to casting. It is quite common for the molten material to contain dissolved gases. The maximum gaseous solubility limit in a liquid metal is generally much higher than the maximum gas solubility limit in the solid material. Solidification of a liquid metal with a high dissolved gas concentration would result in much of the gas being rejected from solution during solidification. Most of this rejected gas would be unable to escape to atmosphere, as a solid skin quickly forms on the casting, and a porous casting would result. Degassing treatments are often necessary to cleanse the liquid metal and allow for the production of sound castings. The type of treatment given will depend on the metal and the dissolved gas, but the following examples will give an indication.

(a) Molten aluminium readily absorbs hydrogen. Removal of hydrogen from aluminium can be affected by bubbling chlorine gas through the melt. Chlorine is insoluble in aluminium, and will combine with hydrogen to form hydrogen chloride, which is also insoluble in molten aluminium, so removing hydrogen from the metal. The gas bubbles will also attach themselves to non-metallic inclusions, such as aluminium oxide particles, and float them to the surface of the melt.

(b) Copper, and many copper alloys, absorb oxygen during melting. One method of oxygen removal is to add phosphorus in the form of a phosphorus–copper alloy, to the melt. Oxides of phosphorus are formed, which separate to the surface of the melt; they can then be removed by skimming. This technique cannot be used for deoxidising high-conductivity copper, as any residual phosphorus in the copper would greatly reduce the conductivity. High-conductivity copper may be deoxidised with calcium boride and lithium.

22.5 Casting

The casting of liquid metal into a shaped mould is a convenient way of making many types of component. The decision on whether or not to produce a component by casting and, if so, by which of the many casting processes, is made after a consideration of both practical and economic factors. The shape and/or size of the component required may be such that a casting process is the only suitable method available even though the crystal structure and mechanical properties may be inferior to those of a wrought product.

The main casting methods available are *sand casting* in which liquid metal is poured into a shaped cavity moulded in a sand, *die casting* in which the mould cavity is machined within metal die blocks, *investment casting* and *centrifugal casting*. Moulding sands have a fairly low thermal conductivity so that the rate of solidification of liquid metal within a sand mould is fairly slow, giving rise to a coarse crystal grain size. In contrast, rapid solidification and a fine grain size will occur with die casting. A broad comparison of the various casting processes is given in Table 22.1.

22.6 Sand casting

The moulds for sand castings are made from particular types of sands. These sands, which may be either naturally occurring or synthetic, must possess a sufficient strength to retain the mould

Table 22.1 *A comparison of casting methods*

	Sand casting	Gravity die casting	Pressure die casting	Centrifugal casting	Investment casting
Alloys that can be cast	Unlimited	Copper-base, aluminium-base and zinc-base alloys	Copper-base, aluminium-base and zinc-base alloys	Unlimited	Unlimited
Approximate maximum possible size of casting	Unlimited	50 kg	15 kg	Several tonnes	5 kg
Thinnest section normally possible (mm)	3	3	1	10	1
Relative mechanical properties	Fair	Good	Very good	Best	Good
Surface finish	Fair	Good	Very good	Fair	Very good
Possibility of casting a complex design	Good	Good	Very good	Poor	Very good
Relative cost for production of a small number off	Lowest	High	Highest	Medium	Low
Relative cost for large-scale production	Medium	Low	Lowest	High	High
Relative ease of changing the design during production	Best	Poor	Poorest	Good	Good

shape both before and during the entry of liquid metal to the mould. They must also be refractory and be permeable enough to allow for the escape of steam and gases.

The naturally occurring material used for mould making is *greensand*. This is a siliceous material containing a proportion of clay. The exact composition of greensands varies with the location of the deposit and the clay content may range from 5 per cent to 20 per cent. The clay, when moist, acts as a binding agent around the silica grains to give the sand a certain strength. The moisture content must be closely controlled to give a mix that is readily mouldable but not so moist that an excessive amount of steam would be generated when liquid metal is poured into the mould, as this would lead to defective castings. Synthetic sands are made by blending a dry silica sand with a clay. A sand mould is only used once, as it has to be broken to retrieve the finished casting, but the sand is re-usable. However, the sand that was at the surface of the mould cavity will have undergone a change. The clay binder in this region will have been heated to a temperature sufficiently high to convert it into a hard non-plastic state, the same process that occurs when clay is fired in a kiln to make bricks or pottery. This altered material should be removed from the sand and not re-used.

Moulds for the production of small castings are normally made from a fine-grained sand as this will ensure a good surface finish. Moulds for large castings are generally made from a coarse-grained sand to give good permeability for steam and gases to escape readily. A thin layer of a fine-grained facing sand would be used at the mould cavity surface to ensure that the casting has a good surface finish.

Cores have to be inserted within the mould cavity if a hollow casting is to be made. A core has to be strong enough to stand largely unsupported within the mould cavity but not so strong that it cannot be removed easily from the finished casting. Core sand may be a greensand with an additional binder. It was common practice to use linseed oil as a binder. Linseed oil-bonded sand cores need to be baked in an oven for several hours to render them sufficiently strong for use. The use of sand-bonded cores has been largely superseded by other types of core that do not require to be baked before use. Cores are now generally made by shell moulding techniques, or by the carbon dioxide process. Shell moulding sand is a mixture of silica sand and a thermosetting resin. The core moulds are made of metal and are heated. When the sand-resin mix is put into the core mould the heat causes the resin to 'cure' and bind the sand grains together. Because of the low thermal conductivity of the sand-resin mixture the curing is effective only for a distance of about 3 mm from the heated metal surface. Surplus sand is then removed leaving a hollow, but strong, shell core (see Figures 22.2 and 22.4). In the carbon dioxide process a greensand or a pure silica sand is mixed with a small amount of sodium silicate solution. After making the core shapes in a core-moulding box, carbon dioxide is passed through the cores. The effect of the carbon dioxide is to convert the sodium silicate into silica gel which firmly binds the sand grains together.

The first stage in the production of sand castings must be the design and manufacture of a suitable pattern. Casting patterns are generally made from hard wood and the pattern has to be made larger than the finished casting size to allow for the shrinkage that takes place during solidification and cooling. The extent of this shrinkage varies with the type of metal or alloy to be cast. For all but the simplest shapes the pattern will be made in two or more pieces to facilitate moulding. If a hollow casting is to be made the pattern design will include extension pieces so that spaces to accept the sand core are moulded into the sand (see Figure 22.3). These additional spaces in the mould are termed core prints.

Sand moulds for the production of small and medium-sized castings are made in a moulding box. The mould is made in two or more parts in order that the pattern may be removed. With a two-part mould the upper half of the moulding box is known as the *cope* and the lower half is termed the *drag*. The various stages in the production of a complete sand mould are described below and illustrated in Figure 22.3.

FIGURE 22.2 Shell mould and cores

The drag half of the mould box is placed on a flat firm board and the drag half of the pattern placed in position. Facing sand is sprinkled over the pattern and then the mould box is filled with moulding sand. The sand is rammed firmly around the pattern. This process of filling and ramming may be done by hand but mould production is automated in a large foundry with the mould boxes moving along a conveyor, firstly to be filled with sand from hoppers and then to pass under mechanical hammers for ramming. When ramming of the sand is complete excess sand is removed to leave a smooth surface flush with the edges of the moulding box.

The completed drag is now turned over and the upper, or cope, portion of the moulding box positioned over it. The cope half of the pattern is placed in position, correct alignment being ensured by means of small dowel pins. Patterns for the necessary feeder, runner and risers are also placed in position. Feeder and runner channels should be placed so as to give an even distribution of metal into the mould cavity. The risers should coincide with the highest points of the mould cavity so that the displaced air can readily escape from the mould. The sizes of risers should be such that the metal in them does not freeze too rapidly. An important function of a riser is to act as a reservoir of liquid metal to feed solidification within the mould. A thin coating of dry parting sand is sprinkled into the mould at this stage. This is to prevent the cope and drag sticking together when the cope half is moulded. The cope is now filled with moulding sand and this is rammed firmly into shape in the same manner as in the making of the drag.

After the ramming of sand in the cope is completed and excess sand has been removed from the top surface the two halves of the moulding box are carefully separated. At this stage *venting* of the mould can be done, if necessary. The venting operation is the making of holes in both cope

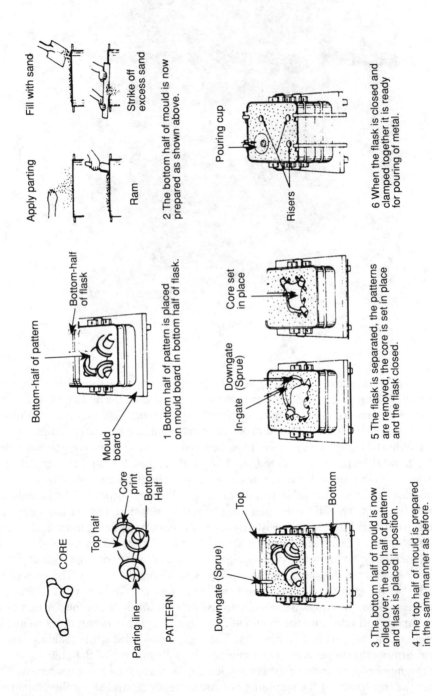

FIGURE 22.3   Stages in making a small sand mould

and drag with the aid of a thin wire. This is to increase the permeability of the mould allowing for easy egress of steam and other gases from the mould when the liquid metal is poured in. After venting the patterns are carefully removed from both cope and drag. Minor repairs are now made to the surface of the mould cavity, if necessary, and a gate or gates are carefully cut to connect the runner channel with the main cavity. Gates should be sited to allow for entry into the mould with a minimum of turbulence. Any loose sand is gently blown away and if a core is to be used it is placed in position. The mould is reassembled by placing the cope upon the drag and it is then ready for use. Liquid metal is poured smoothly into the mould via the feeder. Pouring ceases when liquid metal appears at the top of the risers and the feeder channel is also full.

The moulds for large castings are made in a different way. The mould is generally made in a pit on the foundry floor and the basic outline of the mould is built in brick as a firm basis for the moulding sand.

When the metal that has been poured into a sand mould has fully solidified the mould is broken and the casting is removed. The casting still has the runner and risers attached to it and there will be sand adhering to portions of the surface. Runners and risers are cut off and returned to the melting furnace. Sand cores are broken and removed and adherent sand is cleaned from the surface by vibration or by sand blasting with dry sand. Washing with a high-pressure water jet may also be used with non-ferrous castings. Any fins or metal flash formed at mould parting lines are removed by grinding and the castings are then ready for inspection.

Lack of attention to design or bad practice in either moulding, melting of the metal or in pouring can lead to defects within a casting. Most metals undergo a considerable volume shrinkage during solidification, and it is the function of the risers to provide reservoirs of liquid metal to feed this shrinkage. Adequate provision of risers should largely eliminate the possibility of major solidification shrinkage zones within the casting but finely divided interdendritic porosity is inevitable. Another type of porosity that may occur is gas porosity. If the liquid metal is incompletely degassed before pouring, gases will be rejected from solution as the metal solidifies and will remain as gas-filled blow-holes within the casting. Reaction gas porosity will tend to give blow-holes just under the skin of a casting. This type of defect can occur if the moulding sand is too moist or if the mould has been so hard rammed that it has a very low permeability so that steam generated as hot metal comes into contact with the mould blows back into the metal. Inclusions within a casting may arise from one of two sources. Bad melting practice may give a liquid metal containing slag inclusions, but it is also possible that loose sand may be washed away from the surface of a poorly prepared mould to form sand inclusions within the casting. The removal of sand in this way will also give a casting with unsightly lumps of excess metal on the surface. Another type of defect is the *cold shut*. This is a discontinuity within the casting. The basic cause is that the temperature of the liquid metal is too low so that two streams of metal may meet within the mould but not fuse together completely. Cold shuts may also occur if the pour is irregular. Bad design of a casting may be responsible for the creation of large thermal stresses as the casting cools and contracts under constraint. Such stresses may cause the casting to warp or may cause *hot tears* to occur. Hot tears are fractures within the casting due to the generation of a thermal stress that exceeds the strength of the hot casting just after solidification.

Despite some disadavantages sand casting is suitable for the production of castings in almost any metal and of almost any size from a few grammes up to several hundred tonnes.

Shell moulding has already been mentioned as a method of producing cores for sand moulds. The process can also be used for making moulds for the production of small castings. Shell moulds have the advantage that they are relatively strong and can be transported and stored without damage. This type of mould can be made to close dimensional tolerances and with a good surface finish, ensuring that castings of high quality are made. Prior to casting the two

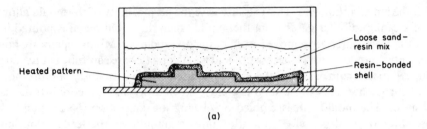

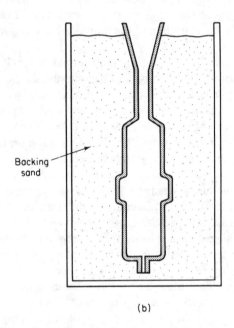

FIGURE 22.4   Shell mould casting: (a) sand/resin mix dumped over heated metal pattern plate, (b) shells
             assembled to receive liquid metal

halves of a shell mould are fixed together, placed in a box and surrounded by sand for support
(see Figure 23.4).

22.7 *Die casting*

Gravity die casting is very similar in principle to sand casting, in that the die design must include
riser heads to feed solidification shrinkage. The main advantages of the process are that a more
rapid solidification rate produces a finer crystal grain structure than in a sand casting, and in
consequence the die casting will possess a higher strength than a corresponding sand casting.
Also, better dimensional control and surface finish will be obtained, as compared with a sand-
cast product. The cost of dies is high and the use of the process is restricted to metals and alloys
with melting temperatures not exceeding values of about 1000–1100°C. Gravity casting into
metal moulds is often referred to as *permanent mould casting*, and the term *die casting* tends to be
reserved for the *pressure die process*.

The dies or moulds are made of steel. There is a variety of die steels available, most being either nickel/chromium/molybdenum steels or chromium/tungsten/cobalt steels. The choice of die steel will be determined by the melting temperature of the metal to be cast and the length of the production run required. Steel cores may be used with gravity dies if it is required to cast hollow shapes of simple design. Hollow shapes of more complex design can be made by gravity die casting with sand-core inserts. An example of a product made in this manner is the body of a domestic water tap.

The injection of liquid metal under pressure into a mould offers several advantages. It means that very thin sections can be cast, that metal can be forced into the recesses of a mould of very complex shape, and that rapid solidificaton under presure will considerably reduce porosity. Also, the need to provide risers in the mould is eliminated. In place of risers all that is necessary is a series of vents, of about 0.1 mm in thickness, at the parting line of both halves of the die. These will be sufficient to allow for the escape of air from the mould cavity as metal enters, but will be too small to permit the passage of metal. In die casting there is little metal wastage and a casting is obtained that requires little, or no, machining. The process is, however, restricted to the lower melting point metals and alloys, the principal ones being zinc, magnesium, aluminium, copper, and their alloys (see Figures 22.5 and 22.6).

There are two main types of process, namely *hot chamber die casting* and *cold chamber die casting*. The former process uses a 'goose-neck' type machine in which the goose-neck is partially immersed in the liquid metal bath in a holding furnace. In each cycle, liquid metal enters the goose-neck and is forced into the die attached to the furnace by either a mechanical piston or pneumatic pressure (Figure 22.5(a)) giving injection pressures of up to 40 MPa. This type of machine is used principally for zinc and magnesium based alloys. In the cold chamber process, the charge of liquid metal is poured into a transfer chamber from a holding furnace immediately prior to injection, under a pressure of up to 150 MPa, into the die (Figure 22.5(b)). The cold chamber process is suitable for the die casting of aluminium and brass as well as for magnesium and zinc.

An interesting feature of die casting is that 'insert' castings can be made, that is, an insert can be placed in the die and the liquid metal cast around it. This is particularly useful for including a threaded bolt or stud in a cast component, as in the examples shown in Figure 22.6. It is, of course, necessary that the part of the thread within the casting be mutilated before insertion into the die to prevent it unscrewing from the casting in use. Pressure die casting is used for the manufacture of such components as automotive parts, including carburettor and petrol pump bodies, door handles and fitments, power tool casings and accurately scaled toys.

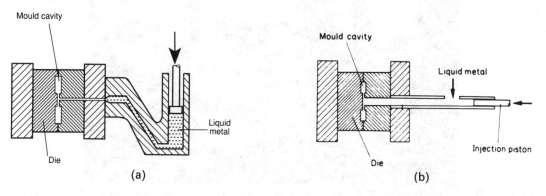

FIGURE 22.5   Principle of pressure die casting: (a) hot chamber machine; (b) cold chamber machine

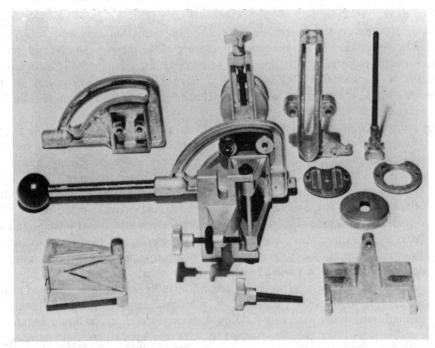

FIGURE 22.6   A saw-grinding attachment making use of nine zinc-alloy die castings. The plate shows the individual castings and the complete assembly. The knobs are die cast on to the adjusting screws (courtesy of Zinc Development Association and Hollands and Blair Ltd)

22.8 Investment casting

Another important casting process is the *investment*, or *lost wax*, process. This process has its origins in the early Chinese and Egyptian civilisations, but it has become a valuable part of twentieth century technology. Some metal components have to withstand very high temperatures and stresses in service and, in the case of such components as aero engine turbine blades, the parts

FIGURE 22.7   Typical examples of parts made by investment casting. The casting shown at the top left is in steel and of 220 mm diameter and 5 kg mass (courtesy of British Investment Casters' Technical Association)

have to be made to very high standards of dimensional accuracy. Many materials sufficiently strong at high temperatures are virtually impossible to machine or to shape at ordinary temperatures. A feasible solution would be to use a precision casting process. Pressure die casting is not possible in these cases, as the melting point of the alloy would be too high, and investment casting can provide the answer.

In the investment process, a master mould is produced in a readily machineable alloy, such as brass, and this mould is used for the production of accurate patterns in wax or a low melting point alloy. The pattern, together with a feeder sprue, is dipped into a ceramic slurry and the coating formed allowed to dry. In the case of small components several patterns may be assembled around a central sprue as a 'tree' before coating with slurry. The dipping process may be repeated several times until the required coating thickness, usually 8–10 mm, is achieved. (For the production of investment castings in metals of relatively low melting point, plaster of Paris may be used in place of a highly refractory material to make the mould). After the completed coating has fully dried out the assembly is heated and the molten wax or low melting point alloy allowed to run out. This is followed by heating to a higher temperature to burn out all traces of wax and to sinter the refractory, forming a perfect ceramic mould. The thin-walled investment

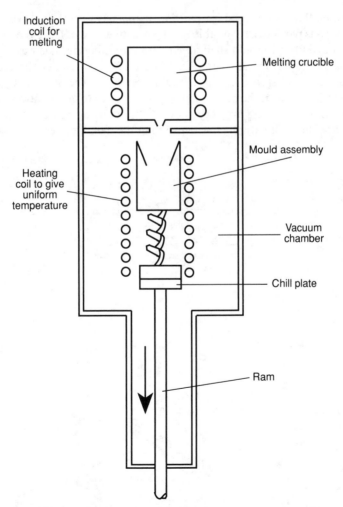

FIGURE 22.8 Arrangement for vacuum casting to give a single crystal product. Solidification commences below spiral. One growing crystal only emerges from spiral section

mould is placed in a mould box, backed by sand for support, and is then ready to accept molten metal. In order to produce precision castings it is necessary to force the metal into the mould under pressure and this can be accomplished by mounting the investment moulds in a centrifuge (Figure 22.9(a)). Examples of components made by investment casting are shown in Figure 22.7.

High performance components, for example aero gas turbine blades, are generally cast into investment moulds in vacuo and using slow directional solidification. This is the technique used for the manufacture of blades in DS eutectic alloys and, with minor modifications, for single crystal blades. In the latter case single crystal solidification is achieved either by using a prepared 'seed' crystal as a nucleus or by using a mould with a spiral extension (see Figure 22.8). Modern gas turbine blades are cast with a network of fine cooling channels within them. These are obtained by inserting a series of soluble ceramic cores in the master mould used to produce the wax patterns. After the castings have been made, the cores are removed by dissolving them out using a strong solution of sodium hydroxide, molten sodium carbonate or other solvents.

22.9 Centrifugal casting

A method of casting that can be used for parts with an axis of rotational symmetry is *centrifugal casting*. This process is particularly suitable for the manufacture of large-diameter pipes for water and gas mains. A cylindrical mould shell, lined with moulding sand, is rapidly rotated about its longitudinal axis and liquid metal poured in (Figure 22.9(b)). Under the action of centrifugal force, the liquid metal will spread evenly along the length of the mould. Rotational speeds are such as to give accelerations of the order of 60 g (g here refers to acceleration due to gravity). The properties of the finished casting are very good as the centrifuging action ensures that any slag particles, which are almost invariably less dense than the metal, segregate to the bore of the pipe

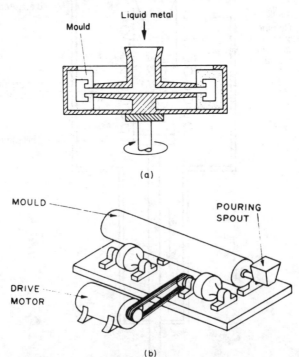

(a)

(b)

FIGURE 22.9   (a) Investment moulds in centrifuge. (b) Centrifugal casting machine from where they may be easily removed by machining. Similarly, any gases will also escape to the bore leaving a sound casting. Centrifugal casting can also be used for producing cylindrical shell bearings in white metal, and in bronze and other copper alloys.

A comparison of the major casting processes is summarised in Table 22.1.

22.10 Ingot casting

In very many instances it is necessary to shape metals by deforming them plastically. The starting point for these working processes is a large regular-shaped casting, termed an *ingot*. Ingots for further working are generally square, rectangular, or circular in cross-section, and may vary in size from about 10 kilogrammes for some non-ferrous alloys, up to 100 tonnes for some large steel forging ingots.

Ingots may be cast by pouring the liquid metal into large permanent moulds made from cast iron. Cast iron moulds for the production of steel ingots are tapered to allow for easy removal of the solidified ingot from the mould, and are termed either *big end up* or *big end down* depending on the direction of the taper. At the end of the steel-refining stage the liquid steel is in a highly oxidised state. Additions are made to the bath to deoxidise the steel before pouring, or *teeming*, the steel. Fully deoxidised steel is termed *killed steel*, and most killed steels are poured into moulds of the big end up variety. In order to minimise the formation of pipe defect as a result of solidification shrinkage, the moulds are generally provided with a refractory collar, or *hot top*, to act as a reservoir for liquid metal to feed the shrinkage (Figure 22.10). Liquid metal poured into a large mould from the top can splash considerably, and the splashes would freeze instantly on the upper mould walls. The splash surface would then rapidly oxidise and not weld satisfactorily to the main body of the ingot when the mould is filled with liquid metal. This would give rise to surface defects in the subsequent rolled or forged product. In order to avoid this some steel ingots are bottom poured. This process, which is considerably more expensive than straight teeming into moulds, is normally used for only the higher-grade steels.

Some grades of low carbon, or mild, steels are not deoxidised before teeming. These are termed rimming steels and they are normally cast into moulds of the big end down type. There is considerable gas evolution during the solidification of rimming steels. When solidification first begins at the mould walls, carbon monoxide is evolved according to the reaction

$$FeO + C = CO + Fe$$

The stream of gas bubbles tend to carry impurities away with it and the outer layers of the ingot solidify as almost pure, clean iron, with the impurities concentrated in the central portion of the ingot. Some of the gas bubbles are trapped between growing crystals of iron giving a widely distributed porosity, and this porosity compensates for the solidification shrinkage of the metal. If the internal surfaces of the gas pores are clean they will readily weld together under rolling pressures, giving a sound product. Rimmed steel is very suitable for the production of mild steel sheet because the pure iron rim gives a ductile skin of good surface quality, and this makes the steel ideal for deep drawing and pressing operations.

Large ingots cast in permanent moulds possess coarse grain structures, and there will also be variation in crystal structure across the ingot. A typical transverse section of an ingot is illustrated in Figure 22.10(e). There are three main crystal zones shown. The initial chill zone, which is of small thickness, possesses small crystals. The second zone is the columnar zone, and this contains large crystals, which have grown preferentially inwards from the mould wall along the direction of the major temperature gradients. Finally, when the temperature of the central

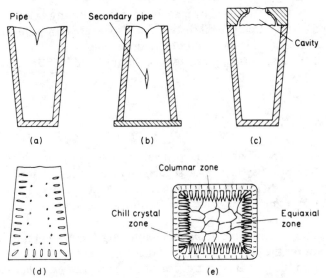

FIGURE 22.10   Ingot casting: (a) big end up type; (b) big end down type, showing secondary pipe; (c) big end up with hot collar; (d) section of rimmed steel ingot; (e) transverse section of ingot

liquid zone has fallen to around the freezing point of the metal and temperature gradients are small, randomly orientated, or equiaxial, crystal grains form.

Before 1940 ingots of all metals were cast, almost exclusively, into permanent cast iron moulds, but considerable attention was given to the development of casting processes that would give ingots of a better and more consistent quality and, in particular, a casting process that would be continuous in operation. The introduction of continuous casting, on a production scale, took place in the aluminium and copper alloy sectors of industry. The continuous casting of steel was a much later development, owing go the greater difficulties that had to be overcome, with molten steel temperatures of about 1550°C, compared with temperatures of about 700°C and 1000°C respectively for molten aluminium and molten brass. Today, almost all large ingots produced in the aluminium industry, and much of the ingot production in the copper industry, are continuously cast. A growing proportion of steel ingot production is also cast in this way.

Figure 22.11 shows the lay-out of a semi-continuous casting unit for the production of aluminium ingots. The mould is shallow, and the distance from the metal surface to the bottom of the mould skirt is about 75 mm. The ingot withdrawal rate must be carefully controlled, and this is done by means of a hydraulic ram. Withdrawal speeds vary considerably with alloy composition and the sectional size of the ingot, and may be between 20 and 200 mm per minute. The liquid metal solidifies very quickly and this means that not only does the ingot possess a fairly fine grain size, but also that segregation of impurities is largely eliminated and there is no possibility of piping defect occurring. In semi-continuous casting the maximum length of ingot possible is determined by the length of stroke of the hydraulic ram. This is usually of the order of 5 m. The fully continuous casting process is similar, but in this case steady withdrawal of the ingot from the mould is accomplished by having a pair of pinch rolls situated about 1 m below the mould. The continuous ingot is cut into suitable lengths by either a high-speed saw, or by flame cutting.

Much material is continuously cast in the horizontal, or near horizontal, mode on belt casting machines (Figure 22.12). This method is suitable for the production of flat strip and plate. Dams

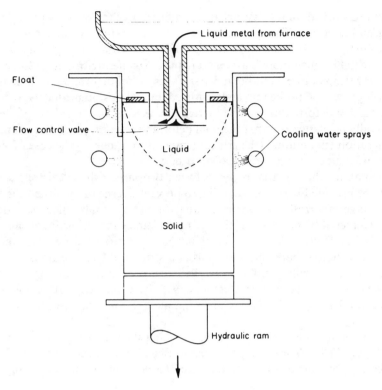

FIGURE 22.11   Semi-continuous casting of aluminium

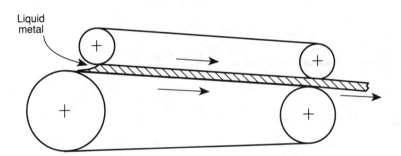

FIGURE 22.12   Moving belt machine for continuous casting

are fitted between the two moving belts to control the width of the cast product. Rod and bar sections can be cast in a similar manner but with shaped channels, rather than flat belts, incorporated in the casting machine. In some cases, particularly in the non-ferrous production industries, continuously cast rod or strip may feed directly into a multi-stand rolling mill.

22.11 Hot working

The term *hot working* refers to plastic deformation carried out at temperatures in excess of the recrystallisation temperature of the metal (refer to Section 8.7). Almost without exception, cast metal ingots are hot worked, in the first instance, and the three major hot working processes are rolling, forging, and extrusion. The ingots are preheated in soaking pits or furnaces to bring them up to the desired temperature, usually about 100–200°C below the melting range. At such a high temperature the metal is much more plastic and requires smaller deformation forces than would be necessary at low temperatures. During the first stages of hot deformation the coarse, 'as cast', structure of the ingot will be broken up and distorted, but, almost immediately, recrystallisation will occcur and further deformation can be carried out. A sequence of rolling passes, or forging blows, will generally be planned so that the final deformation takes place at a temperature close to the recrystallisation temperature (the critical temperature range, in the case of steels) so that recrystallisation without excessive grain growth occurs.

In *rolling* operations the ingot is passed between two large cylindrical rollers, and the roll surfaces have to be provided with flood cooling to prevent overheating of the rolls and possible welding between ingot and rolls. Water, or a water/soluble oil emulsion, is used, depending on the nature of the material being rolled. If the rolls are plain cylinders the ingot can be processed into flat plate material. Rolls with grooves machined into the surface are used for the production of round bars and sectional shapes, such as rails and structural beam sections.

The product of the hot rolling mill may be the finished product, requiring only trimming to size, or it may be an intermediate product that will be deformed further by cold working operations. Structural plate and sections are examples of products used in the hot worked condition.

Generally, hot rolling mills are of the 2-Hi reversing type with a motorised roller table at either side (see Figure 22.13). The rolling mill operator has control over the direction of both rolls and roller tables and, of course, alters the distance between rolls after each rolling pass.

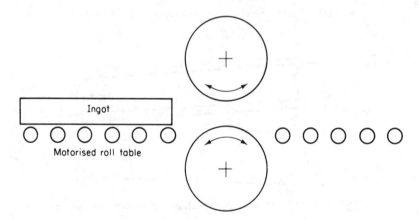

FIGURE 22.13   2-Hi hot rolling mill

In a slab rolling mill the diameter of the cylindrical rolls will be of the order of 1 m and the barrel length may be up to 3 m. The hot ingot, in a highly plastic state, will be passed backwards and forwards between the rolls several times with a major reduction in thickness being brought about with each successive pass through the mill. For example, an aluminium ingot with the dimension length 2 m, width 1.5 m and thickness 0.3 m may enter a hot rolling mill at a temperature of 550°C and eventually emerge, after nine successive reduction passes as plate material with a thickness of 10 mm and approximately 60 m in length. There is very little lateral spread during rolling and virtually all the plastic deformation is in the direction of rolling. At the end of this series of hot rolling operations the temperature of the plate will probably be down to about 250°C.

The rolls, which are made of forged or cast steel, are not perfectly cylindrical but are ground to possess a calculated profile or camber. Massive though they are, the rolls will deflect when operating under load and, if perfectly cylindrical, this deflection will cause the cross-section of the rolled material to be barrel-shaped (see Figure 22.14(a), (b)). If the rolls are given a convex profile or positive camber of the correct amount this will compensate for roll deflection and a parallel-sided rolled product should be obtained (Figure 22.14(c), (d)). This is the case for cold rolling operations but matters are further complicated in hot rolling. The ingots that are hot rolled are always somewhat smaller than the full width of the mill, so that during rolling the central portion of the rolls will become hotter than the edges, giving an uneven expansion. Rolls for hot rolling are surface ground to give a concave profile or negative camber when cold. During

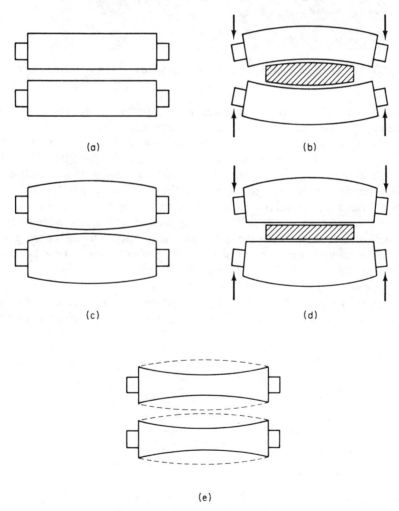

(a)　　　　　　　(b)

(c)　　　　　　　(d)

(e)

FIGURE 22.14　Roll surface contours. Plain cylindrical rolls (a) would deflect under load (b). (c) Rolls with positive camber deflect to give a parallel-sided rolled slab (d). (e) Hot mill rolls ground with negative camber so that a convex profile is obtained when rolls are heated to operating temperatures

rolling operations the greater expansion at the centre will cause the rolls to have a positive camber (Figure 22.14(e)).

The process of hot rolling may be used to produce square, round or shaped sections such as rails and I-sections. For this type of rolling the rolls are not plain cylinders but have matched grooves and shaped profiles to enable the desired section to be produced. Each set of grooves is carefully designed so that the required shape is fully formed but with no excessive fin being produced (see Figure 22.15).

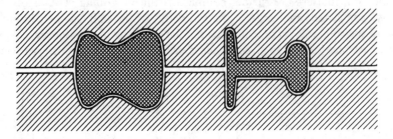

FIGURE 22.15   Roughing and finishing pass for rails from square section

As mentioned earlier there is very little lateral spread during rolling. The metal is accelerated as it passes through the roll gap and leaves the mill at a higher velocity than it enters. An ingot of width w and thickness t_1, entering a rolling mill at velocity v_1, will emerge from the roll gap with a reduced thickness t_2 but at a higher velocity v_2 (see Figure 22.16). Since the volume flow rate of metal must be the same at any point within the mill then this means that at only one point, N, within the roll/metal contact area will the velocity of the metal be equal to the peripheral velocity of the rolls. N is termed the neutral point. The forces that act on the metal within the roll gap are a radial force P_r exerted by the rolls and a friction force F which acts tangentially to the roll surface. The friction force acts to draw the metal into the roll gap between the entry point and the neutral point but its direction is reversed between the neutral point and the point of exit. The vertical component of P_r is the rolling load but as this is equal to the force exerted by the metal on the rolls it is often referred to as the roll separating force.

If the coefficient of friction between rolls and work-piece, μ were zero no rolling would be possible. As μ increases so it becomes possible to take a greater 'bite' on an ingot and hence to

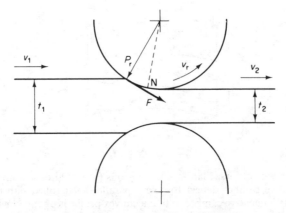

FIGURE 22.16   Rolling mill forces

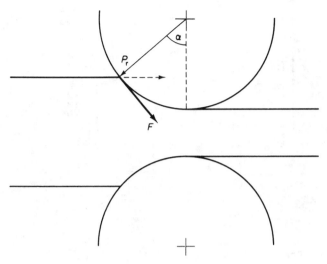

FIGURE 22.17   Effect of friction on bite angle

give a substantial reduction in thickness in one rolling pass. Referring to Figure 22.17 the force tending to pull the work-piece into the roll gap is F and its horizontal component is $F \cos \alpha$. For the work-piece to enter the roll gap $F \cos \alpha$ must be equal to or greater than the horizontal component of the normal force, P_r, which acts in the opposite direction.

Therefore, at the limit $F \cos \alpha = P_r \sin \alpha$
or $F = P_r \tan \alpha$
But $F = \mu P_r$
therefore $\mu = \tan \alpha$

In hot rolling operations the roll surfaces are ground so that μ is of the order of 0.2–0.3 and this, in conjunction with the use of large-diameter rolls, means that a large reduction in thickness per rolling pass can be achieved.

Forging means the shaping of metal by a series of hammer blows, or by slow application of pressure. The simplest example is a blacksmith's forging of a hot piece of metal by hammering the work-piece on an anvil. *Heavy smith's* forging is fundamentally similar, differing only in the scale of the operation. The work-piece may be an ingot of 100 tonnes and the deforming force provided by a massive forging hammer, but the whole process is controlled by the master smith, who decides each time where, and with what force, the blow should take place. Another form of forging is *closed die* forging, in which the hammer and anvil possess shaped recesses which effectively form a complete mould. The hot metal work-piece is caused to flow, under pressure, into the die cavity to produce the desired shape.

Forging operations may be carried out using either forging hammers or forging presses. With a press, as opposed to a hammer, pressure is slowly applied and plastic deformation tends to occur fairly uniformly throughout the material.

Forging hammers are of two basic types with the large hammer either dropping on to the work-piece accelerated by gravity alone or, as in the steam hammer, accelerated by both gravity and steam pressure (see Figure 22.18). A gravity drop hammer can deliver a rapid succession of forging blows but as the drop distance is constant all blows are delivered with the same force. For any given hammer size the power-assisted steam hammer can give a greater forging capacity. It is

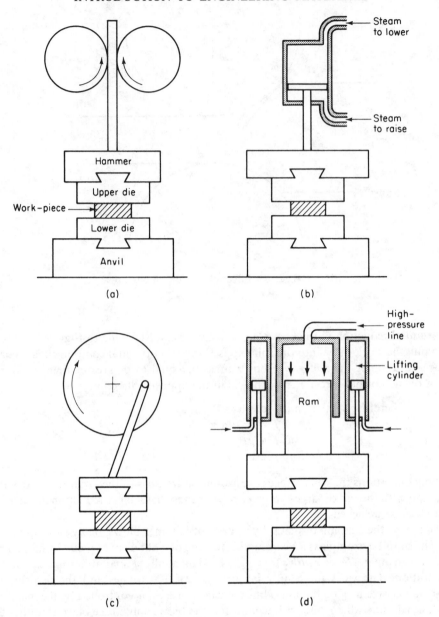

FIGURE 22.18　Forging machinery: (a) gravity or board drop hammer; (b) steam hammer; (c) crank press; (d) hydraulic press

also possible to vary the forging force by varying the steam pressure. Forging hammers are rated by the hammer weight and these tend to vary between 5 MN and 250 MN.

The use of a forging press tends to give forgings of greater accuracy than is normally possible with hammer forging. Forging presses may be of the mechanical or hydraulic type. Most mechanical presses are of the crank-operated type (see Figure 22.18(c)) and have capacities ranging from 3 MN to 120 MN. Hydraulic presses generally have capacities in the range from 5 MN to 200 MN although some very large machines with a 500 MN rating exist. A hydraulic press is more expensive than a mechanical press of the same capacity, but offers greater flexibility in that the force applied can be varied.

Smith's or open-die forging is used for making very large forgings or when the number of forgings of a particular design is small. Open-die forging is also used to pre-form metal prior to final shaping by means of closed-die forging. For Smith's or open-die forging the hammer and anvil are either plain or possess simple shapes.

Closed-die forging involves the use of accurately machined die blocks. The capital cost of a pair of die blocks is high, but this is allowable when it is required to make forgings to close dimensional tolerances and in large quantity, as would be the case for, say, connecting rods for the engine of a popular car. For such a component the die blocks would contain two pairs of cavities. The first pair, roughing or blocking impressions, would produce a forging approximating to the final shape. This would then be placed in the finishing cavity in which it would be forged to the final exact shape. The pre-heated blank or billet for closed-die forging is generally pre-formed to some extent before being placed in the blocking impression.

It is important in closed-die forging that a full and complete component is made and to ensure this the billet used will have a slightly greater mass than the mass of the finished forging. The die blocks are designed to allow for any excess metal to be extruded as a fin or flash (Figure 22.19) and this flash can be easily removed from the finished forging by a blanking operation.

Upset forging, or *upsetting*, is a forging operation in which the direction of plastic flow in the workpiece is changed from that of earlier operations and is carried out so as to obtain the best flow pattern or fibre structure for the component in question (see Section 22.12). Generally, upsetting shortens the length and increases the diameter of the workpiece.

A variant on the forging process is *roll forging*. This is a deformation process for producing straight or tapered sections with a varying shape and section along the length of the workpiece. The forging rolls are grooved, conforming to the shape of the product, and in each cycle rotate through less than a complete revolution, and then reverse. Roll forging is used for forming such components as automotive drive shafts, axles and aircraft propellor blades.

The *Pilger process* is a hot roll forging process for the manufacture of seamless pipe and tube. A Pilger mill is a 2-Hi mill and each roll contains a groove of semicircular section. The size of the groove is not constant and decreases around the roll circumference (see Figure 22.20(b)). A previously pierced ingot is mounted on a mandrel, to control the bore diameter. The rolls rotate through almost 360°, rolling and forging a section of the ingot and reducing the external diameter. The rolls are then reversed, the ingot fed further into the roll gap, and the process repeated roll forging a further section of the ingot. Ingots for tube manufacture are pierced initially by rotary piercing, as for example in the Mannesman process. In this process, a heated steel billet of cylindrical section is rotated between two slightly tapered rolls whose axes of

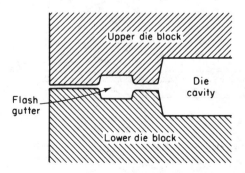

FIGURE 22.19   Flash gutter

rotation are inclined at a small angle on either side of the billet rotation axis (see Figure 22.20(a)). The effect of the rolling is to deform the billet into an elliptical shape and rotation of this cross-section causes shear and the formation of a crack in the centre of the section. As the rolling proceeds, this centre crack is forced over a pointed mandrel, which both enlarges and shapes the central opening.

In the process known as *hot extrusion* the hot ingot is forced to flow, under pressure, through a shaped orifice, or die, in a similar manner to the flow of tooth paste from a collapsible tube (Figure 22.21(a)). It is possible to extrude an almost infinite variety of sectional shapes, including hollow sections, from solid ingots (see Figure 22.25). For many years hot extrusion was restricted to the lower melting point materials, principally aluminium and copper alloys, but in recent years considerable development has taken place in the extrusion of steels.

Extrusion of hot ingots is carried out in large hydraulic extrusion presses and these are generally in the horizontal mode. Extrusion press capacities vary considerably and the ram force may range from 2 MN in a small press up to 150 MN in a very large press. Most hot extrusion is performed by the direct method in which a pre-heated ingot is placed in the chamber of the press and a force applied by the piston, through a pressure pad, causing the ingot to flow plastically and extrude through the die aperture (see Figure 2.21(a)).

In indirect extrusion (Figure 22.21(b)), there is no relative movement between the billet and container. The extrusion die is attached to a hollow ram and is forced into the billet. The extrusion flow is through the die and hollow ram in a reverse direction to the ram movement. This system, also referred to as backward or inverted extrusion, is not used to a great extent.

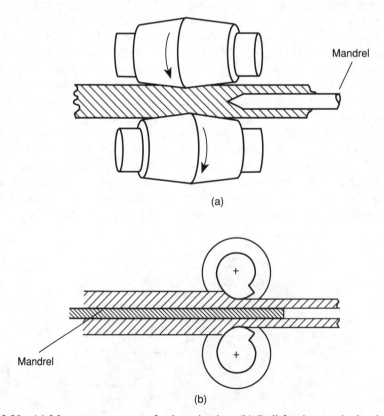

FIGURE 22.20   (a) Mannesman process for hot piercing. (b) Roll forging a tube by the Pilger process

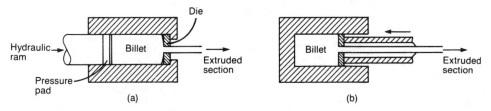

FIGURE 22.21 Principle of extrusion: (a) direct extrusion; (b) indirect extrusion

The dies for extrusion must be extremely hard and high temperature resistant and are made from chromium/vanadium/tungsten steels. A die lubricant must be used to minimise the rate of die wear. For aluminium alloy extrusion molybdenum disulphide is used as a lubricant while for steel extrusion the lubricant is glass. The hot steel ingot or billet is wrapped in a glass fibre mat before insertion in the press chamber and a glass fibre pad is placed against the face of the die. At the temperatures involved the glass softens and acts as a lubricant between the ingot and the chamber walls and also at the die. The glass mat at the die surface softens and a film of glass about 25 µm thick separates the steel extrusion from the die.

A high pressure is needed for extrusion, but in the direct extrusion process the pressure is not uniform throughout the ram stroke. When the hydraulic ram starts moving the pressure on the ingot increases rapidly to a maximum value at which time metal will start to flow through the die aperture. The pressure needed to continue extrusion reduces during the process (see Figure 22.22(a)). This is because friction between the billet and the chamber wall is reduced as the length of the billet within the chamber decreases.

Towards the end of the stroke the pressure for extrusion begins to increase again. Extrusion is normally ceased when about 90 per cent of the billet has been processed. The remaining 10 per cent is not extruded but discarded. This is not only because of the high pressure that would be needed to extrude the last portion but also because of the strong possibility of defect formation within the extrusion. The type of defect that could occur is known as extrusion defect or 'back-end' defect. The outer surface of the billet will be chilled by the chamber wall and so the surface layers of the billet will be less plastic than the core. During extrusion these surface layers may wrinkle as shown in Figure 22.23 and if extrusion is continued these oxidised surface zones could be pushed through the die giving a defective product.

In indirect extrusion, the pressure needed is fairly uniform throughout the whole stroke and the incidence of extrusion defect is almost nil as there is no relative movement between the billet and the container wall.

The range of sectional shapes that can be produced by extrusion is very wide, and includes both solid and hollow sections. Typical examples of shapes produced in aluminium are shown in Figure 22.25. Hollow shapes may be made from solid billets by extruding through a bridge die or porthole die. In a bridge die (Figure 22.24) the plastic metal is split into two streams which flow around the mandrel that gives the bore of the tube. After being split by the bridge the two streams of metal converge and pressure-weld together. As there is no air present to oxidise the metal streams flowing past the bridge, perfect welding occurs within the die to produce a seamless hollow section. A porthole die is similar in principle but the plastic metal is split into three or four streams at the entry side of the die.

A major advantage of any hot working process is that it breaks up the coarse grain structure of a cast ingot. Also, under the pressures involved, clean cavities such as minor shrinkage porosity, will be closed up. Consequently, the strength of a hot worked metal product will be considerably better than that of a casting. Some disadvantages of hot working are that very close

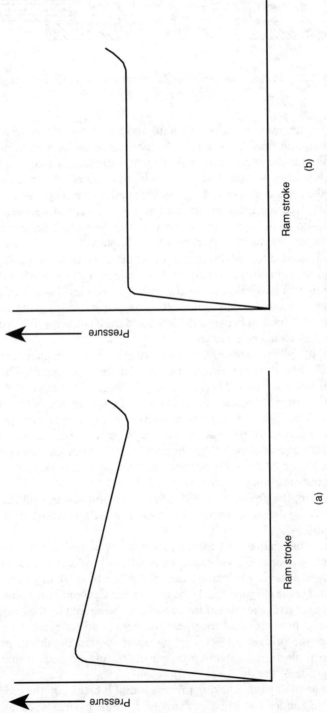

FIGURE 22.22   Force variation during (a) direct extrusion, (b) indirect extrusion

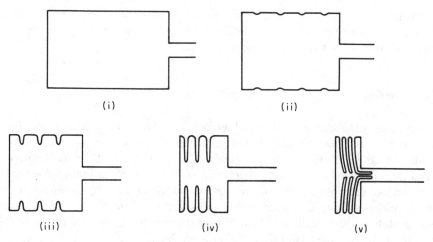

FIGURE 22.23   Formation of extrusion defect

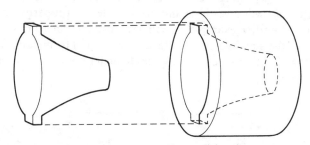

FIGURE 22.24   Bridge die for making hollow extrusions

FIGURE 22.25   Typical examples of extruded sections

control of dimensions are not always possible, and the product will have a poor surface finish owing to the effects of high-temperature oxidation. Subsequent cold working operations will give close dimensional tolerances and a good, clean surface.

22.12 Fibre structure

The final properties of a hot worked product depend to a certain extent on the type of deformation process that was used. Commercial materials are not completely pure and will contain in addition to the alloying and trace elements, small non-metallic particles derived from the fluxes and slags involved in the melting operations. Within the cast ingot these particles will have a fairly random distribution. When the ingot is deformed these foreign particles will flow with the metal and will form fibre lines within the structure. After each hot deformation the flowed metal crystals will recrystallise, but the fibre particles will remain unchanged. In rolled and extruded products all fibre lines will lie parallel and will be in the direction of working, but in a forging the metal flow pattern will be more complex. Figure 22.26(a) shows the fibre structure of two bolts, one of which has been formed by machining from a solid hot rolled bar, and the other of which has been formed by forging the bolt head from small-diameter hot rolled bar stock. The difference in fibre structure will give the components differing properties even though all other factors may be equal. Another example of different production routes giving different fibre structures is the case of the small gear blank (Figure 22.26(b)) formed by (i) blanking from hot rolled plate, (ii) parting off a thin slice from a large-diameter hot rolled bar, and (iii) parting off a

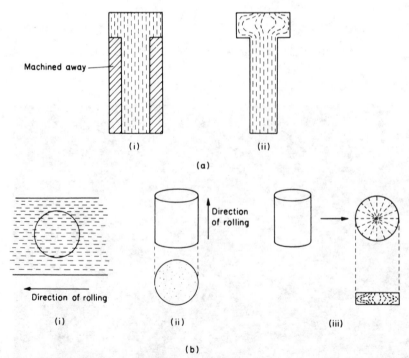

FIGURE 22.26   Fibre lines. (a) Bolt shaped by (i) machining from rolled stock, (ii) forged bolt head. (b) Gear blank formed by (i) blanking from rolled plate, (ii) parting off from rolled bar, (iii) forging from small-diameter bar

length from a small-diameter hot rolled bar and forging this into a thin disc. The fibre lines may constitute lines of weakness within the material, and lines along which fracture may occur. In the case of the gear blank formed from rolled plate the fibre lines are tangential at two points on the disc and teeth in these positions could be subject to failure. In case (iii) all fibre lines within the disc are radial and the risk of failure of the gear is lessened. The term *upsetting* is given to the type of forging that would be used in case (iii), as the former flow pattern is completely reversed and the arrangement of the fibre lines is totally altered.

22.13 Cold working

Cold working is plastic deformation performed at temperatures below the recrystallisation temperature of the metal. During cold working the crystal structure becomes broken up and distorted and the material becomes strain, or work, hardened. The mechanical strength is increased by cold working, and the material becomes harder, but more brittle. The electrical resistivity is also increased. Eventually, the metal becomes so hard that further attempts at cold working would cause fracture. After cold working the material may be softened by heating it to a temperature above the recrystallisation temperature and allowing the distorted crystal grains to recrystallise. This type of process is termed *annealing*. Very good dimensional control can be exercised in cold working processes and good surface finishes can be achieved. The surface of the hot worked product may be covered by an oxide scale that was formed at high temperature. This layer of oxide scale must be removed before cold working operations can be commenced. Oxide scale may be removed from hot worked steel by pickling the steel in hot sulphuric acid.

Cold rolling is used for the production of sheet and strip material (the term 'strip' is applied to any material rolled from coils, irrespective of the width). The starting point for cold rolling is soft annealed material. During cold rolling the material becomes harder and in order to continue rolling it would be necessary to increase the roll separating force. The equipment used for cold rolling is different from that needed for hot rolling (see Section 22.11). It is customary for cold rolling mills to have rolls of comparatively small diameter. In this way the contact area between strip and roll surface is reduced, with a consequent reduction in the total roll separating force. One drawback of small diameter rolls is that they could have a large deflection under load. To obviate this the work rolls are backed by larger diameter rolls. One of the most popular arrangements is the 4-Hi mill (Figure 22.27(a)) in which the small-diameter work rolls are powered and the large back-up rolls are not. Cold rolling mills of the 4-Hi type are used for strip rolling with the material being unwound from a drum on the entry side of the mill and recoiled after passing through the mill. It is then possible to put tension into the strip at both sides of the roll. Both the back and front tensions will be acting in opposition to the friction forces between the strip and the rolls at entry and exit. This will have the effect of reducing the roll-separating force and so reducing the power required for rolling.

Frequently, a number of 4-Hi mills are arranged in tandem for the rolling of continuous strip material. Up to six or seven mill stands may be arranged in a strip mill train.

Another type of rolling mill suitable for the cold rolling of thin hard strip material is the cluster mill in which small diameter work rolls are backed by a series of rolls arranged in arcs (Figure 22.27(b)). The planetary mill configuration (Figure 22.27(c)), in which a set of small work rolls is mounted around a large diameter roll in a manner similar to a roller bearing, will permit very large reductions in thickness to be given in a single pass.

Another difference between cold rolling and hot rolling operations is that the roll surfaces are ground to a finer finish to give a reduced coefficient of friction. A lubricating oil is also sprayed on to the rolls to further reduce friction. If the final cold rolled product is required to have a smooth

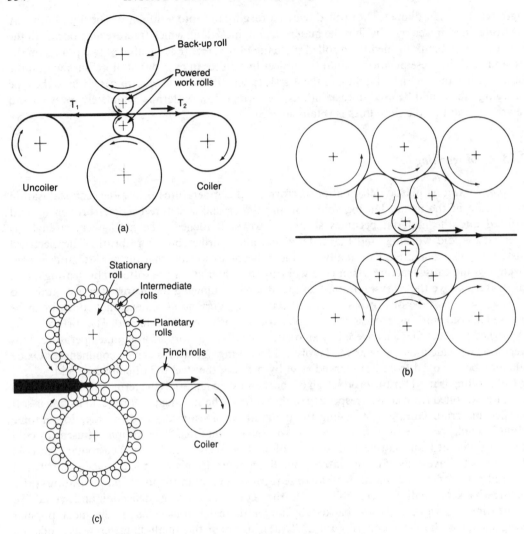

FIGURE 22.27 (a) 4/Hi mill. (b) Cluster mill. (c) Planetary mill

polished surface the last cold reduction passes are through highly polished rolls. When rolling a very thin product such as thin foil the system of pack-rolling, in which two layers of foil are rolled together, is used for the last few rolling passes. Aluminium foils with thicknesses as small as 0.005 mm can be made in this way. Pack-rolled foil is characterised by having one highly polished surface with a matt finish on the other surface, this latter being the interface between the two layers.

As has been mentioned any metal will work harden during cold rolling and one or more intermediate annealing operations may be necessary to soften the material before any further reductions can be obtained. If the finished product is required in a soft and ductile condition a final annealing treatment will be necessary. The annealing of cold rolled material is usually done in controlled-atmosphere furnaces to avoid surface oxidation of the strip. After cold rolling the coiled material may either be uncoiled, levelled and cut into sheets or circular blanks of the desired size, or supplied as coil for some other manufacturing process.

The final properties of cold rolled material may vary between those of the fully work hardened condition and those of the annealed, or soft condition. Properties intermediate between these two extremes may be obtained by carefully controlling the amount of cold reduction given to the material after an annealing operation or by partially annealing fully work hardened material (refer to Table 8.2).

Much rod and bar material is also produced by cold rolling operations. In this case, the rolls possess semicircular grooves in their surfaces and each pass through the rolls is a 'closed' pass. Generally, rod mills contain several stands, up to about 15, arranged in tandem. The product from a rod mill may be marketed as rod but much is used as feedstock for wire drawing.

Another useful cold working process is *drawing*. This is used for the production of wire, rod and tubing. The annealed material is pulled through a die with a reducing diameter. The die materials must be very hard to resist abrasive wear. Drawing dies are made from high carbon or alloy steels. Tungsten carbide is frequently used as a die insert.

Not all metals can be drawn successfully. Some metals may be highly *malleable*, that is they deform readily when subjected to forces that are largely compressive such as hammering, but they are not necessarily highly *ductile*. For good ductility, that is, the ability to be drawn well, it is necessary that the metal work hardens sufficiently rapidly so that the reduced diameter material at the exit side of the die is able to withstand the tensile force that is required to cause plastic deformation of the soft undrawn metal. Low carbon steels are highly ductile materials and a reduction in cross-sectional area of about 40 per cent can be achieved in a single pass through a cold drawing die.

A die for the drawing of round rod or wire will contain a conical hole tapering down to a cylindrical section (Figure 22.28). The diameter of the hole on the inlet side is greater than the diameter of the stock to be drawn and the space between die mouth and stock acts as a reservoir for the drawing lubricant.

Rod material is drawn in straight lengths on a draw bench, but the stock for wire drawing is usually a large coil and the driven coiling drum on the exit side of the die provides the necessary tensile force to pull the wire through the reducing die.

In tube drawing the bore dimensions are controlled by means of a mandrel. Several techniques are used. When relatively large-diameter tubing is to be drawn in a series of straight lengths on a draw bench a fixed mandrel may be used (Figure 22.29(a)). Drawing using a floating mandrel (Figure 22.29(b)) is much more suitable for small-diameter tubing and unlike the fixed-mandrel method there is no limit to the length of tubing that can be drawn. Small-diameter drawn tubing may be coiled on to a drum. The third technique is the use of a moving mandrel

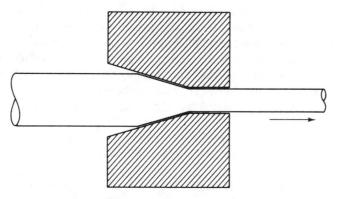

FIGURE 22.28   Principle of wire drawing

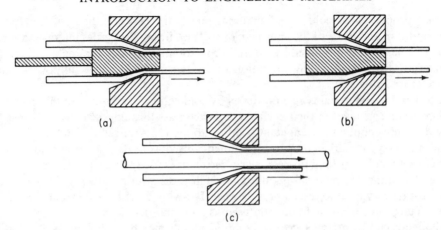

FIGURE 22.29   Tube drawing: (a) with fixed mandrel; (b) with floating mandrel; (c) with moving mandrel

(Figure 22.29(c)). In this case a hard straight rod with a diameter equal to the required bore of the tubing and of the same length as the finished tube length is used as the mandrel. Frictional resistance during drawing is less in this case because the mandrel moves with the tube. This technique may be used for both large-diameter and very small-diameter tubing. For example, the very fine steel tubing used for hypodermic syringe needles is drawn using hard steel wire as a moving mandrel.

Sheet and strip metal may be formed into an infinite variety of shapes. *Roll forming*, *deep drawing*, *pressing*, *rubber forming*, *stretch forming* and *spinning* are some of the processes that can be used for the production of complex and hollow shapes from flat sheet and strip material (Figure 22.30).

Roll forming is the shaping of flat strip metal into sections by passing it through a series of contoured forming rolls. The process can be used for forming angle and channel sections and a wide range of complex shaped sections.

In a true deep drawing operation for the production of a hollow cylindrical shape, such as a saucepan body, the sheet material is drawn through the aperture between punch and die, and there is, generally, a reduction in wall thickness occurring in the process. Pressing is the term given to the production of complex shapes, such as car-body panels. In this type of process the nature of, and the amount of, deformation may vary considerably from one part of the shape to another. It may be simple bending in some parts, and bending with some drawing taking place in other areas of the component. Metal spinning is the most suitable means for making conical shapes or re-entrant shapes, providing there is an axis of rotational symmetry for the latter. A circular sheet metal blank is rotated in a lathe-type machine, and the sheet metal is then forced into shape around a former, in much the same manner as a potter moulds clay shapes on a potter's wheel. Leather faced wooden tools are normally the means of applying pressure to the metal sheet, and these may be hand held. In the manufacture of re-entrant shapes, such as the bodies of tea-pots, the former is segmented in order to facilitate its withdrawal from the moulded shape.

When a large amount of energy is supplied extremely rapidly to a metal, the metal may deform to an extent, and in a manner that would not be possible if force were applied slowly. One process involving high-energy rate forming is *impact extrusion*, and this process is used for the production of aluminium collapsible tubes and cigar tubes, and for zinc dry battery cases. Another high-energy process of note is *explosive forming*, in which an explosive charge is

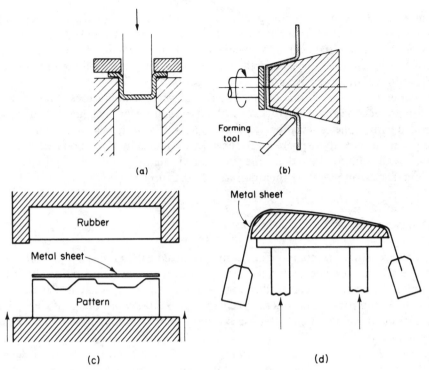

FIGURE 22.30   (a) Deep drawing. (b) Spinning. (c) Rubber forming. (d) Stretch forming

detonated under water and the pressure wave shapes metal sheet or plate by forcing it to flow and follow the contours of a female die. This process may be used for the manufacture of small or large components. It has been used for forming denture plates out of stainless steel sheet, and also for making the large dish ends for the underwater laboratory, Sealab II.

Electro-hydraulic forming is a high-energy rate forming (HERF) process in which electrical energy stored in a bank of capacitors is suddenly discharged within a liquid. The shockwaves generated within the fluid cause a metal blank to be deformed into a shaped die, as in explosive forming. This method is suitable for small scale production and is used for the forming of fairly small items.

22.14 Annealing

In many instances, cold rolled, or cold drawn, material has to be annealed, either as an intermediate stage in the processing, or as a final process. Normally the annealed material is required to have a fine crystal grain size, so that recrystallisation without excessive grain growth is necessary. Many annealing furnaces operate on the batch principle, in which a large batch of material is placed in the furnace and annealed in one lot. For this type of annealing the maximum temperature in the cycle is only marginally above the recrystallisation temperature. The surface layers of the batch furnace load may reach annealing temperature several hours before the inner layers. With only a small temperature excess above the recrystallisation temperature, there will be little grain growth occurring in the outer layers of material. An alternative annealing technique is flash annealing. A flash annealing furnace has a moving conveyor travelling through the hot zone. Single metal components are placed on the conveyor, and these travel fairly rapidly through the hot zone of the furnace and then through a cooling zone. The temperature of the hot zone may be several hundred degrees above the recrystallisaton temperature of the metal, but the components being annealed may be in the hot zone for only two or three minutes, so that full recrystallisaton without excessive grain growth occurs.

With many metals oxidation occurs rapidly at high temperatures, and it is, therefore, necessary to provide the annealing furnace with an oxygen-free atmosphere. Many types of inert, or controlled, atmospheres may be used in conjunction with heat-treatment furnaces. One commonly used furnace atmosphere is produced by burning a hydrocarbon fuel, such as propane or butane, with slightly less than the theoretical quantity of air required for complete combustion. The burnt gases are then passed into the furnace.

22.15 Powder metallurgy

The manufacture of finished components from a particulate material is a well established method for both metals and ceramic materials and it involves preparation and grading of powders, compaction of the powder is a die, followed by sintering the compact at an elevated temperature to give a strong coherent product. (See also Chapter 24.) *Powder metallurgy* is a useful process for the manufacture of parts in metals that possess very high melting points. It is both difficult and expensive to melt these metals. Tungsten, for example, melts at 3410°C, but a powdered tungsten compact may be sintered into a fully coherent part at temperatures of the order of 1600°C. Another useful application of powder metallurgy is in the production of 'hard metals', namely sintered mixtures of hard metallic carbides and powdered cobalt, nickel, or other strong metals. These 'hard metals' are used for the manufacture of cutting tool tips, percussion tools, and die inserts. For many years porous bearings have been made from metal powders. If a fairly low compaction pressure is used, a powder compact with a controlled degree of porosity can be obtained. Porous bronze bearings, with some 40 or 30 per cent porosity are made in this way. The bearings are then soaked in lubricating oil before they are put into service, and they can give long periods of service without requiring attention.

Some metal matrix composites are also made by compaction and sintering.

In recent years powder processes have been used increasingly for the production of small machine components of complex shape. The advantages of fabricating components from powders are considerable. One powder compaction operation produces an accurately dimensioned component that does not require machining and there is no scrap material produced. The only other operation necessary to produce the completed component is the sintering. In order to produce a similar component by conventional methods it may be necessary to commence with stock material produced by hot working and cold working processes and to give this stock several expensive machining operations, with, in consequence, some scrap arising at each production stage. Although the cost of metal powders is high, in comparison with the cost of ingot metal, it may well be cheaper to produce components by powder metallurgy techniques than by conventional machining.

There are several methods available for the production of metal powders, but not all methods apply to all metals. The more brittle metals may be powdered by pulverising in a hammer mill. Softer metals may be converted to powder by atomising a stream of the liquid metal in a jet of air, or an inert gas. Many metals, including aluminium, iron, tin, and zinc, are converted to powder in this way. Magnesium and zinc, metals with comparatively low boiling points are sometimes obtained in powder form by condensing the metal from the vapour phase. Some pure metal powders are obtained by electrolytic deposition of metal from a solution of a salt of the metal.

The electrolytic cell operating conditions are arranged so that the cathode deposit is soft and spongy, rather than a hard coherent plating. This method is particularly suitable for the production of copper powder. Chemical methods for the production of metal powders include the reduction of heated powdered metallic oxide to metal in a stream of a reducing gas, such as hydrogen, and the dissociation of volatile compounds, such as carbonyls. Some metals combine with carbon monoxide to form gaseous compounds, termed *carbonyls*. Heating the carbonyl causes it to dissociate into metal and carbon monoxide. The carbon monoxide may be recirculated. Pure nickel and iron powders can be made in this way.

Powders require to be size graded before compaction. The powder mix for pressing is made up of different size fractions so that small particles fill the void spaces between larger particles and minimum porosity is achieved. The graded powder may be die pressed or isostatically pressed. In the former process, powders are compacted in alloy steel or cemented carbide dies at pressures up to 750 MPa. In isostatic pressing, the dry powder mix is placed in a flexible container, usually made of rubber, and subjected to a hydrostatic pressure of up to 500 MPa. The pressures used in each case will depend upon the nature of the powder, the average particle size, and the degree of porosity required in the compact. Higher and more uniform densities can be achieved by isostatic pressing than by die pressing.

During the compaction of metal powders, there will be some plastic deformation of individual powder particles, allowing for some mechanical interlocking of particles. There will also be some cold pressure welding of particles occurring and the compact will possess sufficient green strength to be handled. The powder compacts are then sintered in an inert atmosphere at some temperature below the melting point of the metal or alloy. The sintering temperature for iron is about 1100°C, and for bronze it is about 800°C. During the sintering operation, there is a diffusion of atoms across interfaces between particles bonding the particles together as one coherent mass. The driving force for the process is the reduction in surface energy achieved by the coalescence of powder particles. (Typically, the surface energy of a grain boundary is about one-half that of a free metal surface.) As the total surface area of a fine metal powder can be of the order of 1000 m^2, the reduction of energy which accompanies sintering is significant.

Pressing and sintering need not be carried out as two separate operations. Hot die pressing

FIGURE 22.31 Typical examples of parts made by powder metallurgy (courtesy of Sintered Products Ltd)

and hot isostatic pressing involve the simultaneous application of heat and pressure. These processes give sintered products of higher density and finer grain size than can be achieved by separate pressing and sintering and products with virtually zero porosity can be made. There is an increased driving force for sintering caused by the stresses established at the particle contact points. Some typical examples of metal parts made by powder metallurgy are shown in Figure 22.31.

Newer developments in powder technology include the production of sheet and strip material by compacting metal powders between rollers. This has been successfully accomplished on a pilot-plant scale, for pure aluminium, some aluminium alloys, and for mild steel. This technique probably has great potential as it offers a short route from raw material to finished product with a considerably lower level of capital investment than in conventional metal processing plant.

Another direct method for the production of sheet metal that is being developed is the direct spraying of atomised liquid metal on to a substrate. The deposited metal is then peeled from the substrate and is further processed by conventional hot and cold rolling.

23

Forming Processes for Polymer Materials

23.1 Introduction

The nature of polymerisation and a brief description of polymerisation methods are given in Chapter 4. The pure polymers which are made are not suitable for shaping and forming into finished articles but have to be further processed to produce moulding compounds.

The polymer formed is ground into powder or broken up into granules, but, in order to produce a suitable moulding material, it is then compounded with a variety of additives. The additives necessary may include lubricants and plasticisers, fillers and reinforcements, pigments and dyes, stabilisers, and flame retardants. It is necessary to add a plasticising agent to many polymers to improve their moulding performance and the toughness of the material. Plasticisers are generally non-volatile solvents that partially dissolve the polymer and increase the flexibility of the material. Another form of plasticising agent is a non-solvent oil, highly dispersed throughout the material. In addition to improving moulding characteristics a plasticiser will also modify the mechanical properties of the plastic material. An extreme example of the effect of plasticiser content on properties is seen with PVC (polyvinyl chloride). Rigid unplasticised PVC is a tough horny solid and is widely used for such products as rainwater pipes and guttering but plasticised PVC, containing up to 50 per cent plasticiser, is a tough, flexible, rubbery material used for the manufacture of rainwear, as simulated leather in upholstery and for adhesive-backed electrical insulating tape. Fillers are used to add bulk to the moulding compound and to cheapen the product but they will also modify properties. Wood flour is widely used as a filler and it will give the material an improved impact strength. Chopped glass fibre is used as a filler and reinforcement (see Chapter 17 for the properties of glass-filled polymers). Pigments and dyes are used to colour the moulding compound and one of the attractions of polymer materials is that they are available in a wide range of colours. A stabiliser is sometimes needed to resist degradation of the material during its service life and stabilisers are generally antioxidants. Polymers, being carbon-based compounds, are often flammable materials and a flame-retarding agent should be incorporated during the compounding stage.

23.2 Forming of thermoplastics

The majority of thermoplastic materials possess comparatively low softening and melting temperatures and are termed *melt processable*. Some, such as PEEK, the polyimides and PTFE are difficult to process and require to be moulded at temperatures of the order of 350°C.

Components in PTFE are generally formed by hot pressing a sintering powder, in a manner similar to that described for powder metal components (Section 22.15).

Casting. Some thermoplastic materials may be cast into shape in moulds and allowed to solidify. Casting is not a widely used process for thermoplastics but is used for some polyamide and acetal products. It is also suitable as a production process when only a limited number of items of any particular design is required and the small number does not justify the high cost of making a set of dies for injection moulding. Sheet material may be made by a process termed casting, but this is not casting in the true sense. It involves pouring a solution of the polymer material on to a moving conveyor. Subsequently, the solvent evaporates, leaving behind a solid sheet of material.

23.3 Injection moulding

Injection moulding is the most important fabrication method for thermoplastics and is analogous to pressure die casting for metals. The compounded polymer material, in the form of granules or pellets, is fed into a cylinder, or barrel, where it is plasticised and the hot melt then injected into a

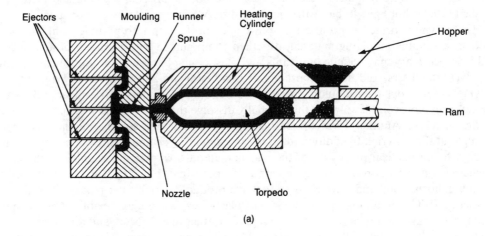

(a)

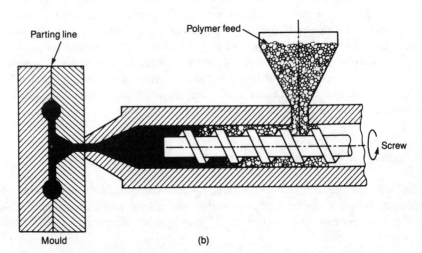

(b)

FIGURE 23.1   Principle of injection moulding: (a) plunger type machine; (b) screw type machine

relatively cool mould or die where it conforms to the shape of the cavity and allowed to solidify. It is then ejected as a finished moulding when it is rigid. The polymer material is moved through the cylinder and injected into the mould by either the action of a simple plunger or by a screw mechanism. The principles of the two types of machine are shown in Figure 23.1. Generally, the cylinder of a simple plunger machine contains a spreader or torpedo to permit a more uniform heating and mixing of the polymer. Most industrial machines are of the screw type. The shearing action of the screw gives a better mixing and plasticising than is possible in the plunger type machine. The sequence of events in a moulding cycle is closing of mould, after which the screw rotates. During this period, plasticised material builds up in front of the screw causing the screw to move axially backwards. Rotation is stopped and the screw moves forward as a ram to inject material into the mould cavity. The mould is then opened and the component ejected.

The screw in an injection moulding machine serves several functions and, generally, comprises three sections, feed, compression and metering. The feed section transports polymer granules from the feed hopper to the heating section of the barrel. The screw flights are of constant dimension in the feed section. In the compression section, the material changes from a granular form into a melt. The volumes of the flights decrease in this section to compensate for density change during the process. The final mixing and heating creating a homogeneous melt takes place in the metering section where again the flight dimensions are constant. There are several designs of screw to cater for various types of polymer material and these are illustrated in Figure 23.2. A highly crystalline polymer, such as a polyamide, has a sharp melting point and the screw best suited for this material has a very short compression section (Figure 23.2(a)). Semi-crystalline polymers, such as polyethylene, melt over a small range of temperature and require a screw with a longer compression section than highly crystalline materials (Figure 23.2(b)). Amorphous polymers such as PVC have no true melting point and the best screw for these materials shows a gradually increasing compression over its whole length (Figure 23.2(c)).

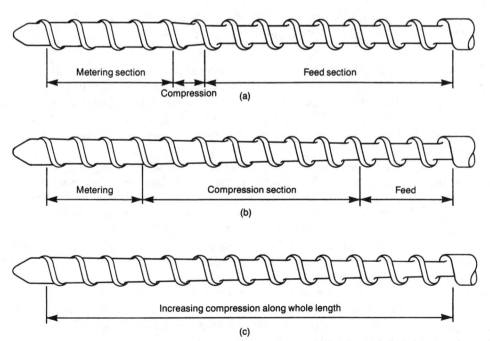

FIGURE 23.2   Types of injection moulder or extruder screw: (a) for highly crystalline polymers; (b) for semi-crystalline polymers; (c) for amorphous polymers

23.4 *Extrusion*

Rod, tubing, sheet, film and solid or hollow sections in a wide variety of shapes can be produced by extrusion. The most common type of plastics extruder is the single screw machine in which a rotating screw conveys granular polymer material along a barrel in which it is heated, plasticised and compressed into a homogeneous melt and then extruded through a shaped die orifice (Figure 23.3). As for screw injection moulding machines, there are several types of screw designed to cater for various polymer types (Figure 23.2), but unlike injection moulding, the screw is in continuous rotation at constant speed during extrusion. Twin screw extruders are also available but are much more complex and expensive and their use is restricted to a few applications, one being the extrusion of large diameter piping in UPVC.

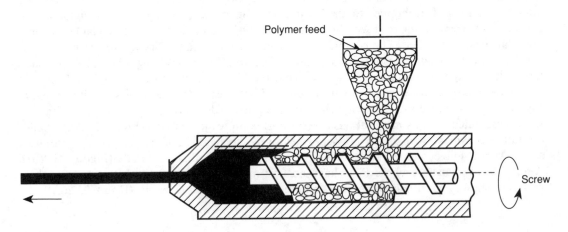

FIGURE 23.3   Principle of screw extruder

The hot extruded plastic must be cooled immediately after emerging from the die to increase its rigidity and prevent loss of shape. This is accomplished by means of cold air jets or water. A necessary part of any extrusion equipment is the 'haul-off' table. This is a moving conveyor, with its speed geared to the rate of extrusion, to give support to the extrudate and to move it uniformly through the cooling zone. For some products, for example, sheet and film, the 'haul-off' is accomplished by a series of rolls.

Sheet and film are produced by extrusion through a flat die with a wide, straight, slit orifice. (Generally, plastic thicknesses of <0.25 mm are referred to as film while thicknesses of 0.25 mm or greater are referred to as sheet.) On emerging from the extrusion die, film or sheet extrudates are usually passed through a series of water cooled rolls with polished surfaces. This not only cools the material but also imparts a good surface finish. Some film extrusion is *co-extrusion* in which two or more layers of film are extruded simultaneously from separate extruders to produce a laminate. This is done for the purpose of improving properties, particularly the degree of impermeability and the barrier resistance of the film.

Filaments are made by extrusion, in this case through a multi-hole die with a number of very fine holes. The extruded filaments are usually quenched in water and then stretched between two sets of rolls to give full molecular orientation within the fibres. This method is used for polyamides, polypropylene and other thermoplastics.

There are several interesting variations on extrusion for the manufacture of a number of products. These are *extrusion coating* and *wire covering, film blowing* and *blow moulding*.

Extrusion coating is a means of applying a molten plastic film, usually polyethylene to a continuous sheet of paper, card or other flexible backing. This produces a waterproof layer and is used in the manufacture of paper sacks and cartons to hold liquids. In some cases a laminate is formed with a thin polyethylene film between two paper layers.

Plastic coated wire and cables are produced by an extrusion process. The extruder nozzle is designed to turn the flowing polymer melt through 90° and to admit the wire to be coated (Figure 23.4(a)).

Thin film may be made by the film blowing technique, the principle of which is illustrated in Figure 23.4(b). During the major deformation which occurs during film blowing there is significant molecular orientation in both the longitudinal and transverse directions giving the film high strength in all directions.

Blow moulding is used for the production of bottles and other hollow shapes. A hollow tube, or *parison*, of plastic material is hot extruded. Immediately after emerging from the extrusion die, the hot parison is clamped in a mould and air is blown into it enlarging the parison so that it conforms to the contours of the mould. After a suitable cooling period, the mould is opened and the moulded article ejected. Materials which have been moulded successfully by this process include polyethylene, polypropylene, polycarbonate, polystyrene, ABS, acetal, polysulphone and PEEK.

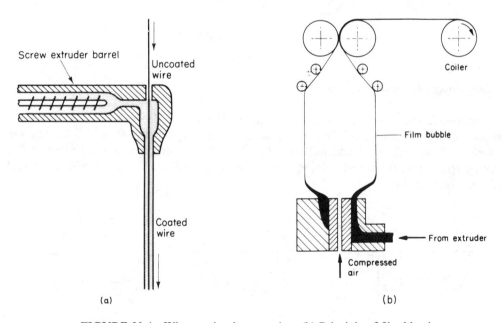

(a)   (b)

FIGURE 23.4   Wire coating by extrusion. (b) Principle of film blowing

23.5 *Calendering*

Calendering is a process for forming thermoplastic sheet material and it is used mainly for PVC and copolymers containing vinyl chloride. Polymer powder and fillers and other additives are first blended together cold and then thoroughly mixed and agitated within an enclosed heated mixer. In one common type of mixer, the Banbury mixer, the two mixing rotors are hollow and contain channels for steam heating and the mixer jacket is also hollow with channels for both steam, for heating the mix, and water, for cooling. The hot and partialy gelled material from the mixer is then fed between a pair of heated blending rolls, in which the peripheral speed of one roller is about 10 per cent faster than the other, emerging as a continuous sheet at a uniform temperature in the range 150–170°C. At this stage, the sheet may still have some inhomogeneities with small areas of imperfectly mixed polymer and filler. The hot sheet from the blending rolls is cut up and extruded through a fine mesh straining screen to remove coarse particles and fed into the calender. A calender consists of either three or four steel or cast iron rolls, often referred to as bowls. Figure 23.5 shows the principle of calendering. The calendering rolls have highly polished surfaces and the barrels contain drilled channels for a heating fluid, oil or water. Sheet with thicknesses ranging from 0.1 mm to several millimetres and about 1 m in width can be produced by this process.

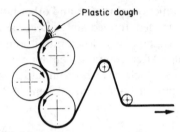

FIGURE 23.5   Principle of calendering

23.6 *Thermoplastic sheet forming*

Flat thermoplastic sheet may be formed into a variety of shapes. The sheet is heated to soften it and it may then be moulded into the required shape. A shaping process which is used very extensively is *vacuum forming* in which the flat sheet is clamped over a mould box, heated, and the intervening space then evacuated causing the hot sheet to conform to the shape of the female mould (see Figure 23.6).

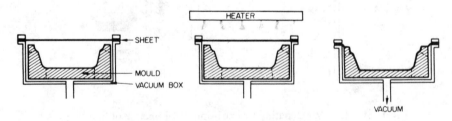

FIGURE 23.6   Principle of vacuum forming

The production of shapes which involve deep formings within female moulds can be aided by giving the heated sheet some preforming before vacuum is applied. This is termed *plug-assist*. A male forming tool, or plug (Figure 23.7), is used to produce an approximate shape immediately before evacuation of the chamber.

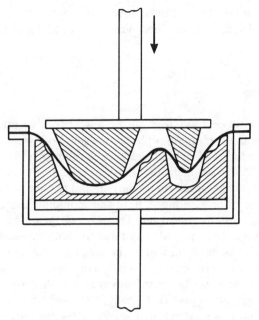

FIGURE 23.7   Preforming sheet by plug-assist

Drape forming is a modified type of vacuum forming in which heated plastic sheet is draped over a male mould and mechanically prestretched, then fully formed under vacuum (Figure 23.8).

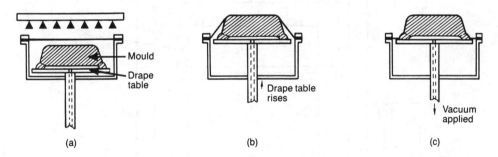

FIGURE 23.8   Principle of drape forming

Bubble-assist is a prestretch technique for sheet in drape forming comparable to plug-assist for vacuum forming. In this case, after the sheet has been heated to the forming temperature, the air pressure in the drape moulding box is increased causing the sheet to stretch upwards in a bubble. The drape table is then raised and vacuum applied. The prestretching reduces the chance of the sheet thinning too much at high or sharp points of the male mould.

Vacuum and drape forming techniques are used for forming a wide range of thermoplastic materials including LDPE, HDPE, polypropylene, polystyrene, ABS, PMMA, CAB, PVC and polycarbonate.

These sheet forming techniques are used to produce a wide range of products, from small to very large, in both thin and thick sheet material. Products made include food packaging trays, refrigerator linings, containers, light fittings, advertising signs, baths and shower trays, boat hulls, lorry and tractor cabs, and so on. Sheet sizes as large as 3 m × 9 m have been formed by these methods.

23.7 *Rotational moulding*

Rotational moulding is a process for forming hollow shapes in thermoplastics. A metered quantity of polymer moulding compound is placed in a split hollow mould. The mould, which is heated, is rotated about two axes at right angles to each other (see Figure 23.9). The polymer in contact with the hot mould surface melts and flows to take up the shape of the hollow female mould. When sufficient time has elapsed for all the material to have melted, a cooling sequence is commenced and biaxial rotation continued until the moulding has cooled sufficiently to be removed from the mould without distortion. The process cycle time may be high, up to 30 minutes for large mouldings. The method is suitable for both large and small mouldings and there is very little molecular orientation within the product, thus minimising the risk of stress cracking. Very large mouldings with sizes up to 4 m × 4 m × 4 m are possible. The female moulds are relatively cheap to make and may be made from mild steel sheet or cast aluminium as they do not have to withstand high pressures. Technically, the method is suitable for use with any powdered thermoplastic but it is used mainly for PVC, polyethylene, polystyrene and CAB.

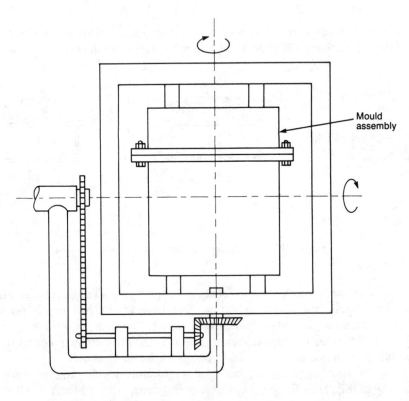

FIGURE 23.9   Principle of rotational moulding

23.8 Expanded plastics

Expanded, or foamed, plastics and rubbers are used to a considerable extent for packaging, heat and sound insulation, and as cushioning materials. Foams may be formed by a number of methods, including fully aerating an uncured resin in liquid form, or mixing the plastic resin with a volatile material that vaporises at the temperature of the curing process. Another method of producing a foam is to add a chemical blowing agent. This is a chemical that decomposes with the evolution of a gas, such as carbon dioxide, when it is heated. This technique is used for the production of expanded polystyrene.

Much polystyrene is used in the form of foam. Expandable polystyrene beads are composed of polystyrene compounded with a gas-releasing ingredient, n-pentane being a common ingredient. Mouldings for such shapes as expanded foam cups, food trays, and packaging shapes are made by in-situ expansion of polystyrene beads within a mould. The mould is heated to about 120°C, at which temperature the polystyrene softens and the expansion agent decomposes.

Foamed polymers may be used as the basis of structural composite materials, as, for example, sandwich structures with a foamed centre. A manufacturing method for this type of composite is *foam reservoir moulding*. In this process, an open cell polyurethane foam is impregnated with a controlled amount of epoxy resin and the foam faced on both sides with sheets of glass or carbon fibre fabric. The assembly is then placed in a mould which is closed under pressure. The resin saturates the facing fabric and the foam and is cured to form a rigid, lightweight composite.

23.9 Forming of thermosets

The thermosetting plastics are produced by condensation polymerisation and the reaction must be stopped before completion, as the raw plastic for moulding is required in a part-polymerised condition. The polymerisation is completed during the moulding process to give a full network rigid structure. The raw plastic material is normally compounded with fillers in order to produce a moulding compound. There are many filler materials that may be used, and the purpose for adding them is to improve the properties of the material, and also to reduce cost, as the fillers are often cheaper than the polymer. Wood flour is a commonly used filler in Bakelite type resins, but cotton fibres, asbestos fibres, and mica, may also be used (refer to Chapter 19).

The principal forming processes for thermosetting plastics are *compression moulding* and *transfer moulding*. In compression moulding, the moulding powder is compressed between the two parts of a heated metal mould, or die. The powder becomes plastic and flows into the recesses of the mould. The pressure is maintained until the polymer has cured, that is, until full polymerisation has occurred and the material has set rigid. Compression moulding is not suitable for the moulding of shapes of thick section, or with large changes of section. The resins possess low thermal conductivities, so that the centre portion of a thick section may not become fully heated and only partially cure. This problem is overcome in transfer moulding. The resin powder is heated in an ante-chamber, and when plastic it is injected into the main mould, where curing occurs (Figure 23.10). It is, however, a more expensive process than simple compression moulding as there is some scrap material formed each time.

The cold setting rigid plastics may be readily formed into shape by casting. Also, these castable materials may be used for encapsulating small electrical components, and for mounting metallic and biological specimens.

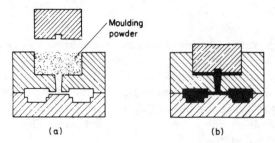

FIGURE 23.10   Transfer moulding: (a) moulding powder in ante-chamber; (b) resin forced into mould for curing

23.10 Reaction injection moulding (RIM)

This is a plastics forming process, developed initially for moulding polyurethanes, in which two rapidly reacting constituents are mixed together just prior to injection into a mould, or co-injected into a mould. The time for curing may be some 30–60 seconds after the initial mixing. Reaction injection moulding is used for some thermoplastic and thermoset materials. Fibre reinforcement may also be added, in which case the process is termed *reinforced reaction injection moulding* (RRIM). Some reacting mixes may include an agent in their formulation which releases a gas to produce a foamed product. In these cases, the foam bubbles collapse at the mould surface to form a solid skin to the moulding.

23.11 Moulding of fibre composites

The majority of the fibre reinforced composites used are based on thermoset or thermoplastic matrix materials with glass, carbon or aramid fibre reinforcement. Continuous fibres are supplied in extremely long lengths wound on a reel, referred to as *rovings* for glass or *tow* for carbon fibre. Glass rovings comprise a number of *strands*, each strand composed of about 200 fine filaments. Carbon fibre tows may contain from several hundreds to several thousands of individual fibres. Strands of glass or carbon fibre may be woven into mats or fabrics. Fibre rovings or tows may also be chopped into short lengths for use as reinforcing fillers, the individual fibre length in chopped fibres generally ranging from about 1 mm up to about 25 mm.

The simplest method for production of FRP shapes is *hand lay-up*. A mould, which may be made of wood, with a varnished surface, plaster or fibreglass, is first given a *gel coat*. This is an initial layer of resin to give a smooth surface free from protrudng fibres. The reinforcing material, in the form of cloth, mat or fibre strands, is placed in position manually while the gel coat is in a tacky condition and further resin applied by pouring on, by brush or by spraying. Full impregnation of the fibres by resin is accomplished by using a roller or squeegee. Successive layers of reinforcement and resin are added until the required thickness is achieved. In the *spray lay-up* technique, a special gun is used to spray both resin and reinforcement onto a mould which has been given a gel coat. Resin and rovings are fed into the gun, the rovings being chopped into short lengths within the gun. The component fabricated by either hand or spray lay-up is then allowed to cure and harden. The curing time may be accelerated by heating. Greater consolidation and densification of a moulding made by hand or spray lay-up can be achieved by means of pressure bag moulding, or vacuum bag moulding. In the former, pressure is applied to the free surface of the moulding via a shaped rubber bag, while in the latter method the mould and moulding are enclosed in an impermeable bag and the air evacuated. The external air pressure then compacts the moulding. *Autoclave moulding* is used for resins that require an elevated temperature for cure. This is similar to pressure bag moulding except that the cure is carried out in a steam autoclave. Autoclave moulding is widely used for the manufacture of structural shells and sections in the aerospace industries.

The methods described above use an open mould but it is possible to produce reinforced components within a closed mould. The two most widely used processes are *resin transfer moulding* and *reinforced reaction injection moulding* (RRIM). In resin transfer moulding, the reinforcing fibres, frequently in the form of *preforms* (see Section 23.12), are placed in the mould before closure and resin is injected to fully impregnate the fibres and fill the mould. Curing occurs within the heated mould. RRIM is similar to RIM (see Section 23.10) but with chopped fibre being mixed in with one of the reactants. RRIM is used mainly in connection with polyurethanes.

Filament winding is used for producing hollow cylindrical or spherical components. Continuous fibre rovings or tow are passed though a bath of resin and then wound around a shaped former, or mandrel. Fibres are wound onto the mandrel in a helical pattern and the helix angle and winding pattern can be designed to give the best strength characteristics for the intended application. Alternatively, *pre-preg* tapes (see Section 23.12) may be wound around the mandrel.

Pultrusion and *pulforming* are forming methods for producing rod, tubing and sections in fibre reinforced plastics. In pultrusion, a continuous length of fibre rovings or tow is passed through a resin bath and then through a series of carding plates to give full penetration of resin into the fibre bundles, and to remove excess resin. The impregnated fibres are pulled through a heated die where full consolidation occurs and the resin gels and then through a tunnel oven to complete the cure (see Figure 23.11).

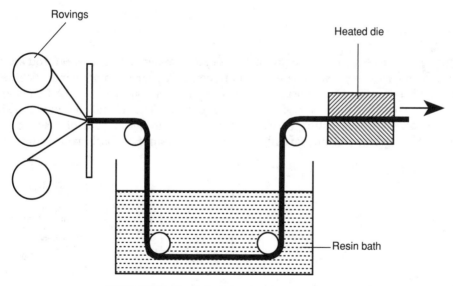

FIGURE 23.11   Principle of pultrusion process

In pultrusion, sections of uniform and constant cross-section are formed. Pulforming is similar to pultrusion but it gives a product with a varying cross-sectional geometry but of constant cross-sectional area. These processes give composites with a high axial strength and stiffness.

23.12 *Preforms, prepregs and moulding compounds*

Various measures can be taken to aid manufacture of fibre reinforced composites, including the use of preforms and prepregs. A preform is used in connection with resin transfer moulding (Section 23.11). Fibres laid within a mould could be moved out of position during the injection of resin. This problem can be obviated by use of a preform, which is a mass of fibres, or fibre fabric, shaped to fill the mould cavity and held in shape by a small amount of a dilute resin as a binder.

Prepreg is the shorthand term in general use describing pre-impregnated fibre/resin intermediate products which can be formed, at a later stage, into the required shapes. Prepregs are available in several forms and can use glass or carbon fibre and either thermoset or thermoplastic matrices. When thermosets are used, the fibres, either as a woven fabric or as continuous unidirectional fibres, are fully impregnated with resin and partially cured. In this form, the resin is slightly tacky and the sheets or tapes are protected with paper coated with a silicone barrier resin for storage. (The term *warp sheet* is used for unidirectional prepreg sheet.) Prepregs may be shaped by conventional processes such as vacuum or drape forming, pressure bag moulding and autoclave moulding, and thermoset prepregs are fully cured during or after the moulding operation.

Thermosetting polyesters with glass reinforcement are available as moulding compounds. These contain uncured resin, chopped glass fibres and an inert filler in approximately equal proportions. They are available in three forms, as sheet moulding compound (SMC), dough moulding compound (DMC) and bulk moulding compound (BMC). SMC, produced in various sheet thicknesses from 3 to 12 mm, can be formed by conventional vacuum and drape forming methods, the resin being cured during the process. DMC and BMC can be injection moulded, the resin curing in the mould.

23.13 *Laminating*

Laminating is a process used for the production of stock material such as sheet, rod and tubing but it is also used for the manufacture of more complex shaped components such as safety helmets. Laminates may be made using a range of thermoset materials as the base material but the majority are based on phenolic resins with paper or cloth. The general manufacturing process is to pass paper or woven cotton or glass fibre fabrics through a bath of resin to give full impregnation of the material and partially cure to gel the resin. For sheet manufacture, the resin

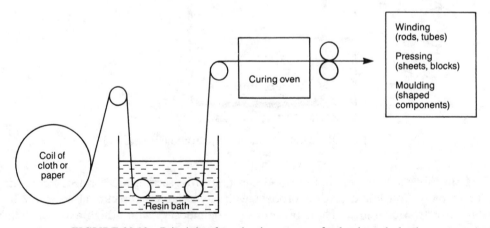

FIGURE 23.12   Principle of production process for laminated plastics

impregnated materials are layered until the required thickness is obtained and the layered sheet is then compressed in a heated press to cure the resin thoroughly. For the manufacture of rod or tubing, impregnated tape is would to a cylindrical shape before curing, while for moulded shapes the required number of layers are assembled and press formed into shape and cured. A proprietary name for paper and cloth laminates is *Tufnol*.

24

Forming Processes for Ceramics and Glasses

24.1 *Forming of clay ceramics*

Ceramics have been used since earliest times, and many ceramic articles are produced today by methods that are basically the same as those used several thousand years ago. Clay materials are normally shaped from a moist soft plastic mass. For articles with rotational symmetry, the clay may be *thrown*, or moulded, on a potter's wheel. *Hand throwing* in the age old manner is still used on a small scale, but for mass production the process is automated and the potter's hands have been replaced by moulds and templates. Plastic clay may also be readily formed by extrusion, and this process is used for the forming of bricks, tiles and pipes.

After shaping, clay articles are dried and then fired in a kiln at temperatures in the range 800–1500°C. The firing causes expulsion of water of crystallisation, and a recrystallisation, creating a rigid, though brittle, material. There is a partial melting of some of the constituents of the clay during firing and a glass phase is formed between the grains, cementing the grains together. The extent of this vitrification during firing depends on both the composition of the clay and also on the firing temperature, higher temperatures increasing the glass content of the fired product. The higher the glass content, the stronger will the body be. The changes which occur during firing are accompanied by a shrinkage. Fired clay products are porous, and this porosity may be as high as 15 or 20 per cent in some earthenware products. In many cases a glaze is applied to the surface of a fired article and this is converted to a hard, vitreous and non-porous layer during a second, or *glost*, firing operation.

Another forming method that may be used for forming clay shapes is *slip casting*. The slip, which is a thick slurry or suspension of clay in water is poured into a porous mould. Water is rapidly absorbed into the mould from the slip, and the surface layers of the slip become solid. When the required wall thickness has solidified, excess slip is poured from the mould, leaving a hollow casting. The casting may then be removed from the mould and fired. Slip casting is suitable for the prouction of complex shapes, such as wash basins and sanitary ware, and it is also used for the production of shapes in some of the newer industrial ceramics.

Another method used for forming ceramics from a slip is *tape casting*. This method is used to produce thin ceramic wafers for use in the electronics industry. The slip is filtered and fed onto a continuous moving belt, the thickness of the deposited film being controlled by an adjustable gate. The continuous tape produced is dried and pieces of the required shape and size can be blanked from the cast tape before firing.

24.2 Pressing

Generally, oxide, carbide and nitride industrial ceramics are formed from powder by processes of compaction and sintering in the same general way as that used for powder metal parts (see Section 22.15). One major difference from powder metallurgy is that the ceramic powders are non-ductile and show no plastic deformation when compressed. To overcome this, ceramic powders are often mixed with a small amount of a binder and a lubricant so that a coherent compact with sufficient green strength to permit handling can be obtained.

Green ceramic pressings may be produced by either *die pressing* or *isostatic pressing*. In die pressing, graded ceramic powder is fed into a metal mould or die and compacted by either one male tool, or plunger, forced into the die under pressure or two plungers applied from opposite directions. After compaction, the green compact is ejected from the die and sintered. The process is used for powder mixes of low plasticity and, although the powder may contain up to about 8 per cent water, it is also referred to as *dry pressing*. Isostatic pressing is capable of producing compacts which possess higher and more uniform densities than is possible with die pressing. The dry ceramic powder, or a powder preform made by die pressing, is contained in a flexible mould, usually made of rubber. There are two major variants, these being 'wet-bag' or hydrostatic pressing or 'dry-bag' pressing. In the former process, the rubber mould is placed in an enclosed pressure vessel and hydrostatic pressure is applied using a water/soluble oil mix as the transmitting fluid. In the latter process, the pressure fluid is a gas. Pressures in the range of 50–500 MPa are applied in isostatic pressing.

Hot pressing and hot isostatic pressing (HIP) are processes in which heat and pressure are applied simultaneously and sintering, under pressure, occurs (see Section 24.3).

24.3 Sintering

Sintering is the term for the process which converts a powder compact into a solid polycrystalline material. It is a thermally activated process and the driving force is the reduction in surface energy which occurs as the powder particles coalesce. In solid phase sintering, which is carried out at temperatures below the melting points of the powder constituents, the first stage in the process is the transport of some material by diffusion and its deposition in the inter-particle spaces immediately adjacent to particle contact points. This deposit is termed a *neck*, and the formation of necks increases the particle contact area. During sintering, the necks grow and merge with one another until, eventually, the original particle structure is replaced by a polycrystalline grain structure with a network of porosity between grains. There is an increase in density, and a volume shrinkage, during sintering and this shrinkage may be as high as 20 per cent.

Sintering under pressure will produce higher density materials with finer grain sizes than is possible with sintering at atmospheric pressure. The stresses at particle contact points caused by the pressure will give an increased driving force for sintering. The two forms of pressure sintering are *hot pressing*, where the compact is sintered under pressure in a heated graphite or ceramic die, and *hot isostatic pressing* (HIP), in which a preformed compact is sintered at high temperature in an autoclave with argon or nitrogen as the pressure transmission medium. HIP is capable of giving a sintered product with uniformity of properties and virtually zero porosity.

Another form of sintering which may be used to produce dense ceramics with little or no porosity is *liquid phase sintering*. A small amount of an additive in the ceramic powder mix may react with the ceramic forming a compound which is liquid at the sintering temperature. An example is the addition of a small amount of magnesia, MgO, or yttria, Y_2O_3, to silicon nitride powder. During sintering in a nitrogen atmosphere, a liquid oxynitride phase is formed. This phase aids densification by means of a solution–diffusion–precipitation mechanism and the final product is a dense silicon nitride with an oxynitride glass as a grain boundary phase. The presence of the glass phase limits the maximum temperature at which liquid phase sintered components can be used.

24.4 Reaction bonding and reaction sintering

Reaction bonding is the conversion of a powder compact into a polycrystalline ceramic by means of a chemical reaction between the powder particles and a gas or a liquid. This occurs in the production of some grades of silicon carbide and silicon nitride ceramics.

Reaction bonded silicon carbide (RBSC) is made by heating a porous powder compact of silicon carbide and graphite in contact with molten silicon. Silicon enters the open pores by capillary attraction and reacts with the graphite to form silicon carbide, which deposits on the surface of the original silicon carbide particles, bonding them together. All the carbon reacts and excess silicon remains, the final structure being of silicon carbide grains with interconnecting silicon-filled pores. This material is used for seals, bearings and engine components.

Reaction bonded silicon nitride (RBSN) is made by heating a compact of silicon powder in an atmosphere of nitrogen at a temperature in the range 1100–1450°C. Silicon is converted into silicon nitride according to the reaction:

$$3Si + 2N_2 \rightarrow Si_3N_4$$

This reaction bonding process is also referred to as nitriding. There is virtually no change in the dimensions of the compact because, although silicon nitride occupies a greater volume than the original silicon, the nitride formed fills some of the pore space of the original silicon compact. However, even after nitriding, the product still has about 20 per cent porosity. A major advantage of this process is that the silicon compact may be partially nitrided into a coherent compact that can be machined to any required shape and to accurate final dimensions using conventional material cutting methods and tooling. Finally, the machined part is reaction bonded to fully nitride it into the very hard silicon nitride. RBSN is used as pouring tubes and pumping equipment for liquid metals and for high temperature engineering components.

Reaction sintering is the term describing the sintering of a powder compact into a polycrystalline ceramic involving a chemical reaction between two solid components of the powder. The product of reaction sintering is a dense single phase or multi-phase ceramic. An example is the reaction sintering of a mixture of alumina, Al_2O_3, and zircon, $ZrSiO_4$, to produce a dense zirconia–mullite composite ceramic.

24.5 Injection moulding

In recent years there has been considerable research and development into the forming of ceramic components by injection moulding. The ceramic powders, which in themselves are non-plastic, are mixed with various plasticisers and binders. Very close control of both the chemical composition and powder particle size is necessary if a powder mix is to possess the rheological properties necessary to give a good injection moulding. Use is being made of injection moulding techniques to produce ceramic components of complex shape. A far greater degree of complexity can be achieved in injection moulding than is possible by die pressing. Injection moulding has been used successfully for the production of ceramic engineering components, including the soluble cores used in the investment casting of superalloy turbine blades with cooling channels.

24.6 Manufacture of glass

The raw materials for the manufacture of common soda–lime–silica glass are soda ash, Na_2CO_3, limestone, $CaCO_3$, and silica sand, SiO_2. These ingredients are crushed and ground to a fine particle size, typically 0.1–0.6 mm, and blended together before charging to a melting furnace, also known as a glass tank. During melting, the carbonates decompose, liberating carbon dioxide, and react with the silica. The gas evolution agitates the melt and aids the mixing to achieve a uniform composition. When fully molten, the glass contains many gas bubbles and the final stage, termed *refining*, involves raising the temperature of the melt to about 1550°C to reduce viscosity and allow the gas bubbles to come to the surface. This stage is helped by making additions of agents such as sodium sulphate, arsenous oxide or antimony oxides to the melt.

Glass tanks are divided into two zones, the melting and refining zone, and the working area, a bridge wall separating the two areas. The glass in the working area is maintained at the appropriate temperature for the type of processing which is to be carried out (see Section 24.7).

24.7 Forming processes for glass

The main forming processes for glass are sheet rolling, tube making, hot pressing, float moulding and blow moulding. Different viscosities are required for these various processes and the working section, or forehearth, of the glass tank is maintained at the appropriate temperature. The range of viscosities for glass working is from 10^3 to 10^6 Pa s. Low viscosities, of the order of 10^3–10^4, are required for float and blow moulding, while for sheet rolling and tube making viscosities in the range 10^4–10^5 Pa s are needed. A high viscosity of 10^5–10^6 Pa s is suitable for forming by pressing.

Much glass sheet, in thicknesses from about 1–10 mm is made by a rolling process in which glass is drawn from the forehearth of the glass tank and flattened by passing through either a pair of rolls or between a pair of moving continuous belts (Figure 24.1(a)). The glass sheet formed by this method contains some imperfections and the surfaces are not perfectly flat and parallel. It is, however, suitable for many applications including use as window glass.

Distortion-free glass sheet and plate is made by the float-glass process, developed by Pilkingtons in the 1950s. In this process, molten glass flows from the forehearth of the glass tank in a continuous strip to float on the surface of a bath of molten tin (Figure 24.1(b)). It is heated sufficiently for the glass to flow so that both surfaces are flat and parallel.

Glass tubing is made by causing glass to flow from the glass tank onto the surface of a rotating mandrel, through which air is blown (Figure 24.1(c)). As the glass cools and its viscosity increases, it is drawn out to smaller dimensions.

Pressing is a process analogous to the closed-die forging of metals. A measured quantity of hot glass is put into a steel or cast iron die and the upper half of the die set pressed in position causing the glass to flow into the contours of the dies.

Blow moulding is used for the manufacture of bottles and other hollow shapes and, as in the blow moulding of thermoplastics, air pressure is used to blow a 'gob' or parison of glass so that it conforms to the shape of a mould. For the production of some shapes a pre-shaped parison is formed by an operation in which a quantity of hot glass is pressed in a die to give a hollow

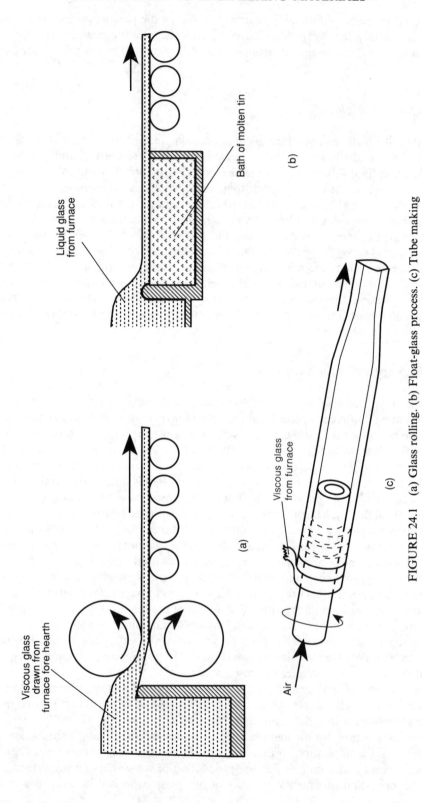

FIGURE 24.1 (a) Glass rolling. (b) Float-glass process. (c) Tube making

preform. This is reheated to give the required viscosity for blow moulding and then blown in a mould.

Hand-blowing of glass is still carried out but this is mainly for art glasswork and not for mass produced glass articles.

Glass products need to be annealed after forming and this is carried out in an oven or *lehr*. An annealing lehr is an oven containing a conveyor belt which carries the glass products through the heated zone. The annealing temperature is that at which the viscosity of the glass is about 10^{12} Pa s (see Section 20.13) and the glass is slowly cooled from the annealing temperature to the *strain point* (viscosity 10^{13} Pa s).

25

Material Removal Processes

25.1 Introduction

The majority of the forming processes described in the three preceding chapters give products in their final, or near final shape. The aim, in many instances, is to produce a finished product with the elimination of machining operations. The term *machining* refers to the forming or generating of shapes by means of a material removal process. The range of machining processes includes both traditional cutting techniques such as drilling, milling, turning and grinding, in which chips are formed, and a large number of non-traditional chipless machining processes, including electro-chemical and electro-discharge machining. During a machining process, amounts of costly material are removed from the part, this often being converted into chips, a form of material scrap which is frequently difficult to handle and recycle. However, although much progress has been made in forming net or near net shapes by precision casting and plastic deformation and moulding techniques, there is still a major requirement for machining operations as it is possible, through their use, to create shapes, surface finishes and dimensional accuracies which may not be achievable by other means.

The majority of machining is carried out on metals and alloys but plastics materials and composites are machineable and some material removal processes are applicable to ceramics.

25.2 The cutting process

Metal cutting is a cold working process in which a cutting tool forms chips. The cutting tool presents a wedge-shaped point to the work. In ideal orthogonal cutting, the cutting edge of the tool is normal to the direction of motion. There is severe deformation by shear in a plane, the shear plane, resulting in the formation of a chip (Figure 25.1). Considerable compression occurs in the deformed material and the thickness of the deformed chip produced will be greater than the depth of cut. There are two aspects of the tool geometry which are of importance, these are the *rake angle* and the *clearance angle*. The rake angle of the cutting face, which is measured from the normal to the workpiece surface, influences the shear angle and the amount of chip compression. The clearance angle is necessary to reduce friction between the tool and the machined surface.

The geometry of a single point cutting tool has a major influence in determining the efficiency of a cutting operation. Increasing the rake angle has the effect of reducing both tool forces and friction and giving a less deformed and cooler chip. However, the provision of a large rake angle gives a smaller tool section, thus reducing the strength of the tool. The most effective rake angle for any cutting operation will be determined by the nature of both the workpiece and cutting tool material. The rake angles used in tools for cutting the softer materials, such as aluminium and its alloys, can be larger than those employed in the machining of steels. Carbide and ceramic cutting

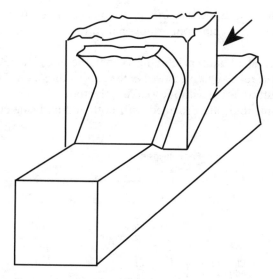

FIGURE 25.1   Orthogonal cutting with a single point tool

tools, although extremely hard, are also brittle. Consequently, the rake angles used in connection with this type of tool are usually either zero or negative to provide added tool strength. The same principle of rake and clearance angles applies to all multipoint tools, such as broaches, milling cutters and saws, in fact, any tool in which the cutting action is one of presenting a wedge-shaped point to the work.

In many practical instances, oblique cutting, rather than orthogonal cutting, is adopted. In this, the cutting edge of the tool is inclined at some angle other than 90° to the cutting direction (Figure 25.2). This offers significant advantages. This chip leaves the workpiece at an angle, equal

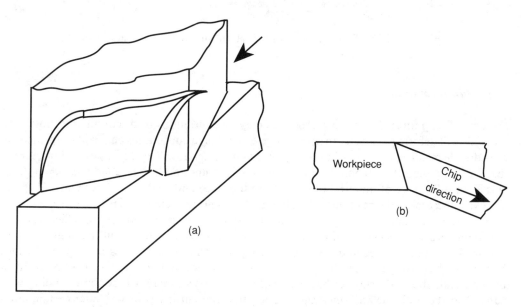

FIGURE 25.2   (a) Oblique cutting; (b) Direction of chip travel

to the cutting edge angle and, in the case of a continuous chip, curls into a helical form rather than the closed spiral shape of a chip from orthogonal cutting. The chip then tends to move away from the work.

The other advantage of oblique machining is that the effective rake angle of the tool is somewhat greater than the rake angle measured in the cutting direction. This higher effective rake angle gives a reduced cutting force without weaking the tool.

Generally, single point tools are provided with both end and side cutting edges and this is shown in Figure 25.3.

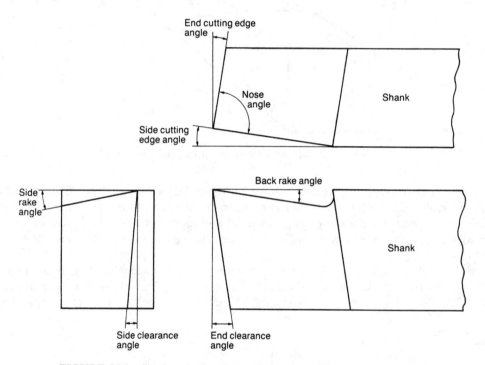

FIGURE 25.3   Single point cutting tool showing rake and clearance angles

25.3 *Chip formation*

The type of chip formed in cutting processes may be continuous or discontinuous. Soft, ductile materials will deform plastically to a very large extent and a continuous chip will be formed. With a less ductile metal, severe work hardening of the cut material will lead to a series of fractures and give rise to a discontinuous chip (Figure 25.4). Under some conditions, a *built-up edge* will form. This is the adherence of material to the cutting tool face and it alters the geometry of the tool, effectively increasing the rake angle (Figure 25.4(c)) and reducing the cutting force.

How does a built-up edge occur and how does its presence affect the efficiency of the machining process? When cutting at relatively low speeds using a cutting fluid, a smooth machined surface is formed and the chip surface is also smooth and slides across the rake face of the tool. As the cutting speed is increased, there is an increase in the temperature of both chip and tool face and, above a certain critical value of cutting speed, there will be some welding of chip material to the tool, the *built-up edge*, altering the tool cutting profile. Under certain circumstances, this may be advantageous in that the larger effective rake angle reduces cutting power

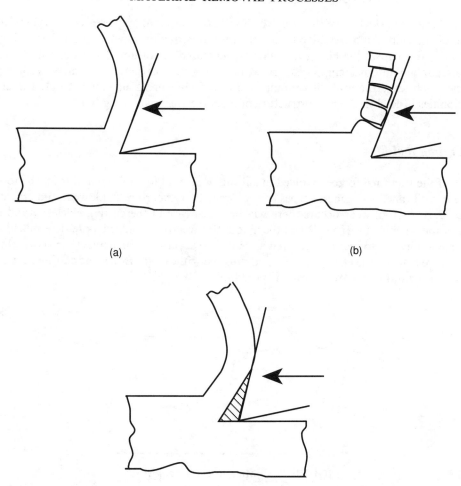

FIGURE 25.4 Chip formation: (a) continuous chip; (b) discontinuous chip; (c) continuous chip with built-up edge

requirements and the built-up edge, if stable, affords some protection to the tool face. Frequently, with a small, stable built-up edge, a machined surface finish of acceptable quality can be produced. However, in many cases, the built-up edge is not stable, may break off from time to time and a machined surface of poor quality results. As the cutting speed is increased further, and even higher chip temperatures occur, a second critical speed is reached at which a built-up edge does not form and a machined surface with a good finish is again achieved. The critical speeds are a function of several factors, the nature of workpiece, the cutting tool material and the type of cutting fluid. For the machining of mild steel with a high speed steel cutting tool, the two critical speeds are of the order of 320 mm/s and 650 mm/s respectively.

In high speed machining, particularly with automatic machine tools, it is desirable to have discontinuous chip formation as these can be removed easily from the cutting zone. Long continuous chips pose considerably greater problems. They are more difficult to remove from the cutting zone and they could wrap around workpiece or tool creating a potentially hazardous situation. The situation can be improved by incorporating a *chip breaker* into the tool. This is a device for breaking up a continuous chip into shorter lengths. One form of chip breaker is a groove cut into the rake face of the tool. This forces the chip into an increased curvature causing fracture of the chip. An alternative is to attach an 'obstruction-type' chip breaker to the rake face. Again, this causes the chip to assume an increased curvature. The so-called *free-cutting* materials contain internal chip breakers within their microstructures, namely second phase particles which are structural discontinuities and aid chip break-up, for example the discrete small globules of lead in the microstructure of a leaded free-machining brass.

25.4 Cutting tool life

Because of the frictional forces involved in cutting, wear and loss of material from the tool face will occur with time. The general types of tool wear that occur are flank wear and crater wear (Figure 25.5). As tool wear occurs there will be a decrease in the cutting efficiency and in the quality of the machined surface. When the amount of wear has reached a pre-determined value, based on inability to produce the required dimensional accuracy or machined surface quality, the tool will have reached the end of its 'life' and must be withdrawn. Either the cutting face may be re-ground to permit re-use or the tool disposed of.

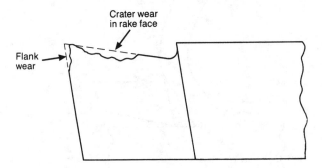

FIGURE 25.5   Main forms of tool wear

F W Taylor developed an empirical law relating cutting tool life, T, to cutting speed, V, and other parameters. The general form of Taylor's tool life equation is:

$$VT^n = C$$

where C is a constant based on feed rate and other parameters and the exponent n is characteristic of the cutting tool material. The relationship can also be written in the form $\log V + n \log T = \log C$, and a graphical plot of $\log V$ against $\log T$ gives a straight line with a slope equal to $-n$. Typical values of the exponent n are: 0.14 for high speed steel, 0.25 for carbide cermets, 0.3 for coated carbides and 0.4 for sintered alumina and some other ceramics.

The extent of the reductions in cutting tool life brought about by an increase in cutting speed is better indicated by expressing Taylor's equation in terms of tool life, T. This gives: $T = C^{1/n} \div V^{1/n}$. It can be seen clearly now that, for a high speed steel cutting tool ($n = 0.14$), the tool life reduces according to the seventh power of cutting speed whereas, for cemented carbide tools, the life reduces according to the fourth power of speed and, for ceramic tools, it reduces according to $V^{2.5}$.

25.5 Cutting fluids

Although, in a cutting operation, the majority of the heat energy liberated is carried away in the chip, there is a significant rise in the temperature of the cutting tool, particularly at high cutting speeds. The majority of machining operations are carried out with use of a cutting fluid and such cutting fluids fulfil a number of functions. These can be listed as:

(a) to act as a coolant and conduct heat away from tool and work,
(b) to act as a lubricant and reduce friction between the chip and tool,
(c) to carry chips away from the cutting point,
(d) to aid production of a good surface finish,
(e) to reduce energy consumption, and
(f) to extend the life of the tool.

Two main types of cutting fluid are used, coolants and lubricants. In high speed machining, it is the cooling and heat removal aspects which are the most important and water/soluble oil emulsions are widely used. They are good coolants and give some lubrication. Some complex machining operations, such as gear cutting, require a greater lubrication than can be provided by soluble oil emulsions and a suitable oil will give a good surface finish and a prolonged tool life. Most of the lubricant type cutting oils are based on mineral oils, often with chlorinated or sulphurised additives to increase lubricity.

25.6 Cutting tool materials

The properties required in a cutting tool material are high hardness, the ability to retain hardness at elevated temperatures and to possess a degree of toughness sufficient to minimise the incidence of chipping and cracking. The main categories of cutting tool materials in use are: carbon tool steels, high speed tool steels, cemented carbides, coated carbides, ceramics and diamond.

Carbon tool steels

Carbon tool steels contain between 0.6 and 1.5 per cent carbon and are used in the hardened and tempered condition (refer to Sections 16.9 and 16.10). They lose much of their hardness, through tempering when heated above 300°C. These days, carbon tool steels are used mainly for the cutting of plastics and timber and have been almost entirely superseded for metal cutting operations by other materials.

High speed steels

High speed steels are highly alloyed steels containing tungsten, chromium, vanadium and molybdenum, the most widely used composition being 18 per cent tungsten, 4 per cent chromium, 1 per cent vanadium and 1 per cent carbon. Steels of this type contain hard tungsten and chromium carbides in their microstructures and require a three-stage heat treatment comprising oil quenching from a high temperature and two tempering treatments (see Section 16.14). These steels retain their hardness at reasonably high temperatures and do not soften appreciably until they are heated above 600°C. High speed steels are used for the manufacture of many metal-cutting tools, including drills, reamers, milling cutters, hacksaw blades and single point tools for general purposes. They are not used to a great extent in connection with high rate production operations using automatic machine tools, having been largely replaced by carbide and ceramic materials for this type of application.

Cemented carbides

Cemented carbides, also known as cermets and hardmetal, are composed of carbides of tungsten, titanium, niobium and other metals, in a matrix of cobalt (see also Section 21.7), and are made by powder metallurgy processing involving compaction and sintering. Cemented carbides are produced as both tool tips, which are then brazed to a steel tool shank, or as 'throw-away' inserts which can be secured to a tool shank by a set screw.

Coated carbides

These are cemented carbides provided with a thin coating, typically about 5 μm thick, of a harder ceramic material. The coating may be applied by chemical vapour deposition (CVD) or by sputtering. Titanium carbide, titanium nitride and alumina are among the hard coatings used and in some cases a multi-layer coating may be applied, for example, titanium carbide, followed by alumina and titanium nitride. These coatings give the tool tip a wear resistance approaching that of a ceramic tool tip whilst retaining the greater toughness of a cemented carbide.

Ceramic tools

A number of ceramic materials are used as disposable tool tips. These include alumina (Section 20.7), cubic boron nitride (Section 20.11), sialons (Section 20.9) and polycrystalline diamond (PCD). These ceramics are harder than cemented carbides but are of lesser toughness. They are capable of cutting very hard materials such as *superalloys*, and of operating at much higher cutting speeds than is possible with cemented carbides (Section 25.4). PCD, although extremely hard, interacts with iron at high temperatures and is not suitable for cutting steels. The ceramic materials are formed by powder compaction and sintering into tool tips and 'throw-away' inserts.

25.7 Machinability

Machinability is the relative ease with which a material may be machined by a cutting process. A very soft and ductile metal may spread under the tool cutting pressure, with the result that the tool tends to become buried in the workpiece and a tearing, rather than a clean cutting action, may occur. A similar effect can occur with soft thermoplastics. In the case of a less ductile metal, the severe work hardening effect will lead to the creation of a series of fractures giving discontinuous chip formation. It is this type of chip formation which is desirable for high-speed machining with automatic machine tools.

For the best machining characteristics, the requirements are for low hardness, coupled with low ductility. The machinability of a metal can be improved by any means that will decrease the ductility and increase the susceptibility to fracture. This includes work hardening and alloying to give solution or dispersion strengthened materials. Of course, the material must not be made too hard, as then it will increase the difficulties of cutting and lead to very rapid tool wear. Multi-phase alloys often machine well because the additional phases provide discontinuities within the microstructure and assist in the formation of discontinuous chips. The machining characteristics of some common materials is given below.

Plain carbon steels

The hardness and ductility of plain carbon steels is affected by both carbon content and the type of heat treatment (see Chapter 16). Low carbon steels in the annealed or normalised condition are of high ductility and have poor machinability. In the work hardened condition, the machinability of these materials is improved considerably. A very large number of engineering components are manufactured from low carbon steels by machining rod and bar stock and there is a major requirement for low carbon steels possessing good machinability. The so-called *free-cutting* steels are low carbon steels with either an increased sulphur content or added lead. Their microstructures contain either manganese sulphide (MnS) or lead as globular second phase particles which act as internal chip breakers. Medium carbon steels, with their higher carbon content, are harder and of lower ductility and are best machined in the annealed or normalised condition. High carbon steels are best machined after being spheroidise annealed, as otherwise their hardness would be too great and result in rapid tool wear.

Alloy steels

The low alloy pearlitic steels are similar to plain carbon steels of corresponding carbon content in their machining characteristics. Many high alloy steels, including stainless steels are of low machinability.

Cast irons

Grey cast irons are, in general, free-machining materials as the graphite flakes act as internal chip breakers. The graphite also acts as a dry lubricant. White and mottled irons, however, are difficult to machine owing to the very high hardness of cementite.

Aluminium and its alloys

Pure aluminum is extremely soft and ductile and it is difficult to obtain a good machined surface unless the material is in a work hardened condition. A tearing of the metal surface rather than a clean cut may occur. This also applies to the softer alloys. The strong heat treatable alloys have good machining characteristics.

Copper and its alloys

Pure copper and some of the α phase alloys are soft and ductile and are best machined in the work hardened condition. Free-machining alloys are available, for example, tellurium copper (Section 15.5) and leaded brasses.

Magnesium and its alloys

These materials possess excellent machinability as they possess both a low hardness and low ductility. The main problem is the hazard posed by fine magnesium chips, as these could ignite. It is essential that there is no interruption to the flow of cutting fluid during the machining of these materials.

Plastics materials

Most plastics materials can be machined using conventional wood-working or metal cutting equipment. The main problems with thermoplastics are that their properties are highly heat sensitive and temperature build-up must be avoided. This problem is somewhat aggravated by their low thermal conductivities. The soft thermoplastics, with T_g values below room temperature, give continuous chip formation but discontinuous chips are formed in materials, such as polystyrene and PMMA, with high T_g values. Polymers possess low elastic moduli and can be deformed relatively easily by the cutting tool, making the machining less precise than metal cutting. A small depth of cut is necessary to keep the tool force to a minimum. Glass-filled thermoplastics and most thermosets, including the phenolic laminates, although much harder than unfilled thermoplastics, can be readily machined with discontinuous chip formation.

Ceramics

Ceramics are of very high hardness and extremely difficult to shape by machining. Generally, of the traditional machining operations, only abrasion and grinding processes are suitable. The non-traditional methods of electron beam and laser beam machining can be used in conjunction with ceramics. An exception is silicon nitride. Partially nitrided silicon compacts (see Section 24.4) can be machined using conventional metal cutting equipment.

25.8 Forming and generating

Forming is the production of a shape by a cutting process when the cutting tool has the reverse shape or contour to that required in the machined product. A form tool is a lathe or milling tool with a profiled cutting edge. In *generating* a surface, the shape and accuracy of the machined surface is determined by the movements of the machine tool slides rather than by the shape of the cutting tool. For example, the obtaining of a cylindrical or conical shape by lathe turning using a single point tool is generating. In practice, many machined surfaces are produced by a combination of generating and forming. A good example is the cutting of a screw thread on a lathe. The thread profile is *formed*, as it conforms to the contour of the single point tool, but the helical thread is *generated* along the length of the workpiece by the movement of the tool.

25.9 Single point machining

The main machining processes involving use of a single point tool are *shaping* and *planing*, which generate flat surfaces, and *turning* on a lathe to generate curved surfaces (Figure 25.6), although other operations such as *boring* and *facing* can also be carried out on a lathe.

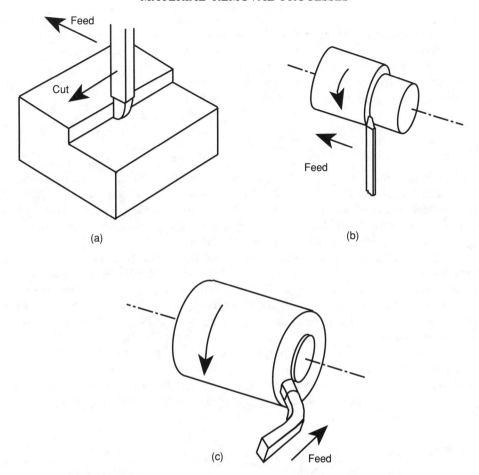

FIGURE 25.6   (a) Shaping or planing; (b) Turning; (c) Facing

Shaping and planing

Both shaping and planing are machining processes for producing flat surfaces by the straight line cutting action of a single point tool. Most shaping is in the horizontal plane with a horizontal push-cut, the cutting tool being pushed across the work by a ram. During the return stroke of the ram, the tool slides over the work surface. Workpiece feed between cutting strokes is at right angles to the cutting direction. In planing, it is the workpiece which moves in a horizontal plane relative to a stationary cutting tool. Planing is used for workpieces which are too large to fit onto shapers. Both planing and shaping are slow processes and productivity is low in comparison with other types of machine tools. Their use has been largely replaced by other more efficient cutting processes such as milling.

Turning, boring and facing

Turning is a rotational cutting process in which the work rotates and tool feed is parallel to the axis of rotation. Turning is only one of several cutting operations which may be made on a lathe. Boring is the generation of the internal surface of a hollow cylindrical part using a single point tool. The tool may be stationary and the work rotated, or vice versa, with the feed parallel to the axis of rotation. Facing is the generation of a flat surface on a component which is rotated in a lathe. The cutting tool is fed towards the centreline and the cut follows a spiral path on the work, giving a flat surface normal to the lathe centreline (Figure 25.6(c)).

Lathes

Centre lathes are probably the most widely used types of machine tools. There are many different types and sizes of lathe but basically a lathe provides a rotary motion to the work. The cutting tool, in its holder, can move along sideways in two directions, parallel to and normal to the axis of rotation (Figure 25.7). In addition to turning, facing and boring, other operations such as drilling, reaming and screw-cutting can be conducted on a lathe. A centre lathe requires an operator who is a skilled craftsman and this machine is used for prototype manufacture and small production runs. The *capstan lathe* is a production lathe for small to medium sized components and can be operated by semi-skilled labour. The machine carries a hexagonal capstan or turret in place of the tailstock and this carries up to six tools or special attachments. The operator can move the turret head to and from the work very quickly and the provision of stops enables each tool to cut to the required depth and length. The turret indexes automatically on withdrawal from the workpiece to bring the next tool into position.

In addition to manually operated lathes, there are many types of automatic or autolathes, some with turret heads and with single spindles or multi-spindles. Fitted with cams and a variety of rapid-acting electro-mechanical devices to control spindle speed and tool movement, they are

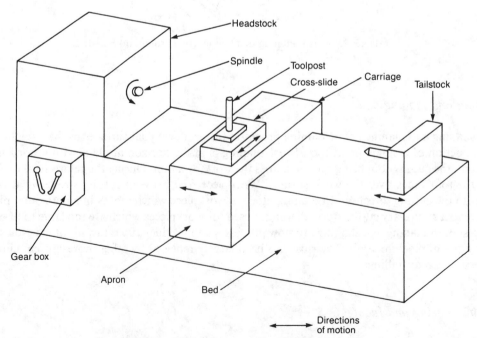

FIGURE 25.7   Schematic representation of a centre lathe

used to machine accurately large batches of components. Special types include screw-cutting lathes and copying lathes in which tool movement is controlled by the movement of a stylus following a template or copy of the required product.

25.10 Multi-point machining

In multi-point machining, two or more cutting edges of the same tool are in operation at the same time. The main multipoint machining processes are *drilling*, *milling*, *broaching* and *sawing*.

Drilling

The provision of cylindrical holes in components is a major requirement and the majority of machined holes are produced by drilling. The spade drill, with a relatively simple shape, was the traditional type of drill in use for thousands of years for drilling holes in wood. The most common type of drill in use today is the twist drill. A twist drill has two cutting edges with two helical grooves, termed flutes (Figure 25.8(a)). The flutes allow for the access of cutting fluids and also a path for the removal of chips. The helix angle for most twist drills made from high speed steel is 24° but smaller angles may be used for drilling plastics and soft metals, whilst large angles can be used for materials which can be drilled very rapidly. Drills with straight flutes are generally used for drilling holes in thin sheet. The spade drill design has been revived and this type, with cemented carbide cutter inserts (Figure 25.8(c)), is being used increasingly in automatic mass production work.

Milling

Milling is a metal cutting process in which a multi-point tool rotates and the work feeds past the tool. In horizontal milling the cutter rotates about an horizontal axis, parallel to the work piece surface. The cutter is mounted on a shaft which is supported at both ends. In vertical milling, the

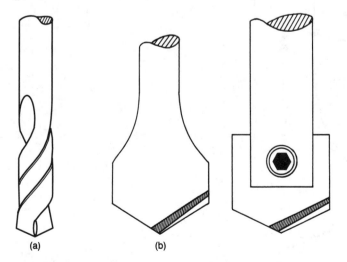

(a)   (b)

FIGURE 25.8   (a) Twist drill; (b) Simple spade drill; (c) Spade drill with insert cutter

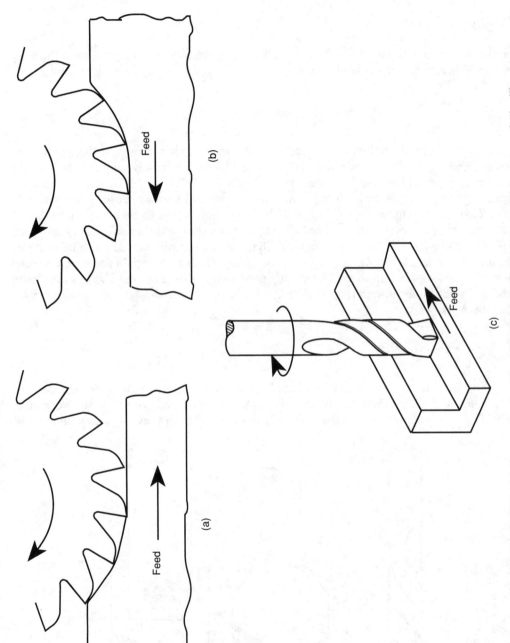

FIGURE 25.9   (a) Conventional milling; (b) Climb milling; (c) End milling on a vertical mill

cutter rotational axis is perpendicular to the workpiece surface and the cutter is supported at one end only. Horizontal milling may be of two types. In *conventional* or *up-milling*, the workpiece feed is in the opposite direction to the motion of the cutting tool (Figure 25.9(a)) while in *climb milling*, also known as *down-milling*, the feed is in the same direction as the direction of cutter rotation (Figure 25.9(b)). In the former method, the cutter teeth engage in the work at minimum depth and starting cutting forces are lower than in climb milling. However, the machined surface may have a greater degree of waviness than that generated by climb milling. In climb milling, the chip thickness is greatest at the start of the cut and decreases as the cut proceeds. The initial cutting forces are high but a good machined surface quality can be achieved. Formerly, much milling was of the conventional type until the introduction of more advanced and rigid machine tools allowed climb milling to be carried out with ease. In vertical milling operations, there are two main types of cutter used, the *face mill*, with cutting edges on the surface normal to the axis of rotation, and the *end mill*, with cutting edges on the cylindrical surface as well as on the surface perpendicular to the axis (Figure 25.9(c)).

CNC machining

For modern day mass production, machine tools are computer controlled and NC is the universally used abbreviation for numerical control. In CNC (computer numerical control), the milling machine is controlled by a dedicated mini or microcomputer attached or assigned to it. Instead of using relatively simple milling machines, much production work today is carried out on *machining centres*. These are CNC milling machines with an extended capability and possessing the ability to perform operations such as boring, drilling and tapping, in addition to milling. They may possess more than one machining head and be capable of performing machining operations on five sides of the workpiece.

Broaching

This is a metal cutting operation in which a tapered multi-toothed tool, a broach, is pushed or pulled through (internal broaching) or over (external broaching) the work. A broach has a series of cutting edges projecting from a rigid bar with successive teeth protruding further as one proceeds along the length of the tool. The rise, or step, between each successve cutting edge determines the chip size. Figure 25.10 is a representation of an internal broach. An internal broach may have any cross-section and can be used to cut holes of any profile, for example a circular hole with an internal keyway, to a high degree of accuracy.

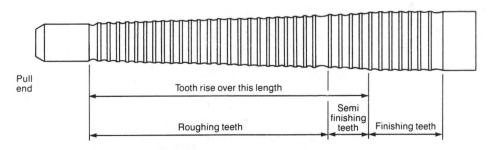

FIGURE 25.10   Internal broach

25.11 Abrasive machining and finishing

Abrasives are very hard materials and they are used for the removal of material by scratching, grinding or polishing. The abrasive machining processes enable components to be produced to close dimensional tolerances and with good surface finishes. Until the development of some of the non-traditional machining techniques, the only methods available for the machining of extremely hard metals and ceramics were abrasive processes.

Abrasives

Many different materials are used as abrasives, the common feature being that they are all of high hardness. The materials used include silica, alumina, silicon carbide (carborundum), cubic boron nitride and diamond. Corundum is a naturally occurring form of alumina and an impure variety of this is known as *emery*. Other abrasives used include magnesia, chromic oxide and ferric oxide (jeweller's rouge). Abrasive powders may be consolidated and bonded into a solid shape, such as a grinding wheel, made into *coated abrasives*, that is, bonded to a flexible paper or cloth backing, or used as a paste or powder within a carrying fluid. The abrasive powders are sized and graded to produce a wide range of wheels, papers and pastes capable of giving high or low rates of material removal and enabling a variety of surface finishes, from coarse to fine polished, to be achieved.

Grinding

This is material removal by abrasion and may be effected by using a grinding wheel or a coated abrasive. Each abrasive particle acts as a mini-single point cutting tool, producing very small chips. There are several types of grinding operation. Surface grinding is used for generating a flat, precise surface using either the cylindrical surface of a grinding wheel or by bringing the surface of the work into contact with a rapidly moving flat belt of coated abrasive. Cylindrical grinding is used for the generation of cylindrical surfaces and may be carried out either by rotating the work between centres or by centreless grinding. In both cases, the work rotates at a slower rate than the grinding wheel. The principle of centreless grinding is shown in Figure 25.11. The small control wheel is made of abrasive particles bonded with rubber and the rotational speed of the work is essentially that of the slower moving control wheel. This process is often automated and used for mass production. The principle can also be applied to the internal grinding of cylindrical bores.

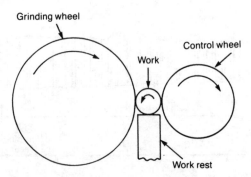

FIGURE 25.11   Principle of external centreless grinding

Form grinding is grinding using a profiled grinding wheel and is used for producing formed shapes, for example screw threads.

Grinding wheels

There are many different types of grinding wheel available as there are many variable parameters, including type of abrasive, grain size and packing density of abrasive, type of bonding and amount of bonding material. The abrasives used may be silica, alumina (or the natural form corundum), silicon carbide, cubic boron nitride or diamond, while the bonding may be a vitreous bonding, shellac, rubber or a polymer resin. The majority of grinding wheels are vitreous bonded and made by mixing the abrasive with a clay, pressing, drying and then firing in a kiln to give a glass bond around all the particles. Bonding with rubber gives a wheel with some flexibility while the use of a shellac bonding tends to give a ground surface with a polished finish. Both rubber and shellac have now largely been superseded by a range of polymer resins as bonding agents. The efficiency of a grinding wheel will reduce during use due to a rounding of the abrasive grains and the filling of spaces between grains with metal chips and other wear debris. Also, general wear may alter the grinding wheel profile. These problems can be rectified by dressing and trueing. Dressing is a cleaning operation, using a hardened steel tool, which removes or fractures rounded grains and removes foreign particles, thus exposing sharp cutting edges. A diamond tool may be used to restore the wheel profile, an operation termed trueing.

Abrasive jet machining

This is a process which is used for cutting holes or slots in very hard materials by means of a jet of high pressure air or carbon dioxide carrying entrained dry abrasive particles.

Ultrasonic machining

This is an abrasive machining process using an abrasive slurry in connection with a form tool. The form tool holder is vibrated by a piezoelectric transducer at a frequency of about 20 kHz and the tool is fed slowly into the work. The abrasive particles at the tool/workpiece interface are accelerated by the vibrating tool and perform the cutting. This process is suitable for machining holes, slots and shaped cavities in hard metals, ceramics and composite materials.

Barrelling

This is a general term covering several abrasive cleaning and finishing processes. In *barrel cleaning*, or *tumbling*, components are loaded into a cylindrical drum together with an abrasive such as sand, alumina, or granite chips and the drum rotated. Flashings, fins and oxide scale are removed by this process. *Barrel finishing* is similar but employs finer abrasive particles and is used to remove surface scratches and produce an even finish. In some cases, components may be passed successively through several finishing barrels with a finer grade of abrasive in each drum. In *barrel burnishing* or *barrel finishing*, the components are tumbled in a drum together with steel balls or shot. This is usually done wet, with water containing soap or other cleaning agent. A smooth even surface is produced.

Honing, lapping and polishing

These are abrasive finishing processes and only a small amount of metal removal occurs. *Honing* is a technique for sizing and finishing bored holes. The honing head carries four or six honing stones. These are made of abrasive particles bonded by a resin or wax. The honing head moves in the hole with both a rotating and a reciprocating motion, the stones being held against the work by a light pressure. Surface finishes as good as 0.075 µm CLA can be achieved. *Abrasive flow machining* is a finishing process used for internal surfaces but, in this case, unlike honing, it can be used for small holes and for non-cylindrical holes. A semisolid medium, charged with fine abrasive particles, is forced through a hole. One of the mediums used is methyl polyborosiloxane, or 'potty putty'. *Superfinishing* is a process similar to honing but used for external surfaces. *Lapping* is a finishing process in which the surface is in relative sliding motion with a lap, the abrasive being a slurry containing very fine particles. Laps may be made from a variety of materials, including pitch, plastics, lead, copper and cast iron, but it is always softer than the surface to be finished. The lap surface may be flat or curved. Lenses and curved mirrors are finished by lapping on a lap with a curved surface. *Polishing* is an abrasive finishing process in which very fine abrasive particles are embedded in a soft backing, or polishing cloth, and there is relative motion between the surface and cloth with a fluid lubrication. Very little material is removed and the main effect is to produce a smooth surface of high reflectivity. In the polishing of a fine ground metal surface, the effect of frictional heating during polishing is to cause a softening of 'peaks' causing them to flow into 'troughs'.

25.12 Gear and thread manufacture

Gears and screw threads are complex shapes which are required to be manufactured in great quantity to an accurate size with close dimensional tolerances. Both types of component may be made by a variety of methods, many of which involve material removal.

Gear making

Manufacturing methods for gears include metal cutting, plastic deformation, casting, polymer moulding techniques and processing from powder. The major metal cutting processes used are:

(a) Gear hobbing, which is a continuous cutting process using a multi-point tool, the hob, and is the fastest of the gear generating processes. The hob and blank are connected by means of gearing so that they rotate in mesh. The hob is fed into the blank in full depth and then feeds across the face of the blank, generating involute teeth. This is a very versatile process suitable for straight and helical gears, splines and sprockets.

(b) Form cutting using a profiled form tool in conjunction with a milling machine or a machine tool with a reciprocating cutting action.

(c) Broaching, in which a blank is pulled or pushed through a pot broach, a female tool as opposed to the usual male broach.

(d) Cutting on special purpose gear cutting machines whose action involves both forming and generating. This is the only cutting method which can be used for bevel gears, whether straight, helical or hypoid.

Gear forms in metals may be made by sand, die or investment casting. Sand casting is used for the large gears in some low speed machinery and finish machining operations are usually required. Gears in some plastics materials are made by casting but, generally, injection, compression or transfer moulding techniques are used. The plastic deformation processes of forging and roll forming of metals can be used for the production of most gear forms and spur gears can be made by parting off lengths from an extrusion having a spur gear cross-section. Spur gears for low load applications, as in analogue instruments and clocks are made by blanking from cold rolled metal strip. Also, many small gears are made from powder by compaction and sintering.

Gear finishing

For many high performance applications, it is necessary to use one or more finishing processes, either to give improved meshing surfaces or to manufacture to very close dimensional tolerances. Many gears are made from hardenable steels and gear cutting is carried out prior to hardening and tempering. It may be necessary, then, to use more than one finishing process, one before and one after hardening and tempering. The main finishing processes used are shaving, grinding, burnishing and lapping.

(a) *Gear shaving* is the most widely used finishing process and is applied prior to hardening. The shaving tool, which is a meshing gear, the teeth of which have a number of grooves or slots, is run in mesh with the gear. The shaving tool acts as a broach and a small amount of metal, of the order of 0.025 to 1 mm, is removed from the gear faces.

(b) *Gear grinding* is a finishing process for hardened gears and may take one of two forms. One form involves a straight sided grinding wheel and an accurate ground surface is generated. The other process uses a profiled grinding wheel and is a forming operation. Gear grinding is a slow and expensive process.

(c) *Gear burnishing* is a rolling process and an unhardened gear is rolled in mesh and under pressure against three hardened gears. There is some plastic deformation of the gear surface.

(d) *Gear lapping* is a finishing process for hardened gears in which the gear is run in mesh with a cast iron lapping gear in conjunction with a slurry of fine abrasive powder in oil.

Screw thread manufacture

The main principles involved in the manufacture of thread forms, both external and internal are cutting, plastic deformation and direct casting or moulding to shape. The main thread cutting techniques used are:

(a) *Use of taps and dies.* A solid tap is a multi-point cutting tool, similar in appearance to a threaded bolt but with a number of flutes, usually four, to give the cutting edges, and is used for cutting internal threads in pre-drilled holes. Taps may be used manually or on lathes, drilling machines and other machine tools. Also, collapsible taps are available for use with fully automatic machine tools and the cutting heads collapse inwards after the full thread depth has been cut to facilitate rapid withdrawal of the tool. A threading die is similar to a threaded nut with several, usually four, longitudinal grooves which expose multiple cutting edges. For manual thread cutting, dies are held in a die-stock. After cutting an external thread, the die is removed by unscrewing. Dies for automatic production work are of the self-opening type with a series of multi-point cutters arranged either radially or tangentially within a die head, and designed to open automatically when the required length of thread has been cut. This permits the die head to be removed rapidly from the work.

(b) *Screw cutting on a lathe* is suitable for both internal and external threads. The cutting tool is a single point tool which conforms to the thread profile. Longitudinal movement of the tool is obtained through use of a lead screw which permits the correct movement of the lathe carriage in relation to workpiece rotation.

(c) *Cutting on a special purpose machine.* A number of types of automatic machines dedicated to the cutting of either external or internal screw threads have been designed. These may be single-spindle or multi-spindle machines.

(d) *Milling.* External screw threads may be cut to a very high degree of accuracy by a milling operation using specially designed form cutters. Milling is often used for the production of threads on large diameter stock.

(e) *Grinding* is used to produce accurate external thread forms, particularly in very hard materials. The grinding wheel is shaped and is, essentially, a form cutting tool. The grinding wheel may be shaped with one rib, in which case the process is similar to thread cutting on a lathe with a single point cutting tool, or it may have several ribs and the action is analogous to thread milling with a multi-point form tool. It is also possible to use a centreless grinding technique for forming threads and this last method is particularly useful for the manufacture of headless set screws.

A very high proportion of external threads required in large quantity are made by *thread rolling*, a plastic deformation process. The rolling between grooved dies or rolls, with consequent work hardening of the material, gives a smooth thread profile with a hard, wear resistant surface. Internal threads in ductile metals may also be made by a plastic deformation process using a tool which is basically a tap without flutes. There is no cutting action and the thread is formed by plastic deformation in a manner analogous to the rolling of external threads.

In some instances, threads may be formed directly in die-cast or investment cast products and many threaded components in thermoplastic materials are produced by injection moulding.

25.13 Non-traditional machining processes

There is a range of material removal processes which do not involve the formation of chips. These non-traditional, or chipless, machining processes are suitable for use with a variety of materials, including very hard metals which, otherwise, would be difficult to machine, and some can be used in connection with ceramics.

Chemical machining (CM)

CM is the removal of material by controlled chemical attack. One form of chemical machining, engraving, has been in use for hundreds of years. The basic principle of chemical machining is to cover those parts which are not to be attacked by a mask or resist, so that only portions of the surface are etched away. The term chemical milling is used when there is overall removal of material and the formation of etched pockets, contours or thinned down sections, as in the manufacture of integrally stiffened structural aircraft panels. Chemical blanking is the term used when the process involves etching completely through thin sheets, as in the manufacture of printed circuits. This latter process is also known as *photochemical machining* when used in connection with a photo-sensitive resist. The thin sheet or foil surface is coated with a light-sensitive emulsion, placed in contact with a film negative of the master pattern, and exposed to light. After development and washing away of unexposed emulsion, the photosensitive resist or mask remains. Lines as thin as 2 or 3 μm can be produced in the chemically blanked product using this type of resist.

Electrochemical machining (ECM)

ECM is a metal removal process based on the solution of anode material in an electrolytic cell. The workpiece is made the anode and a tool, shaped so that it is a negative of the required component shape, is made the cathode of the cell. The tool is fed into the work at a constant rate, and equal to the rate of metal removal, thus maintaining a gap of about 0.01 mm between the tool face and the work. The electrolyte is not static, but is pumped around the system at high velocity, 30 m/s or greater. The high electrolyte flow rate is necessary as a major function of the electrolyte is to conduct heat away from the work. ECM is an expensive process but is of particular use in the machining of very hard complex alloys, such as the superalloys used for high temperature components in gas turbines. Surface finish values as good as 0.1 μm CLA can be achieved. There are two other variants of ECM, these being electrochemical hole drilling, suitable for the production of round or shaped holes in alloys which are difficult to machine, and electrochemical grinding (ECG). ECG, which is also known as electrolytic grinding, is used widely for grinding carbide tool tips. If offers several advantages over conventional grinding, giving higher material removal rates and lower wear rates of the abrasive material. The process is similar to conventional grinding, but the grinding wheel is electrically conductive and is made the cathode, with the workpiece made the anode of an electrolytic cell. A continuous stream of electrolyte is directed into the interface between wheel and work. During the process, about 90 per cent of material removal is by electrolytic action, the remainder by abrasive cutting. Surface finishes as good as 0.1 μm CLA are possible.

Electrodischarge machining (EDM)

EDM, also referred to as *spark erosion*, is a machining process based on the erosive effect of an electric spark. Of the two electrodes used to produce a spark, one is the tool and one is the workpiece. The work is made positive because erosion of the positive electrode is greater than that of a negative electrode, other things being equal. The tool is produced to a shape which is a negative of the required machined component profile. The work and electrode tool are contained in a tank filled with a dielectric fluid, such as kerosene, white spirit, transformer oil or a glycerol/water mixture, and sparks at a high frequency are generated between tool and work. The gap between work and tool is kept at about 0.05 mm and the spark frequency may be between 400 Hz and 200 kHz, depending on the nature of the tool and workpiece, rate of metal removal and surface finish required. As erosion occurs, the tool is moved into the work by a closed-loop servocontrol mechanism so as to maintain a constant gap size. EDM is widely used for the manufacture of dies for casting, forging, stamping and extrusion, and surface finishes as good as 0.2 μm CLA can be achieved.

Electron beam machining (EBM)

In this process, a high energy electron beam is focused onto the workpiece, which is contained in a vacuum chamber. The heat energy generated when the beam strikes the work is sufficient to melt and vaporise the material. EBM is used with extreme precision for making small diameter holes and slots in relatively thin materials. Holes with diameters as small as 2 μm can be made in very thin materials. The process may be used to drill holes in sapphire and other ceramics for use as bearings in delicate instruments.

Laser beam machining (LBM)

LBM is similar to EBM in that the intense heat generated when a finely focused laser beam strikes a material is sufficient to melt and vapourise the material. A laser beam can be focused down to a very small spot size, about 0.02 mm, and can be used for the drilling of very small holes in almost any metal in thicknesses of up to 2.5 mm. Unlike EBM, a laser beam does not have to be operated in vacuo. The energy of a laser beam can be used for the cutting of materials and almost any material can be cut. For the cutting of many metals, a jet of an inert gas or oxygen is directed at the area of cut to blow molten metal away. In the case of an oxygen jet, oxidation effects tend to accelerate the cutting process.

Plasma jet machining

A plasma is a highly ionised gas produced by heating a suitable gas, such as argon, to a very high temperature in a confined space. Plasmas can be created by passing the gas through a constricted electric arc. The plasma jet produced in a plasma arc torch can be used for metal cutting. The extremely high temperatures and jet action of the plasma will give relatively smooth cut surfaces. Very high cutting speeds are possible and cutting rates of 2.5 m/min have been achieved in steel plate of 12.5 mm thickness.

Jet machining

This is a slitting process suitable for cutting soft materials, including paper and paper products, some plastics, leather, rubber and GRP material, using a very high velocity fluid jet as the cutting medium.

26

Joining Processes

26.1 Introduction

The permanent assembly of individual manufactured components is an important aspect of fabrication and there are many ways of accomplishing this. Many assemblies involve the use of fastenings such as rivets, bolts and screws but this form of assembly will not be discussed here. Many parts are joined by welding and the allied processes of soldering, brazing and diffusion bonding. Adhesives also play a major part in joining technology.

Welding methods can be broadly divided into *fusion* and *pressure* processes. In the former, some part of the materials to be joined is melted during the process and a molten filler of similar composition to the parent material is generally needed to complete the joint. In pressure welding processes, no filler material is used. Most metals and thermoplastic materials can be joined by some form of welding. *Soldering* and *brazing* are techniques used for joining some metals and a material of a different composition from the metals to be joined is used to effect the joint. The solder or brazing alloy is of lower melting point than the metals to be joined and no portion of the parent metals is melted during the process.

26.2 Soldering and brazing of metals

Soldering and brazing processes are similar in principle in that the filler material melts at a comparatively low temperature and this liquid filler is drawn by capillary action into the small gap between the parts being joined. One major advantage of soldering and brazing operations is that the joint is made at fairly low temperatures so that there is little heat distortion of parts and little change to the microstructure of the parent metal. Soft solders are comparatively weak but hard solders (brazes) may have tensile strengths in the range 400–500 MPa. Figure 26.1 shows the soldering process diagrammatically.

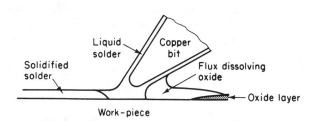

FIGURE 26.1   The soldering process

Soft soldering

Soft solders are alloys of tin and lead or tin/lead with antimony, cadmium or bismuth and have melting temperatures in the range 70–300°C (see Table 26.1). For soldering to be effective the liquid solder must 'wet' the surface of the metal to be soldered. This means that some alloying must take place between a constituent of the solder and the metal to be joined. The type of alloying that occurs may cause formation of a solid solution or the formation of an intermetallic compound. In soft solders tin will form intermetallic compounds with both copper and iron. A liquid solder will only wet a metal surface that is clean and grease free so all surfaces to be soldered must be perfectly clean. A thin oxide film will quickly form on a fresh metal surface and so a soldering flux is necessary to dissolve this oxide layer. The flux used is either zinc chloride solution or of the resin type. When the former flux is used the part must be washed after soldering to remove any remaining flux, otherwise corrosion could occur. When a resin flux is used any flux residue left after soldering is not corrosive. Soft solder wire with a resin core is invariably used for electrical jointing work.

Table 26.1 *Some soft solders*

Composition (per cent)						Melting range or point (°C)	Comments
Pb	Sn	Sb	Bi	Cd	Ag		
4.5	95	0.5				183–223	Electrical instruments
34	65	1				183–186	High strength – quick setting
38	62					183	Tinman's solder
49.5	50	0.5				183–214	General-purpose solder
60	40					183–236	Soldering of tin cans
35	40		25			96–160	Soldering of glass to metal
66	32	2				185–245	Cable wiping solder
25	12.5		50	12.5		70	Very low melting point solder
97.5					2.5	304	High-strength joints in copper

The most commonly used heat source for soft soldering is a soldering iron with an electrically heated copper bit. The size of the heated bit may vary from large to very small depending on the type of work to be undertaken. A gas torch may be used as a source of heat for some soldering operations, for example when making pipe joints using 'Yorkshire' fittings. 'Yorkshire' fittings are capillary fittings with a solder insert.

Pre-tinning of a component to be soldered is frequently carried out. This will ensure that when the soldered joint is made a rapid penetration of the joint with solder will occur.

If too high a temperature is used in soldering, or if a soldered joint is maintained at elevated temperature during service, an excessive amount of intermetallic compound may be formed and the joint could become brittle. To avoid this type of brittleness in soft-soldered joints in copper, a lead/silver solder is used instead of a lead/tin alloy.

Hard soldering

Hard soldering, or *brazing* makes use of copper-base alloys with melting temperatures ranging from 620°C to 900°C for making the joint (see Table 26.2). Again, the materials to be joined must be clean and a flux must be used. The most commonly used flux is borax. Borax melts at 750°C

Table 26.2 *Some brazing alloys and silver solders*

Composition (per cent)						Melting range (°C)	Comments
Cu	Zn	Ag	Si	Ni	Cd		
50	50					860–870	General-purpose alloy
60	39.5		0.5			880–890	High-fluidity filler
59	40	1				880–890	High-fluidity filler
50	38		0.5	10		860–900	High-strength brazing
16	4	80				740–795	High electrical conductivity
29	10	61				690–735	Braze with good conductivity
15	16	50			19	620–640	Low melting point hard solder
37	20	43				700–775	General-purpose hard solder

and is a good solvent for many metal oxides. For brazing at lower temperatures than this alkali metal fluorides are used as fluxes.

As with soft soldering the parts to be joined are prepared and fitted together. During brazing the molten filler metal is drawn by capillary action into the joint. The most commonly used heat source for brazing is the oxy-gas torch. The gas used is propane or acetylene. Generally, a neutral or slightly reducing flame is used, but when brazing copper a slightly oxidising flame should be used to reduce the possibility of hydrogen embrittlement.

Furnace brazing

Furnace brazing is a technique used for the mass production of small assemblies (see Figure 26.2). The assemblies are made up and the filler placed in the joint, either as a powder or a pre-formed ring or disc. The parts are then placed in a controlled atmosphere furnace and heated to a temperature just above the melting temperature of the brazing filler.

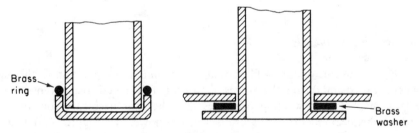

FIGURE 26.2   Assembly for furnace brazing

Hard solders

Brasses, alloys of copper and zinc, are widely used as hard solders, hence the term brazing. The brasses used contain more than 40 per cent zinc but some zinc is lost during the operation and the final jointing metal possesses an α or $(\alpha + \beta)$ structure. The brazing alloys may contain small amounts of silicon or silver to increase the fluidity of the filler, or nickel and manganese to increase the strength of the joint. Silver solders, while much more expensive, are very strong jointing materials and have the additional advantage that they possess much lower melting temperatures than brasses (see Table 26.2).

Braze welding

This is a metal joining process in which the filler metal has a lower melting point than the metals to be joined but, unlike brazing, the molten filler is not drawn into the joint by capillary action. There is a wider gap between parts to be joined than in a brazed joint and the filler metal, typically either a brass or a bronze, is melted by means of an oxy-gas torch or an electric arc. When the filler material is a bronze the process is termed *bronze welding*. Cast irons may be successfully joined by braze or bronze welding. Because of the relatively low melting point of the filler, the iron is not melted and the tendency for thermal stress crack formation in the cast iron is minimised.

Aluminium soldering

Lead–tin alloys will not 'wet' an aluminium surface and it is not possible to solder aluminium using soft-solders. There are two main groups of aluminium solders, these being:

(a) aluminium–zinc alloys, with varying quantities of zinc and small additions of tin or cadmium, and possessing melting ranges close to 400°C,
(b) aluminium–silicon alloys; the 7.5 per cent silicon alloy has a melting range 580–620°C and the eutectic alloy melts at about 580°C.

Aluminium can be soldered successfully with these alloys using a chloride flux. For the manufacture of aluminium cans the canning material is commercially pure aluminium sheet with a thin layer of aluminium silicon alloy roll-bonded to the surface. This layer is put on during an early stage in manufacture by placing a cladding plate of aluminium–silicon alloy on the surface of an aluminium ingot and hot rolling them together.

26.3 Fusion welding of metals

As stated earlier, in fusion welding a filler material of similar composition to the metals being joined is generally used and during the process a portion of the metals being joined is also melted. There are very many fusion welding processes and the heat necessary for welding may be obtained from a chemical reaction or by using electrical power (see Figure 26.3). One of the first successful fusion welding processes to be developed was gas welding using heat energy from the combustion of acetylene in oxygen in a special torch.

Gas welding

The main type of gas welding is oxy-acetylene, in which acetylene (ethine) and oxygen are mixed and burnt in a torch. The full and complete combustion of acetylene is

$$2C_2H_2 + 5O_2 = 4CO_2 + 2H_2O \ (53.4 \ MJ/m^3 \ acetylene)$$

In this reaction the ratio oxygen volume/acetylene volume is 5/2. In practice the volume ratios of oxygen/acetylene entering the torch are much less than this and normally range from 0.85/1 to 1.7/1. For a ratio of 1/1 the reaction equation would be

$$C_2H_2 + O_2 = 2CO + H_2 \text{ (18.3 MJ/m}^3 \text{ acetylene)}$$

The carbon monoxide and hydrogen reaction products of this equation then burn with atmospheric oxygen at the outer envelope of the flame. By controlling the relative amounts of oxygen and acetylene, flames with differing characteristics can be obtained. A chemically neutral flame is obtained when the oxygen/acetylene ratio is about 1.1/1, with a flame temperature of about 3250°C. If the oxygen/gas ratio is increased a shorter, fiercer, oxidising flame with a temperature of about 3500°C is produced, while an oxygen/gas ratio of about 0.9/1 will give a chemically reducing flame with a flame temperature of about 3150°C.

Gas welding is less widely used than the electric arc processes but it is still an important process. The gas torch can be used to pre-heat the joint area and to reduce the rate of cooling after welding. This is a great advantage when welding hardenable steels, as it reduces the possibility of brittle martensite formation in the weld zone.

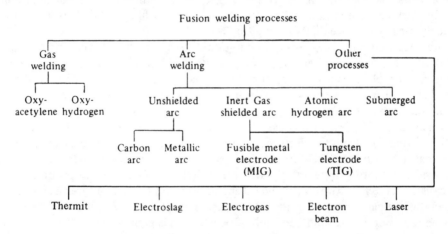

FIGURE 26.3   Fusion welding processes

The chemically neutral flame is used for the welding of plain carbon steels and most ferrous and non-ferrous alloys, but an oxidising flame is used for welding copper, bronze, brass and nickel-silvers. This is to avoid any possibility of hydrogen absorption which would lead to brittleness. A reducing flame is used for welding aluminium alloys and some stainless steels.

Flame cutting

For flame cutting, the torch-nozzle design is different from that used for welding. The cutting nozzle consists of a series of small holes arranged around a large central jet. The oxy-acetylene mixture emerges through the small holes while a high-velocity jet of pure oxygen passes through the central hole. When cutting steel the oxy-acetylene flame heats up the surface of the steel to a sufficiently high temperature (900–1000°C) so that the iron will ignite and burn in the strong oxygen jet according to the reaction.

$$3Fe + 2O_2 = Fe_3O_4 \text{ (4.9 MJ/kg iron)}$$

The heat generated by this strongly exothermic reaction melts some of the surrounding metal and this is blown away by the strong oxygen jet. In this way steel thicknesses of up to 1 m can be cut. In addition to cutting, this principle can be used for planing, turning and boring large steel sections.

Arc welding

Arc welding in one form or another is the most widely used form of welding. The electrical supply is low voltage but high amperage and may be either alternating or direct. In many cases the supply for welding operations is obtained from the secondary of a step-down transformer connected to the mains supply, with rectification of the secondary output if direct current is required. Also, many portable welding units are available in which current is generated by a direct-current compound-wound generator driven by a petrol engine. The latter are particularly useful for working on construction sites.

The earliest forms of arc welding used carbon electrodes but nowadays the arc is struck between a metal electrode and the work-piece. The electrode may either be of tungsten or be a consumable metal electrode that melts, acting as a source of filler metal. This latter form is the most widely used type of welding (see Figure 26.4).

An alternating-current arc is broken and re-established at each half cycle and this leads to arc instability although the use of arc-stabilising agents in the flux coating of electrode wires can overcome this problem. When direct current is used there is a choice of polarity, with the electrode being either positive or negative with respect to the work-piece. The choice of polarity is important and both types of polarity are used in practice. The most widely used system is with the electrode negative (DCEN). In this case the electron flow is from electrode to work-piece. A concentrated arc issues from the electrode tip and heating of the work-piece is largely confined to the very small area beneath the electrode. The tip of the electrode and the metal directly under the arc melt and mix thoroughly with considerable turbulence. The pressure produced by the arc stream causes a crater to form in the liquid metal pool. As the arc progresses along the joint, liquid metal flows back into the crater. The depth to which melting of the work-piece occurs is called penetration of the arc, and the depth of penetration is greatest with DCEN welding. If the electrode is positive with respect to the work-piece (DCEP) the electron flow is from work-piece to electrode. The emission of electrons from the work-piece has the effect of breaking up and

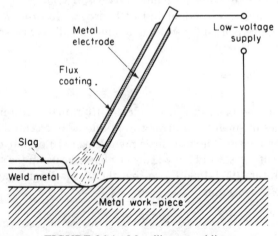

FIGURE 26.4 Metallic arc welding

scattering any tenacious oxide film that may be present on the work-piece metal. Also, greater heat is generated at the electrode tip giving more rapid electrode melting than with DCEN. The extent of heat penetration into the work-piece is less with DCEP than with DCEN, but there is a smoother transfer of liquid metal from the electrode. DCEP polarity is the preferred system for the arc welding of those metals that possess tenacious oxide films, namely, aluminium alloys, stainless steels and some copper alloys.

Transference of molten metal from the electrode tip to the weld pool may occur in one of several ways, these being *dip transfer, free-flight arc* and *spray transfer*. In dip transfer, when an arc is established the metal droplet at the end of the electrode wire begins to grow and the arc gap closes. When the drop touches the weld pool it creates a short circuit and the arc is extinguished. The high short circuit current causes resistance heating in the electrode wire until a molten drop leaves the electrode tip, re-establishing the arc. This sequence repeats with a relatively constant frequency. The conditions for effective dip transfer are dependent on arc voltage, wire feed rate and wire size. This form of transfer is used mainly for the welding of steels with carbon dioxide shielding, but it can be used also with other metals.

The arc condition in which the droplets form and detach themselves from the electrode tip to transfer to the weld pool without bridging and short circuiting is termed free-flight. Changes in the voltage and arc current will affect the droplet transfer frequency and as the transfer frequency increases, so the droplet size reduces. At a high transfer frequency, the metal transfer takes the form of a stream of small droplets projected axially from the electrode wire. This last condition is termed spray transfer.

Electrode coatings

Uncoated welding rods can be used for arc welding, but it is more usual to use flux-coated electrodes. When steels are welded using uncoated electrodes, oxides and nitrides can form and remain in the weld with a consequent loss of toughness.

Flux-coated electrodes are widely used. The composition of the coatings is complex and a variety of different coatings are used to cater for different types of welding application. However, in all cases the coating is formulated to satisfy three objectives. These are: (1) to form fusible slags, (2) to stabilise the arc, and (3) to produce an inert gas shield during welding.

Manual metal-arc welding uses flux coated electrodes in short lengths, ranging from 250 to 500 mm. For automatic flux-shielded arc welding, continuous electrode wire is required and this must be constructed in such a way that electrical contact can be made between the flux coated wire and the electrode holder. One method of doing this is to wrap the core electrode wire with helical wire spirals, with the spaces between the spirals being filled with flux. The electrical contacts in the welding head touch the outer wire spiral, where it protrudes through the flux coating, and current is conducted to the core wire. Another solution to the problem is to manufacture the electrode in tubular form and have the welding flux filling the tube. Flux coated welding wires are much more expensive than bare electrode wire and there are two types of automatic flux shielded arc welding in which a bare metal electrode wire is used. These are submerged arc welding and magnetic flux/gas-arc welding.

Submerged arc welding

Submerged arc welding (see Figure 26.5) is an automatic form of metallic arc welding and is suitable for making long straight-line welds in the horizontal mode in plate thicknesses ranging

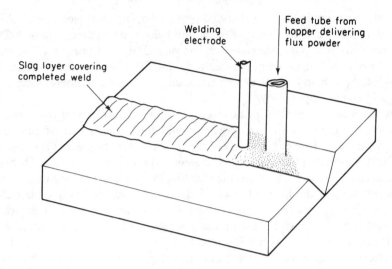

FIGURE 26.5   Submerged arc welding

from 6 to 30 mm. The surfaces to be joined are prepared with chamfered edges for a vee-butt joint. The welding head consists of a hopper and delivery tube for powdered flux together with the wire feed mechanism for the welding electrode wire. Powdered flux is delivered into the prepared joint ahead of the welding wire. The melting tip of the electrode wire is below the flux surface. The heat generated by the arc melts not only the electrode wire and some of the work-piece but also the flux, which forms a molten protective slag over the weld. When cool, the solidified slag can easily be chipped away from the metal surface. Submerged arc welding is used mainly in connection with low carbon and low alloy steels but the process can also be used for welding copper, aluminium and titanium alloys.

Magnetic flux/gas-arc welding

This is an automated continuous welding technique in which a magnetisable powdered flux is carried in a stream of carbon dioxide through a nozzle surrounding a bare electrode wire. When current is flowing, the electrode wire has a magnetic field surrounding it. This attracts the flux, which adheres to the wire. This method which can be used for making both vertical and horizontal weld runs in mild steel, has advantages over submerged arc welding. These are that the weld is visible and that welding is not confined to the horizontal mode.

Inert gas shielded arc welding

It is possible to produce a complete blanket of an inert gas around a weld area and by this means arc welding without fluxes, and using uncoated electrodes, can be readily accomplished. The most widely used shielding gas is argon, but nitrogen can be used as a shield for the welding of copper and carbon dioxide can be used with steels. The welding electrode may be of tungsten and non-consumable, in which case a separate filler rod has to be fed into the arc to provide a source of weld metal, or a consumable metal electrode can be used. The former is termed the TIG (tungsten-inert gas) process (Figure 26.6) and the latter is known variously as MIG (metal electrode-inert gas), MAGS (metallic-arc gas shielded) or GMA (gas metallic arc) welding (Figure 26.7).

In both types of electrode holder (welding torch) the electrode is surrounded by a tube and nozzle, through which inert gas is fed and blown over the weld area. Inert gas welding techniques lend themselves to automation and automatic TIG and MIG welding processes are used to produce welds of consistent high quality.

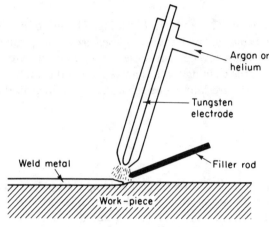

FIGURE 26.6   Principle of TIG welding

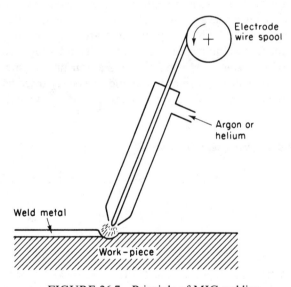

FIGURE 26.7   Principle of MIG welding

When carbon dioxide, CO_2, is used as a shielding gas, dissociation of the CO_2 can occur at high temperatures according to the reaction

$$2CO_2 = 2CO + O_2$$

When using CO_2 as a shielding gas for the welding of steels the oxygen produced by the above reaction could oxidise some of the iron and the iron oxide thus formed could then react with some carbon in the molten steel weld pool: $FeO + C = Fe + CO$. This generation of carbon monoxide could lead to porosity in the final weld.

The CO_2 welding process is widely used for the welding of mild and low-alloy steels, but in order to produce consistent high-quality welds the electrode wire is of a modified composition. The welding wire must contain relatively high proportions of manganese and silicon as de-oxidising agents to remove FeO from the weld pool. In this way the CO_2 process can be both effective and economic.

There are some variants on TIG and MIG welding. The use of a pulsed d.c. in conjunction with an inert gas shielded tungsten arc, known as *pulsed arc welding*, can offer advantages for the welding of thin sheet material. A low current is used to maintain the arc and a square wave pulse is superimposed on this. As the welding torch is moved along the joint, a series of weld nuggets is made and these may be made overlapping or separate, depending on the speed of travel of the torch.

Plasma arc welding

This can be regarded as a specialised form of TIG welding. The plasma is a mass of extremely hot ionised gas formed by passing argon, or some other suitable gas, through a constricted electric arc. A constricting orifice in the plasma torch gives a stable, high energy, concentrated arc. When the arc is squeezed, the voltage gradient, current density and heat transfer intensity all increase. An annular layer of gas flows around the arc and this relatively cool, non-conductive layer gives electrical insulation, stabilises the arc, and protects the water-cooled constricting nozzle. The plasma arc process gives a high energy concentration and, consequently, higher temperatures than in conventional arc processes. This gives deep penetration with a narrow heat affected zone, less distortion and permits fast welding speeds. Most metals and alloys can be welded by means of this process. The variant *plasma GMA welding* uses a consumable metal wire electrode fed through a nozzle which also guides a plasma stream and can be regarded as a specialised form of MIG welding. As with plasma arc welding, this process permits fast welding speeds.

26.4 Other fusion welding processes for metals

There are numerous other fusion-welding processes, and some of these are briefly described below.

Electroslag welding

Electroslag welding was first developed in the USSR as a means of joining large castings or forgings to produce large complex assemblies, for example, the housing for a large rolling mill or forging press, and the method has been used to successfully join pieces with sectional thicknesses of up to 1 m. The process, which can be used only in the vertical mode, is now used for welding material with thicknesses of 50 mm and greater (see Figure 26.8).

The two pieces to be joined are placed close together and water-cooled copper shoes are placed across the joint to make the third and fourth sides of a mould. A special flux is placed in the bottom of the joint and two welding wires dip below the surface of the flux. This is not an arc

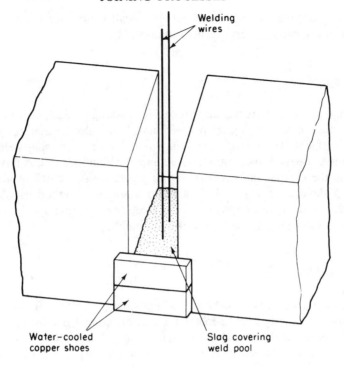

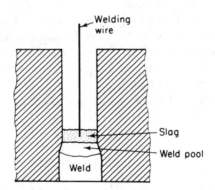

FIGURE 26.8   The electroslag welding processs

process. A circuit is completed through the flux when the electrode wires are energised. The heat generated in overcoming the resistance of the flux is sufficient to melt the flux into a protective slag and to start melting the ends of the electrode wires. As the process proceeds a pool of weld metal is produced beneath a protective slag layer. There is sufficient heat generated to begin melting the edges of the work-pieces so that total fusion occurs. As the weld metal solidifies in the lower portion of the joint so the water-cooled copper shoes are moved slowly up the sides of the joint. It is customary when welding very thick sections to oscillate the wires across the weld zone to achieve uniformity of temperature and weld deposition.

Electroslag welding can be used for welding plate thicknesses from about 10 mm upwards, but is used chiefly for sectional thickness greater than 50 mm. For the lower end of the thickness range *electrogas welding* is preferable and gives faster welding speeds. The main applications for electroslag welding are for the fabrication of thick-walled pressure vessels and the joining of castings or forgings to produce large structural assemblies.

Electrogas welding

This is an automated shielded arc welding process for making vertical, or near vertical, weld runs in steel plate of thicknesses ranging from 12–75 mm. It can also be used for joining plates with slightly curved surfaces. There is a superficial similarity to electroslag welding in that water cooled copper shoes are used to contain the weld metal within the joint. A flux-cored electrode is used to maintain the arc and this provides a thin protective slag cover. Additional shielding is given by carbon dioxide or argon. It is a rapid automatic method capable of consistently producing good welds. Typical applications for electrogas welding are for the construction of large gas and petroleum storage tanks and for shipbuilding.

Electron beam welding

The heat source here is a finely focused beam of high-energy electrons. The electron beam can be finely focused and the extent of heat penetration is confined to a narrow zone. The energy of an electron beam is dissipated rapidly when not in a vacuum. For much electron beam welding, the assembly to be welded is placed in a chamber which is then evacuated before the beam is energised and focused on the joint area (Figure 26.9). In high vacuum electron beam welding, the chamber pressure is reduced to between 10^{-4} and 10^{-5} torr (0.0133–0.00133 Pa). The advantages of electron beam welding are that a deep penetration fusion weld with low distortion and a very thin heat affected zone is made. Almost all metals can be welded. The absence of contaminating gases means that reactive metals can be welded. Dissimilar metals may be joined to one another and it is also possible to weld some ceramic materials—alumina, magnesia and quartz can be joined, both to themselves and to metals.

Electron beam welding in a partial vacuum and in a non-vacuum (in air or inert gas) has also

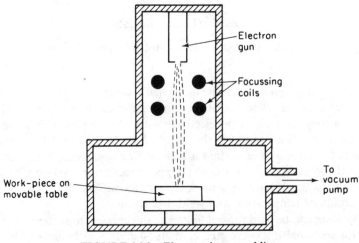

FIGURE 26.9   Electron beam welding

been developed. Cycle times can be reduced using a partial vacuum chamber working under pressures in the range of 0.05 to 0.3 torr (6–40 Pa). A high voltage set has to be used if out-of-vacuum electron beam welding is to be carried out and the workpiece must be sited very close to the source of the beam.

There are many industrial applications for electron beam welding and these include nuclear fuel elements, gas turbine rotors, finned superheater tubing, gears, bellows, diaphragms and honeycomb structures.

Laser beam welding

A disadvantage of electron beam welding, namely the rapid dissipation of the beam in air, is not present when a high intensity laser beam is used as the energy source. The power outputs of lasers used for welding are in the range from 1 to 20 kW. The laser beam can be very finely focused and, like electron beam welding, this is a precision process with a deep penetration and little lateral spread of energy. In consequence, there is a very small heat affected zone. Welding may be carried out in air or under an inert gas atmosphere and high welding speeds are possible. It is possible to weld virtually all metals and joints between dissimilar metals can be made, for example, plain carbon steels can be welded to stainless steels, nickel or titanium. It can also be used for making welds in very small components and with very fine wires. Applications of the process include high precision watch and instrument components, bellows, diaphragms and dental bridgework.

Flash butt welding

Flash butt welding is a fusion process which was developed from resistance butt welding (see Section 26.5). The parts to be joined are made part of an electrical circuit, brought lightly together, and then separated slightly to strike an arc. Arcing is continued until the surfaces are uniformly heated and beginning to melt. The current is switched off and the parts are forged together under pressure. Molten metal is expelled to form a ragged fin around the joint (Figure 26.10). The process is used for joining bars, tubing and sections. Dissimilar metals can also be joined by this process, as for example in the joining of a mild steel stem to an alloy steel head in the manufacture of internal combustion engine valves.

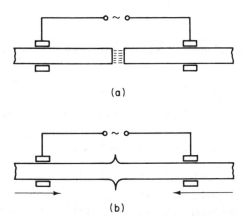

FIGURE 26.10   Flash butt welding: (a) arc struck; (b) parts brought together under pressure

A further development of this process is MIAB (magnetically impelled arc butt) welding and this is particularly suitable for the butt welding of hollow sections. A small concentrated arc is created between the two parts and, under the influence of a rotating magnetic field, this arc moves uniformly around the complete section at a frequency of 50 Hz. The arc is ceased when the surfaces begin to melt and the two parts are brought together under pressure to forge the weld. The decrease in length due to upsetting when the two parts are brought together is less, and more predictable, than in flash butt welding. Also the size of fin formed at the joint is less. This process is used as a fabrication technique for the manufacture of such assemblies as rear axle housings and propellor shafts for road vehicles.

Arc stud welding

It is frequently necessary to attach studs and fixing attachments to plate or sheet material. Arc stud welding is similar to flash butt welding in that an arc is struck between a stud and a plate for a preset time, usually between 150 and 500 ms. This is sufficient to melt the end of the stud and to form a small crater in the plate. The two are then brought into contact under pressure and a weld is made. The process is used for the placing of studs and fixing attachments with diameters ranging from 3 to 25 mm in steels, including stainless steels, brass and aluminium alloys for a wide range of applications in civil engineering, ship building and general engineering. No elaborate surface preparation is necessary other than shot blasting or wire brushing.

Percussion welding

This process, which is also known as *spark discharge welding* or *capacitor discharge welding*, is used for joining wires to larger parts and for welding studs of up to 6 mm in diameter to sheet metal. The energy source is the sudden discharge from a bank of storage capacitors. For stud welding, the stud is shaped with a small cylindrical tip projecting from the end. Within the first millisecond of current discharge, the very high current (about 1000 A) causes the stud tip to vaporise and an arc is initiated. The arc causes some melting of the stud and sheet surfaces. The stud moves forward under a percussive force to make the weld and the whole process is complete in about 10 ms.

Thermit welding

In this form of welding the source of heat is the chemical reaction between iron oxide and aluminium:

$$8Al + 3Fe_3O_4 = 9Fe + 4Al_2O_3 \; (15.4 \, MJ/kg \, Al)$$

The thermit reaction can be used for the site welding of steel sections such as rails, in which a few kilogrammes of weld metal are required, and for the joining of very heavy sections where several tonnes of weld metal might be needed. It can also be used for the repair of large castings and forgings (see Figure 26.11).

The thermit powder, when ignited, produces pure liquid iron at a very high temperature (3500°C). In order to keep temperatures down and produce a liquid steel of similar composition to the steel being joined, steel scrap and alloying additions are made to the thermit mixture. When the reaction is complete the liquid steel produced is allowed to run into the prepared mould space.

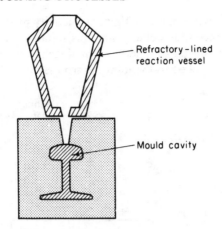

FIGURE 26.11   Thermit welding a rail

26.5 *Pressure welding of metals*

The oldest form of pressure welding is forge welding in which two pieces of wrought iron are heated to red heat and hammered together to make a weld. The slag within the wrought iron acts as a flux. Similarly, mild steel can be forge welded with a little silica sand used as a flux. Most of the pressure welding processes used industrially today are of the electric resistance type. The range of pressure welding processes is shown in Figure 26.12.

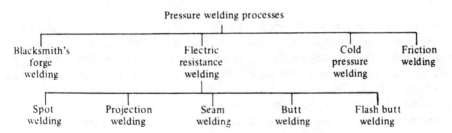

FIGURE 26.12   Pressure welding processes

Spot welding

In this process the parts to be joined are tightly clamped between a pair of electrodes and a large current passed through the metal for a short, but accurately controlled, interval of time (see Figure 26.13). The electrodes, which are made from a strong chromium–copper or beryllium–copper alloy, are hollow and watercooled to prevent the possibility of welding between the electrode and the workpiece. The pressure exerted by the electrodes is generally in the range 70–100 N/mm², and the welding current density is up to 75 A/mm² at a voltage of 10–20 V.

The maximum resistance in the welding circuit occurs at the interface between the two parts to be joined. The total heat generated is given by I^2Rt, where I is the current, R the total resistance and t the welding time. R is made up of the resistance of the work-piece and interface resistances. Some of the heat generated is dissipated from the joint area by conduction through the work-piece. Spot welding is far more suited to steels and other materials with low electrical and thermal conductivities than to high-conductivity metals such as aluminium and copper.

A type of resistance spot welding which is suitable for welding metals of high conductivity and those which are difficult to weld by normal techniques is *magnetic force spot welding*. In this method, the force is applied to the joint electro-magnetically, using the welding current.

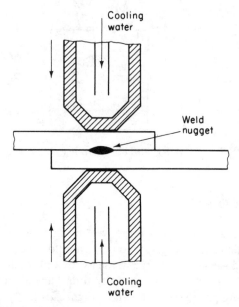

FIGURE 26.13   Principle of spot welding

A spot-welding cycle comprises three parts, firstly the bringing together of the electrodes on the work-piece, creating a pressure load within the material. Secondly, the welding current flows for a pre-set time. This is frequently of the order of 0.5 s. The welding time is accurately controlled by means of an electronic timer. The timers are generally capable of giving pre-set times of 0.1–2 s with an accuracy of ±0.01 s. During this part of the cycle the weld is made and some plastic deformation of the hot metal occurs as a result of electrode pressure. This shows as a surface depression on both sides of the work-piece. Finally, electrode pressure is maintained for a short time after the welding current has ceased. This aids cooling of the weld area and ensures formation of a fine crystal grain size within the weld nugget.

Spot welding is mainly used for the joining of thin-gauge sheet metal, generally mild steel, although other materials can also be joined by this method. It has become a major fabrication technique for the assembly of such items as pressed-steel radiator panels, car-body panels, washing machines, refrigerator cabinets and many others. Much factory-assembly work employs specially designed multi-spot welding machines capable of making up to 100 simultaneous spots.

While spot welding is generally used in connection with light-gauge metal sheet, it is also possible, using heavy-duty equipment, to produce satisfactory spot welds in plate thicknesses of up to 25 mm and this has been used as an alternative to arc welding or rivetting for the assembly of girder frameworks for civil-engineering constructions.

Projection welding

This is a modified form of spot welding in which one of the parts to be joined is embossed before welding. Projection welding offers advantages over spot welding when the two parts to be joined differ in composition or thickness (see Figure 26.14).

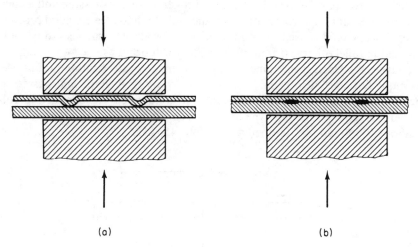

(a)   (b)

FIGURE 26.14   Projection welding: (a) electrical circuit completed through embossed projections; (b) after welding current ceases pressure causes plastic deformation, flattening the projections

Seam welding

This is a process for making a continuous seam between two overlapping sheets of metal. In this case the two electrodes are discs, between which the sheets to be joined are passed (see Figure 26.15). As the overlapping sheets move between the discs an intermittent current flows between the electrode discs through the work, creating a series of overlapping spot welds.

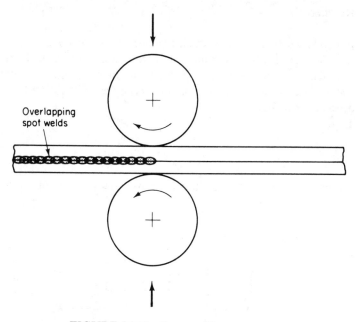

Overlapping
spot welds

FIGURE 26.15   Seam welding

Butt welding

In butt welding the two parts to be joined are held together under pressure and a current passed through the joint area until the temperature is sufficiently high to create a good pressure weld (see Figure 26.16). This method is used for joining tubes and rods. The surfaces to be joined must be clean, plane and parallel otherwise imperfect welding will occur.

Continuous seam-welded steel tubing is frequently made by the continuous butt weld process. In this technique flat steel strip is cold formed into cylindrical shape and passed through the welding head where a continuous welding current is supplied via conducting rolls. A high frequency source is required for this type of resistance butt welding. The electrical supply is at a voltage of about 100 V and at a frequency of 400 to 450 kHz.

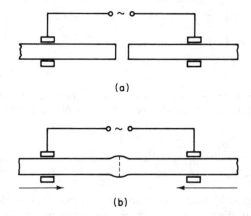

FIGURE 26.16   Butt welding: (a) parts ready for welding; (b) parts brought together and current passed

Induction welding

This is a solid state welding process which has been used for the creation of a continuous butt weld in seam welded steel tubing manufactured from cold formed steel strip. The necessary heat for welding is generated by the effects of electrical currents induced within the material. This process, while effective technically, is probably less efficient than the high frequency resistance welding process described above.

Friction welding

In this process one of the parts to be joined is rotated at a high angular velocity in a chuck, while

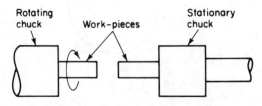

FIGURE 26.17   Friction welding

the other part is held in a stationary chuck. The two parts are then brought together under pressure and seizure and welding occurs between the two surfaces owing to the frictional heat generated (see Figure 26.17). Friction welding has been successfully used for joining shafts of up to 0.3 m diameter. The method is also very suitable for joining dissimilar metals, for example, steel to titanium. This process is also known as *inertia welding* and *spin welding*.

Explosive welding

Explosive welding was first developed in the late 1950s. It has been mainly used for cladding the surface of one metal with another, but it can also be used for making butt and lap welds. In the process one portion of metal hits the other at a very high velocity at an oblique angle causing welding between the two surfaces (see Figure 26.18). For explosive cladding the moving or 'flyer' plate is inclined at a small angle to a target plate resting on a firm base. A sheet of explosive is

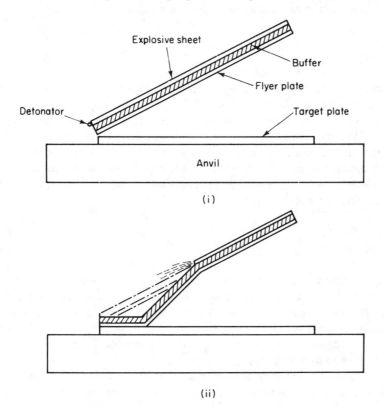

FIGURE 26.18   Principle of explosive welding

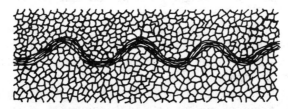

FIGURE 26.19   Sketch showing the microstructure of the ripple interface in an explosive weld

separated from the flyer plate by a buffer sheet. The buffer sheet, made of a material such as PVC, protects the surface of the flyer plate from damage. When the explosive is detonated the flyer sheet is deformed and strikes the target at velocities of up to 300 m/s, generating pressures of up to 5000 N/mm^2. At these very high pressures the flyer plate and target plate surfaces behave as fluids and a weld interface with a ripple-type surface is produced (see Figure 26.19).

Examples of the explosive-cladding method are the bonding of a thin layer of aluminium, titanium or stainless steel to a mild steel base.

Cold pressure welding

Cold pressure welding is a solid phase welding process carried out at ambient temperature by the application of pressure across the interface. The process is applicable for soft, ductile metals, particularly aluminium and copper. Both lap joints and butt joints can be made. This process is used for sealing aluminium cans and for joining wires and bar stock.

Ultrasonic welding

This is a method for producing spot welds in which the two pieces to be welded are clamped between an anvil and a probe, or *sonotrode*, which is vibrated at frequencies within the range 10 to 50 kHz. This causes a small lateral movement between the mating, or faying, surfaces which disrupts surface oxide films, giving an increase in temperature and creating a weld. Most metals can be joined and joints between dissimilar metals are possible. This method is used mainly for delicate work, including the making of electrical connections to electronic devices using fine aluminium or noble metal wires. The welding time may range from 5–10 ms for very fine wires up to about 1 s for thicker gauges.

26.6 Diffusion joining

Diffusion joining is a generic title which covers the three solid-state joining processes of *diffusion bonding*, *diffusion brazing* and *diffusion welding*. A common feature of these closely related processes is that the joint is made by diffusion across the interface at elevated temperatures in vacuo or in an inert atmosphere. The mating surfaces have to be prepared carefully by grinding or abrasive polishing. These methods were developed primarily for the fabrication of some of the newer high technology materials such as beryllium, nickel and cobalt based superalloys, and titanium alloys, particularly for aerospace and nuclear engineering applications.

Diffusion bonding is the process in which a very thin layer of another metal separates the two parts to be joined. This layer may be electroplated or sputtered onto the surfaces to be welded or may take the form of a thin foil. When the assembly is held at high temperature, this intermediate layer diffuses into both parts of the assembly and the final joint shows little or no evidence of the layer.

Diffusion brazing is similar to diffusion bonding in that an intermediate layer of another metal is placed as a filler in between the two parts to be joined. During the heating stage, there is an interdiffusion of filler and parent metal to the extent that the properties of the material in the joint approach those of the parent material.

Diffusion welding is the joining of two mating parts by high temperature diffusion but without the presence of a film or shim of another metal in the joint.

Dissimilar metals, some ceramics and some ceramic-metal joints can be made by the diffusion joining processes described above.

26.7 The welding of plastics

Welded joints can be made in most thermoplastics and many can be joined using adhesives. Fully cured thermosets cannot be welded and any joints required are made using either adhesives or mechanical fastenings. The main welding methods for thermoplastics are given below.

Hot gas welding

Hot gas welding is a fusion welding process for thermoplastics which resembles the oxy-gas welding of metals but the heat source is a jet of heated gas rather than a flame. Air or an inert gas is passed through a torch, where it is heated by either gas or electrical heating, and used to melt the joint surfaces and a thermoplastic 'welding filler rod'. At the same time, pressure is applied at the joint helping to 'forge' the joint. The filler rod is of the same material as the parts to be welded. The bond strengths which are achieved are, typically, about 70 per cent of that of the parent material.

Hot plate welding

This method, also known as *heated tool welding*, relies on the application of heat and pressure. The parts to be joined are heated by an electrically heated resistance strip or a heated metal plate, the softened areas brought together and cooled under pressure. In some cases, a strip or rod of the thermoplastic can be used as a filler material in order to improve the joint quality. Materials most commonly joined by hot plate welding are polyethylene, polypropylene, PVC, ABS and ABS/polycarbonate alloys. The *heat sealing* of thin polymer film and film laminates, as for example in the welding of polyethylene film in the manufacture of bags, is a type of hot plate welding.

Spin welding

This is another name for *friction welding* (see Section 26.5). Generally, the term friction welding is reserved for the joining of metal parts and the welding of thermoplastics is referred to as spin welding. It is a welding technique whereby the frictional heat generated by the rapid rotation of one thermoplastic component against a second, stationary thermoplastic part causes localised surface melting of both parts. Usually, pressure is applied to the joint as soon as the spinning stops, thus producing a good weld. A typical example of the use of the process is the welding of rims to can bases in the manufacture of polypropylene industrial paint pots.

Vibration welding

The frictional heat generated at the mating surfaces of thermoplastic parts when one of the parts is subject to a low-frequency oscillation is sufficient to cause localised melting and the two parts fuse together. Vibrational frequencies of the order of 90 to 120 Hz are used and the method is used for the manufacture of butt joints. When very much higher frequencies are used, the process is termed ultrasonic welding.

Ultrasonic welding

This is similar in principle to vibration welding, described above, but vibrational frequencies in the range 10 to 50 kHz are employed. Ultrasound energy is transmitted into the thermoplastic component by means of a transducer assembly or horn and the small relative lateral movement at the mating surfaces causes local heating and melting of the thermoplastic. Two other processes involving the use of ultrasonic vibrations are *ultrasonic spot welding* and *ultrasonic staking*. The first process is used for making spot welds in thermoplastic sheet material. The most common type of equipment consists of a high-energy pistol with a vibrating pointed probe. In the welding of sheet the probe is forced through the top sheet and into the lower sheet. A flow of fused plastic from the top sheet into the cavity formed in the base gives a spot weld. The best results are obtained with polyethylene, polypropylene and high impact polystyrene. *Ultrasonic staking* is a method which can be used for the assembly of metal parts to polymers. A thermoplastic component with moulded studs is assembled to a metal part so that the plastic studs protrude through holes in the metal. The studs are then headed using ultrasonic energy, resulting in a rivet type joint.

High frequency welding

Thermoplastics are dielectrics and dielectric heating occurs when they are placed in a high frequency electromagnetic field. Although, theoretically, frequencies in the range 1 to 200 MHz are feasible, permissible wave bands have been specified by international agreement and, of these, the only practicable frequency available is 27.12 MHz. This form of welding is suitable for polyamides, PPVC, PVA and ABS.

Induction welding

The principle of electromagnetic induction heating can be used for the welding of thermoplastics. Although thermoplastics are insulators, weldable materials can be made by incorporating fine magnetic particles within the polymer. Generally, these are of the order of 1 μm in size. When subject to a high-frequency field the particles are heated causing a heating and melting of the surrounding plastic. The heated surfaces are then brought together under pressure to create the weld.

Solvent welding

Some thermoplastics can be softened and joined with the aid of specific solvents. The solvent partially dissolves the surfaces to be joined allowing them to fuse into one another when pressed together. Evaporation of the solvent leaves a sound bond. Acrylics, polystyrene and PVA are the main polymers which are joined in this manner.

26.8 Adhesive bonding

Adhesives, or glues, are polymeric materials which can be used to make joints between similar or dissimilar materials. The use of adhesive bonding offers a number of advantages, the main ones being that it can offer a low cost assembly method, the joint is made at ambient or only slightly elevated temperatures, and it may be the only method for joining dissimilar materials. The nature of adhesion between an adhesive and a metal surface is not fully understood, but there is no alloying and it must rely on secondary bonding forces. The adhesive must fully wet the surface of the adherend, and surface preparation of a metal adherend may be necessary. Surface films and loosely bonded surface oxide need to be removed but a strongly bonded oxide coating, particularly if of a porous nature, can improve adhesion. A roughened surface may also help adhesion as it provides a greater surface contact area and can permit some mechanical interlocking. Surface treatment with some chemical compounds, termed adhesion promoters, can aid adhesion. In the case of porous materials, such as many ceramics, and fibrous materials, such as timber, the adhesive can penetrate into the pores or between fibres giving a good mechanical interlocking. Many adhesives tend to be relatively strong in both tension and shear but weak in cleavage. The design for adhesive joints needs to take account of this.

Animal and vegetable glues, made from hoof, bone and vegetable starches, have been used from earliest times. Today there is a very wide range of adhesives, many of them specific in application, based on both thermoset and thermoplastic polymer materials and they fall into five general categories. These are: solvent adhesives, dispersion adhesives, hot melt adhesives, thermosets, and monomers which polymerise in situ.

Solvent adhesives have been covered already under the heading *solvent welding* in Section 26.7.

Dispersion adhesives are dispersions of a finely divided thermoplastic material in a fluid and the fine particles bond as the carrying fluid evaporates. This group includes water borne emulsions of PVA and plastisols, emulsions of PVC in an organic fluid. The former are only suitable for use with porous adherends, for example paper, while the latter are used mainly for bonding PVC.

Hot melt adhesives are thermoplastics of low softening temperature. The polymer is heated to some temperature above its T_g and applied to the surfaces to be joined. Rapid cooling under pressure establishes a bond.

Several *thermoset* resins including phenolics, amino resins, polyesters and epoxies are used as adhesives. A resin and hardener mix will react together within the assembled joint. Epoxy materials possess excellent adhesion to most materials and this, coupled with their high strength, make them excellent structural adhesives.

There are several liquid *monomers* which can be used as adhesives, these polymerising *in situ*. Typical of this type are the cyanoacrylates which are low viscosity liquids which polymerise rapidly in the presence of moisture, which is usually present on the surface of an adherend. The so-called 'super-glues' are based on the cyanoacrylate family.

26.9 Metallurgical considerations for welding

In general, the conventional oxy-gas and arc welding processes are used for joining parts of the same or very similar compositions. In order to obtain a good joint the filler metal should normally be of the same composition as the parts being joined so that, for example, a plain carbon steel filler or electrode is used for welding a plain carbon steel while for welding a 1 per cent chromium low-alloy steel a 1 per cent chromium steel filler would be used. However, for oxy-gas and unshielded arc welding the carbon content of the filler or electrode should be slightly higher than the carbon content of the steel being welded to compensate for the small carbon loss that occurs through oxidation during welding.

Weldability of metals

Not all materials can be easily welded by all processes. Mild steels are readily weldable by almost all processes but high carbon steels are difficult to weld with oxy-gas or unshielded arc techniques. Some of the carbon content of the steel may oxidise producing carbon monoxide which may be trapped within the weld pool giving porosity. A shielded arc process is much better for these steels. It is possible to weld stainless steels with an oxy-acetylene torch, providing that a reducing flame and a suitable flux are used, but these days it is very unusual to use anything other than MIG or TIG welding for these materials. Similarly, aluminium and its alloys are generally welded by means of the TIG process even though they are readily weldable with oxy-gas using a reducing flame and a fluoride flux. If aluminium is gas welded it is most important that all traces of residual flux are removed from the material after welding as the fluxes are highly corrosive.

Structure of weldments

A fusion weld is a small casting and the actual weld area will tend to have the structure of a casting. During the welding process the temperature of the surrounding material will be raised and this may cause microstructural alterations in the area adjacent to the weld, the heat affected zone. Some examples are given below.

(a) Weld between two pieces of normalised mild steel (see Figure 26.20). The weld itself will solidify to give an initial coarse cast structure but rapid cooling through the critical temperature range will produce very fine ferrite and pearlite with a high strength. The heat affected zone will be that area adjacent to the weld in which the temperature rose to or above the critical temperatures. Here, with subsequent rapid cooling, a fine pearlite and ferrite structure will be formed. This is generally finer than the original normalised structure. The

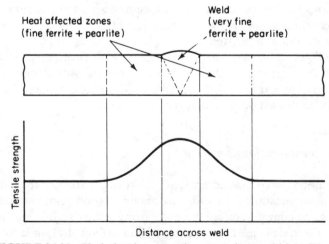

FIGURE 26.20 Variation in properties across butt weld in mild steel

result is that the heat affected zone and the weld will usually have better mechanical properties than the parent metal.

(b) Weld in a low-alloy steel. Low-alloy steels possess a much greater hardenability than plain carbon steels and the rate of cooling experienced after welding may exceed the critical cooling velocities for these materials giving formation of brittle martensite in the weld and heat affected zones. A good flux cover will reduce cooling rates to minimise martensite formation but a post-weld heat treatment may be necessary.

(c) Weld in austenitic stainless steel. A different type of problem may occur with this type of steel. The normal heat treatment condition for an austenitic steel is water quenching from 1000°C to ensure that all carbon is in full solution in austenite. During welding grain-boundary precipitation of carbides may occur in that portion of the heat affected zone where the temperature was raised to 500–700°C, and this carbide precipitation can lead to subsequent grain-boundary corrosion of the type known as weld decay. This type of corrosion can be avoided by fully heat treating the component after welding. Alternatively, a more expensive welding quality stabilised stainless steel should be used (see Section 16.14).

(d) Weld in a non-ferrous material such as aluminium or copper (see Figure 26.21). The weld zone will normally possess the coarse grain structure of a small casting. Within the heat affected zone grain growth may have occurred. This means that both the heat affected zone and the weld zone will be softer and weaker than the basis material. The relative effects will be even greater if the original material was in the work hardened condition.

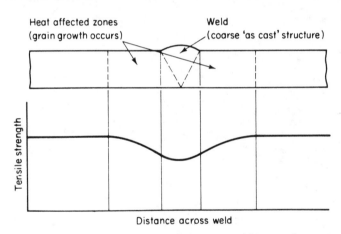

FIGURE 26.21   Variation in properties across weld in non-ferrous metal

26.10 Defects in welds

With the correct choice of welding method and good welding practice, defects within welds should not occur. However, nothing in this world is perfect and defective welds are sometimes made. The most common types of welding defect are:

(a) Lack of penetration. Cause: bad welding practice.
(b) Oxide and slag inclusions. Cause: incorrect choice of welding method or poor execution of weld.
(c) Gas porosity. Cause: incorrect choice of method or poor execution of weld.
(d) Distortion and residual stresses. Distortion of parts during welding may be caused by poor design of parts for welding or it may be caused by bad welding practice. Residual stresses

may frequently occur in weldments after cooling. Occasionally, owing to poor design for welding, the stresses may be sufficiently large to cause immediate cracking. For many welds, particularly in components of complex shape, it is necessary to give the welded article a post-weld stress-relief heat treatment.

PART V
BEHAVIOUR IN SERVICE

27

Failure, Fatigue and Creep

27.1 *Failure*

A material, component or structure is deemed to have failed when its ability to fully satisfy the original design function ceases. This may be due to a variety of causes. It may be due to fracture, either partial or complete, plastic buckling, dimensional change with time, loss of material by corrosion, erosion or abrasive wear or an alteration of properties and characteristics with time due to environmental or other effects. These various modes of failure may be dependent on stress, time, temperature, environment or a combination of any of these.

The conditions for failure by brittle fast fracture were considered in Chapter 10. Failure by fatigue is stress/time dependent and the effect of a cyclical stress system is to cause nucleation and propagation of cracks within the material. Final failure is by fast fracture when one or more cracks have grown to critical size.

Creep is a continued straining of a material under constant stress conditions and is stress/time/temperature dependent. The creep process will lead to eventual fracture but a component subject to creep may be deemed to have failed when its dimensions have changed by some pre-determined value.

27.2 *Fatigue*

If a material is subjected to repeated, or cyclic, stressing, it may eventually fail even though the maximum stress in any one stress cycle is considerably less than the fracture stress of the material, as determined by a short-term static test. This type of failure is termed *fatigue failure*. (The term fatigue is sometimes used to describe the failure of a plastic material in a long-term static load test although this would be more correctly described as creep).

Very many components are subjected to alternating or fluctuating loading cycles during service, and failure by fatigue is a fairly common occurrence. The mechanism of fatigue in metals has been thoroughly investigated. When a metal is tested to determine its fatigue characteristics, the test conditions usually involve the application of an alternating stress cycle with a mean stress value of zero. The results are plotted in the form of an *S-N* curve (Figure 27.1), where *S* is the maximum stress in the cycle, and *N* is the number of stress cycles to failure. Most steels show an *S–N* curve of type (i), with a very definite *fatigue limit*, or *endurance strength*. This fatigue limit is usually about one-half of the value of the tensile strength, as measured in a static test. It means that if the maximum stress in the stress cycles is less than this fatigue limit, fatigue failure should never occur. Many non-ferrous materials show *S–N* curves of type (ii) with no definite fatigue limit. With these materials it is only possible to design for a limited life, and a lifing of 10^6 or 10^7 cycles is often used.

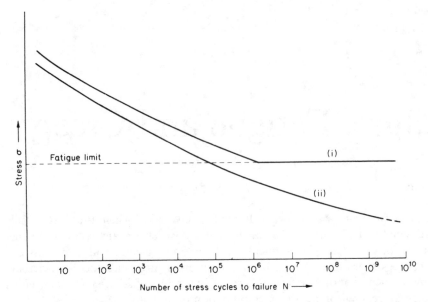

FIGURE 27.1   *S–N* curves for (i) metal showing fatigue limit, (ii) metal showing no fatigue limit

There are, essentially, two forms of fatigue failure, *low-cycle fatigue* and *high-cycle fatigue*. In the former case, the maximum stress in any cycle, while below the tensile strength of the material, is greater than the yield stress and the number of stress cycles to failure is low, generally less than 1000. High cycle fatigue, which requires some 10^5–10^6 cycles, is the result of a stress regime in which the maximum stress in a cycle does not exceed the yield stress of the material.

Failure by fatigue is the result of processes of crack nucleation and growth or, in the case of components which may contain a crack introduced during manufacture, the result of crack growth only brought about by the application of cyclical stresses. This is evident from the appearance of a typical fatigue fracture surface. The appearance of a fatigue fracture surface is distinctive and consists of two portions, a smooth portion, often possessing conchoidal, or 'mussel shell', markings showing the progress of the fatigue crack up to the moment of final rupture, and the final fast fracture zone (Figure 27.2). The striations visible in the smooth

FIGURE 27.2   Fatigue failure of large shaft showing the two zones

portion, sometimes referred to as 'beach marks', indicate stages in the growth of the crack. The final fracture is usually of the cleavage type.

There are several types of stress cycle and cycles are described as *alternating*, when the mean stress, σ_m, is zero, *repeating*, when the minimum stress, σ_{min}, is zero, and *fluctuating*, when the mean stress, σ_m, has some value other than zero. These cycle types are shown in Figure 27.3. The stress range of a cycle, $\Delta\sigma$, is $(\sigma_{max}-\sigma_{min})$, the cyclic stress amplitude, $\sigma_a = \frac{1}{2}(\sigma_{max}-\sigma_{min})$, and the mean cycle stress, $\sigma_m = \frac{1}{2}(\sigma_{max} + \sigma_{min})$.

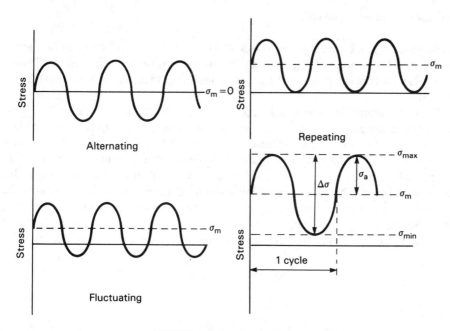

FIGURE 27.3   Types of stress cycle

Much fatigue testing is carried out using alternating stress cycles in which σ_m is zero but, in many practical situations where components or structural assemblies are subject to cyclical stresses the mean stress, σ_m, is not zero. Several *laws* of an empirical nature have been proposed to give the relationships between stress ranges, mean stresses and fatigue life. These are:

The modified Goodman equation: $\dfrac{\sigma_a}{\sigma_{FL}} + \dfrac{\sigma_m}{\sigma_{TS}} = 1$

The Gerber parabolic equation: $\dfrac{\sigma_a}{\sigma_{FL}} + \left(\dfrac{\sigma_m}{\sigma_{TS}}\right)^2 = 1$

The Soderberg equation: $\dfrac{\sigma_a}{\sigma_{FL}} + \dfrac{\sigma_m}{\sigma_y} = 1$

In the above equations, σ_{FL} is the fatigue strength as determined in tests with a mean stress of zero, σ_{TS} is the tensile strength, and σ_y is the yield strength of the material. These empirical laws are not very precise but they are used for some preliminary design calculations.

There are also many practical situations in which a component or structure may be subjected to more than one type of loading system during its lifetime with the various stress cycles having different stress ranges, $\Delta\sigma$. Miner's law of cumulative fatigue is an empirical rule which can be used to estimate the fatigue life of a component when subjected to a series of different loading cycles. For example, if stressed for n_1 cycles in a regime which would cause failure in a total of N_1 cycles and for n_2 cycles in a regime where failure would occur after N_2 cycles, then n_1/N_1 of the fatigue life of the component would be used up in the first case and the fraction n_2/N_2 of the fatigue life used in the second instance. Miner's rule can be expressed by stating that failure will occur when $\Sigma(n/N) = 1$.

27.3 Crack nucleation and growth

In low cycle fatigue, the maximum stress in a cycle is greater than the yield stress and some plastic deformation occurs. The process of slip creates slip bands which roughen the surface. This effect creates points of stress concentration and a crack is initiated which, in the first instance, grows along a slip plane.

Although the maximum stresses in high cycle fatigue conditions are nominally below the elastic limit of the material, it has been established that some plastic slip takes place. During continued cyclic stressing, slip bands appear on the material and some extrusive and intrusive effects are associated with the slip bands (Figure 27.4). The extrusions and intrusions so formed are extremely small, being of the order of 1 μm in size, but once an intrusion has formed on a highly polished surface it can act as a point of initiation for a fatigue crack. The intrusion, with a very small root radius, acts as a point of stress concentration and a crack can propagate through the material until, eventually, the remaining sound portion of the section is too small to sustain the maximum load in a stress cycle and sudden fast fracture of the component occurs.

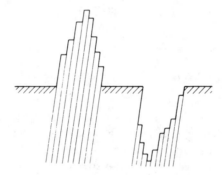

FIGURE 27.4   Intrusions and extrusions formed in early stages of fatigue

Cracks, once initiated, propagate in materials by one of two methods, either by ductile tearing or by cleavage. In the former process, which applies to ductile metals and many thermoplastic materials, a lot of energy is absorbed in plastically deforming the material just ahead of the advancing crack. The plastic deformation tends to blunt the crack tip, so reducing its stress concentrating effect. However, during the severe plastic deformation, small cavities tend to form in the deformed area ahead of the crack and these eventually link up, so extending the crack length. In non-ductile materials such as ceramics, glasses, brittle metals, thermoplastics at temperatures below their T_g values, and thermosets, little, if any, plastic deformation can occur ahead of a crack tip to blunt the crack. When the local stress at the crack tip exceeds the fracture stress of the material, the interatomic bonds will be broken and the crack develops catastrophically as a fast brittle fracture.

An important factor is the cyclic stress intensity, ΔK, this being:

$$\Delta K = C \, \Delta\sigma \sqrt{\pi a}$$

where $\Delta\sigma$ is the cyclic stress range, a is the depth of a surface flaw or crack (or depth of a trough in a machined surface profile) and C is a material constant. The value of ΔK increases with time, as crack depth a increases and catastrophic failure will occur when ΔK reaches a critical value, this value being a function of the material. The rate of crack growth per cycle, da/dN, can be related to the cyclic stress intensity by an expression: $da/dN = A \, \Delta K^m$, where A and m are constants of the material.

27.4 Factors affecting fatigue

There are many factors that affect the fatigue strength of a material, and these include surface condition, component design, and the nature of the environment. Specimens for fatigue testing are usually prepared with a highly polished surface, and this condition will give the best fatigue performance. The fatigue limit for highly polished steels is approximately one-half of the tensile strength. If the surface of the specimen contains a scratch or notch, or is a ground rather than a polished surface, the fatigue limit of the material will be reduced. The presence of scratches or notches will provide small defects from which fatigue cracks can be initiated. Similarly, a sharp change in section with a small fillet radius can act as a stress raiser, and fatigue cracks can commence from such points. Keyways and oil holes in shafts are often points at which fatigue commences. The effect of a notch or scratch is not the same for all materials. As in the case of brittle failure, a ductile metal is much less sensitive to the presence of surface flaws than a brittle material. If conditions are such that corrosion can occur, not only is the fatigue limit very greatly reduced but also the rate of corrosion is increased. For some materials, including some steels, there is no fatigue limit in a corrosive environment, and failure will eventually occur, even when the stress levels are very low.

The resistance of a metal to failure by fatigue will be highest when the surface of the metal is highly polished and free from scratches and other defects. The fatigue life may be improved by treating the metal surface by *peening*. Peening involves lightly hammering the surface of the material with a round-nosed hammer, or by bombarding the surface with small steel pellets. The result of the peening treatment is to induce residual compressive stresses within the surface layers of the material.

The rate of stress cycling has no effect on the fatigue strength for frequencies up to about 150 Hz, but at higher frequencies there appears to be a small increase in the fatigue strength. This increase is of the order of 10 per cent at frequencies of up to 15 000 Hz.

Materials other than metals are also subject to failure by fatigue, but comparatively little work has been done in this area. For concrete and polymers, as with metals, the number of stress cycles necessary for failure is increased as the maximum stress in the loading cycles is decreased, but there does not appear to be a definite fatigue limit with these materials. There are difficulties in the fatigue testing of polymers because, owing to the low thermal conductivities and high damping capacities of these materials, there is an increase in the temperature of a polymer test-piece during a test.

Composite materials are also subject to fatigue. The growth of a crack in a composite material in a direction normal to the line of reinforcing fibres may be stopped at a fibre. The high stress concentrated at the crack tip may cause a rupture of the bond between matrix and fibre, thus blunting the crack tip so that fracture cannot proceed into the fibre. This does not happen if crack development is parallel to the fibre direction. Aligned (uni-axial) fibre composites may give *S–N* curves with a fatigue limit and often show, under cyclic loading conditions, a behaviour which is superior to metals such as aluminium. The greatest resistance to fatigue damage is provided by carbon and boron fibre reinforcement, rather than glass fibre reinforcement, since the higher elastic moduli of the former limits the amount of matrix strain. Random fibre composites usually exhibit larger matrix strains in the direction of applied stress and, hence, do not perform so well. True fatigue effects are only observed in polymer matrix composites at cyclic frequencies less than about 30 Hz, since cyclic frequencies of greater than 30 Hz lead to heating and thermal degradation of the polymer.

27.5 *Creep*

Creep is the continued slow straining of a material under the influence of a constant load. The amount of strain developed in a material becomes a function of time and temperature, as well as of stress. The phenomenon of creep can occur in all types of materials but there is a threshhold temperature, below which creep does not become a factor in stress–strain relationships. In the case of thermoplastic materials, creep is of significance at all temperatures above the T_g value. In metals and ceramics, creep is considered as a high temperature phenomenon, although lead creeps at room temperature, and occurs at temperatures above about 0.3–$0.4\ T_m$ in metals and above 0.4–$0.5\ T_m$ in ceramics, where T_m is the melting temperature in Kelvin.

A typical creep curve of strain v. time at constant stress is shown in Figure 27.5. This shows three stages of creep.

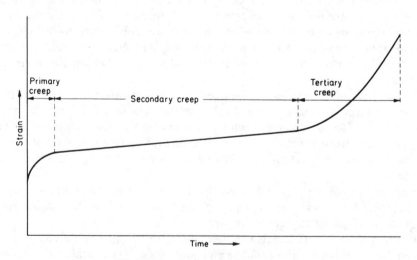

FIGURE 27.5 Typical creep curve

In the first stage, *primary* or *transient creep*, the rate of creep strain is initially high but steadily reduces to the constant rate of the second stage, *secondary* or *steady-state creep*. Eventually this moves into the final stage, *tertiary creep*, where the creep rate increases and leads rapidly to failure. A number of mechanisms have been postulated for creep in polycrystalline materials but they fall into two main categories—those involving the lattice and those involving grain boundaries. Lattice mechanisms involve vacancy diffusion, dislocation climb and mobility (see

Section 8.11) while grain boundary sliding may also be important, particularly where second phases are present at grain boundaries.

At a constant stress level, the rate of secondary creep increases with increased temperature according to an Arrhenius type relationship:

$$d\varepsilon/dt = K \exp(-B/T)$$

where $d\varepsilon/dt$ is strain rate, T is temperature (Kelvin) and K and B are constants for the material.

The effect of increasing temperature on the creep rate is shown in Figure 27.6. It will be seen that there is primary creep only, with a secondary creep rate of zero, at a relatively low temperature, but at a very high temperature the primary stage merges in with tertiary creep.

At a constant temperature the creep rate is stress-dependent and follows a power law, the relationship being:

$$d\varepsilon/dt = C \sigma^n$$

where C is a constant for the material and n is another constant, the time exponent. For many materials, the value of the exponent, n, lies between 3 and 8, while for glasses it is of the order of 10.

The above two relationships can be combined, to give a general creep equation:

$$d\varepsilon/dt = A \sigma^n \exp(-B/T)$$

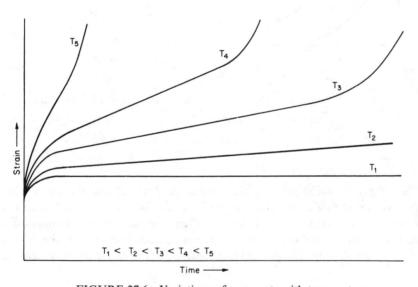

FIGURE 27.6   Variations of creep rate with temperature

The change from secondary to tertiary creep is caused by the intervention of some other effect such as the formation of small internal cavities or localized necking of the material. When such effects occur, the effective stress in the fissured or necked zone will increase and cause an increase in the rate of strain. This is the onset of tertiary creep and, usually, there is a rapid acceleration in the creep strain rate leading to failure.

A creep limit, as such, does not exist. However, for design purposes, it is useful to know the stress/temperature combinations which will give some limiting value of total creep strain, say 10^{-3} in a time of 1000 hours or 10 000 hours, or which will give some limiting rate of creep strain, say 10^{-5} or 10^{-6} per hour. Summary curves, showing data derived from very many creep tests, are used to present information in these forms (Figure 27.7).

When creep occurs in ceramics under a tensile load the overall strain is usually less than 1 per cent and failure often occurs during steady state creep, before the tertiary creep stage is reached. However, under compressive stress, tertiary creep, with an increasing creep rate, is observed before final failure. In glasses, creep is due to the viscous flow of a Newtonian fluid of high viscosity. Creep of thermoplastic polymers also occurs by a process of viscous flow. The intermolecular bonding forces between the molecular chains are relatively weak van der Waal's forces and, even at low levels of stress, creep can occur in many thermoplastics. The creep resistance of polymers is increased substantially by incorporating glass or carbon fibre reinforcement.

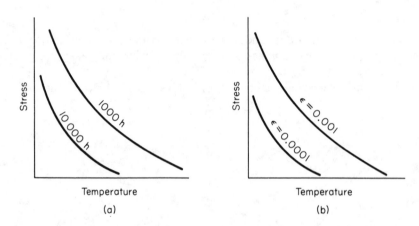

FIGURE 27.7   Summary creep curves: (a) stress-rupture; (b) limiting strain

27.6 Relaxation

Relaxation is a phenomenon related to creep and is the slow reduction of stress with time in a material which is subject to a constant value of strain. This will occur in metals and ceramics at elevated temperatures but in thermoplastics it will often take place at room temperature. The rate of relaxation is exponential and the *relaxation time* is the time required for the level of stress to fall to $1/e$ of the original value. The general relationship for relaxation at a constant temperature is:

$$\sigma_t = \sigma_0 \exp(-At)$$

where σ_0 is the initial stress, σ_t is the stress after time t, and A is a constant for the material.

An example of relaxation occurring within a material during service is the level of stress in a bolt. The bolt is strained to a particular level when the nut is initially tightened. During service, the total strain in the bolt remains constant but, if the temperature is such as to permit creep, then stress relaxation will occur within the bolt and retightening of the nut will be required at intervals.

27.7 Questions

27.1 A steel has the following properties: tensile strength 640 MPa, tensile yield stress 510 MPa and a fatigue limit of 320 MPa. The steel is to be used for a component which will be subject to a static tensile stress of 225 MPa with a superimposed cyclic stress. Estimate the maximum range of cyclic stress on the basis of: (a) Gerber's law, (b) the modified Goodman law, and (c) Soderberg's law.

27.2 The cyclic stress stress intensity, ΔK, for a steel is given by the relationship: $\Delta K = 1.12$ $\Delta\sigma (\pi a)^{\frac{1}{2}}$, where $\Delta\sigma$ is the cyclic stress range and a is the depth of a surface defect. For a steel, for which $\Delta K = 5$ MPa m$^{\frac{1}{2}}$, calculate the maximum safe cyclic stress range for the following surface conditions:

(a) polished with surface roughness of 1 μm,
(b) turned surface with roughness of 10 μm,
(c) rough turned surface with roughness of 100 μm, and
(d) polished with a surface defect 1 mm deep.

27.3 (a) Define Miner's law of cumulative fatigue, clearly identifying each of the parameters. (b) A steel with a tensile strength of 900 MPa and a fatigue strength (for 10^6 cycles) of 450 MPa is subjected to a series of alternating stress cycles as follows: 10^4 cycles at $\sigma_{max} = 575$ MPa and 10^4 cycles at $\sigma_{max} = 550$ MPa. Estimate the number of additional stress cycles the steel can endure at a value of $\sigma_{max} = 500$ MPa.

27.4 Given that the growth rate of a fatigue crack, da/dN, occurs according to the relationship $da/dN = A\, \Delta K^m$, where $\Delta K = 1.12\, \Delta\sigma (\pi a)^{\frac{1}{2}}$, estimate the number of cycles required for a crack in a steel component to grow in depth from 2 mm to 10 mm when subjected to an alternating cyclic stress with a stress range of 225 MPa. The constants for the steel are: $A = 4 \times 10^{-38}$, and the exponent $m = 4$.

27.5 Sketch the form of the creep curve for a metal under conditions of constant stress and temperature and explain the significance of the various sections of the curve.

Creep tests are conducted on a material at a stress of 100 MPa. At 900 K, the steady state creep rate is 1.0×10^{-4} s^{-1} while at 750 K it is 7.5×10^{-9} s^{-1}. Calculate the maximum working temperature for the material if the limiting creep rate is 1.0×10^{-7} s^{-1}.

27.6 The steady state creep strain rates for an alloy at several temperatures and at a constant stress of 105 MPa were determined and are given in the Table below.

Temperature (°C)	180	220	260
Strain rate (s^{-1})	4.72×10^{-8}	2.08×10^{-7}	7.34×10^{-7}

Estimate the life of the material when stressed at 105 MPa at a temperature of 160°C if the total creep strain is not to exceed 0.01.

27.7 The data in the Table below are those for the creep of an aluminium alloy at 200°C.

Time (hr)	Strain at 60 MPa	Strain at 35 MPa
0	320×10^{-6}	108×10^{-6}
20	532×10^{-6}	215×10^{-6}
60	815×10^{-6}	332×10^{-6}
90	1000×10^{-6}	400×10^{-6}
120	1200×10^{-6}	477×10^{-6}
140	1320×10^{-6}	532×10^{-6}
170	1477×10^{-6}	610×10^{-6}
220	1800×10^{-6}	723×10^{-6}

From the data, plot creep curves and calculate the value of stress that would give a creep rate of $4.8 \times 10^{-6}\,hr^{-1}$.

27.8 The steady state creep rates for a steel, when tested at 510°C at several stresses, are: $1.69 \times 10^{-5}\,hr^{-1}$ at 180 MPa; $3.22 \times 10^{-5}\,hr^{-1}$ at 200 MPa; $7.55 \times 10^{-5}\,hr^{-1}$ at 240 MPa; $1.41 \times 10^{-4}\,hr^{-1}$ at 265 MPa.

(a) Does steady state creep at 510°C obey a simple power law and, if so, evaluate the exponent?
(b) Estimate the maximum stress to which the steel can be subjected at 510°C if the limiting strain rate is $1.65 \times 10^{-4}\,hr^{-1}$.

28

Oxidation, Corrosion and Other Effects

28.1 Oxidation of metals

Metals possess an affinity for oxygen and tend to oxidise, but with most metals the amount of oxidation that takes place at ordinary temperatures is not serious. In fact, in many instances, the oxide layer that rapidly forms on a freshly exposed metal surface tends to protect the metal from further oxidation.

The direct reaction between a metal and oxygen can occur at any temperature, but the rate of the oxidation reaction is temperature dependent and increases rapidly with an increase in temperature. The reactivity of a metal with oxygen varies very considerably from one metal to another, with some metals such as copper and lead showing low reactivities, and other metals such as aluminium and magnesium possessing very high reactivities with oxygen. However, the rate at which oxidation occurs at the surface of a metal is largely controlled by the nature of the oxide that is formed. This may form as a coherent and almost non-porous layer on the surface of the metal, in which case it can act as a protection against further oxidation, or it may form as a broken or porous layer, allowing oxygen relatively free access to an exposed metal surface allowing oxidation to continue at a constant rate. In the case of the metal molybdenum there is a worse phenomenon. At temperatures above 900°C there is a catastrophic oxidation of the metal. No surface oxide layer can be formed as molybdenum oxide is a vapour at such temperatures.

The oxide layer that forms on the surface of a metal is termed either a film or a scale. There is no clear distinction between these names, apart from the fact that scale refers to a thick layer and films are very thin. Generally, the term film is used for an oxide thickness of 10^{-3} mm or less, while the term scale is used for greater thicknesses than this.

The Pilling–Bedworth (P–B) ratio defines the characteristics of an oxide layer on a metal surface and is expressed as:

$$\text{P–B ratio} = \frac{M_o \times \rho_m}{n M_m \times \rho_o} = \frac{M_o \times V_o}{n M_m \times V_m}$$

where ρ_m and ρ_o are the densities of metal and oxide respectively, V_m and V_o are the respective volumes, M_m and M_o are the atomic mass numbers of metal and oxide, and n is the number of metal atoms in the oxide molecule. If the P–B ratio is less than 1, the oxide occupies a smaller volume than the volume of metal from which it was formed and, in this case, the oxide film is porous. When the P–B ratio is $\simeq 1$, the volumes of oxide and metal are nearly equal and the tendency is for the formation of a tenacious, non-porous, protective oxide film. When the P–B ratio is greater than 1, the oxide film formed is initially protective but, as the thickness of the oxide layer increases, stresses are developed within the oxide layer and these may cause the oxide to flake off from the surface.

Porous oxide films or scales are non-protective as there is free access of oxygen to the metal surface. In such cases oxidation will continue at a constant rate at any given temperature and the thickness of the oxide layer will increase linearly with time according to the expression $y = kt$, where y is the thickness of the oxide coating after time t and k is a constant. The value of k, however, varies with temperature according to the Arrhenius relationship, $k = A \exp(-B/T)$, where A and B are constants and T is the temperature (K).

When a non-porous protective film forms on the exposed surface of a metal it prevents the access of gaseous oxygen to the surface. Consequently, the diffusion of ions and electrons must take place through the oxide film if there is to be any further growth in the thickness of the film. Such diffusion processes are aided by the presence of defects within the crystalline structure of the oxide. The reactions that occur are as follows: (i) at the oxide film/air interface formation of oxygen ions takes place, $O_2 + 4e^- = 2O^{2-}$; (ii) at the metal/oxide interface ionisation of the metal occurs, $M = M^{2+} + 2e^-$. The electrons liberated by the ionisation of metal atoms will be conducted fairly freely through the oxide layer to combine with oxygen molecules forming negatively charged oxygen ions at the oxide/air interface. Both metal and oxygen ions can diffuse through the oxide film, but usually the rate of diffusion of metal ions is greater than that for oxygen ions as many species of metal ions possess smaller diameters than the oxygen ions. The metal ions meet the oxygen ions close to the oxide/air interface and there combine forming oxide and the oxide layer grows outwards (see Figure 28.1). (In those cases where the oxygen-ion diffusion rate is greater than that for the metal ions the film growth occurs from the metal/oxide interface). As the thickness of the oxide layer increases and diffusion distance increases there will be a corresponding decrease in the rate of oxide formation. The variation of film thickness with time follows a parabolic law expressed like $y^2 = kt$, where y is the thickness of the oxide layer after time t and k is a constant. As for linear growth, the value of the constant, k, varies with temperature according to the Arrhenius expression. This type of oxide growth is observed in copper, iron, nickel and cobalt.

Some metals have highly protective oxide films with a growth rate that decreases much more rapidly than given by the parabolic law. In these cases the growth rate follows a logarithmic law expressed as: $y = k \log(at + 1)$. Again, y is the film thickness at time t and k and a are constants. Chromium and zinc are of this type. In the case of aluminium the oxide film possesses an extremely high value of electrical resistance and when the film thickness is about 100 atoms thick electron flow through the film ceases.

Alloys to resist oxidation are often based on an alloy system in which the solute has a much higher affinity for oxygen than the parent metal. A good example of this are alloys of copper and aluminium (aluminium bronzes). When these alloys are oxidised at high temperatures cuprous

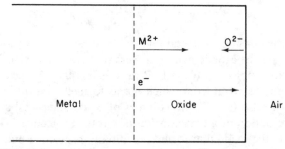

FIGURE 28.1   Oxidation of a metal through a protective film

oxide, Cu_2O, forms very rapidly and copper ions cross the metal/oxide interface into the oxide layer. The concentration of aluminium on the metal side of the interface therefore increases until it is sufficiently high to form a continuous layer of protective aluminium oxide. This layer is highly impermeable to copper ions, which can no longer enter the cuprous oxide layer to increase its thickness. The higher the aluminium content of the alloy, the quicker will the oxidation rate of copper be reduced (see Figure 28.2). This type of protection may not occur at low temperatures. Aluminium oxide is still formed but it tends to form as particles within the Cu_2O layer.

There is a similar behaviour in copper–beryllium alloys where a protective layer of beryllium oxide forms at elevated temperatures. The principle of protection against oxidation by formation of a continuous oxide layer also applies to stainless steels where protection is afforded by chromic oxide, Cr_2O_3, in all materials containing more than 12 per cent Cr. In ferrous materials containing several per cent of silicon the oxide that forms is a double oxide of iron and silicon, namely a silicate. The silicate layer has a glass-type structure, is relatively impervious to the passage of ions and is largely protective.

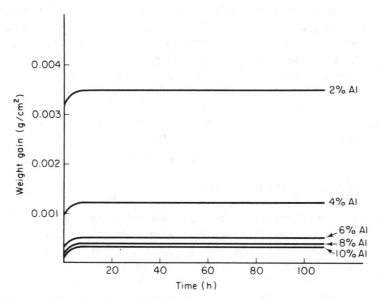

FIGURE 28.2   Oxidation rate of Cu–Al alloys at 800°C

28.2 Degradation of polymers

When polymeric materials are heated, there is a tendency for them to depolymerise, or *degrade*. These degradation reactions commence at comparatively low temperatures, and for this reason, accurate control of temperature is necessary in the plasticising cylinders of injection moulding and extrusion equipment. Polymers that are formed by condensation reactions involving the formation of water may depolymerise if heated for lengthy periods in an atmosphere containing steam. In a dry atmosphere they will tend to degrade when heated to a fairly high temperature, but in this case the reaction is essentially charring, with carbon formed as one of the degradation products.

The term *degradation* is used to refer to any change in a polymer which leads to some breakdown of its chemical structure or, in other words, causes it to depolymerise. This may be brought about by oxidation, hydrolysis, namely a reaction with water, heat, radiation, stress, or any combination of these. Most thermoplastics and elastomers can be affected by these factors and the stability of the materials is improved by incorporating antioxidants and stabilisers into the polymer moulding compounds.

Many polymers will age slowly in air with a consequent deterioration in some properties, particularly toughness and, in the case of transparent materials, optical clarity. This is the result of oxidation. The process is accelerated in the presence of sunlight, especially if there is radiation in the ultra-violet sector of the spectrum. This is termed *photo-oxidation*. The high energy incident photons stimulate the chemical reactions involved in the degradation. An increase in temperature increases the rate of degradation in the presence of oxygen, but heat alone can also bring about polymer degradation. Ozone is more reactive than oxygen and a number of elastomer materials, including natural rubber and SBR, are attacked by ozone in the atmosphere. The ozone attacks unsaturated C=C bonds in the molecular chains leading to a splitting of the chains.

The term *weathering* is used to describe the deterioration of polymer materials in an outdoor environment. Weathering includes the effects of water, together with oxygen, heat and ultra-violet radiation, and leads to general deterioration in appearance and properties with possible embrittlement and cracking together with degradation.

Environmental stress cracking (ESC) is a phenomenon associated with some polymer materials when stressed and in contact with certain active substances. For example, polyethylene is subject to this form of cracking if, when stressed, it is in contact with a surface-active substance such as an aqueous solution containing a detergent. Other materials which are prone to ESC are polystyrene, ABS and UPVC.

28.3 Oxidation of ceramics and composites

Very many ceramic materials are pure or mixed oxides and, as such, do not oxidise as they are already in a fully oxidative state. However, carbide and nitride ceramics can oxidise to some extent if heated in air at temperatures in excess of 1000°C. At high temperatures, in the presence of atmospheric oxygen, a layer of oxide can form on the surface of a ceramic. In the case of silicon carbide or silicon nitride, the oxide layer formed on the surface is silica, SiO_2. The exact nature of the oxide formed is determined by the composition of the carbide or nitride ceramic. Also, the composition and nature of the oxide layer is affected by presence of other constituents, for example, any sintering additives used to regulate the type of grain boundary phases.

The thermosetting resins used as matrix materials for many fibre reinforced composites are highly resistant to oxidation and chemical attack and, thus, protect the fibre reinforcement against environmental conditions. However, some composite materials designed for use at high temperatures may suffer from oxidising effects. Examples are nickel, reinforced with carbon fibre and ceramic matrix composites. In the former material, oxygen can diffuse through the nickel matrix to the fibres while, in the latter example, the presence of microcracks in the ceramic matrix will permit the ingress of oxygen to the fibres.

28.4 Corrosion

A metal in contact with an electrolyte ionises to a small extent, a typical reaction being

$$Zn \rightleftharpoons Zn^{2+} + 2e^-$$

The free electrons produced by the forward reaction remain in the metal and the ions pass into the electrolyte. At equilibrium, therefore, the metal will possess a certain electrical charge. Different metals ionise to differing extents and will possess different values of electrical charge when in equilibrium with their ions. The potential difference between a metal electrode and a standard hydrogen electrode is termed the *Standard Electrode Potential** for the metal. Table 28.1 gives the standard electrode potentials for some metals.

Table 28.1 *Standard Electrode Potentials*

Base metals	Metal	Ion	Electrode potential V	Anodic
	Sodium	Na^+	−2.71	
	Magnesium	Mg^{2+}	−2.38	
	Aluminium	Al^{3+}	−1.67	
	Zinc	Zn^{2+}	−0.76	
	Chromium	Cr^{2+}	−0.56	
	Iron	Fe^{2+}	−0.44	
	Cadmium	Cd^{2+}	−0.40	
	Cobalt	Co^{2+}	−0.28	
	Nickel	Ni^{2+}	−0.25	
	Tin	Sn^{2+}	−0.14	
	Lead	Pb^{2+}	−0.13	
	Hydrogen	H^+	0.000	
	Copper	Cu^{2+}	+0.34	
	Mercury	Hg^{2+}	+0.79	
	Silver	Ag^+	+0.80	
	Platinum	Pt^{2+}	+1.20	
Noble metals	Gold	Au^+	+1.80	Cathodic

Much metallic corrosion is of the galvanic or electrochemical type. If two dissimilar metals, for example copper and zinc, are connected via an electrolyte and are also electrically connected external to the electrolyte (Figure 28.3) a galvanic cell will be set up. The potential difference of this cell will be 1.10 V, with the zinc being the anode and copper being the cathode of the cell. At the anode, the reaction $Zn \rightleftharpoons Zn^{2+} + 2e^-$ will be thrown out of equilibrium, as the electrons

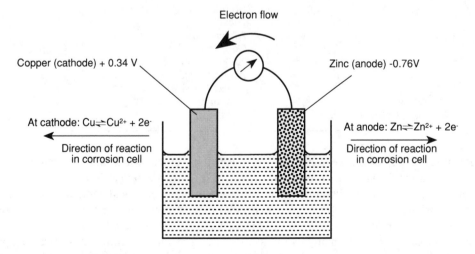

Electron flow

Copper (cathode) + 0.34 V

Zinc (anode) -0.76V

At cathode: $Cu \rightleftharpoons Cu^{2+} + 2e^-$

Direction of reaction in corrosion cell

At anode: $Zn \rightleftharpoons Zn^{2+} + 2e^-$

Direction of reaction in corrosion cell

FIGURE 28.3   Galvanic cell between copper and zinc

* Standard Electrode Potentials are determined with electrodes immersed in a Normal solution of hydrogen ions (1 gramme equivalent of hydrogen ions per litre).

produced in the reaction flow through the external circuit to the copper, and ionisation of zinc will be accelerated in an attempt to restore equilibrium, that is, the zinc will be dissolved or corroded. Conversely, at the cathode, the ionisation of copper will be suppressed owing to the constant arrival of free electrons from the anode. Corrosion cells may be established in many ways in practice.

The order in which the various metals are placed, in Table 28.1, is termed the electromotive series. It should be noted, however, that the potential difference between metals in equilibrium with their ions may differ from the standard values if the metals are not immersed in a standard 'normal' electrolyte but in some other electrolyte, for example, sea water. Also, the order of metals in the series may differ from the order of the standard electromotive series when metals are immersed in other electrolytes. The order of metals, when determined for some electrolyte other than a normal solution of hydrogen ions, is termed a galvanic series. The galvanic series for some metals and alloys in sea water is given in Table 28.2.

Usually, in a galvanic cell, it will be the anode material which is corroded and the cathode material is protected against corrosion. There are a few exceptions to this and cathodic corrosion, a type of corrosion associated with the formation of alkaline reaction products, may occur in some instances.

Table 28.2 *Galvanic series in sea water*

Anodic end:	Magnesium
	Magnesium alloys
	Zinc
	Galvanised steel
	Aluminium
	Cadmium
	Mild steel
	Wrought iron
	Cast iron
	Austenitic stainless steel (active)
	Lead
	Tin
	Muntz metal
	Nickel
	Alpha brass
	Copper
	70/30 Cupronickel
	Austenitic stainless steel (passive)
	Titanium
Cathodic end:	Gold

Galvanic, or corrosion, cells may be established in several ways, and the main types of galvanic cells are: (a) *composition* cells; (b) *stress* cells; (c) *concentration* cells. An obvious example of a composition cell is two dissimilar metals that are in direct electrical contact and also joined via a common electrolyte. For this reason, dissimilar metals should not be in contact in situations where corrosion conditions may prevail. It is also possible to have composition cells formed within one piece of material. In any two-phase alloy, or a metal containing impurities, one phase may be anodic with respect to the other and galvanic microcells may be established. Examples of this are pearlite in steels, and precipitate particles of chromium carbide in austenitic stainless steels. In pearlite, the ferrite is anodic with respect to cementite, and in austenitic stainless steels the grain boundary precipitate of chromium carbide, which is formed if the steel is incorrectly heat treated, causes the austenite solid solution adjacent to the precipitate particles to be depleted in chromium. These low-chromium zones are anodic with respect to the normal composition austenite. This latter fact is responsible for the weld decay type of corrosion referred to in section 16.14. From the above, it follows that, normally, pure metals of single-phase alloys will possess better resistance to corrosion than impure metals or multi-phase alloys.

Atoms within a highly stressed metal will tend to ionise to a greater extent than atoms of the same metal in an annealed condition. Consequently, the stressed material will be anodic with respect to the unstressed metal. A stress type of galvanic cell may be established in a component, or in a structure, where the stress distribution is uneven. (This also applies to an uneven distribution of residual stresses with a cold worked metal). Some alloys are particularly prone to stress corrosion failure, which is often a grain-boundary (intercrystalline) type of corrosion. This may occur in cold worked α brasses, if they are not properly stress relieved. Randomly arranged grain-boundary atoms tend to ionise more rapidly than atoms within a regular crystal lattice, and so a grain boundary tends to be anodic with respect to a crystal grain. Much corrosion is of an intercrystalline nature. It also means that a coarse-grained metal tends to possess a better resistance to corrosion than a fine-grained sample of the same metal.

When a metal is in contact with a concentrated electrolyte solution it will not ionise to as great an extent as when it is in contact with a dilute electrolyte. In other words, if one piece of metal is in contact with an electrolyte of varying concentration, those portions in contact with dilute electrolyte will be anodic with reference to portions in contact with more concentrated electrolyte. Galvanic cells of the concentration type may be encountered in situations involving flowing electrolytes in pipes and ducts. Another type of concentration cell can occur if there are variations in dissolved oxygen content throughout an electrolyte. But, before considering the oxygen concentration cell, let us consider the various reactions which can take place at anodes and cathodes.

At the anode:

(a) metal may go into solution

$$M \rightarrow M^+ + e^- \tag{28.1}$$

(b) anions, such as $(OH)^-$, will be attracted to the anode giving rise to the formation of an oxide, hydroxide, or an insoluble salt

$$2M + 2(OH)^- \rightarrow M_2O + H_2O + 2e^- \tag{28.2}$$

$$M + (OH)^- \rightarrow MOH + e^- \tag{28.3}$$

$$M + X^- \rightarrow MX + e^- \tag{28.4}$$

At the cathode:

(a) metal may be deposited

$$M^+ + e^- \rightarrow M \tag{28.5}$$

(b) hydrogen may be evolved

$$2H^+ + 2e^- \rightarrow H_2 \tag{28.6}$$

(c) oxygen may be reduced

$$O_2 + 2H_2O + 4e^- \rightarrow 4OH^- \tag{28.7}$$

A most important type of galvanic corrosion cell is the oxygen concentration cell in which the electrolyte has variations in its dissolved oxygen content. Metal in contact with electrolyte low in dissolved oxygen will be anodic with respect to metal in contact with electrolyte rich in dissolved oxygen. A principal reason for this is that in an area with water rich in oxygen the following reaction will take place: $2H_2O + O_2 + 4e^- \rightarrow 4(OH)^-$. This is a cathode-type reaction, in that it absorbs electrons. The electrons will be supplied from another part of the metal where there is less oxygen and so such areas act as anodes. This is one of the main features of the *rusting* of iron.

Rust is the hydrated oxide of iron, $Fe(OH)_3$. Both oxygen and water are required for the formation of rust and iron and steels will not rust in a dry atmosphere or when immersed in oxygen-free water.

Consider the situation with a drop of water on the surface of a piece of steel (Figure 28.4). There will be a greater amount of oxygen available near the edge of the water drop and this area will become cathodic with hydroxyl ions formed according to the reaction: $2H_2O + O_2 + 4e^- \rightarrow 4(OH)^-$. At the anode area under the centre of the drop electrons will be generated by the reaction: $Fe \rightarrow Fe^{2+} + 2e^-$. The Fe^{2+} cations will move through the electrolyte toward the cathode and the $(OH)^-$ anions will tend to move toward the anode. The very much smaller cations have much greater mobility than the large hydroxyl ions and the two ion types meet close to the cathode areas forming insoluble ferrous hydroxide, $Fe(OH)_2$, which settles as a deposit near the cathode area. This subsequently oxidises to rust, $Fe(OH)_3$.

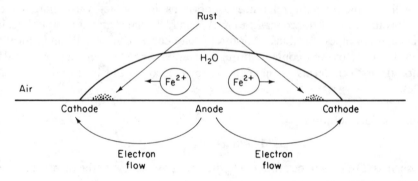

FIGURE 28.4   Oxygen concentration cell forming rust

This type of galvanic cell can form in many situations and can give rise to severe effects. Waterline corrosion is a good example of this. The oxygen concentration at a water/air interface will be higher than at lower levels in the water, making the steel at the waterline cathodic and with anode areas existing just below the water surface. The most rapid corrosion will take place just below the waterline with solution of the anode areas, while the greatest rust deposit will form at the waterline.

Once a rust deposit has formed on the surface of a steel, the area under the porous deposit will become oxygen deficient, and hence anodic, compared with bare metal. This means that dissolution of iron can continue, unobserved, below the covering of rust. Similarly, any covered or hidden area to which moisture as electrolyte can gain admittance, for example, a dirt-covered surface, a crack or crevice, or the gap between overlapping rivetted plates, can become an anode (Figure 28.5). A water-tight caulking of seams, in the latter case, can minimise this type of corrosion.

Corrosion due to this type of cell will be more severe in salt water than in fresh water as the presence of other dissolved ions greatly increases the conductivity of the electrolyte.

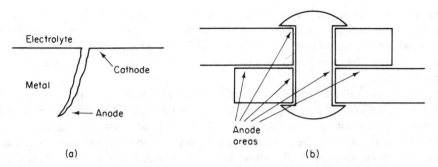

Electrolyte

Cathode

Metal

Anode

Anode
areas

(a)

(b)

FIGURE 28.5   Formation of anode and cathode areas in (a) crack and (b) rivetted joint

The extent to which a metal corrodes depends not only on its electrode potential relative to other materials present but also on the nature of the electrolyte and in particular on its pH value. Many different types of reaction may occur during corrosion. A reaction may be of the type that involves only electrons as, for example, $Fe^{2+} \rightarrow Fe^{3+} + e^-$. This type of reaction will be dependent on potential. A second type of reaction may involve hydrogen ions, but not electrons as, for example, $Fe^{2+} + H_2O \rightarrow Fe(OH)_2 + 2H^+$. Reactions of this type will be dependent on the pH of the electrolyte. Both electrons and hydrogen ions may be involved in a reaction, such as for example, $Fe^{2+} + 3H_2O \rightarrow Fe(OH)_3 + 3H^+ + e^-$. This type of reaction will be dependent on both potential and pH value. *Pourbaix* collected thermodynamic data on reactions between metals and water, together with data on the solubilities of oxides and hydroxides. All this information is combined to produce Pourbaix diagrams which indicate stable situations as a function of electrode potential and pH value. A simplified Pourbaix diagram for iron is shown in Figure 28.6.

The two dotted lines in the diagram are (i) the hydrogen line, below which hydrogen will be evolved, and (ii) the oxygen line, above which oxygen will be evolved. The electrode potential of iron in water of pH 7 is given by point X in Figure 28.6, and it will be noted that this is within the corrosive zone. The diagram indicates ways in which the corrosion of iron may be reduced. By making the electrolyte alkaline, say to pH 9, the iron is moved into the passive region. One of the functions of inhibitors is to control pH values and maintain a value that will keep the metal in a non-corrosive regime. Another way would be to reduce the electrode potential value while

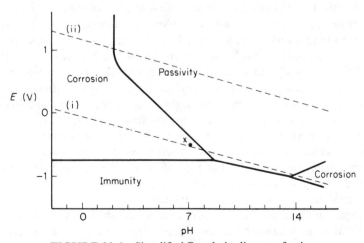

FIGURE 28.6   Simplified Pourbaix diagram for iron

keeping the electrolyte pH value constant. This would place the iron within the region of immunity on the diagram. This is the principle of cathodic protection which is achieved using an anodic metal such as aluminium in conjunction with iron. Thirdly, the iron could be placed in the region of passivity by increasing the potential and this could be done by applying a suitable external e.m.f.

Once a galvanic cell has been established there is a tendency for *polarisation* to occur, namely a reduction in the current flowing within the corrosion cell. When an anode starts dissolving, metal ions go into solution in the electrolyte. If the diffusion rate of these ions away from the anode area is relatively slow the concentration of such ions in the anode area increases and exerts a back-e.m.f. effect, suppressing the ionisation of the anode material. A different effect occurs at the cathode. Here cations are neutralised by electrons and the concentration of cations in the vicinity of the cathodes is greatly reduced if ion diffusion rates are low. Both anode and cathode effects combine to reduce the current flow in the cell and, hence, the corrosion rate. Since such polarisation is controlled by diffusion rates it follows that any change that will increase ion diffusion, such as an increase in temperature, an increase in concentrations or agitation of the electrolyte, will reduce the polarisation effect and maintain high rates of corrosion.

Passivation is the reduction in corrosion rate that occurs when products of corrosion reactions form a largely impervious layer over an anode surface. Passivity is due to the formation of a very thin impervious oxide layer on the surface and this often occurs in highly oxidising conditions. Materials such as aluminium, titanium and stainless steels readily form passive films but these may be destroyed in a reducing atmosphere and the metals will then become active for corrosion.

Corrosion of a metal can easily lead to failure. When the effective thickness of the metal is reduced the level of stress within the material is increased. This increase could accelerate corrosion in a localised area until the level of stress is sufficient to cause failure of the component or structure. The following paragraphs contain a brief description of some of the types of corrosion that could lead to early failure.

Pitting corrosion is a highly localised form of corrosion and is usually associated with a local breakdown of an otherwise protective film. Pits can start where there is a discontinuity in the passive oxide layer and they can also occur, owing to a differential aeration effect, where an external deposit has settled on a metal surface. Frequently the source for corrosion pits is a non-homogeneous metal and impurity inclusions within the structure may be the cause of a break in the oxide coating. A corrosion product often forms over the surface of a pit, but corrosion continues beneath this. Pitting can occur on any metal but it is much more common with metals and alloys that possess a high-resistance passive film. The presence of the film tends to prevent the spread of corrosion to a wider area and attack is concentrated in the series of small localities. Stainless steels are susceptible to this form of corrosion in the presence of chloride solutions.

It has already been stated that a metal is more susceptible to corrosion when it is highly stressed. *Stress corrosion* is a phenomenon in which there is the initiation of a crack and then propagation of the crack through the material as the result of the combined effects of a stress and a corrosive environment. Stress corrosion is generally found only in alloys, not in pure metals, and it occurs only in certain specific environments, for example, the cracking of brass in ammonia or the cracking of aluminium alloys in chloride solutions. The stress within the material must have a tensile component, and both stress and specific corrosive environment must be present for cracks to form and propagate. The stress corrosion cracks are generally of an intercrystalline type although transcrystalline cracking occurs in some cases.

Corrosion fatigue failure is the development and propagation of cracks in a material that is subjected to alternating or fluctuating cycles of load. The presence of a corrosive environment, even if it is only mildly corrosive, will accelerate the formation and growth of fatigue cracks, thus reducing the fatigue life of the material.

Impingement attack is a form of corrosion attack found in marine condensers and occurs when large bubbles in the turbulent cooling fluid burst on the tube surface. The effect of this is to produce large horseshoe-shaped pits. Where a large bubble impinges at high velocity it will remove any corrosion products that are causing a polarising effect, and hence re-activate the corrosion cell. The bursting of a bubble will also provide a good supply of oxygen. The problem may be minimised in two ways. Careful design to reduce turbulence to a minimum and ensure that any bubbles formed will be small is necessary. Also, the materials used in the construction of marine condensers must be highly corrosion resistant. Cupronickel alloys are about the best materials available for this application.

Cavitation attack occurs on ships' screws and steam-turbine blades and it can result in the formation of pits. As the blade moves through the fluid air bubbles will be formed at, or close to, those parts of the blade where there is a low fluid pressure. The later collapse of these bubbles at the surface of the blade will produce a compression wave. There is evidence to show that the pitting that occurs is due to a combination of the mechanical effect of collapsing bubbles and the chemical corrosion effect of aerated cell formation, as in impingement attack. Selection of suitable corrosion-resistant alloys and careful attention to blade design will minimise the problem.

Fretting occurs as a result of the minor amounts of relative motion between two parts in an assembly. Fretting is often associated with fatigue. When the surfaces abrade each other the very fine metal debris formed rapidly oxidises and this oxide debris can itself act as an abrasion agent. As the process continues the metal surface gradually wears away, forming more fine oxide. With steels the corrosion product is a fine brown powder known as 'cocoa'. The continued wastage of metal results in severe wear and may proceed to complete failure. Fretting is more pronounced in a dry atmosphere. The presence of moisture tends to reduce fretting, the moisture possibly acting as a lubricant.

28.5 Corrosion protection

Many methods are used to minimise the possibility of corrosion and these may be grouped into the following categories: (1) the use of protective coatings, (2) cathodic protection, and (3) the use of inhibitors. There are many types of protective coating available ranging from paints and bituminous coatings, through plastic and vitreous enamel coatings to a variety of metallic coatings applied by hot dipping, electroplating, metal spraying or other methods. Also, by careful material selection and attention to design the formation of composition-type galvanic cells may be avoided.

The object of a protective coating is to isolate the metal from the corrosive environment. One of the more common forms of protection is a paint layer. Paints are easily applied, either by brush or spray, but are seldom fully effective. Most paints age and weather, particularly in strong sunlight, and the layer either becomes porous or flakes off the surface. Also paint surfaces are fairly soft and may be easily scratched and damaged. Once oxygen and moisture can reach the metal surface, corrosion can begin and it can even occur beneath portions of the paint-covered surface.

Metal parts are sometimes covered with a thermoplastic coating. This is frequently used for the protection of steel wire trays, as used in refrigerators and freezers. The heated-metal assembly is dipped into a container filled with powdered thermoplastic. The plastic adjacent to the dipped part melts and forms a continuous protective layer over the component.

Vitreous enamelling is used for the protection of metal parts that will be subjected to moderate heating during their service life, for example, gas and electric cookers. An enamel coating is hard but brittle. It is a wear-resistant coating but it is easily damaged by mechanical shock treatment.

Metallic coatings are very widely used for the protection of metals, particularly steels. The metal coating could be of a metal that is anodic with respect to the steel such as aluminium, zinc or chromium but is itself resistant to corrosion because of a passive oxide film. This type of coating may still afford some protection to the steel even if there are breaks in the coating, as the coatings will be anodic and will corrode in preference to the cathodic steel. Other coatings, such as tin, are cathodic with respect to steel and if the coating is broken the exposed steel will become the anode and corrosion will take place rapidly (Figure 28.7). Cadmium, which has an electrode potential value very close to that of iron, is very often used as a protective plated layer on parts such as nuts, screws and bolts. The plated layer could be broken by abrasion during the tightening of a nut on a bolt, but if this happens the e.m.f. of a galvanic cell will be small (0.04 V) and the rate of corrosion will be very low.

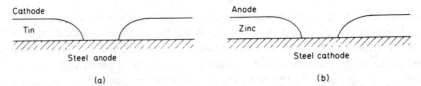

FIGURE 28.7   Effect of break in protective coating on steel. (a) Tin (cathodic) coating. Exposed steel becomes anodic and is corroded. (b) Zinc (anodic) coating. Coating is anodic and corrodes, thereby protecting the steel

Cathodic protection is widely used to protect structures in which the effects of severe corrosion could be serious or where the renewal of a corroded structure could prove very expensive. It is particularly useful for the protection of ships' hulls, steel piers and harbour installations, buried steelwork in building foundations and buried pipe-lines. A block of a highly anodic material, magnesium, aluminium or zinc, is placed adjacent to the structure to be protected and connected electrically. A galvanic cell is deliberately created and the sacrificial anode will corrode, protecting the structure. The sacrificial anodes are inspected periodically and renewed when necessary. For the protection of the hulls of ships the anode material is strapped to the hull near the stern and below the waterline (Figure 28.8). A variant on the use of a sacrificial anode of another metal is to use an impressed voltage. The steelwork will be protected as long as it receives a steady supply of electrons. When using a direct-current supply as a means of protection it is common to bury scrap iron or steel adjacent to the structure, with connections that render the scrap metal anodic and the steel requiring protection cathodic.

Underground pipework is usually given a thick bituminous coating before installation but this coating may be damaged during pipe laying and cathodic protection is used to minimise corrosion of any exposed surface. Uncoated pipes are not used as the very large surface area needing protection would result in too rapid a solution of sacrificial anode material. The risk of corrosion will vary considerably with soil type. In low-risk conditions anodes may be widely separated but in high-risk conditions anodes would be placed closer to one another. Corrosion would be minimal in dry sandy soils but there could be a high risk in wet clays. The presence of

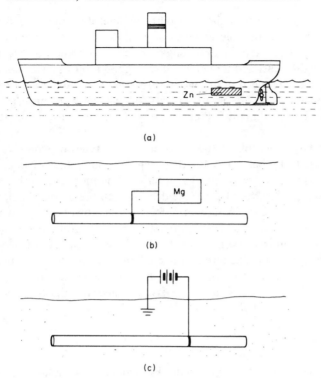

FIGURES 28.8   Cathodic protection. (a) Protection of ship's hull below waterline. (b) Protection of buried pipe-line with magnesium anode. (c) Protection of buried pipe-line using impressed voltage

bacteria in soils or clays with a sulphate content could result in the breakdown of sulphates and the liberation of oxygen, and this could lead to the establishing of oxygen concentration cells. There is also a major corrosion risk at the junction of two different soil types.

Inhibitors and *passifying agents* are widely used for the reduction of corrosion in closed cooling systems where cooling water is recirculated. They may take the form of pH regulators or they may help to form a protective film on metal surfaces within the system. Chromate salts are frequently used as inhibitors within car radiator systems. Any iron ions generated at a steel-anode surface will combine with chromate ions to form an insoluble iron chromate which will create a protective coating over the steel surface. Certain phosphates and tungstates may also be used as inhibitors.

28.6 *Microbial attack*

Micro-organisms do not attack metals or ceramics and glasses. Thermoplastic and thermosetting materials are highly resistant to attack by micro-organisms, but some thermoplastics, because of their very low hardness, may suffer attack from borers. Some materials of animal or vegetable origin, including casein and some cellulose derivatives, are subject to some forms of bacterial or fungal attack. One of the few synthetic materials which may suffer microbial attack is PPVC and in this it is the plasticiser which is affected rather than the PVC itself. This high resistance of plastics materials to organisms poses a major problem from the point of view of waste disposal and research is taking place into the possibility of microbial methods for the breakdown of plastic refuse. Some bio-degradable varieties of plastics materials are currently available.

Timbers are subject to bacterial and fungal attack, and numerous organisms feed on moist wood fibre. Thoroughly seasoned timber is not such good food value as moist wood, and the rate of attack is very greatly reduced in dry wood. Preservatives may be used to retard the decay of timber. A common and cheap preservative is creosote, as this is poisonous to the wood-attacking organisms. There are also many other wood preservatives available that contain specific fungicides.

There are also larger organisms that feed on certain types of wood. A number of types of beetle cause damage to timber. They lay their eggs on the surface or in small crevices and the grubs that hatch tunnel their way into the interior. The most common types that affect well-seasoned timber are the furniture beetle, the death-watch beetle and the house longhorn beetle. The grubs of the furniture beetle, woodworm, feed on the sapwood of softwoods used in building structures, as well as in furniture. The death-watch beetle is common in old hardwood timber, particularly if moist or subject to some fungal decay. In some parts of the world, attack on timber by termites can be a problem. The teredo is a small marine animal that attacks timber immersed in sea water, and the lower hulls of old sailing ships were coated with copper sheeting to afford protection, hence the use of the phrase 'copper-bottomed' to refer to a sound proposition.

28.7 *Effects of radiation*

Radiation may have an effect on materials. Electromagnetic radiation, of various frequency ranges, will have an effect on some materials. A good example of this is the effect of radiation in the infra-red, visible light, ultra-violet, and X-ray ranges on photographic emulsions. The radiation causes dissociation of silver from the silver salts in the emulsion. Similarly, visible light, ultra-violet light, or X and γ radiation will stimulate numerous chemical processes. An example is the effect of X or γ radiation on a linear polymer, such as polyethylene. The radiation increases the energy of the polymer and promotes branching and cross-linking reactions, thus producing thermoplastics of greater rigidity. This process is utilised in the production of some plastic components. Bombardment of a thermoplastic with electrons (β rays) will also produce similar effects. The effects of high-frequency electromagnetic radiation on some compounds is to excite electrons to higher states of energy. When the excited electrons subsequently fall back to lower energy states, energy is emitted, often in the visible-light range, producing fluorescence and phosphorescence. A similar effect is noticed when some compounds are subjected to β radiation (electrons), and this is the principle used in cathode ray tubes. High-frequency radiation or electron bombardment will not affect metals, but the radiation will be partially absorbed by the material. A considerable amount of radiation will pass through the metal, and this principle is utilised in the radiography of metals and in electron transmisson microscopy. Because the wavelengths of X and γ radiation approach the sizes of individual atoms, there will also be diffraction effects and diffraction patterns are used in the identification and analysis of crystals.

The irradiation of a material with neutrons is invariably damaging. A neutron is a small particle possessing no electrical charge, and a fast neutron will not be attracted to or deflected by charged particles, such as electrons and protons. A fast neutron may travel through a large number of atoms before it collides with an atomic nucleus. When such a collision occurs the atom, or ion, that has been hit is displaced from its normal position, and the path of the neutron will be deflected. A neutron will lose some energy at each collision and, eventually, a lower-energy neutron will be 'captured' by an atomic nucleus. When a nucleus captures a neutron, the nucleus will become unstable and will emit α, β, or γ radiation, and in doing so may be changed into a radioactive isotope of another element (refer to Section 2.10).

The irradiation of polymers with neutrons at low neutron flux densities will cause some cross-linking and branching to occur, in a similar manner to activation by X rays, but at higher neutron flux densities the effect of neutron collisions is to cause degradation of polymers. The effect of a fast neutron colliding with the nucleus of an atom in a metallic crystal lattice is to cause the atom to be knocked out of its equilibrium position within the lattice. The knocked-out atom moves into an interstitial lattice position, leaving behind a vacancy. The lattice defects created by neutron collisions make the material harder and increase the yield strength. They also make the material more brittle. In the case of those metals that show ductile-brittle transition behaviour, the transition temperature is increased. (The neutron irradiation of steels will raise ductile-brittle transition temperatures by up to 100°C.) This type of radiation damage may be removed by annealing the material. Irradiation with slow neutrons will give a different type of radiation damage. Slow neutrons may be captured by atomic nuclei and the irradiated metal will become radioactive or 'hot'. Also the transmutation product may alloy with, or react chemically with, the other atoms of the metal.

28.8 Questions

28.1 Estimate the thickness of the oxide film formed on the surface of a metallic alloy assuming growth according to a parabolic law, $x^2 = kt$, under the following conditions: (a) at 350°C after 1 hour, after 150 hours and after 1000 hours; (b) at 600°C after 1 hour, 150 hours and after 1000 hours. Assume that k varies with temperature according to $k = \exp(-16 \times 10^3/T)$ cm^2/s where T is temperature (K).

28.2 Define the Pilling–Bedworth ratio and discuss its significance in relation to the degree of protection offered to a metal surface by an oxide layer.

Cadmium oxidises to the oxide, CdO. Given the following data, deduce whether or not the oxide layer formed is likely to be protective: M for Cd = 112.4, M for O = 16.00, ρ for Cd = 8640 kg/m^3, ρ for CdO = 8100 kg/m^3.

PART VI
EVALUATION
OF
MATERIALS

29

Property Testing

29.1 Introduction

Testing is an essential part of any engineering activity. Inspection and testing must take place at many stages in the complex process of producing engineering materials, be they metals, polymers, ceramics or composites, and during the forming of these materials into components and assembling the components to create an engineering product to satisfy some specific requirement. The requirement for testing does not automatically cease when the product has been manufactured. It is frequently necessary to check and test the article during its service life in order to monitor changes, such as the possible development of fatigue or corrosion damage.

The types of test used can be broadly classified into two categories:

(a) tests to establish the properties of the material, and
(b) tests to determine the integrity of the material or component.

Those tests in the first category are generally of a destructive type. They are performed on samples of a material and the test-piece is damaged or broken in the process, as is the case when determining the tensile strength of a material in a tensile test to destruction. If the sample test-piece is correctly chosen and prepared, the results should be indicative of the properties of the bulk material represented by the sample.

The tests in the second category are of a non-destructive nature and are used to detect the presence of internal or surface flaws in a material, component or finished product. By their very nature, these tests do not damage the parts being tested and sampling is not required as, if necessary, every item can be checked. The tests of this type are described in Chapter 30.

Many test procedures have been devised for the determination of some of the characteristic properties of materials, for example, the tensile test. For the results of any test to have value, it is important that the tests be conducted according to certain set procedures. To this end, standardised test procedures have been evolved. In the United Kingdom, it is the British Standards Institution which publishes the standards and codes of practice which cover most aspects of the testing and utilisation of materials. All the major developed countries possess standards organisations and while standards for a particular type of test may differ slightly on points of detail from one country to another, the broad principles will be similar.

29.2 Hardness tests

The hardness of a material may be determined by using either a scratch test, or by making a surface indentation. This latter type of test may be used only for materials that are capable of plastic deformation, namely metals and thermoplastic materials. The resistance of a material to indentation is not necessarily the same as its resistance to abrasion, but a hardness measurement obtained from an indentation test can be used as an empirical check for abrasion resistance. In general, metals possessing high hardness will have a high resistance to abrasive wear. Indentation-type hardness tests are widely used for the checking of metal samples as they are easy to make and yield information on heat treatment condition. There are also empirical relationships between the hardness of a metal and its tensile strength.

The hardness of a ceramic material may be determined using a scratch test. Mohs' scale of hardness, which was devised for assessing the relative hardness of minerals and rock, is based on ten naturally occurring minerals ranging from very soft talc to diamond (Table 29.1). In this test an attempt is made to scratch the surfaces of the standards with the material under test. The hardness of the unknown lies between the number of the mineral it just fails to scratch and that of the mineral it just scratches.

Table 29.1 *Mohs' scale of hardness*

Number	Mineral
1	talc
2	gypsum
3	calcite
4	fluorite
5	apatite
6	orthoclase felspar
7	quartz
8	topaz
9	corundum
10	diamond

Of the various types of indentation test available for the hardness testing of metals the most commonly used are the Brinell test, the Vicker's diamond test, and the Rockwell test.

The Brinell test

The Brinell test was the first static indentation test to come into general use. In its original form, it utilised a hardened steel ball indentor of 10 mm diameter forced into the surface of the metal being tested under a static load of 3000 kg (29.43 kN) and the load maintained for 10 to 15 seconds. A load of 3000 kg, in conjunction with a ball indentor of 10 mm diameter, is suitable for use with steels and cast irons. For softer non-ferrous metals and alloys, lower values of static load are used (see Table 29.2). The diameter of the resulting impression is then measured with the aid of a calibrated microscope. The Brinell hardness number, H_B, is given by:

$$H_B = \frac{\text{applied load (kg)}}{\text{surface area of the impression (mm}^2)}$$

The hardness number obtained from a Brinell test is not independent of the load used and if two tests are made on the same material, one using a large static load and one using a small load the hardness value obtained may differ. The recommended values of F/D^2, where F is the load in kg and D is the diameter of the ball indentor in mm, to be used in the testing of various materials are given in BS 240 (1986) and ASTM E 10-84 (see Table 29.2).

Table 29.2 *Recommended ratios of F/D^2 for the Brinell hardness test on various materials*

Material	F/D^2
Steels and cast irons	30
Copper alloys and aluminium alloys	10
Pure copper and aluminium	5
Lead, tin and tin alloys	1

After making a Brinell hardness impression, the parameter that is measured is the diameter, d. Accurate Brinell hardness values will only be achieved if the impression diameter, d, is between the limits of 0.25 D and 0.5 D, where D is the diameter of the ball. The surface area of an impression of diameter, d, made by a spherical indentor of diameter, D, is given by:

$$\text{Area} = \pi D/2 \{D - \sqrt{(D^2 - d^2)}\}$$

and so

$$H_B = \frac{2F}{\pi D\{D - \sqrt{(D^2 - d^2)}\}}$$

The Brinell test is not suitable for the testing of very hard materials. As the hardness of the metal approaches that of the ball indentor, there will be a tendency for the indentor to deform. The Brinell test will produce reliable results up to H_B values of around 400 and it is not recommended that the test be used on metals which would give values of H_B greater than 500. A comparison between H_B values and hardness values obtained from other tests is given in Table 29.4. H_B values and Vickers diamond hardness values, H_D, are comparable up to $H_B = 300$ but for higher hardness values there is a divergence between the two scales, with the H_B values being less than the H_D values for a given material.

The Vickers diamond test

In the Vickers diamond test, the indentor used is a pyramidal shaped diamond, and as in the Brinell test, the indentor is forced into the surface of the material under the action of a static load for 10 to 15 seconds. As in the Brinell test, the Vickers Diamond hardness number, H_D, is given by:

$$H_D = \frac{\text{applied load (kg)}}{\text{surface area of impression (mm}^2)}$$

The Vickers diamond hardness test is covered by BS 427 Part 1 (1981) and ASTM E 92-82. The standard indentor is a square pyramid shape with an angle of 136° between opposite faces. One advantage of the Vickers test over the Brinell test is that the square impressions made are always geometrically similar, irrespective of size. The plastic flow patterns, therefore, are very similar for both deep and shallow indentations and, in consequence, the hardness value obtained is independent of the magnitude of the indenting force used.

After an impression has been made, the size of the impression is measured accurately using either a calibrated microscope or by projecting a magnified image of the impression onto a screen and measuring this image. Both diagonals of the impression are measured and the mean value of D, the diagonal length, is used in the determination of the hardness number.

$$H_D = \frac{2F \sin \theta/2}{D^2} \quad \text{where } \theta = 136°$$
$$H_D = \frac{1.8544F}{D^2}$$

where F is the applied load in kg and D is the mean diagonal length in mm.

Convenient loads to use for some common materials are:

steels and cast irons	30 kg
copper alloys	10 kg
pure copper, aluminium alloys	5 kg
pure aluminium	$2\frac{1}{2}$ kg
lead, tin, tin alloys	1 kg

The Rockwell test

The Rockwell test machine is a rapid action direct reading machine. This provides a very convenient method for speedy comparative testing. In this test, or rather series of tests, it is the depth of the impression which is measured and directly indicated by a pointer on a dial calibrated, inversely, into 100 divisions (1 scale division = 0.01 mm of impression depth). Consequently, a low scale number indicates a deep impression, hence a soft material, and vice versa. There are a series of Rockwell hardness scales because there are several indentors and several indenting loads available and covered by standards (BS 891 (1989) and ASTM E 18-89a refer to Rockwell hardness testing). The indentors used are hardened steel balls of various diameters or a diamond cone with an included angle of 120°. The standard ball indentors are of $\frac{1}{16}$ inch, $\frac{1}{8}$ inch, $\frac{1}{4}$ inch and $\frac{1}{2}$ inch diameter. The standard indenting loads are 60 kg, 100 kg and 150 kg. Each separate scale of hardness is designated by a letter—A scale, B scale and so on. The various scales and their applicability are given in Table 29.3.

There are other Rockwell scales of hardness available. These scales, known as the N and T scales, are variants of the Rockwell A and B scales but involve smaller indenting forces and are suitable for hardness tests on thin samples. BS 4175 (1989) and ASTM E 18-89a are the relevant standards for these tests. The scales using the diamond cone indentor are known as the 15N, 30N and 45N scales and those using a $\frac{1}{16}$ inch diameter steel ball indentor are known as the 15T, 30T and 45T scales. In each case, the initial number refers to the total indenting force in kg.

Microhardness testing

There are systems available for microhardness testing and the two most widely used methods are the *Vickers diamond* test and the *Knoop Diamond* test. The principle of the Vickers diamond

Table 29.3 *Rockwell Hardness Scales*

Scale	Symbol	Indentor	Total Indenting Load	Material for which the scale is used
A	H_{RA}	Diamond cone	60 kg	Thin hardened steel strip
B	H_{RB}	$\frac{1}{16}$-inch diameter steel ball	100 kg	Mild steel and non-heat treated medium carbon steels
C	H_{RC}	Diamond cone	150 kg	Hardened and tempered steels and alloy steels
D	H_{RD}	Diamond cone	100 kg	Case hardened steels
E	H_{RE}	$\frac{1}{8}$-inch diameter steel ball	100 kg	Cast iron, aluminium alloys and magnesium alloy
F	H_{RF}	$\frac{1}{16}$-inch diameter steel ball	60 kg	Copper and brass
G	H_{RG}	$\frac{1}{16}$-inch diameter steel ball	150 kg	Bronzes, gun metal and beryllium copper
H	H_{RH}	$\frac{1}{8}$-inch diameter steel ball	60 kg	Soft aluminium and thermoplastics
K	H_{RK}	$\frac{1}{8}$-inch diameter steel ball	150 kg	Aluminium and magnesium alloys
L	H_{RL}	$\frac{1}{4}$-inch diameter steel ball	60 kg	Soft thermoplastics
M	H_{RM}	$\frac{1}{4}$-inch diameter steel ball	100 kg	Thermoplastics
R	H_{RR}	$\frac{1}{2}$-inch diameter steel ball	60 kg	Very soft thermoplastics

microhardness test is basically the same as for the standard Vickers test but the indenting loads used are measured in grammes rather than kilogrammes. This type of hardness testing is performed on a metallurgical microscope adapted for the purpose. The small pyramidal diamond indentor is embedded in the surface of a special objective lens. The surface of the test sample is prepared to a high polish and etched for micro-examination. When viewed under the microscope with a high magnification, usually some value between × 200 and × 2000, any particular micro-constituent or feature can be centred in the field of view and a micro-sized diamond indentation made using a small indenting load.

The *Knoop* test was developed in the United States of America and uses a diamond pyramid indentor designed to give a long thin impression, the length being seven times greater than the width and about thirty times greater than its depth. This shape offers an advantage over the square pyramid of the Vickers test for microhardness work in that the length, l, of a Knoop impression is about three times greater than the diagonal, D, of a Vickers impression and can be measured with a greater degree of accuracy.

The Knoop hardness number, H_K, is given by:

$$H_K = \frac{\text{load}}{\text{projected area of impression}}$$

From the geometry of the indentor

$$H_K = \frac{10F}{l^2 \times 7.028}$$

The range of loads used with the Knoop indentor is similar to that used for Vickers microhardness tests. Knoop hardness test results are very similar to those obtained from the Vickers test but are consistently 20 or 25 numbers above Vickers values for the same material. Standards BS 5441 Part 6 (1988) and ASTM E 384-89 cover microhardness testing.

The Shore scleroscope test

This is a dynamic test and involves allowing a small diamond tipped weight to fall freely through a known height onto the test-piece surface and measuring the height of rebound of the weight. In the test apparatus, the small weight of a half ounce (14.2 g) falls freely through a height of 10 inches (250 mm) in a graduated glass tube onto the surface of the material being tested. The tube is graduated into 140 equal divisions. The rebound height of the weight is estimated, by eye, against the graduations on the tube. The Shore test is particularly useful for measuring the hardness of very hard metals and, because the test equipment is small and very portable, it is also very useful for the *in situ* testing of parts such as gears and the surfaces of the large rolls used in metal working operations. The relationship between Shore hardness values and the results of static indentation tests is given in Table 29.4.

A dynamic test, such as the Shore test, can be used to gauge the elastic recovery response of a rubber type material. While this is not the same as hardness, when the test is used for metals, the rebound values obtained with rubbers and thermoplastics is termed their Shore hardness value and this parameter is a good indication of the quality of the material. The Shore value for rubber and plastic materials is generally determined using a small instrument known as a *Durometer*, rather than using the standard Shore falling weight apparatus. A typical durometer is a compact hand-held device in which a rounded indentor is pressed into the material surface under the action of a spring or weight and a pointer registers a hardness value on a graduated scale. Various designs are available to cover the range of elastomers and plastics from the very soft to the very hard over the range Shore A to Shore D. The hardness of rubbers and plastics is often quoted on the IHRD scale (International rubber hardness degrees). The IHRD scale closely approximates to the Shore scale.

29.3 Relationships between hardness and other properties

As a general rule it appears that, for metals, as the hardness value increases, so also do properties such as tensile, compressive and shear strengths. There is no specific relationship between hardness and strength which holds for all metallic materials but some empirical equations have been used to estimate the tensile strengths of some metals from a hardness value. One such relationship, which is valid for annealed and normalised steels, is:

$$\text{tensile strength (MN/m}^2) = H_D \times 3.4$$

This equation does not hold for either heavily cold worked steels or for austenitic steels. However, hardness testing, which is a rapid and relatively simple process, is used frequently as an approximate means for assessing the tensile strength of materials.

Similarly, there is a general trend between the hardness of metals and other properties, such as ductility and toughness. Most materials of high ductility tend to be relatively soft and the ductility decreases as the hardness of the material increases. Toughness, or its antithesis, brittleness also tends to vary with hardness with very hard materials having a tendency to be very brittle, but these are only tendencies and there are no empirical equations which enable a hardness number to be converted into, say, an impact strength value.

A work hardening index for a metal, the *Meyer index*, can be derived from a series of Brinell hardness tests. For this a series of hardness measurements is made using varying indenting loads but with the same sized ball indentor. The Meyer relationship is:

$$F = a \; d^n$$

where F is the indenting load (kgf), d is the diameter of indentation (mm) and a and n are constants of the material and its condition. The resistance to indentor penetration is represented by a and n is the work hardening index. The expression may be written:

$$\ln F = \ln a + n \ln d$$

A graphical plot of $\ln F$ against $\ln d$ will give a straight line from which both a and n can be evaluated.

Table 29.4 *Approximate Hardness Conversions*

H_D	H_B	H_{RB}	H_{RC}	Shore	H_D	H_B	H_{RB}	H_{RC}	Shore
20	19				520	482		51.1	66.5
40	38				540	497		52.4	68.0
60	57				560	512		53.7	69.5
80	76	31.9			580	527		54.8	71.5
100	95	52.5			600	542		55.7	73.0
120	114	66.3			620	555		56.7	74.5
140	133	76.1			640	568		57.6	76.0
160	152	83.4			660	580		58.5	77.5
180	171	89.2			680	592		59.3	79.0
200	190	93.8	14.0	31.5	700	602		60.1	80.5
220	209	97.5	18.0	34.5	720			60.9	82.0
240	228		21.8	38.0	740			61.7	83.5
260	247		25.1	40.5	760			62.5	85.0
280	266		28.2	43.0	780			63.3	86.5
300	285		30.0	45.5	800			64.0	88.0
320	304		33.4	48.0	820			64.8	89.0
340	323		35.7	50.0	840			65.5	90.5
360	342		37.8	52.0	860			66.3	92.0
380	361		39.8	54.0	880			67.0	93.5
400	380		41.7	55.5	900			67.7	94.5
420	399		43.5	57.5	920				96.0
440	418		45.1	59.5	940				97.5
460	437		46.7	61.0	960				98.5
480	452		48.2	63.0	980				100.0
500	467		49.7	64.5	1000				101.0

29.4 Tensile, compressive and shear testing

A knowledge of the mechanical properties of materials in tension, compression, shear and bending is needed by design engineers and a series of tests, many of which have been standardised, have been developed for the determination of these properties. In the case of metals and many plastics materials, tensile testing is used to a much greater extent than compression or shear testing, mainly because the problems associated with this form of testing tend to be less. Materials such as concrete, ceramics and glasses are generally much stronger in compression than in tension (this is due mainly to the presence of porosity and other internal flaws) and tend to be used in situations where the service stresses are largely compressive. Compressive testing is widely used for this type of material. When it is required to know the tensile strength of such brittle materials, it is often determined by testing the material as a beam in bending using a three-point loading arrangement or by testing a cylindrical test-piece in diametral compression (see Section 29.10).

29.5 Testing machines

There are many different types of testing machine available. Some are designed to perform one type of test, for example, tensile, while others are of the 'Universal' type and as such are suitable for uniaxial testing in both tension and compression and also for three-point bend tests on beam type test-pieces. Some designs are small 'table-top' machines with maximum force capacities ranging between 500 N and 20 kN, while at the other end of the scale large machines with load capacities of 1 MN or greater exist.

There are certain features which must exist in any testing machine be it for tensile, compressive or shear tests, and irrespective of size. These are:

(a) a system for locating and holding the test-piece in a satisfactory manner,
(b) a mechanism for applying a force to the test-piece and for varying the force at a controlled rate,
(c) a system for accurate measurement of the applied force.

In addition, some designs of testing machines incorporate systems for the accurate measurement and recording of changes in test-piece dimensions but, generally, such measurements are made using separate devices, such as extensometers and torsionmeters, which can be attached to the test-piece (see Section 29.6).

The system used within a testing machine for the application of force may be either mechanical or hydraulic. The mechanical system normally comprises a screw, or screws, attached to the load-applying cross-head, and the screw is moved by means of a rotating nut. In a hydraulic system, the load is applied by a hydraulic ram moving in an oil-filled cylinder. In many test situations, it is important that the rate of strain is kept constant. Most types of machine have the facility for operating at several specific strain rates but some are capable of infinitely variable rates of strain between fixed upper and lower limits. The results obtained during the testing of metal samples is not affected by the rate of strain employed but polymeric materials are generally strain-rate sensitive. Specific rates of strain for these materials are recommended in the relevant standards.

The force applied to the test-piece may be measured in one of several ways. The oldest system, but highly accurate, is that involving a mechanical lever with a moveable jockey weight. The major disadvantage of the lever system is that it is bulky and occupies a lot of space, particularly for a large capacity machine. Some machine types in which the load is applied hydraulically use a hydraulic system for measuring the load on the test piece. In this arrangement, the increasing oil pressure, as the force is applied to the test-piece, moves a piston in a small calibrated pressure cylinder against a weighted pendulum lever, moving the pendulum out of the vertical position. As the force, and hence the oil pressure, increases so the pendulum is moved further from the vertical. Movement of the pendulum causes a pointer to move around the face of a calibrated dial scale. This is also a measurement system capable of high accuracy. A third system of load measurement utilises the deflection of a spring. This is the principle used in the small Monsanto Hounsfield Tensometer table top machine in which the elastic deflection of a steel beam under load causes a piston to move in a small cylinder containing mercury. The mercury is forced into a glass tube mounted alongside a calibrated scale. A range of beams of varying degrees of stiffness is available to give a series of load ranges. A fourth system of load measurement involves the use of load cells. A load cell may be either a transducer, or a carefully prepared piece of material fitted with sensitive strain gauges. Application of a load creates an electrical output signal from the cell and this is amplified and presented as a digital meter read-out and/or a graphical display on a pen-recorder. The standards of accuracy for tensile testing machines are given in BS 1610 (1985) and ASTM E 4-89.

29.6 Measurement of strain

Many testing machines are fitted with autographic recorders which give a graphical display of the force–deflection behaviour of the test-piece. However, the graphs produced must be treated with caution. For example, in a tensile test while an autographic record shows the correct shape of the force–extension diagram for the material and also shows accurate values of load applied, the readings on the extension axis refer to the separation distance between the test-piece holding grips and not the extension relative to the gauge length of the test sample. Some of the latest generation of testing machines are interfaced with a microcomputer and after completion of a test a full tabulated and graphical print-out of test results can be obtained including, if required, a comparison with previous test results.

The accurate measurement of test-piece dimensional change and, hence, strain is generally achieved by attaching a sensitive measurement device to the test-piece. The devices used for the measurement of longitudinal strain are termed *extensometers* and those designed to measure strain during a torsion test are termed *torsionmeters*.

One of the most commonly used extensometers is the Lindley type (Figure 29.1). This is a robust, yet sensitive device. The Lindley extensometer is attached to a test-piece by tightening two screw grips, which are set 50 mm apart. When a force is applied to the test-piece and strain occurs, relative movement between the gripping points is transmitted through a lever to a dial gauge. The dial gauge is calibrated in steps equivalent to an extension of 0.001 mm and the maximum amount of extension which may be measured is 2.5 mm. Another type of extensometer using a mechanical principle is the Monsanto Hounsfield extensometer. In this instrument, a small relative movement between the gripping points causes a pair of electrical contacts to open. The contacts may be closed again, allowing a small bulb to light up, by turning a calibrated screw. Extensions of 0.01 mm can be read using this instrument, but the applied force has to be held at a constant value while each extensometer reading is taken. This type of extensometer also operates on a gauge length of 50 mm (Figure 29.2).

Electronic extensometers are an integral feature of some of the newer generation testing machines, the extensions being displayed as a digital LED or LCD display. Some extensometers operate on an optical principle. One such is Marten's extensometer (Figure 29.3). A change in the gauge length of the material will cause an angular movement of the mirror. A scale is viewed

FIGURE 29.1 Lindley type extensometer positioned on a test-piece

FIGURE 29.2 Monsanto Hounsfield type extensometer positioned on a tensile test-piece

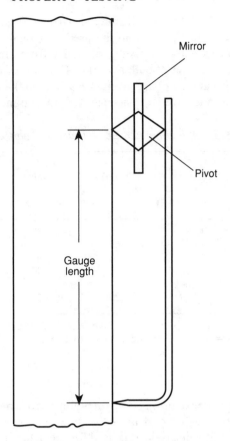

FIGURE 29.3   Principle of Marten's mirror extensometer

through the mirror by means of a telescope. Alternatively, a light source may project, by way of the mirror, a spot of light onto a graduated scale. Optical extensometers are highly sensitive and are often used for the measurement of small strains during long term creep tests.

Another device which may be used for the measurement of strain is the electric resistance strain gauge. The resistivity of some conductors is very sensitive to variations in elastic strain. A resistance strain gauge consists of a zig-zag of very fine wire mounted between waterproof sheets. The gauge is securely bonded to the surface of the material under test and the ends of the wires connected to a bridge network so that changes in resistance can be measured.

In torsion testing, a cylindrical shaped test-piece is subjected to axial torsion and the strain measurement which needs to be taken is the angle of twist. A common type of torsionmeter is similar to the Lindley type extensometer with angle of twist transmitted via a lever to a dial gauge. Such an instrument is capable of giving an angular reading to an accuracy of 0.001 radian over a gauge length of 50 mm.

29.7 The tensile testing of metals

The metal test-pieces used for a standard tensile test are shaped in such a way that fractures will occur within the desired portion, that is, within the gauge length, and definite standards are laid down for their dimensions. In some instances, particularly in some fundamental experimental programmes, very small test-pieces may be used, but these small sizes are not quoted in the relevant testing standards as the results obtained from small test-pieces may not be truly representative of the properties of bulk material. The recommended dimensions for metal tensile test-pieces (Tables 29.5 and 29.6) are given in BS 18 (1987) and ASTM E 8-89b and E 8M-89b. (The testing of cast iron is covered in a separate publication, BS 1452 (1990).)

Table 29.5   *Round bar tensile test-pieces. Table of standard dimensions*

S_0 (mm^2)	d (mm)	L_0 (mm)	L_c (mm)	r wrought materials (mm)	r cast materials (mm)
200	15.96	80	88	15	30
150	13.82	69	76	13	26
100	11.28	56	62	10	20
50	7.98	40	44	8	16
25	5.64	28	31	5	10
12.5	3.99	20	21	4	8

d = Diameter of test-piece, L_0 = gauge length, L_c = parallel length, S_0 = original cross-sectional area, r = radius at shoulder (refer to Figure 29.4(a)). For proportional test-pieces $L_0 = 5.65 S_0$, $L_0 \simeq 5d$.

The shape of the load–extension curve for a non-ferrous metal is shown in Figure 29.5(a). The initial strain is elastic, but beyond point E, the elastic limit, strain is plastic. Point U is the maximum load, and this value of load is used for the determination of the tensile strength of the material. Point F marks the point of fracture. Although the applied load on the test-piece decreases beyond point U, the true stress acting on the test-piece, taking into account the reducing cross-sectional area, continues to increase until fracture occurs. (Refer to Section 29.8.) In the commercial testing of metals, it is the load–extension curve, rather than a true stress–strain curve, that is plotted and strengths are calculated on the basis of the original cross-sectional area of a test-piece. The following information is determined in a routine tensile test.

Tensile strength (T.S.) (formerly known as ultimate tensile strength). This is based on the maximum load sustained by the test-piece, when the latter is tested to destruction, and corresponds to point U in Figure 29.5. The numerical value of tensile strength is calculated as a nominal stress and is given by:

$$T.S. = \frac{\text{maximum load applied}}{\text{original cross-sectional area}}$$

The units in which tensile strength is normally quoted are megapascals (MPa), meganewtons per metre^2 (MN/m^2) or newtons per millimetre^2 (N/mm^2). Numerically these three values are equal. (In the construction industry, the units used generally are N/mm^2.)

Yield point or *yield stress*. There is a sharp discontinuity in the load–extension diagram for wrought iron and many steels (Figure 29.5(b)) and the material will suddenly yield with little or no increase in the applied load necessary (point Y in the figure). The extent of this sudden yielding is about 5–7 per cent of the original gauge length. The yield point Y is close to, or coincident with, the elastic limit. The yield stress of the material is given by:

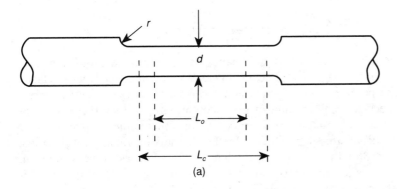

d = Diameter
r = Radius
L_o = Gauge length
L_c = Minimum parallel length

(a)

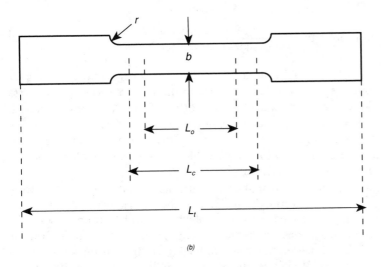

r = Radius
b = Width
L_o = Gauge length
L_c = Minimum parallel length
L_t = Minimum total length

(b)

FIGURE 29.4   Types of tensile test piece:
(a) round test-pieces; (b) flat test-pieces

Table 29.6   *Flat tensile test-pieces. Table of standard dimensions*

b (mm)	L_0 (mm)	L_c (mm)	L_t (mm)	r (mm)
25	100	125	300	25
12.5	50	63	200	25
6	24	30	100	12
3	12	15	50	6

b = width of test-piece, L_0 = gauge length, L_c = parallel length, L_t = total length, r = radius at shoulder (refer to Figure 29.4(b)).

Tables 29.5 and 29.6 are extracts from BS 18 and are reproduced by permission of the British Standards Institution, 2 Park Street, London W1A 2BS, from whom complete copies of the standard can be obtained.

$$\text{Yield stress} = \frac{\text{applied load at yield point}}{\text{original cross-sectional area}}$$

Proof stress or *offset yield stress*. Most metals, unlike many steels, show a smooth transition from elastic to plastic deformation behaviour. It may be difficult to determine the value of elastic limit with exactitude and the parameter *proof stress* is determined. This is the level of stress required to produce some specified small amount of plastic deformation. Often the amount of plastic strain specified is 0.1 per cent (nominal strain = 0.001) but other values of plastic strain may be used also. Frequently, proof stress is referred to by the alternative term *offset yield stress*, or just *yield stress*. In Figure 29.5(c), the load–extension diagram for some test-piece is shown in curve OEPUF. If OA corresponds to some percentage x of the original gauge length, and AP is drawn parallel to OE, then the point P denotes the x per cent proof load. The value for the x per cent proof stress is given by:

$$x \text{ per cent proof stress} = \frac{x \text{ per cent proof load}}{\text{original cross-sectional area}}$$

In the United Kingdom, values are normally quoted for either the 0.1 per cent proof stress, or the 0.2 per cent proof stress, but in the USA it is customary to quote 0.2 per cent proof stress values. There will be a fairly wide difference between the value of proof stress and tensile strength for an annealed metal, but in a work hardened metal, or, in the case of a fairly hard and brittle material, the proof stress and tensile strength values will be fairly close to one another. The extent of the separation between proof stress and tensile strength values gives a measure of the amount of cold work that may be performed on the material.

Limit of proportionality and *elastic limit*. The limit of proportionality is the value of the stress at which the stress–strain curve ceases to be a straight line, that is, the point at which Hooke's law ceases to apply. The elastic limit may be defined as the level of stress at which strain ceases to be wholly elastic. In most cases, these two values will be the same, but some polycrystalline metals do not completely obey Hooke's law. Such materials are termed anelastic, but the stress–strain curve does not vary greatly from linear. It is very difficult to determine a value for the limit of proportionality, as the values obtained depend on the sensitivities of the load-measuring system and the extensometer used. In general commercial testing, the limit of proportionality is not usually reported. In its place, for metals which do not show a marked yield point, a *proof stress* is quoted.

Modulus of elasticity, E, (Young's modulus). The modulus of elasticity may be calculated from the slope of the straight line portion of the load–extension curve. E is given by (gauge length/cross-sectional area) × slope. The units in which E is quoted may be gigapascal (GPa), giganewton/metre^2 (GN/m^2) or kilonewton/millimetre^2 (kN/mm^2).

Percentage elongation. A definite length, the gauge length, is marked off on the test-piece before testing. After fracture, the two portions of the test-piece are placed together and the distance between gauge marks is remeasured. The amount of extension, expressed as a percentage of the original gauge length, is then quoted as the elongation value:

$$\text{Percentage elongation on gauge length} = \left(\frac{L - L_0}{L_0}\right) \times 100$$

where L is the length between gauge marks after fracture and L_0 is the original gauge length. For an elongation figure to have any validity, the fracture must occur in the central section of the gauge length and the gauge length must be specified, for example, *the percentage elongation on*

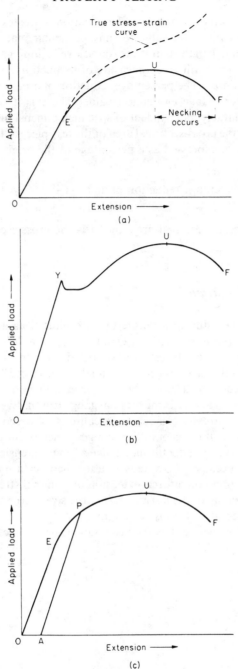

FIGURE 29.5   (a) Load–extension curve and its relation to a true stress–strain curve.
(b) Load–extension curve with sharp yield point (mild steel).
(c) Determination of proof load from load–extension diagram

50 mm is 20 per cent. As the amount of plastic deformation of the test-piece is greatest nearest to the point of fracture, the elongation value for any particular material will be much higher if measured over a short gauge length than if measured over a long gauge length. The percentage elongation value for a material will give a measure of its ductility.

Percentage reduction of area. The percentage reduction of area is often quoted for round-bar specimens instead of a percentage elongation value. There is a certain merit in this as the reduction of area value is largely independent of specimen dimensions and gauge length. It is the difference in area between the cross-sectional area of the test-piece at the point of fracture and the original cross-sectional area, expressed as a percentage of the original cross-sectional area.

$$\text{Percentage reduction of area} = \left(\frac{A_0 - A}{A_0}\right) \times 100$$

where A_0 is the original cross-sectional area and A is the cross-sectional area at the point of fracture.

29.8 True stress and true strain

Generally, the data obtained during a tensile test are plotted as a *force–extension* curve and sometimes this type of curve is referred to, erroneously, as a *stress–strain* curve. The force–extension data obtained from a tensile test may be plotted in the form of stress and strain to give a nominal stress–nominal strain curve in which nominal stress, $\sigma_n = F/A_0$, where F is the force and A_0 is the original cross-sectional area of the test-piece, and nominal strain, $\varepsilon_n = (L - L_0)/L_0$, where L is the extended length and L_0 is the original gauge length. Even so, such a nominal stress–strain curve is not a true stress–strain curve because it is based on the original test-piece dimensions and does not take into account the changes which occur during a test to destruction. A nominal stress–strain curve for a ductile metal, like a force–extension curve, shows a maximum before fracture. This is because, in a tensile test, there is a point of plastic instability, corresponding to the maximum on a force–extension or nominal stress–nominal strain curve, at which there is a sudden necking of the test-piece. If a true stress–true strain curve were drawn this would show that stress increases steadily to fracture.

The true stress, $\sigma_t = F/A_i$, where A_i is the instantaneous cross-sectional area, and true strain,

$$\varepsilon_t = \frac{dL}{L} = \ln\left(\frac{L}{L_0}\right)$$

but

$$\frac{L}{L_0} = (1 + \varepsilon_n)$$

so

$$\varepsilon_t = \ln(1 + \varepsilon_n)$$

where L_0 is the initial gauge length and L is the instantaneous gauge length. Assuming there is no volume change during plastic deformation, $A_i L = A_0 L_0$ or

$$A_i = \frac{A_0 L_0}{L}$$

so

$$\sigma_t = \frac{F L}{A_0 L_0}$$

but

$$\frac{L}{L_0} = (1 + \varepsilon_n)$$

So

$$\sigma_t = \frac{F}{A_0}(1 + \varepsilon_n)$$

For most metals the relationship between true stress and true strain during plastic deformation can be written:

$$\sigma_t = k \, \varepsilon_t^{\,n}$$

where k and n are constants for the material. The constant n is the strain hardening exponent and the true stress–true strain equation is an empirical expression for work hardening. Instability and necking of a ductile metal occurs when $\varepsilon_t = n$ and at this point, the maximum on the nominal stress–strain curve, the maximum value of σ_n, the nominal tensile strength of the material, is given by:

$$\sigma_{max} = k \, n^n (1 - n)^{(1 - n)}$$

29.9 The tensile testing of plastics

The tensile testing of plastics materials is conducted in a generally similar way to the tensile testing of metals. There are differences, however, and these are necessary because of the different nature of plastics materials and metals. Many thermoplastic materials do not show Hookean elasticity even at low stress levels but are viscoelastic. In other words, the strain developed is not dependent on the level of stress alone, as would be the case for an elastic solid, but also depends on the length of time for which the stress is applied. One consequence of this is that variations in the rate of strain can give differences in the force–extension values and, hence different test results (see Figure 29.6). It is recommended in the relevant standards publications that testing be conducted at comparatively high rates of strain (refer to BS 2782, Part 3, Methods 320A to 320F (1986) and ASTM D 638-89 and D 638M-89).

Unlike metals, many thermoplastic materials cold draw during a tensile test, and in these cases values are quoted for the yield stress and the drawing stress, these being:

$$\text{Yield stress} = \frac{\text{yield load}}{\text{original c.s.a.}} \qquad \text{Draw stress} = \frac{\text{drawing load}}{\text{original c.s.a.}}$$

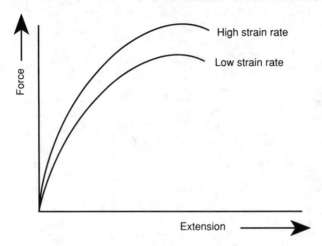

FIGURE 29.6   Influence of rate of strain on the force–extension values for a thermoplastic material

The phenomenon of cold drawing can occur in some thermoplastic materials and this is illustrated in Figure 29.7. At a stress corresponding to point Y in Figure 29.7(a) the test-piece necks down considerably and thereafter further strain takes place at a constant stress, usually slightly lower than the yield stress, with undrawn material being drawn into the necked zone. This drawn material is much stronger than the original undrawn plastic material due to the alignment of polymer molecules which occurs during drawing. The drawn state, with a more crystalline structure, is also referred to as stress induced crystallinity.

The load–extension curves that are obtained for many plastic materials either show no initial straight line portion, or show a departure from Hooke's law at very low values of applied load. In

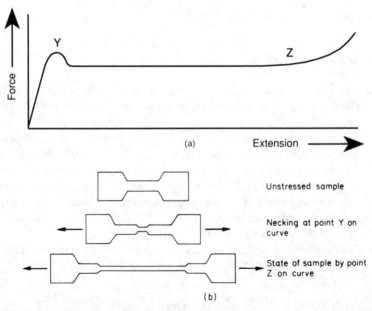

FIGURE 29.7   (a) Typical load–extension curve for a thermoplastic which cold draws.
(b) Stages in cold drawing of a test-piece

these cases, it would be very difficult, or impossible, to obtain a value for Young's modulus. The value that is quoted as the modulus of elasticity, E, for plastics is in fact a *secant modulus*, and is obtained by determining the stress at a value of 0.2 per cent strain (see Figure 29.8).

Many thermoplastics extend to very considerable extents during a tensile test, but there is an almost instantaneous recovery of much of this strain as soon as the test-piece fractures. The elongation value quoted for thermoplastics is known as the *percentage elongation at break*, and is determined by noting the distance between the gauge marks at the moment of fracture, and not by placing the fractured portions together after the test, as is the case with tests on metals.

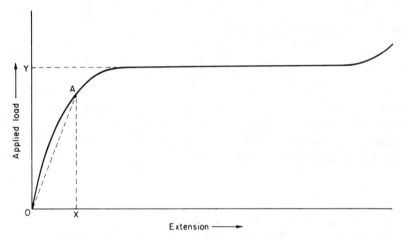

FIGURE 29.8   Load–extension diagram for a thermoplastic. Y is the yield load. OX is an extension
               corresponding to a strain of 0.2 per cent. The slope of the straight line OA is used to
               determine the modulus of elasticity for the plastic, the secant modulus

29.10 Determining the tensile strength of brittle materials

The axial tensile testing of brittle materials, such as ceramics and glasses, would be extremely difficult, if not impossible, because of the problems of preparing suitable shaped test-pieces and those of effectively holding them within the testing machine. It is customary to determine the fracture strength from a three-point bend test. The flexural strength value determined in this type of test is also known as the *modulus of rupture* of the material. When a sample is subjected to bending, as shown in Figure 29.9, a compressive stress is generated in the upper surface and a tensile stress is generated in the lower surface. The tensile strength of these materials is less than the compressive strength. When a sample is tested in bending in this way fracture commences at the tensile surface so the breaking load, F, is related to the tensile strength of the material. The magnitude of the direct stress, σ, is related to the bending moment, M, by the general bending equation:

$$\frac{\sigma}{y} = \frac{M}{I}$$

where y is the distance from the neutral surface (half the thickness for a sample of symmetrical section) and I is the second moment of area of the section.

$$I = \frac{BD^3}{12} \text{ for a rectangular section of width } B \text{ and thickness } D$$

$$I = \frac{\pi D^4}{64} \text{ for a circular section of diameter } D$$

The maximum value of bending moment for a symmetrical three-point loading system as in Figure 29.9 is $FL/4$ and so

$$\sigma = \frac{My}{I} = \frac{FLD}{8I}$$

For a sample of rectangular section, therefore

$$\sigma = \frac{3FL}{2BD^2}$$

The modulus of rupture value for a material is approximately double the true tensile strength. This mode of testing is also used to determine the fracture strength of rigid thermoset materials including laminated plastics such as 'Tufnol', a phenol–formaldehyde resin laminated with either cloth or paper, and is covered by BS 2782, Part 3, Method 335A (1989) and ASTM D 790-86 and D 790M-86.

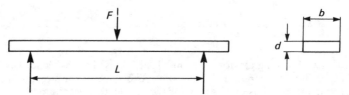

FIGURE 29.9   Principle of modulus of rupture test by three-point loading

Another indirect method which may be used to determine a tensile strength is the split cylinder test. This type of test, shown diagrammatically in Figure 29.10, is widely used for determining the tensile strength of concrete and is detailed in BS 1881; Part 117 (1983) ASTM C 496-86. As the compression force is imposed, circumferential tensile stresses are induced in the material, these having maximum value at the central horizontal plane. The cylinder will eventually fail and split at this plane. The value of the tensile stress, σ, at failure is given by:

$$\sigma = \frac{2F}{\pi LD}$$

where F is the maximum force applied, L, is the cylinder length and D is its diameter.

One test result alone is not sufficient for a brittle material, such as a glass or ceramic. There is a major variation in strength for such materials because of the presence and distribution of cracks, pores and other flaws of differing sizes. Many test results are necessary in order to obtain a statistical mean. Weibull devised a method for dealing with the statistical nature of the fracture strength of such brittle materials and he defined the probability of survival, $P_s(V_0)$, as the fraction of identical samples of volume V_0 which survive after the application of a tensile stress of σ. The Weibull expression is:

$$P_s(V_0) = \exp\left(-\sigma/\sigma_0\right)^m$$

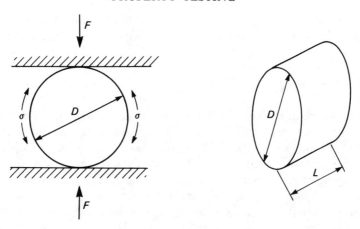

FIGURE 29.10   Principle of split cylinder test for tensile strength

where σ_0 is the value of tensile stress at which $1/e$, that is 37 per cent, of the samples do not fail and the power m is the Weibull modulus.

The Weibull expression can be written in logarithmic form:

$$\ln P_s(V_0) = (-\sigma/\sigma_0)^m$$

A plot of σ/σ_0 against $\ln P_s(V_0)$ should give a straight line from which the modulus m can be evaluated. The higher the value of m, the less will be the variation in strength of a group of similar samples (see Figure 29.11). The value of the Weibull modulus for ceramic materials such as brick and pottery is about 5, while the value for some engineering ceramics, such as alumina and silicon

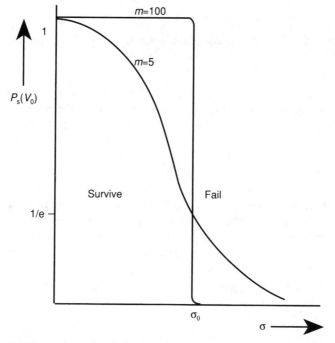

FIGURE 29.11   Weibull curves showing probability of survival

nitride, is about 10. Most metallic materials possess Weibull moduli of around 100. With such a high value of m, the material can be considered as having a single failure strength.

29.11 Compression testing

The compression testing of metals is little used. A tensile test sample is fairly long and waisted down to give a parallel sided central portion, within which the tensile characteristics of the material can be accurately assessed. In compression testing, a long sample cannot be used because buckling failure would occur rather than direct axial compression. Test-pieces have to be short and the length not be greater than three times the diameter to avoid buckling. The other problem is that of friction between the ends of the test-piece and the platens of the testing machine. As compressive strain occurs, reducing the test-piece length, so there is a corresponding increase in the diameter of the sample. Friction hinders lateral expansion of the diameter at the test-piece ends and once plastic deformation occurs the material tends to deform into a barrel shape (Figure 29.12(a)). This type of deformation generates tensile stresses at the surface of the material and failure is based more on the tensile strength of the material than anything else. Brittle materials, when tested in compression, normally fail by shear at 45° to the direct stress axis. The type of fracture may be either of the double cone type or straight shear (Figure 29.12(b)). These failure modes apply to all brittle materials, whether metallic or non-metallic.

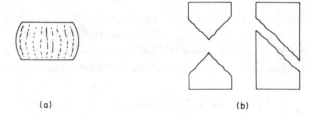

(a)   (b)

FIGURE 29.12   (a) Ductile compressive failure; (b) Brittle compressive failure

Compressive test-pieces tend to be small and of cylindrical shape, as mentioned above but compression test samples of cement or concrete are often cast into cube shapes with edge lengths of either 100 or 150 mm. (See BS 1881. Part 116 (1983) and ASTM C 39-86.)

29.12 Testing in shear

There are two main types of test which may be used for the determination of the properties of a material in shear. One is based on the application of a torque to a cylindrical sample causing the test-piece to twist while the other involves direct shear. There are problems associated with both types of test. The former method causes the development of an almost pure shear stress within the material but the shear stress is not uniform throughout and varies from zero at the central axis to a maximum value at the surface of the cylinder. For pure torsion on a cylinder the general torsion equation applies within the elastic range.

$$\frac{\tau}{r} = \frac{T}{J} = \frac{G\theta}{L}$$

where τ is the shear stress at radius r from the central axis, T is the applied torque, J is the second polar moment of area, G is the modulus of rigidity of the material, and θ is the angle of twist in radians over the gauge length, L, of the test-piece.

The value of J for a solid cylindrical section is given by $\pi D^4/32$ and for a hollow cylindrical section by $\pi(D^4 - d^4)/32$ where D and d are the outside and inside diameters, respectively.

As the applied torque, T, is increased so the angle of twist, θ, increases in direct proportion within the elastic range. The fact that the level of shear stress varies across a diameter is relatively unimportant when all stresses and strains are elastic. However, where the surface shear stress reaches the elastic limit and plastic strain commences results become unreliable. Plastic deformation tends to spread inwards in an irregular manner. Materials such as mild steel with a pronounced yield point, tend to show greater irregularities than other metals. One method used to reduce the problem is to employ hollow cylindrical test-pieces. In this way, differences between the value of shear stress at outer and inner surfaces, is reduced, but a further problem is encountered. The tendency for plastic buckling of the tubular specimen increases rapidly as the wall thickness is reduced.

The results of a torsion test may be plotted graphically in the form torque, T, against angle of twist, θ. The value of the modulus of rigidity, G, may be obtained from the slope of the linear elastic portion of the curve (T/θ).

Direct shear tests are sometimes used but it is very difficult to obtain accurate quantitative data from them. Several systems have been adopted, the main ones being: double shear for round bar test-pieces using a fork and eye device, double knife shear for samples with a rectangular section, and the shearing of a disc from sheet material using a punch and die. The test equipment generally takes the form of attachments which can be used in conjunction with a universal testing machine. These three methods all suffer from the same disadvantage. Even when all parts of the attachments are made to close dimensional tolerances from very hard materials, there will be some bending of the test-piece and the stresses within the material will not be of the pure shear type but will include bending stresses. The results of such tests, while not providing fundamental values of shear strength, can be used to give a qualitative assessment and indicate the behaviour patterns of the material for fabrication processes involving blanking and shearing.

29.13 Notch impact testing

In this type of test a bar specimen with a milled notch is struck by a fast-moving hammer, and the energy that is absorbed in fracturing the test-piece is measured. This type of test has the advantage of revealing a tendency to brittleness that is not revealed by the standard tensile test, or by a hardness test. As mentioned in Chapter 10, the ductile-to-brittle transition temperature is considerably higher for impact loading conditions than for slow straining conditions. It is also possible for two samples of a material to possess very similar properties, as determined in the standard tensile test, but to possess widely different properties when tested under impact loading conditions. This is illustrated in Table 29.7 for samples of a low-alloy steel subject to temper brittleness.

One major use of the notch impact test is to determine whether heat treatment of a material has been carried out successfully, as this type of test gives the most information on this. The impact strength should not, however, be judged alone but viewed in conjunction with the results of other tests, including the standard tensile and hardness values.

The types of notched bar impact test that are most widely used in this country are the Charpy test and the Izod test. In both types of test a heavy pendulum is released and is allowed to strike a test-piece at the bottom of its swing. A proportion of the energy of the pendulum is absorbed in

Table 29.7 *Material with similar tensile properties but different impact loading properties, according to condition*

Composition (per cent)	Condition	Tensile strength (MN/m²)	Elongation on 50 mm (per cent)	Charpy impact value (J/mm²)
C 0.3 Ni 3.2 Cr 1.0	Hardened, tempered, and cooled rapidly after tempering	855	28.6	1.31
	Hardened, tempered, and cooled slowly after tempering	836	26.5	0.15

fracturing the test-piece. The height of the follow-through swing of the pendulum is measured, and the energy absorbed in fracture determined. In the Charpy test the test-piece is tested as a simply supported beam and the sharp edge of the pendulum strikes at mid-span directly behind the milled notch. An Izod test specimen is tested in a cantilever mode. The test-piece is firmly clamped in a vice with the prepared notch level with the edge of the vice. The impact blow is delivered on the same side as the notch (Figure 29.13). Impact strengths are generally quoted in J/mm² and based on the cross-sectional area of the test-piece below the notch.

The impact strengths determined in both types of test are roughly comparable, provided that the same type of notch is used (see BS 131, Parts 1, 2 and 3 and ASTM E 23-88).

A miniature Charpy-type testing machine is available for the impact testing of plastics.

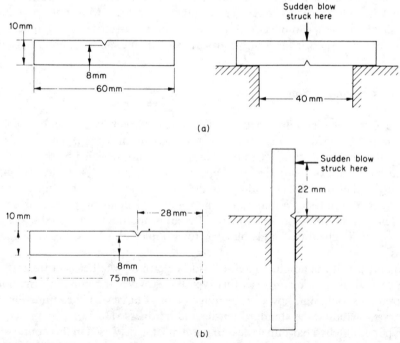

FIGURE 29.13 (a) Charpy test specimen and position of impact. (b) Izod test specimen and position of impact

29.14 Fatigue testing

A component or structure in service may be subjected to fluctuating or alternating cycles of stress, but rarely can it be found that one constant type of loading cycle applies during the whole of the life of a component. Laboratory tests to determine fatigue strength are usually based either on an alternating stress cycle, with a mean stress of zero, or on a fluctuating stress with some positive value of mean stress, rather than making any attempt to duplicate exactly the variable conditions that may occur during the life of the component.

The types of fatigue test that are often employed are those that involve axially loading the test-piece alternately in the tensile and compressive modes, and rotary bending tests. Of the latter, the two most common types are the Wöhler system, in which the test-piece is loaded as a cantilever (Figure 29.14(a)) and the four-point loading system (Figure 29.14(b)). With both of these loading systems, the test-piece is rotated so that, during one complete revolution, any element of the test-piece surface goes through a complete stress cycle of tension and compression. The four-point loading system offers the advantage of a constant value of bending moment over the central section of the test-piece. Because notches and sharp changes in cross-section act as points of high stress concentration, specimens for fatigue testing are prepared with polished surfaces, and the design is such that changes in section are gradual and well radiused. The various recommended types of fatigue test-piece, and their dimensions are quoted in BS 3518, Parts 1 and 2 (1984) and ASTM E 466-82 and E 468-90.

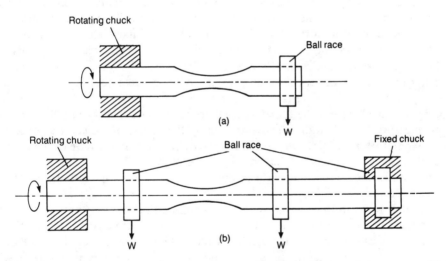

FIGURE 29.14 Principle of rotary bending fatigue tests: (a) Wöhler cantilever system; (b) four-point loading

The rate of cycling in fatigue testing is normally up to 150 Hz, but if a very rapid assessment is required for a material a high-frequency vibration technique may be used. With some high-frequency machines, tests may be carried out at up to 500 Hz. At this rate it takes only about five hours to complete 10^7 cycles. The results obtained, however, show fatigue limits as being slightly higher than those obtained from lower-frequency tests.

In certain cases there may be a need to conduct fatigue tests to failure on a complete engineering structure. Following the pressure-cabin failures in early Comet airliners in 1951, a complete Comet structure was tested to destruction in a special test rig. The test simulated the cabin-pressure changes that occurred in each flight, and also wing flexing during flight. Since that date a similar test to destruction has been made on one air-frame of every civil-airliner type destined for service. This has been a requirement of the airworthiness certification procedure.

29.15 Creep testing

Almost all creep testing is conducted in the tensile mode and the test specimens are similar in form to those used in tensile testing (see Section 29.7). Test-pieces for tensile creep testing may be of either circular or rectangular cross-section but there are no standard sizes. The actual dimensions of the test-pieces used will depend upon the type of creep testing machine used. The basic requirements for a creep testing machine are:

(a) that it must possess means for applying and maintaining a constant tensile load,
(b) there must be a furnace capable of keeping the temperature of the test-piece at the desired value to within very close limits,
(c) there should be means for the accurate measurement of test-piece extension.

This last requirement is not necessary if the equipment is to be used only for the determination of stress-to-rupture data.

Tensile creep test machines are designed so that the test-piece is mounted vertically and, generally, the axial load is applied to the specimen holder by dead weights and a lever system. It is essential that the temperature of the test-piece be very closely controlled for the duration of a test and that the temperature be uniform along the length of the specimen. The usual arrangement is to have an electric resistance tubular furnace mounted on the frame of the testing machine and moveable in the vertical plane. BS 3500 (1987) and ASTM E 139-83 cover the standard conditions for creep testing and specify the very close temperature tolerances required. Temperatures must be maintained to within $\pm 2°C$ for tests at temperatures up to 600°C, to $\pm 2.5°C$ for tests at temperatures between 600°C and 800°C and to $\pm 3°C$ for temperatures between 800°C and 1000°C.

Accurate measurement of strain is necessary during the course of a creep test, except in the case of creep stress-to-rupture tests, and generally the extensometers used are of the mirror type (see Figure 29.3) capable of measuring extensions of the order 10^{-3} or 10^{-4} mm. Extension rods made from a heat resistant alloy connect the gripping points at the test-piece to the extensometer outside the furnace.

When the time to rupture at some particular combination of temperature and stress is required, and there is no need to monitor the progress of creep strain, it is possible to achieve some economy in the use of creep testing machines and test a number of samples simultaneously within the same machine. A series of test-pieces can be mounted in line as a long string within a furnace and the required load applied (Figure 29.15). As in other forms of creep testing, individual thermo-couples are attached to each specimen and the temperature of test must be controlled within the close limits listed above. When one of the test-pieces breaks, the load on all the remaining specimens will be released. Usually the equipment is arranged so that load release accompanying a fracture will open a micro switch and switch off both the furnace and the clock which records the test duration. The broken test-piece can then be removed and the test continued. It is a requirement of BS 3500 Part 1 that the time to rupture be measured to an accuracy of $\pm 1\%$.

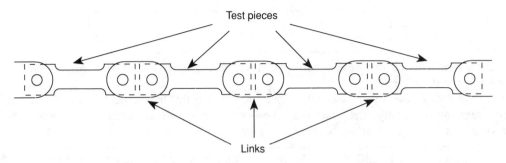

FIGURE 29.15   String of test-pieces for stress–rupture testing

29.16 Relaxation testing

The principle of relaxation testing is to stress a material by straining the sample to some predetermined value. The strain is then maintained at a constant level and the variation of stress with time is recorded. As in creep testing, close control of temperature is required. A test-piece, with thermocouples attached, is located in a furnace positioned within a rigid straining frame. When the specimen is at the required control temperature, tensile strain is applied by turning a straining screw at one end of the frame. The tensile force developed may be measured by means of a load cell and the changes in load on the sample with time are monitored over the duration of the test. The general requirements for the conduct of relaxation tests are given in BS 3500, Part 6 (1987), ASTM E 328-86 and D 2991-84.

29.17 Questions

29.1   Why it is necessary to use different loads when making Brinell or Vickers hardness impressions on two different materials, for example, pure copper and bronze?

Brinell hardness impressions were made on samples of pure copper and a phosphor bronze using a 2 mm diameter ball indentor. The results were:

Pure copper—20 kg load—impression diameter = 0.772 mm
Phosphor bronze—40 kg load—impresson diameter = 0.750 mm

Calculate the Brinell hardness of the materials

29.2   Vickers and Brinell hardness tests are made on a sample of a steel and the following results were recorded:

Vickers diamond test with 30 kg load—the mean lengths of impression diagonals were: (a) 0.527 mm, (b) 0.481 mm, (c) 0.497 mm
Brinell test using 10 mm ball indentor and 3000 kg load—impression diameters were: (a) 4.01 mm, (b) 4.005 mm, (c) 4.02 mm

Calculate the diamond and Brinell hardness numbers for the steel and explain any variations in the results.

29.3   Brinell hardness impressions on a sample of annealed copper, using a 5 mm diameter ball indentor, gave the values in the Table below.

Load (kg)	125	250	375
Impression diameter (mm)	2.20	2.70	3.10

Determine whether this data satisfy the Meyer relationship and, if so, evaluate the Meyer constants for the material.

29.4 The data in the Table below were determined in a full tensile test. Evaluate the tensile properties and identify the material. Test-piece dimensions: width $= 12.61$ mm, thickness $= 3.47$ mm, gauge length $= 50$ mm.

Force (N)	Extension (mm)	Force (N)	Extension (mm)	Force (N)	Extension (mm)
25	0.018	125	0.121	225	0.293
50	0.040	150	0.153	250	0.355
75	0.064	175	0.192	275	0.425
100	0.090	200	0.238	300	0.520

Maximum force in test $= 1290$ N; Length between gauge marks at break $= 97$ mm

29.5 From the data in the Table below plot a force–extension diagram and determine values for: (a) tensile strength, (b) modulus of elasticity, (c) 0.2 per cent proof stress, (d) percentage reduction of area, and (e) the true stress at a nominal strain of 8 per cent.

Force (kN)	39.4	67.5	84.4	90.0	95.6	112.5	123.8	131.1	131.1	123.8
Extension (mm)	0.25	0.40	0.50	0.60	0.75	1.75	3.00	5.00	6.50	8.00

The test-piece original diameter was 11.26 mm and the diameter after fracture was 9.34 mm.

29.6 A metal sample of 10 mm diameter was stressed in tension and the following information obtained. A force of 30 kN gave a total extension of 2.0 mm and a force of 34 kN gave a total extension of 11.0 mm over an original gauge length of 40 mm.

Assuming that the true stress–true strain relationship, $\sigma_t = k\varepsilon_t^n$, applies determine:

(a) the values of the coefficient k and the exponent n,
(b) the total extension at the maximum load,
(c) the maximum load applied, and
(d) the tensile strength of the material.

29.7 Figure 29.16 gives details of a tensile test on a material. From the data determine values for: (a) E, (b) 0.1 per cent proof stress, (c) tensile strength, and (d) percentage elongation, and identify the type of material tested.

29.8 (a) A glass test-piece with the dimensions length $= 95$ mm, thickness $= 3$ mm and width $= 15$ mm is simple supported over a span of 75 mm and a force applied at mid-span. The load required to break the glass is 35 N. Calculate the tensile strength (modulus of rupture) of the glass. (b) A concrete cylinder of 100 mm diameter and 100 mm length is subjected to a diametral compressive force. Calculate the tensile strength of the concrete if the maximum force at the failure point is 47.5 kN.

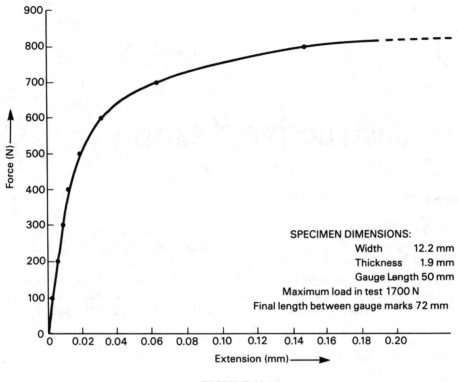

FIGURE 29.16

30

Non-destructive Testing

30.1 Introduction

Defects, such as cracks, porosity and inclusions, which may be potentially damaging may be introduced into materials or components during manufacture, and other defects, such as fatigue cracks, may be generated during service. It is necessary to be able to detect and identify such defects and to ascertain their position and size so that decisions can be taken as to whether specific defects can be tolerated or not. A range of non-destructive test methods is available for the inspection of materials and components and the most widely used techniques are:

dye penetrant,
magnetic particle inspection,
eddy current systems,
ultrasonics, and
radiography.

Some of the features and applications of the main test methods in use are given in Table 30.1. All these NDT systems co-exist and, depending on the application, may either be used singly or in conjunction with one another. There is some overlap between the various test methods but they are complementary to one another. The fact that, for example, ultrasonic testing can reveal both internal and surface flaws does not necessarily mean that it will be the best method for all inspection applications. Much will depend upon the type of flaw present and the shape and size of the components to be examined.

30.2 Visual inspection

Often the first stage in the examination of a component is visual inspection. Examination by naked eye will only reveal relatively large defects which break the surface but the effectiveness of visual inspection for external surfaces can be improved considerably through use of a hand lens or stereoscopic microscope. Generally, high magnifications are not necessary for this type of inspection. Optical inspection probes, both rigid and flexible, which can be inserted into cavities, ducts and pipes have been developed for the inspection of internal surfaces. An optical inspection probe comprises an objective lens system at the working end and a viewing eyepiece at the other end, with a fibre optic coherent image guide linking the two (see Figure 30.1). Illuminating light is conveyed to the working end of the probe through an optical fibre light guide, and both the optical and illumination systems are contained within either a stainless steel tube, for rigid probes, or a flexible plastic or braided metal sheathing in the case of flexible probes. Inspection

Table 30.1

System	Features	Applicability
Visual inspection probes	Detection of defects which break the surface, surface corrosion, etc.	Interior of ducts, pipes and assemblies
Liquid penetrant	Detection of defects which break the surface	Can be used for any metal, many plastics, glass and glazed ceramics
Magnetic particle	Detection of defects which break the surface and sub-surface defects close to the surface	Can only be used for ferro-magnetic materials (most steels and irons)
Electrical methods (eddy currents)	Detection of surface defects and some sub-surface defects. Can also be used to measure the thickness of non-conductive coatings, e.g. paint on a metal	Can be used for any metal
Ultrasonic testing	Detection of internal defects but can also detect surface flaws	Can be used for most materials
Radiography	Detection of internal defects, surface defects and to check correctness of assemblies	Can be used for most materials but there are limitations on the maximum material thickness

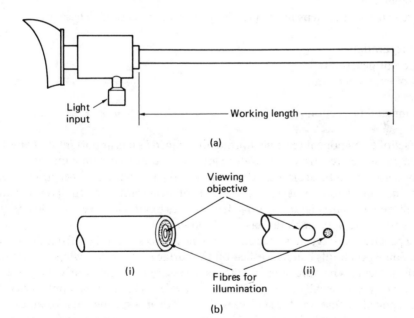

FIGURE 30.1 (a) Rigid optical inspection probe. (b) Probe ends (i) for direct viewing, (ii) for angled viewing

probes are made in many sizes with, for rigid probes, diameters ranging from about 2 mm up to about 20 mm. The minimum diameter for flexible probes is about 4 mm. Probe lengths may vary considerably also, and the maximum working length for a 2 mm probe is about 150 mm. The maximum permissible working length increases as probe diameter increases and may be up to 5 m for a 20 mm diameter probe.

Inspection probes can be designed to give either direct viewing ahead of the probe end, or to a view at some angle to the line of the probe. It is possible to mount a miniature TV camera in place of the normal eyepiece lens system and display an image on a monitor screen.

30.3 Liquid penetrant inspection

Liquid penetrant inspection is a technique which can be used to detect defects in a wide range of components, provided that the defect breaks the surface of the material. The principle of the technique is that a liquid is drawn by capillary attraction into the defect and, after subsequent development, any surface-breaking defects may be rendered visible to the human eye. In order to achieve good defect visibility, the penetrating liquid will either be coloured with a bright and persistent dye or else contain a fluorescent compound. In the former type, the dye is generally red and the developed surface can be viewed in natural or artificial light but in the latter case the component must be viewed under ultra-violet light if indications of defects are to be seen.

Liquid penetrant inspection is an important industrial method and it can be used to indicate the presence of defects such as cracks, laminations, laps and zones of surface porosity in a wide variety of components. The method is applicable to almost any component, whether it be large or small, of single or complex configuration, and it is employed for the inspection of wrought and cast products in both ferrous and non-ferrous metals and alloys, ceramics, glass ware and some polymer components.

There are five essential steps in the penetrant inspection method. These are:

surface preparation,
application of penetrant,
removal of excess penetrant,
development,
observation and inspection.

The surface of a component must be thoroughly cleaned as it is important that any surfaces to be examined must be free from oil, water, grease or other contaminants if the successful indication of defects is to be achieved. After surface preparation, liquid penetrant is applied by brush, spray or immersion so as to form a film of penetrant over the component surface. Penetrant will seep into fairly large flaws in a few seconds but it may take up to 30 minutes for the liquid to penetrate into very small defects and tight cracks.

After the penetration stage, it is necessary to remove excess penetrant from the surface of the component. Some penetrants can be washed off the surface with water, whilst others require the use of specific solvents. The development stage is necessary to reveal clearly the presence of any defect. The developer is usually a very fine chalk powder. This may be applied dry, but more commonly is applied by spraying the surface with chalk dust suspended in a volatile carrier fluid. A thin uniform layer of chalk is deposited on the surface of the component. Penetrant liquid present within defects will be slowly drawn by capillary action into the pores of the chalk. There will be some spread of penetrant within the developer and this will magnify the apparent width of a defect. When dye penetrant is used, the dye colour must be in sharp contrast to the uniform white of the chalk covered surface. The development stage may sometimes be omitted when a fluorescent penetrant is used. If it takes a long time for penetrant to be drawn into a tight crack then it follows that a similar length of time will be needed for liquid to be drawn from the defect by the developer. In consequence, development times of between 10 and 30 minutes are generally required to ensure that all defect indications are visible.

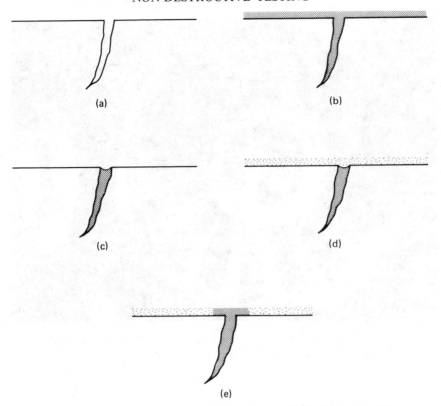

FIGURE 30.2   Stages in penetrant testing. (a) Material surface clean and grease-free. (b) Penetrant
absorbed into defect. (c) Excess penetrant removed, but liquid remains in defect. (d)
Developer applied to surface. (e) Penetrant absorbed into developer giving indication of
defect

After an optimum developing time has been allowed, the component surface is inspected for
indications of penetrant 'bleedback' into the developer. Dye-penetrant inspection is carried out
in strong lighting conditions, whilst fluorescent-penetrant inspection is performed in a suitable
screened area using ultra-violet light. The latter technique causes the penetrant to emit visible
light, and defects are brilliantly outlined. The five essential operations are shown in Figure 30.2.

The liquid penetrant process is comparatively simple as no electronic systems are involved,
and the equipment necessary is cheaper than that required for other non-destructive testing
systems. The obvious major limitation of liquid penetrant systems is that it can detect surface-
breaking defects only. The method is not suitable for use with naturally porous materials such as
unglazed ceramics but is used to detect glazing defects. Some thermoplastics may be affected by
the penetrant fluid, an organic solvent.

The range of applications of liquid penetrant testing is extremely wide and varied. The system
is used in the aerospace, automotive and general manufacturing industries for the quality control
of production and by users during regular maintenance and safety checks. Typical components
which are checked by this system are turbine rotor discs and blades, pistons, cast cylinder heads,
wheels, forged components and welded assemblies. Figure 30.3 shows the inspection of the inlet
assembly of an aircraft engine. The operator is looking for surface cracks under ultra-violet light,
and after the assembly has been processed with fluorescent penetrant.

FIGURE 30.3   Inspection of inlet assembly of aircraft gas turbine using ultra-violet light after application of fluoroescent penetrant. (Courtesy of Magnaflux Ltd)

30.4 Magnetic particle inspection

Magnetic particle inspection is a sensitive method of locating surface and some sub-surface defects in ferromagnetic components. When a ferromagnetic component is magnetised, magnetic discontinuities that lie in a direction approximately perpendicular to the field direction will result in the formation of a strong 'leakage field'. This leakage field is present at and above the surface of the magnetised component, and its presence can be detected using finely divided magnetic particles. The application of dry particles or wet particles in a liquid carrier, over the surface of the component, results in a collection of magnetic particles at a discontinuity. This 'magnetic bridge' indicates the location, size, and shape of the discontinuity.

Current passing through any straight conductor such as a wire or bar creates a circular magnetic field around the conductor. When the conductor is a ferromagnetic material, the current induces a magnetic field within the conductor as well as within the surrounding space. Hence, a component magnetised in this manner is given a circular magnetisation, as shown in Figure 30.4(a). An electric current can also be used to create a longitudinal magnetic field in components. When current is passed through a coil of one or more turns surrounding a component, a longitudinal magnetic field is generated within the workpiece as shown in Figure 30.4(b).

The effectiveness of defect indication will depend on the orientation of the flaw to the induced magnetic field and will be greatest when the defect is perpendicular to the field. This is shown schematically in Figure 30.5. In this Figure, defect A will give a strong indication. Defects B and C are sub-surface defects. An indication will be obtained in respect of defect B as it is normal to the magnetic field but defect C will not be indicated. In the case of defect C, it is too deep within the material to give an indication but, even if such a defect were close to the surface its alignment with the magnetic field would render detection unlikely.

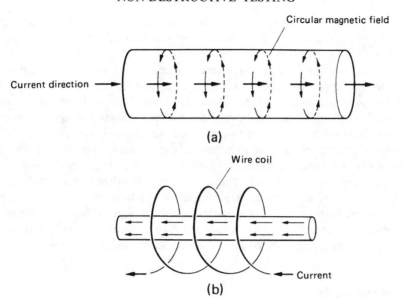

FIGURE 30.4   (a) Current passes through workpiece inducing circular magnetisation. (b) Longitudinal magnetisation induced by placing workpiece within a coil

Generally, in order to indicate the presence of all flaws a component will need to be magnetised more than once. For components of relatively simple shape this is achieved by, firstly, inducing circular magnetisation to indicate longitudinal defects. The component is then demagnetised before magnetising for a second time within a coil to induce longitudinal magnetisation which will enable transverse defects to be located.

Magnetisation of a component could be accomplished using permanent magnets but generally magnetic fields are induced by passing a heavy current through the component, by placing a coil around or close to the component under test, or by making the component part of a magnetic circuit, for example by means of a hand yoke. The actual method used will depend on the size, shape and complexity of the parts to be inspected and, for components to be inspected *in situ*, the accessibility to such components.

The magnetic particles which are used for inspection may be made from any ferromagnetic material of low remanence and they are usually finely divided powders of either metal oxides or metals. The particles are classified as *dry* or *wet*. Dry particles are carried in air or gas suspension while wet particles are carried in liquid suspension. Very high sensitivities are possible with wet

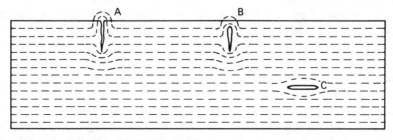

FIGURE 30.5   Magnetic flaw detection. Detectable surface leakage fields produced by defects A and B. Defect C is likely to remain undetected

particle inspection, particularly when a fluorscent chemical is adsorbed on the particles and inspection is made under ultra-violet light. In the best conditions, it is possible to detect cracks with a width of as little as 10^{-3} mm.

Magnetic particle inspection is a sensitive means of detecting very fine surface flaws. It is also possible to obtain indications from some discontinuities that do not break through the surface provided that they are close to the surface. The major limitations of the technique are that it is only suitable for ferromagnetic materials, that for the best results the induced magnetic field should be normal to any defect and thus two or more magnetising sequences will be necessary, and a demagnetising procedure will need to be carried out for many components after inspection. The sensitivity of magnetic particle inspection is very good generally, but this will be reduced if the surface of the component is covered by a film of paint or other non-magnetic layer.

The principal industrial uses of magnetic particle inspection are for in-process inspection, final inspection and the inspection of components as part of planned maintenance and overhaul schedules. The methods are well suited to the inspection of castings and forgings and components such as crankshafts, connecting rods, flywheels, crane hooks, axles and shafts.

30.5 Electrical test methods

The basic principle underlying the electrical test methods is that electrical eddy currents and/or magnetic effects are induced within the material or component under test and, from an assessment of the effects, deductions can be made about the nature and condition of the test-piece. These techniques are highly versatile and, with the appropriate equipment and test method, can be used to detect surface and sub-surface defects within components, determine the thickness of surface coatings, provide information about structural features such as crystal grain size and heat treatment condition, and also to measure physical properties including electrical conductivity, magnetic permeability and physical hardness.

If a coil carrying an alternating current is placed in proximity to a conductive material, secondary or eddy currents will be induced within the material. The induced currents will produce a magnetic field surrounding the coil. This interaction between fields causes a back e.m.f. in the coil and, hence, a change in the coil impedance value. If a material is uniform in composition and dimensions, the impedance value of a search coil placed close to the surface should be the same at all points on the surface, apart from some variation observed close to the edges of the sample. If the material contains a discontinuity, the distribution of eddy currents, and their magnitude, will be altered in its vicinity and there will be a consequent reduction in the magnetic field associated with the eddy currents, so altering the coil impedance value.

Eddy currents flow in closed loops within a material and both the magnitude and the timing or phase of the currents will depend on a number of factors. These factors include the magnitude of the magnetic field surrounding the primary coil, the electrical and magnetic properties of the material and the presence or otherwise of discontinuities or dimensional changes within the material. Several types of search coil are used, two common types being the flat or pancake type coil which is suitable for the examination of flat surfaces and the solenoid type coil which can be used in conjunction with solid or tubular cylindrical parts. For tubes, a solenoid type coil may be placed around the tube or inserted into the bore.

If a component contains a crack or other discontinuity, the flow pattern of eddy currents will be altered and this will cause a change in the magnetic field and, hence, a change in coil impedance. A schematic representation of the effect of a discontinuity on eddy current pattern is shown in Figure 30.6. The impedance of a coil can be determined by measuring the voltage across it. In eddy current test equipment, changes in coil impedance can be indicated on a meter or a chart recorder or displayed on the screen of a cathode ray tube. At component edges, eddy current flow is distorted, because the eddy currents are unable to flow beyond this limiting barrier. The magnitude of this edge effect is usualy very large, and hence, inspection is inadvisable close to edges. In general, it is recommended to limit inspection to an approach of 3 mm from a component edge.

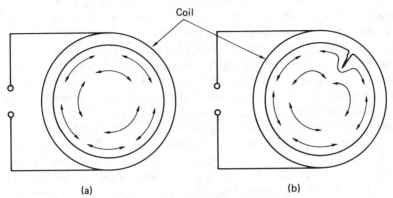

FIGURE 30.6   Cross-section of a bar within a solenoid type coil showing eddy current pattern. (a) Defect free section—uniform eddy currents. (b) Eddy current pattern distorted by the presence of a defect

Eddy currents are not distributed uniformly throughout a part under inspection. They are most dense at the component surface, immediately beneath a coil, and become progressively less dense with increasing distance from the surface. At some distance below the surface of large components, eddy current flow is negligible. This phenomenon is commonly termed '*skin effect*'. When the thickness of a test-piece is small, the distribution pattern of the eddy currents will become distorted and the extent of such distortion will vary with the thickness of the material. This effect is shown schematically in Figure 30.7. It follows then, that for materials of thin section a change in thickness will alter the impedance value of a test coil. An eddy current system,

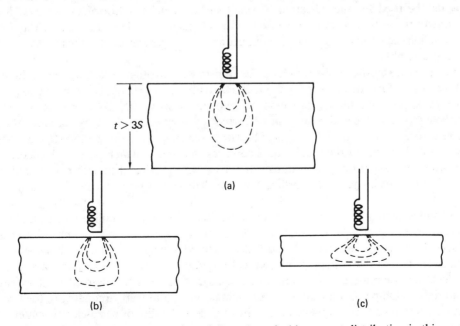

FIGURE 30.7   Schematic representation of distortion of eddy current distribution in thin sections

therefore, when calibrated against known standards, can be used successfully for the accurate measurement of the thickness of thin materials. Thickness measurements can also be made using ultrasonic techniques but, in this case, the degree of accuracy possible diminishes when the material is very thin. The reverse is true for thickness measurements made using eddy current techniques and so the two different methods become complementary to each other.

The inspection frequencies used in electrical techniques range from 20 Hz to 10 MHz. Inspection of non-magnetic materials is performed at frequencies within the range 1 kHz–5 MHz, whilst frequencies lower than 1 kHz are often employed with magnetic materials. The actual frequency used in any specific test is often a compromise. Sensitivity for the detection of small surface flaws is greatest at high frequencies but detection of sub-surface flaws requires low frequencies. Inspection of ferromagnetic materials demands very low frequencies because of the relatively low penetration depth of eddy currents in these materials.

The electrical eddy current system is a highly versatile system which can be used to detect not only cracks, but several other conditions, including corrosion. The test equipment may be small and portable and, with a suitable selection of test probes, can be used in many situations. However, the correct interpretation of signal indications does require considerable skill and experience on the part of the operator. The eddy current equipment which is used for the quality control inspection of the products of many material manufacturing processes is often completely automatic and is highly sophisticated. Production of tubing, bar stock and wire are checked in this way by many manufacturers. The range of applications in which eddy current inspection is used successfully is almost infinite. Some typical uses are the routine in-service inspection of many components in aircraft, including undercarriage wheels, routine inspection of railway track, accurate determination of thickness, including wall-thinning due to corrosion and measurement of the thickness of surface coatings on metals.

30.6 Ultrasonic inspection

Ultrasonic techniques are very widely used for the detection of internal defects in materials, but they can also be used for the detection of small surface cracks. Ultrasonics are used for the quality control inspection of part processed material, such as rolled slabs, as well as for the inspection of finished components. The techniques are also in regular use for the in-service testing of parts and assemblies.

Sound waves are elastic waves which can be transmitted through both fluid and solid media. The audible range of frequency is from about 20 Hz to about 20 kHz but it is possible to produce elastic waves of the same nature as sound at frequencies up to 500 MHz. Elastic waves with frequencies higher than the audio range are described as ultrasonic. The waves used for the non-destructive inspection of materials are usually within the frequency range 0.5 MHz to 20 MHz. Sound waves in fluids are of the longitudinal compression type in which particle displacement is in the direction of wave propagation but shear waves, with particle displacement normal to the direction of wave travel, and elastic surface waves, or Rayleigh waves, can also be transmitted in solids.

Ultrasound is generated by piezo-electric transducers. Certain crystalline materials show the piezo-electric effect, namely, the crystal will dilate or strain if a voltage is applied across the crystal faces. Conversely, an electrical field will be created in such a crystal if it is subjected to a mechanical strain, and the voltage produced will be proportional to the amount of strain. The original piezo-electric material used was natural quartz. Quartz is still used to some extent but other materials including barium titanate, lead metaniobate and lead zirconate are used widely. When an alternating voltage is applied across the thickness of a disc of piezo-electric material, the disc will contract and expand and in so doing will generate a compression wave normal to the disc in the surrounding medium. A transducer for sound generation will also detect sound. An ultrasonic wave incident on a crystal will cause it to vibrate, producing an alternating current across the crystal faces. In some ultrasonic testing techniques, two transducers are used, one to transmit the beam and the other acting as the receiver, but in very many cases only one transducer is necessary. This acts as both transmitter and receiver. Ultrasound is transmitted as a series of pulses of extremely short duration and during the time interval between transmissions the crystal can detect reflected signals.

When a beam of longitudinal compression sound waves reaches a boundary between two media, a proportion of the incident waves will be reflected at the interface and a proportion will be transmitted across the interface. At an air/metal interface, reflection of sound waves will be almost 100 per cent. A fluid, oil or water, is needed as a couplant between an ultrasound transmitter and metal to permit some of the sound energy to be transmitted across the interface into the metal. For normal incidence waves, the transmission across the interface will be of the compression wave type. When the incident beam is at some angle other than normal that portion of the beam which is transmitted across the interface will be refracted. However, there may be two refracted beams transmitted into the metal, because part of the transmitted energy is converted into the shear wave mode. One refracted beam will be of the compression type while the other will be a shear wave, as shown in Figure 30.8.

Much ultrasonic inspection uses normal incidence waves but, in many instances, waves at some angle other than normal are required. The presence of two types of wave with differing velocities within the material would give confusing results and so the angle of incidence is

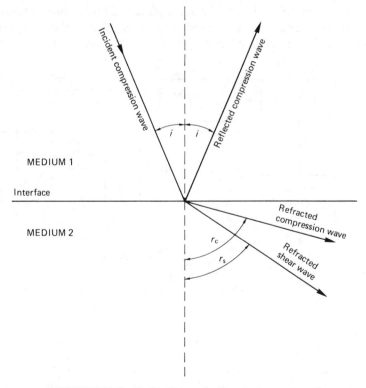

FIGURE 30.8   Reflection and refraction at an interface

adjusted to be greater than the critical angle for compression wave refraction so that only a shear wave is transmitted into the material. This critical angle for total reflection of an incident compression wave at a perspex/steel interface is 27.5°.

If it is required to generate a surface, or Rayleigh, wave, the angle of incidence should be adjusted to a second critical angle to produce a Rayleigh wave at a refracted angle of 90°. The value of this critical angle for a perspex/steel interface is 57°.

An ultrasonic beam transmitted through a metal will be totally reflected at the far surface of the material, a metal/air interface. It will also be wholly or partially reflected by any internal surface, namely cracks or laminations, porosity and non-metallic inclusions, subject to the limitation that the size of the object is not less than one wave length. From this it follows that the sensitivity and defect resolution will increase as the frequency of the beam is increased. The wavelengths of sound at frequencies of 1 MHz and 5 MHz are 6.19 mm and 1.24 mm respectively in aluminium and 5.81 mm and 1.16 mm respectively in steel. At an inspection frequency of 1 MHz defects in an aluminium component with a size smaller than 6.2 mm would not be detectable but would be observed if higher frequencies were employed.

Some metallic materials can only be inspected satisfactorily with a relatively low frequency sound beam because the use of high frequencies could cause a mass of reflections from a normal structural constituent which would mask the indications from defects. This situation arises in the inspection of grey cast iron components where the flakes or nodules of graphite in the iron structure may have a size of several millimetres.

In most ultrasonic test equipment, the signals are displayed on the screen of a cathode ray oscilloscope. The basic block diagram for a typical flaw detector is shown in Figure 30.9.

The information obtained during an ultrasonic test can be displayed in several ways. The most commonly used system is the '*A*' *scan display* (see Figure 30.10). A blip appears on the CRT screen at the left hand side corresponding to the initial pulse and further blips appear on the time base corresponding to any signal echoes received. The height of the echo is generally proportional to the size of the reflecting surface but it is affected by the distance travelled by the signal

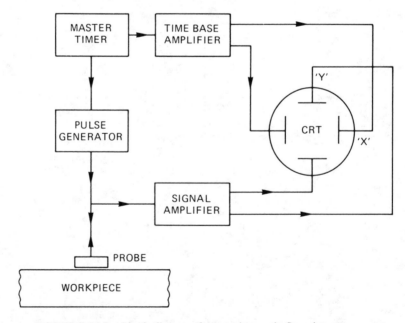

FIGURE 30.9   Block diagram for an ultrasonic flaw detector

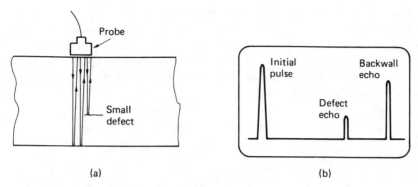

FIGURE 30.10 'A' scan display: (a) reflections obtained from defect and backwall; (b) representation of 'A' scan screen display

and attenuation effects within the material. The linear position of the echo is proportional to the distance of the reflecting surface from the probe, assuming a linear timebase. This is the normal type of display for hand probe inspection techniques.

The presence of a defect within a material may be found using ultrasonics with either a reflection or a transmission technique. The most widely used inspection technique is *normal probe reflection*, as illustrated in Figure 30.10. The pulse is wholly or partially reflected by any defect in the material and received by the single probe, which combines as transmitter and receiver.

There are certain testing situations in which it is not possible to place a normal probe at right angles to a defect and the only reasonable solution is offered by angle probes. A good example of this technique is in the inspection of butt welds in parallel sided plate. The transmitter and receiver probes are arranged as in Figure 30.11(a).

If there is any defect in the weld zone, this will cause a reduction in the received signal strength. Distance AB is known as the skip distance and for the complete scanning of a weld the

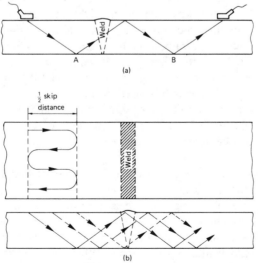

FIGURE 30.11 Angle probe transmission method: (a) probe positions and skip distance; (b) scanning method for complete inspection of butt weld

probes should be moved over the plate surface as shown in Figure 30.11(b). In practice, both probes would be mounted in a jig so that they are always at the correct separation distance.

A Rayleigh, or surface, wave can be used for the detection of surface cracks (see Figure 30.12). The presence of a surface defect will reflect the surface wave to give an echo signal in the usual way. Surface waves will follow the surface contours and so the method is suitable for shaped components such as turbine blades.

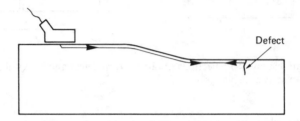

FIGURE 30.12   Crack detection using a surface wave probe

Ultrasonic test methods are suitable for the detection, identification and size assessment of a wide variety of both surface and sub-surface defects in materials, provided that there is access to one surface. Using hand held probes, many types of component can be tested, including *in situ* testing. This latter capability makes the method particularly attractive for the routine inspection of aircraft and road and rail vehicles in the search for incipient fatigue cracks. There are automated systems also which are highly suitable for the routine inspection of production items at both an intermediate stage and the final stage of manufacture.

30.7 Principles of radiography

Very short wavelength electromagnetic radiation, namely X or γ-rays will penetrate through solid media but will be partially absorbed by the medium. The amount of absorption which will occur will depend upon the density and thickness of the material the radiation is passing through and also upon the characteristics of the radiation. The radiation which passes through the material can be detected and recorded on either film or sensitised paper, viewed on a fluorescent screen or detected and monitored by electronic sensing equipment. Strictly speaking, the term *radiography* implies a process in which an image is produced on film.

The basic principle of radiographic inspection is that the object to be examined is placed in the path of a beam of radiation from an X-ray or γ-ray source. A recording medium, usually film is placed close to the object being examined but on the opposite side from the beam source (as shown in Figure 30.13).

X or γ-radiation cannot be focused as visible light can be focused and, in many instances, the radiation will come from the source as a conical beam. Some of the radiation will be absorbed by the object but some will travel through the object and impinge on the film producing a latent image. If the object contains a flaw which has a different absorptive power from that of the object material the amount of radiation emerging from the object directly beneath the flaw will differ from that emerging from adjacent flaw-free regions. When the film has been developed, there will be an area of different image density which corresponds to the flaw in the material. Thus the flaw will be seen as a shadow within the developed radiograph. This shadow may be of lesser or

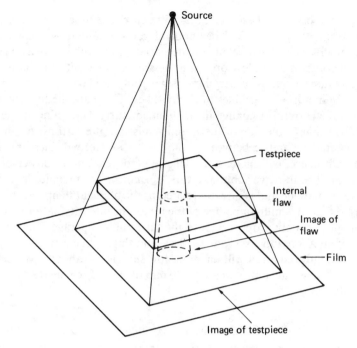

FIGURE 30.13   Schematic diagram showing a radiographic system

greater density than the surrounding image depending on the nature of the defect and its relative absorptive characteristics.

X-rays and γ-rays are indistinguishable from one another. The only difference between them is the manner of their formation. X-rays are formed by bombarding a metal target material with a stream of high velocity electrons within an X-ray tube. γ-rays, on the other hand, are emitted as part of the decay process of radioactive substances.

The major components of an X-ray tube are a cathode to emit electrons and an anode target, both being contained within an evacuated tube or envelope. The general arrangement is shown in Figure 30.14. Electrons, liberated from the cathode by thermionic emission, are accelerated

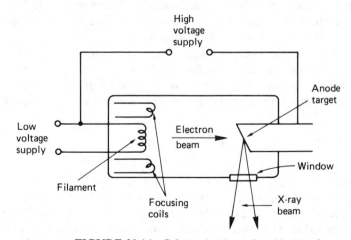

FIGURE 30.14   Schematic view of an X-ray tube

towards the target anode by a large potential difference, the tube voltage. X-ray tube voltages generally range from 50 kV to 1 MV. The anode target is usually tungsten, embedded in a water cooled copper block. The envelope of an industrial X-ray tube is usually of ceramic/metal construction and contains a window opposite the anode to permit X-radiation to emerge from the tube. A 3 to 4 mm thickness of beryllium is generally used as a window material.

There are three important variables in X-ray tubes. These are the filament current, the tube voltage and the tube current. A change in the filament current will alter the temperature of the filament which will change the rate of thermionic emission of electrons. An increase in the tube voltage, the potential difference between cathode and anode, will increase the energy of the electron beam and, hence, will increase the energy and penetrating power of the X-ray beam which is produced. The third variable, the tube current, is the magnitude of the electron flow between cathode and anode and is directly related to the filament temperature. The tube current is usually referred to as the milli-amperage of the tube. The intensity of the X-ray beam produced by the tube is approximately proportional to the tube milli-amperage. The upper practical limit for the voltage of an X-ray tube is about 1000 kV and this will give an X-ray spectrum in which the shortest wavelength radiation will have photon energies of about 1 MeV. Radiation of this high frequency will penetrate approximately 140 mm of steel. The penetrating ability of X-rays is given in Table 30.2.

Table 30.2 *Penetration ability of X-rays*

Tube voltage (kV)	Photon energy for lowest wavelength radiation (MeV)	Penetration ability (mm of steel)
150	0.15	up to 25
250	0.25	up to 70
400	0.4	up to 100
1000	1.0	5 to 140

γ-radiation is emitted during the decay of radioactive nuclei. Unlike the broad 'white' spectrum of radiation obtained from an X-ray tube, a γ-ray emitter will give one or more discrete radiation wave-lengths, each one with its own characteristic photon energy. Radium, a naturally occurring radioactive element, has been used as a source of γ-radiation for radiography but it is far more usual to use radio-isotopes produced in a nuclear reactor.

The specific isotopes which are generally used as γ-ray sources for radiography are caesium-137, cobalt-60, iridium-192 and thulium-170. (The numbers are the atomic mass number of the radioactive nuclei.) There is a continuous reduction in the intensity of radiation emitted from a γ-ray source as more and more unstable nuclei decay. The rate of decay decreases exponentially with time according to:

$$I_t = I_0 \, e^{-kt}$$

where I_0 is the initial intensity, I_t is the intensity of radiation at time t and k is a constant for a particular disintegrating atomic species. An important characteristic of each particular radio-isotope is its *half-life* period. This is the time taken for the intensity of the emitted radiation to fall to one-half of its original value. After two half-life time intervals, the intensity will fall to $\frac{1}{4}$ of the original value, after three half-life time intervals, the intensity will fall to $\frac{1}{8}$ of the original value, and so on. Another characteristic of a γ-ray source is the source strength. The source strength is the number of atomic disintegrations per second and is measured in curies (one curie is 3.7×10^{10} disintegrations per second). The source strength decreases exponentially with time, and the source strength at any given time can be determined using the expression:

$$S_t = S_0 \, e^{-kt}$$

The radiation intensity, usually measured in roentgens per hour at one metre (rhm) is given by source strength (in curies) × radiation output (rhm per curie). (One roentgen is the amount of radiation that will produce ions carrying one electrostatic unit of energy in 1.293 mg of air.) The value of radiation output is a constant for any particular isotope. Another term used in connection with γ-ray sources is *specific activity*. This characteristic, expressed in curies per gramme, is a measure of the degree of concentration of the source.

The characteristics of the commonly used γ-sources for radiography are given in Table 30.3. Commercial radioactive sources are usually metallic in nature, but may be chemical salts or gases adsorbed on carbon. The source is encapsulated in a thin protective covering. This may be a thin sheath of stainless steel or aluminium. Containing the radioactive material in a capsule of this type prevents spillage or leakage of the material and reduces the possibility of accidental mishandling. The encapsulated source is housed in a lead-lined steel container.

Table 30.3 *Characteristics of γ-ray sources*

Source isotope	Half-life period	Photon energy (MeV)	Radiation output (rhm/curie)	Effective penetrating power (mm of steel)
Caesium-137	33 years	0.66	0.39	75
Cobalt-60	5.3 years	1.17, 1.33	1.35	225
Iridium-192	74 days	12 rays from 0.13 to 0.61	0.55	75
Thulium-170	128 days	0.084	0.0045	12 (aluminium not steel)

The obtaining of satisfactory radiographs requires a very high degree of skill and expertise on the part of the radiographer as there are very many factors which affect the formation of an X-ray or γ-ray image. These include the composition, density and dimensions of the object, the type and intensity of the radiation, and the characteristics of the radiographic film. The correct exposure for a particular application may be determined by a process of trial and error or by using an aid such as an exposure chart which relates to a specific grade of film. Often screens are used to improve contrast and reduce exposure times. A lead screen, which is a very thin film of lead bonded to thin card, is placed in contact with both sides of the radiographic film and serves two functions. It absorbs low-energy scattered radiation and interacts with incident high-energy radiation with a resulting emission of electrons. The emitted electrons activate the film emulsion giving improved developed film densities and enhanced contrast. Another type of intensifying screen is the fluorescent screen. Small crystals of calcium tungstate on a thin card base will fluoresce, emitting visible light, when subjected to radiation. They can intensify a radiographic image by a factor of up to 100, thus reducing exposure times, but they do not have a filtering effect on scattered radiation. A third type of screen is the fluoro-metallic screen, comprising both lead and fluorescent crystals, thus combining the advantages of both the former types.

Devices termed *Image Quality Indicators* or *penetrameters* are used as a means of assessing the sensitivity of radiographic inspection. Several designs of image quality indicators have been devised by the various standards organisations and they are generally made in the form of steps or wires of varying thickness in the same or a similar material to that being inspected. The relevant British Standard (BS 3971) covers both step-hole and wire IQI types, and these are shown in Figure 30.15. A straight, six-stage step-hole IQI is shown in Figure 30.15(a) while Figure 30.15(b) shows the wire type IQI. Identification markers must be put on to indicate the type of material and the thickness range of the IQI. In Figure 30.15(a), the coding 8 AL 13 indicates that the IQI is in aluminium with the thinnest step being No. 8 (0.630 mm) and the thickest step being No. 13 (2.00 mm) while in Figure 30.15(b) the coding 9 CU 15 means that there are seven copper wires ranging in diameter from wire No. 9 (0.200 mm diam.) to wire No. 15 (0.80 mm diam.). With the aid of penetrameters, the image quality, or sensitivity, can be expressed as a percentage. The sensitivity is the thickness of the thinnest wire or step or hole visible in the developed radiograph expressed as a percentage of the thickness of the test-piece. In the USA the ASTM standards E 142-86, E 747-90 and E 1025-89 cover the use of IQIs and penetrameters.

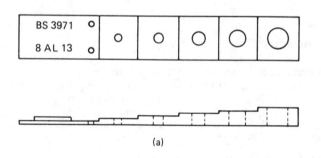

(a)

(b)

FIGURE 30.15　Penetrameters to BS 3971: (a) step-hole type; (b) wire type

The placement of penetrameters is important. They should be placed on the source side of the test-piece and at the edge of the area, namely, in the outer zone of the radiation beam with the thinnest step or wire being outermost.

A radiograph is valueless unless the developed image can be sensibly interpreted, and correct interpretation needs a person who possesses a considerable amount of knowledge, skill and experience. The interpreter, therefore, needs to have a thorough knowledge of the principles of radiography and to be fully aware of the capabilities and the limitations of the techniques and equipment. In addition, the interpreter should have knowledge of the components to be inspected and the variables in the manufacturing processes which may give rise to defects. For example, in the inspection of castings, it would be beneficial if the interpreter is aware of the way in which defects such as gas porosity, shrinkage and cold shuts can occur and the most likely areas in the particular casting where they may be found. The radiographic interpreter is looking for changes in image density in the radiograph. Density changes may be caused by one of three factors, namely a change in the thickness of the test-piece, including visible surface indentations or protuberances, internal flaws within the component, and density changes which may be induced by faulty processing, mishandling or bad film storage conditions and it is important that the interpreter can assess the nature and cause of each density difference observed. The conditions in which radiographs are viewed, therefore, are highly important, and the films should be correctly illuminated by means of a purpose built light source which will give good illumination without glare or dazzle. The radiograph should be viewed in a darkened room so that there will be no light reflections from the surface of the film and the image is seen solely by means of light transmitted through the film. Viewing in poor conditions will cause rapid onset of eye fatigue and so it is also important that the interpreter is in a comfortable position and has no undue distractions. Ultimately, the efficiency of flaw detection is determined by the skill and experience of the interpreter and a highly experienced radiograph interpreter may locate defect indications which could be missed by a less experienced person.

X- and γ- radiation can cause damage to body tissue and blood, but any damage caused is not immediately apparent. The effects of any small doses of radiation received over a period of time is cumulative and so all workers who may be exposed to even small quantities of radiation should have a periodic blood count and medical examination. Strict regulations cover the use of X-and γ-rays and the quantity of radiation to which workers may be exposed. The extent of radiation which may be received by classified workers in the field of radiography must be monitored and this is best achieved by recording the dosage received on a radiation monitoring film (film badge) or by using a pocket ionisation chamber.

Radiography is capable of detecting any feature in a component or structure provided that there are sufficient differences in thickness or density within the test piece. Large differences are more readily detected than small differences. The main types of defect which can be distinguished are porosity and other voids and inclusions, where the density of the inclusion differs from that of the basis material. Generally speaking, the best results will be obtained when the defect has an appreciable thickness in a direction parallel to the radiation beam. Plane defects such as cracks are not always detectable and the ability to locate a crack will depend upon its orientation to the beam. The sensitivity possible in radiography depends upon many factors but generally if a feature causes a change in absorption of 2 per cent or more, compared with the surrounding material then it will be detectable.

Radiography and ultrasonics are the two methods which are generally used for the successful detection of internal flaws that are located well below the surface, but neither method is restricted to the detection of this type of defect. The methods are complementary to one another in that radiography tends to be more effective when flaws are non-planar in type whereas ultrasonics tends to be more effective when the defects are planar. Radiographic inspection techniques are frequently used for the checking of welds and castings and in many instances radiography is specified for inspection of these components. This is the case for weldments and thick-wall castings which form part of high pressure systems. Radiography can also be used to inspect assemblies to check the condition and proper placement of components. One application for which radiography is very well suited is the inspection of electrical and electronic component assemblies to detect cracks, broken wires, missing or misplaced components and unsoldered connections. Radiography can be used to inspect most types of solid material but there could be problems with very high or very low density materials. Non-metallic and metallic materials, both ferrous and non-ferrous, can be radiographed and there is a fairly wide range of material thicknesses that can be inspected. The sensitivities of the radiography processes are affected by a number of factors including the type and geometry of the material and the type of flaw.

30.8 Acoustic emission inspection

When strain energy is released as a consequence of structural changes taking place within a material, some of the energy is emitted as high-frequency waves with frequencies in the range from 50 kHz to 10 MHz. Phase transformations, plastic yielding and crack growth generate acoustic signals which may be detected and analysed. Basically, there are two types of acoustic emission from materials, continuous and an intermittent or 'burst' type. Continuous emission is usually of low amplitude and is associated with dislocation movement and plastic deformation, while burst type emissions are of high amplitude and short duration which result from the development and growth of cracks.

Acoustic emission inspection can offer several advantages over other forms of non-destructive inspection. For example, it can assess the dynamic response of a flaw to an imposed stress. There is a marked increase in the intensity of emission when a crack approaches critical size and, thus, a warning is given of instability and catastrophic failure. It is also possible to detect growing cracks of about 2×10^{-4} mm in length. This is a much smaller size than is detectable by other techniques.

Acoustic emissions are sensed by a transducer close-coupled to the component and the signal from the transducer is amplified, filtered and processed to give an audio and/or video output. In some applications, for example, inspection of pressure vessels, it is necessary to detect both the nature of and the precise location of the source of acoustic signals. In such cases, several transducers are used, spaced over the surface of the vessel, and computer-assisted monitoring of time of arrival of signals to the various transducer locations permits analysis of both the source type and its location.

A wide variety of materials can be inspected, including metals, ceramics, polymers, timber and composites. Although the emission sources from each material type may differ, the characteristic signals received can be correlated with the integrity of the material.

30.9 Vibration testing

There is a natural frequency of vibration for any object. This natural frequency is dependent on the density and the elastic constants of the material. A cracked bell will have a different natural frequency (that is, emit a different sound) from a sound bell of the same dimensions and made from the same material. In this case the presence of a crack has changed the dimensional factors determining the natural frequency. This principle of stimulating vibrations and listening to the sound produced has been used as a non-destructive test technique, for example, in the checking of railway rolling stock wheels by tapping them with a hammer.

Nowadays, vibration testing is used in fundamental research as an extremely accurate method for the determination of the elastic constants of materials. For bars of uniform cross-section the relationships between the frequency of vibration and the constants of the material are as follows:

For bar of circular or square cross-section, clamped centrally, and stimulated into longitudinal vibration:

$$f = \frac{n}{2l}\sqrt{\frac{E}{\rho}}$$

where f is the frequency; n is an integer; l is the length of the bar; E is Young's modulus; ρ is the density of the material.

For a bar of circular cross-section, clamped centrally, and stimulated into torsional vibration:

$$f = \frac{n}{2l}\sqrt{\frac{G}{\rho}}$$

where G is the modulus of rigidity.

Poisson's ratio, v, is a function of a ratio of the longitudinal and torsional vibration frequencies for the same bar.

$$\frac{f_L}{f_T} = \sqrt{[2(v + 1)]}$$

It is possible to determine the elastic constants of a material to an accuracy of up to one part in a hundred thousand in a vibration test.

The damping capacity of a material can also be determined using a vibration test technique The damping capacity of a material is its ability to attenuate vibrations. The damping capacity, ζ, is given by:

$$\zeta = \frac{1}{n}$$

where n is the number of vibrations of a test bar in free attenuation from an amplitude A to an amplitude of A/e, that is, $0.3684A$ (e is the exponential function).

The damping capacity of a material is highly dependent on its structure, and minute changes in structure are reflected in changes in the damping characteristics. This feature is used in much fundamental research in materials.

30.10 Questions

30.1 The velocity of compression waves in aluminium is 6190 m/s. Determine the wavelength of ultrasound waves in aluminium at a frequency of 1.5 MHz. What would be the minimum frequency which could be used if it was a requirement that all defects of size 3 mm and upwards be detected in samples of aluminium using ultrasonic inspection?

30.2 Explain how the sensitivity of radiographic inspection can be assessed.

In a radiograph of an aluminium component with a maximum sectional thickness in the line of the radiation of 28 mm, the thinnest portion of a step-hole IQI which is clearly visible has a thickness of 0.500 mm. Determine the sensitivity of the radiograph.

30.3 The dimensions of a metal bar of circular cross-section are measured accurately and are: length—100.12 mm, diameter—10.06 mm, mass—62.58 g. The bar is then clamped midway along its length and stimulated into longitudinal vibration. The first two resonant frequencies are found to be 25.75 kHz and 51.5 kHz. Determine the value of Young's modulus for the material.

31

Macro- and Micro-examination

31.1 Macro-examination

Macro-examination, examination with the naked eye or at a low magnification, is a useful technique for the inspection of fracture surfaces, and for determining some of the characteristics of metal structures.

Much useful information can be obtained from visual examination of a fracture surface with the naked eye. The difference between a tough fibrous fracture and a brittle cleavage fracture can be readily observed. A fatigue fracture also possesses a characteristic appearance (refer to Chapter 27 and Figure 27.3). It may also be possible to detect the presence of slag inclusions and porosity on a fracture surface, and such defects may have been the points of initiation of the failure. This type of surface detail will be revealed with greater clarity if the surface is further examined with the aid of a low-power (up to × 50 magnification) stereo microscope.

In order to obtain information about the structure of a metal it is necessary to section the material. The cut surface is then ground flat and etched with a chemical reagent. A high degree of surface finish on the specimen is not essential for macro-examination, and grinding can be finished with a grade 0 emery paper. During the etching treatment the surface layer of metal is removed and there is a preferential attack on certain constituents and inclusions. There are very

Table 31.1 *Some etchants for macro-examination*

Composition	Method of application	Uses
Hydrochloric acid 140 ml Sulphuric acid 3 ml Water 50 ml	Immersion in solution for 15–30 minutes at 90°C	A deep etch for steels
Copper ammonium chloride 9 g Water 91 ml	Immersion in solution for 0.5–4 hours	To reveal dendritic structures in steels
Ferric chloride 25 g Hydrochloric acid 25 ml Water 100 ml	Immersion	For copper and its alloys
Ammonium persulphate 10 g Water 100 ml Ammonium hydroxide 50 ml	Immersion	For copper and its alloys
Hydrofluoric acid 20 ml Water 80 ml	Swab surface of sample with etchant	For aluminium and its alloys

many macro-etching agents and the choice of agent will be dependent on the nature of the metal and the type of feature that it is desired to reveal. Details of some macro-etchants are given in Table 31.1. Macro-examination after etching will reveal such detail as defects, segregation effects, and fibre structure in wrought metals.

Sulphur printing is a macro-examination technique that is suitable for the examination of plain carbon steels. Steels contain small sulphide inclusions, and the distribution of these inclusions within the steel is a guide to the distribution of all non-metallic inclusions. The sample of steel to be examined is sectioned and the cut surface is ground flat. A piece of bromide photographic paper is soaked in a 3 per cent solution of sulphuric acid for two minutes and it is then carefully placed, with the emulsion side downward, on the prepared steel surface. The sulphuric acid reacts with sulphide inclusions forming hydrogen sulphide gas. This gas reacts with the silver bromide in the emulsion forming a brown deposit of silver sulphide. The bromide paper should be in contact with the steel surface for about three minutes. The bromide paper is then removed from the steel, washed in water, and the print is 'fixed' by immersion in a 'hypo' solution for a few minutes. Dark-brown areas on the print indicate areas containing sulphide inclusions in the steel section.

31.2 Micro-examination of metals

Microscopic examination of metals is used to reveal fine details of structure. Metals may be examined at magnifications of up to × 3000 with the aid of an optical microscope. In order to examine the structure of a metal at high magnification it is necessary that the metal specimen be carefully prepared and, because metals are opaque to light, incident illumination must be used. Oblique incident illumination is perfectly satisfactory for macro-examination, but normal incident illumination must be used for examination at magnifications of × 50 and greater. The construction of a metallurgical microscope differs from that of a biological microscope (using transmitted light) to cater for this. The metallurgical microscope possesses a built-in light source. A partially reflecting mirror is situated in the microscope tube, and this will reflect the illuminating light through the objective lens on to the specimen surface (Figure 31.1). Alternatively, a small reflecting prism may be used in the microscope tube.

A metal specimen for micro-examination is prepared by cutting a small, but representative, sample from a metal component, followed by grinding and polishing a surface of the specimen to a mirror finish. This is achieved by a series of successive hand-grinding operations using progressively finer grades of paper. The specimen should be rotated through 90° at each change of paper. The final grinding should be on grade 500 or 600 silicon carbide paper. Paraffin should be used as a lubricant for the grinding of very soft metals. After grinding is complete the very fine surface scratches are removed by polishing the surface to a mirror finish. The polishing powders that may be used are alumina, jeweller's rouge, or magnesia. These polishing powders are suitable for the polishing of steels and many other materials, and are used in water suspension in conjunction with a Selvyt or broad-cloth polishing cloth. The polishing cloth is mounted on a flat rotating disc and the specimen is held against this under light pressure. The proprietary metal polishes 'Brasso' and 'Silvo' are highly suitable for polishing copper and copper alloy specimens. Pastes containing very fine diamond dust (particle sizes ranging from $\frac{1}{2}$ to 3 microns) are also widely used for metal polishing. Alternatively, the finely ground surface may be polished electrolytically. This entails anodic solution of the ground surface in a suitable electrolyte.

The flat polished surface of the specimen should be examined under the microscope. The highly polished metal surface will appear bright and will show no structure, as it acts as a mirror to the normal incident light, but the following features, if present, can be observed:

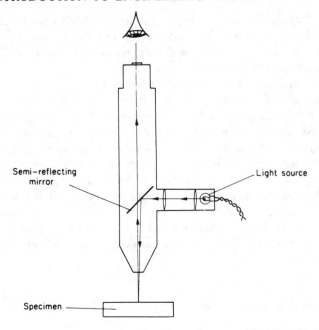

FIGURE 31.1   Principle of metallurgical microscope for normal incident illumination

(a) cracks and porosity,
(b) non-metallic inclusions (slag inclusions, and graphite in cast irons),
(c) hard constituents, which stand out in relief from the matrix.

   In order to reveal the complete structure of the specimen it is necessary to etch the polished surface in a dilute chemical reagent. The choice of etchant will depend on the nature of the material and the type of feature being investigated. Details of some micro-etchants are given in Table 31.2. The action of the etching agent will be selective and there will be a preferential attack at crystal grain boundaries and at constituent boundaries. Examination of the etched specimen under the microscope will now show the grain boundaries revealed (Figure 31.2). The effect of an

Table 31.2 *Some etchants for micro-examination*

Etchant	Uses
Nital: 2 ml nitric acid in 100 ml industrial alcohol	Excellent general etchant for irons and steels, other than stainless steels
Ammonium persulphate 10 g Ammonium hydroxide 20 ml Water 80 ml	Good general etchant for copper and its alloys
Ferric chloride 10 g Hydrochloric acid 30 ml Water 100 ml	Good etchant for copper and its alloys; provides more contrast than ammoniacal ammonium persulphate solution
Hydrofluoric acid 5 ml Nitric acid 1 ml Water 100 ml	Etchant for aluminium alloys; surface of the specimen should be swabbed with cotton wool soaked in etchant
Sodium hydroxide 1 g Water 100 ml	General etchant for aluminium alloys

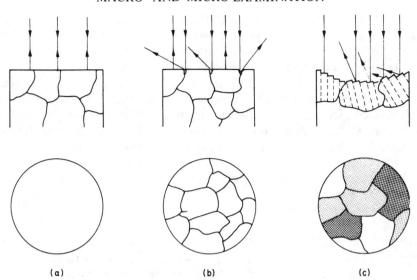

FIGURE 31.2   (a) Polished metal surface. No detail revealed. (b) Etched surface revealing grain boundaries. (c) Etched surface revealing grain boundaries, but with grain shading

etching agent is not usually a general solution of the metal surface. Often the attack produces a faceted surface, and the orientation of the surface facets may vary from grain to grain. In this case the amount of reflected light recaptured by the objective lens of the microscope will vary from one crystal grain to another. This will lead to some crystals appearing to be light in colour and other crystals appearing dark, even though all the crystals may be identical in composition and type. The photo-micrographs of copper and α bronze (Figures 15.4 and 15.7) show this effect. In each of these alloys there is only one phase present, even though there are many shades present in the micrographs.

31.3 Micro-examination of non-metals

Unlike metals, most non-metallic materials are transparent in thin section. For micro-examination, ceramics and polymer materials are prepared as very thin sections mounted on a glass slide and illuminated for viewing by transmitted light. The preparation method for a ceramic material for microscopic examination involves, first, cutting a small thin sample using an abrasive cutting wheel, smoothing one surface to give a fine surface finish and cementing this sample, smooth surface downwards, onto a glass slide. The cement generally used is Canada balsam. After mounting, the exposed surface of the ceramic specimen is then ground down by means of carborundum, emery, or diamond abrasives to a thickness of about 30 μm. Finally, the prepared surface is covered with a thin glass strip, again cemented with Canada balsam.

While considerable information can be gained from viewing a thin slice of material using ordinary transmitted light, further information, leading to positive identification of mineral constituents, will be obtained if the specimens are viewed using polarised light. Polarising microscopes are fitted with two Nicol prisms, one, termed the *polariser*, mounted below the specimen stage and the other, the *analyser*, mounted in the microscope tube. A Nicol prism is a rhomb-shaped prism composed of two pieces of clear calcite bonded with Canada balsam and when incident ordinary light is transmitted into one end of the prism, a ray of plane polarised light emerges from the other end. Many materials show the property of *birefringence*, meaning that they can divide a ray of polarised light into two rays. There are two refractive indices, one for each ray, and the two rays possess different velocities in the material. In the micro-examination of a transparent material in polarised light, the incident ray is divided into two in passing through a birefringent material. After passing through the analysing Nicol the two rays emerge polarised in the same plane but out of phase with one another (as they possess differing velocities) and the resulting interference produces an interference colour. This effect is of diagnostic value. In addition to its use in the examination of ceramics and minerals, polarised light is used in microscopic studies to determine the fibre orientation in composite materials.

31.4 Transmission electron microscopy

The wavelength of visible light is of the order of 5×10^{-5} m and this means that it is impossible to resolve objects of smaller size than about 1 micron using an optical microscope. For magnifications in excess of about $\times 2000$ it is necessary to use the electron microscope. Electrons possess wave characteristics and the wavelength of electrons in a 100 kV electron microscope is about 3.5×10^{-12} m. In an electron microscope a beam of electrons is produced by an electron 'gun'. The electron beam passes through an evacuated tube and is focused on to the specimen by means of magnetic lenses. The specimen is inserted into the vacuum chamber via an air-lock. The electrons pass through the specimen and through the magnetic objective and projector ('eye-piece') lens systems of the microscope to be projected on a fluorescent screen. Magnifications of up to $\times 100\,000$ may be obtained with electron microscopes, and objects of sizes of the order of 10^{-9} m may be resolved. Because the electron microscope uses a transmission technique and metal specimens readily absorb electrons, metal samples can be examined directly only if in the form of very thin foils (foil thicknesses of the order of 10^{-6} m). These thin foils of metals have to be prepared very carefully. This is usually done by thinning a small piece of the metal by chemical or electrolytic solution. Transmission electron microscopy of metal foils has been used as a research technique since the mid-1950s. Before that date metal samples were not viewed directly and replica techniques were used. A replica of an etched metal surface can be obtained by depositing a thin film of a plastic on the surface. The replica can then be stripped from the metal surface and viewed in an electron microscope. Both replica and thin foil techniques are used today.

31.5 Scanning electron miscroscopy

Another type of electron microscope that is used as a research tool is the scanning electron microscope. This type of microscope can accommodate a fairly large specimen and is used for examining the surface topography of a material, for example a fracture surface. A very fine electron beam scans the surface of the specimen, in a similar manner to the scanning of a television picture. The electron beam is scattered from the specimen surface and scattered electrons are picked up by a collector. The signal produced from the collected electrons is used to modulate the scanning beam of a cathode ray tube and this produces a picture of the surface area of the specimen under examination. One major advantage of the scanning electron microscope is that it possesses a large depth of focus. Magnifications of up to $\times 40\,000$ are possible with this type of instrument.

31.6 Other analytical techniques

An electron beam will suffer diffraction by the regularly spaced rows or planes of atoms and the diffraction pattern is characteristic of the type of crystal structure. This diffraction effect can be used for the analysis of crystal structures in the same manner as for X-ray diffraction, as described in Section 5.8.

Spectrographic analysis

When a material is excited in some way, the energy emission which follows excitation is characteristic of the chemical elements present in the material. Each chemical element gives rise to a unique line spectrum emission. This is used as the basis of a rapid and accurate method of chemical analysis. A line spectrum, produced by spark excitation, is analysed. In one method, the spectrum is recorded on a photographic plate and the image examined. A qualitative analysis is obtained by observing the number of characteristic spectra present. From measurement of the relative intensities of certain spectral lines, a quantitative analysis can be obtained. In a direct reading spectrograph, the complete spectrum is not recorded but photomultiplier tubes are set in positions to receive emissions corresponding to specific lines for various elements, and the quantity of light received is directly related to the amount of each element present, thus enabling a quantitative chemical analysis to be given.

Electron probe microanalyser

A fine electron beam can be directed at a specific area or areas of a test-piece. The beam stimulates an X-ray emission from the excited area, this emission being characteristic of the chemical element or elements in the scanned area. From an analysis of the emission, both a qualitative and quantitative assessment of composition can be obtained.

32

Materials Selection

32.1 Introduction

At some stage in the process of converting a design idea into 'hardware', decisions must be taken on the choice of material and the manufacturing route. These decisions should be taken at the earliest possible moment as there are a series of complex interactions between the three elements of materials, manufacturing and design, as illustrated in Figure 32.1

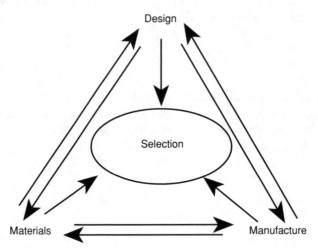

FIGURE 32.1   Inter-relationships in material selection

Let us look at some simple examples of these interactions. The rough size and shape of a component design will have an influence on both the process route and the choice of material. If die-casting or injection moulding are considered as the best methods to produce the desired shape then it follows that the choice of possible materials is restricted to the lower melting point metals and melt processable thermoplastics.

The type of material considered will also influence design decisions. Details of design will differ between, for example, metals and polymers. Also, specific characteristics of a material may permit differing approaches to certain design features. An example of this is the low stiffness and good 'spring-back' of plastics which permits design for 'clip-fit' assembly. Similar components made from metals would require the use of spot welds, rivets or screws for assembly.

The properties of a material also exert major influences on methods of manufacture. The material properties may preclude the use of some process methods or make processing difficult and costly. The use of extremely hard metals could give rise to the necessity for expensive machining processes to obtain the required dimensional tolerances, although, as an alternative, precision casting using the investment process may be feasible. Conversely, the method of manufacture will affect the properties of the material. Wrought metal products are often stronger than castings made from metals or alloys of similar composition.

In the selection of a suitable material to satisfy a particular design and product requirement, it is necessary to look at many aspects to ensure that the component or assembly can be manufactured within the resources available, that the completed product will function satisfactorily throughout its design life, and that all this can be achieved at an acceptable cost.

32.2 *Parameters to be considered*

The parameters which need to be considered when selecting a material to fit a design specification include:

Material properties—mechanical		elastic moduli and stiffness yield and maximum strengths fatigue strength creep strength fracture toughness hardness ductility abrasion resistance
	physical	density T_m and T_g values electrical conductivity magnetic properties thermal conductivity thermal expansion thermal stability
	chemical	resistance to chemicals and solvents corrosion resistance oxidation resistance weathering resistance
Manufacturing characteristics—		castability formability machinability
Cost and availability—		material cost manufacturing cost availability price stability

With reference to material properties, it is often very useful, when comparing the relative merits of a range of materials, to look at *specific properties*, that is, the ratio of the particular property to the density of the material. Figures 32.2 and 32.3 are bar charts showing the ranges of specific modulus and specific tensile strength respectively for materials. It will be seen from these figures that the specific modulus, that is, specific stiffness, is very similar for all the main

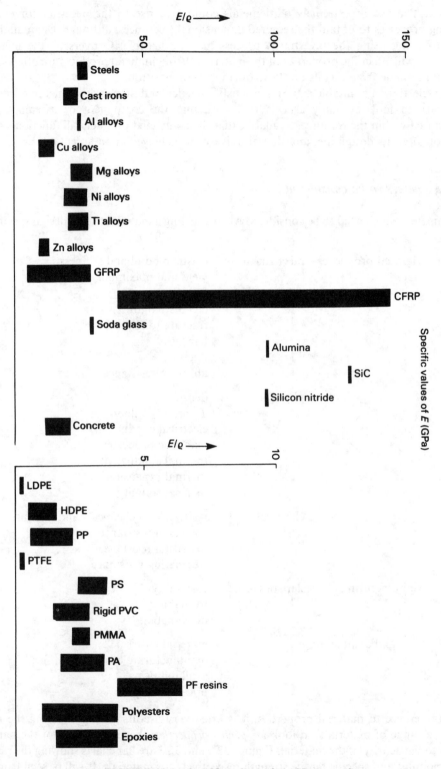

FIGURE 32.2   Specific values of E (GPa/g cm^{-3})

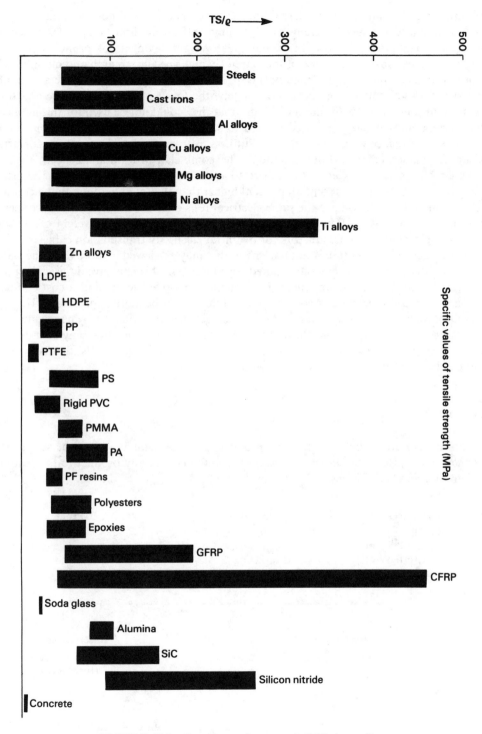

FIGURE 32.3   Specific tensile strength (MPa/g cm^{-3})

engineering metals, and very much less than it is for carbon fibre composites and a number of ceramics. Similarly, the specific strengths of titanium and carbon fibre composites are considerably higher than those of most other materials. Weight saving is of prime importance in aerospace applications and, because of the extremely high specific strength and modulus offered by carbon fibre composites, these materials were investigated extensively for use as a fan blade material for turbofan aero engines. Unfortunately, the lower abrasion resistance and fracture toughness, in comparison with titanium, meant that they could not be used for this application. Solid titanium alloy blades were used instead to satisfy this design requirement but the latest development, to reduce weight and increase stiffness, is to manufacture composite construction titanium fan blades, composed of a titanium honeycomb clad with titanium sheet.

Similar effects can be seen with respect to other properties, for example, the electrical conductivity of pure copper is some 60 per cent higher than that of aluminium, but a conducting wire of aluminium made to have the same electrical resistance per unit length as a copper wire, although of larger cross-sectional area, is approximately one-half of the weight of the copper, hence the use of aluminium conductors for overhead electricity transmission cables.

Material cost is another important parameter that may be viewed in more than one way. It is customary to see the cost of materials quoted per unit mass. This may give a misleading picture as, frequently, the size of a component is determined by other factors and a cost comparison between materials on the basis of cost per unit volume would be more appropriate. The relative positions of materials in a league table of costs change when the criterion is changed from £ per unit mass to £ per unit volume, as shown in Table 1.5. Yet another method may be used to good effect when attempting to compare the relative costs of competing materials, that is to establish tables on the basis of cost per unit of property.

32.3 Costs

Cost and quality are all-important. For a product to be marketable, it must be supplied at an *acceptable cost* and be of an *acceptable quality*. The full cost of any product to the customer is made up of a number of items, as shown in Figure 32.4

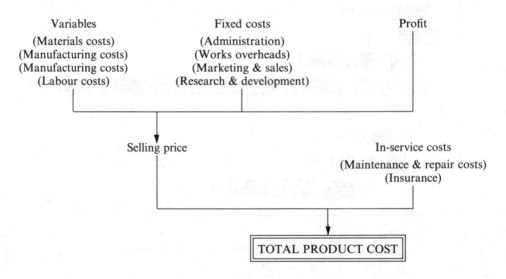

FIGURE 32.4   Product cost build-up

It is difficult to generalise, but in many manufacturing industries the cost of materials accounts for about 50 per cent of the total works cost of the finished product. From this, it follows that the use of a cheaper raw material will have a significant effect on the final selling price. While this is generally true, it is not always so. In some cases, the high cost of a more expensive material may be offset by the material having better processing characteristics. For example, maraging steels are more expensive than high strength low alloy steels but the former materials may be machined to final dimensions and tolerances when in a soft condition, with final high strength and hardness being developed by a low temperature ageing heat treatment. The low alloy steel, on the other hand, may require to be machined to final dimensions after all hardening heat treatments have been completed. If the component is of complex geometry, the use of costly machining processes on a very hard material may make the total cost of the low alloy steel product higher than that of one in maraging steel.

Labour, general manufacturing costs and overheads account for about 50 per cent of total works cost and anything which can reduce this aspect of costs will be worthwhile. Major savings could accrue if the final component shape can be produced with the minimum number of process operations. The production of 'near-net' shapes by a single direct process, such as die-casting, injection moulding, compaction and sintering from powder, etc., should lead to a lower works cost than use of a processing route involving many intermediate stages and costly machining processes.

A few examples of the use of intelligent design and materials selection to effect cost reduction are given below:

(a) Figure 22.6 shows the individual castings and the complete assembly of a saw grinding attachment. All the parts are precision die castings in a zinc alloy. The knobs of the adjusting screws are cast direct on to the screws. The whole of this unit was specifically designed on the basis of production by die casting in a zinc alloy, but the cost of producing a component to fulfil the same purpose using other processes involving machining operations would probably be about twice the cost of producing this design.

(b) Production of parts by powder metallurgy processes can frequently lead to the unit cost of components being about 50 per cent less than the cost of producing parts by rolling or forging followed by machining operations. Some parts produced by powder techniques are shown in Figure 22.31.

(c) Figure 32.5 shows two examples of replacement of complex assemblies by a single injection moulding. Part (a) shows a combined gear cam, formerly made by assembling five separate metal parts, made as a one-piece moulding in nylon 6.6. Part (b) shows a car accelerator pedal made in polypropylene. This is also a one-piece moulding, the hinge being part of the moulding.

(d) Design changes may frequently be progressive. A pulley for use in a washing machine was formerly made as an iron casting, which needed a considerable amount of machining. The works cost of the complete pulley was 14.4p per component. The part was redesigned as an aluminium die casting, but some machining was still required. This change resulted in a reduction of 4.7p in the cost per component. A further design change making the pulley wheel as a two-piece die casting in a zinc alloy, requiring assembly of the two parts, but no machining, gave a further saving in works cost of 4.1p. Thus the two design changes resulted in a total reduction of 61 per cent of the works cost per component*.

Energy is a costly resource and the general trend is for energy costs to rise continually. A very

* Note that the cost figures given in this example relate to design changes made between 1965 and 1970.

(a)

(b)

FIGURE 32.5   (a) Combined gear/cam for an office calculating machine, showing the original component made from five metal parts and the new component which is a single moulding in nylon 6.6 (courtesy of ICI Plastics Division and Bell Punch Co. Ltd). (b) Polypropylene accelerator pedal with integrally moulded hinge (courtesy of ICI Plastics Division and Plastics Division of Hills Precision Die Castings Ltd)

considerable quantity of energy is consumed in the processing of materials, as for example in the extraction of metals from their ores and the working and shaping processes for materials. The approximate energy content of some common materials is given in Table 32.1. The energy content of specific materials is not fixed and for some, particularly some metals, may rise considerably as mineral reserves diminish and ores of lower metal content have to be worked.

The rising cost of energy and the rapid decline in reserves of non-renewable sources makes it more important than ever that, wherever possible, materials are recycled. The additional energy content of metals produced from secondary sources, namely recycled scrap, is comparatively small. Also, greater recovery and use of metals from secondary sources will extend the lifetimes of already scarce mineral resources.

Table 32.1 *Approximate energy content of some materials (GJ/tonne)*

A Materials derived from primary sources			
Titanium bar	560	Zinc (castings)	70
Magnesium extruded bar	425	Mild steel (rod)	60
Aluminium (ingot)	280	Cast iron (castings)	50
Aluminium (sheet)	300	Glass	20
PVC (rod or tube)	180	Reinforced concrete	12
Nylon 66 (rod)	180	Cement	8
Polyethylene (sheet)	110	Brick	4
Stainless steel (sheet)	110	Timber	2
Copper (pipe)	100	Gravel	0.1

B Additional energy content of metals from secondary sources			
Aluminium	45	Mild steel	20
Copper	30	Cast iron	17

32.4 The selection process

There are some instances where the choice of a material is so severely restricted by service requirements that there is either no choice or only an extremely limited choice of possible materials. For example, if it is considered essential to the design that a component possess an electrical resistivity of less than $2.0 \times 10^{-8}\Omega m$, then the only choices are silver, high-conductivity copper, or a few copper alloys. However, perhaps the design requirement, by quoting a maximum resistivity value, is too restrictive. On a weight basis, aluminium is a more efficient electrical conductor than copper, and is used for many high conductivity applications. It is another factor, not conductivity, which prevents aluminium replacing copper for all electrical applications, namely, the fact that copper can be soft soldered with ease, permitting connections to be made with greater ease than is possible with aluminium.

In most cases, the problem is not simple as there may be very many materials which could be considered as possible contenders to meet a design specification. In many cases, there may be conflicting requirements, for example, high strength coupled with high ductility, and compromises have to be made. Frequently, there will be no perfect solution, but several alternative viable solutions can be achieved. The solution eventually arrived at will be the result of an optimisation process.

How can we come to a decision? In order to take an informed decision two things are necessary.

(a) Reliable property data on materials.
(b) Full appreciation of the service conditions that the component or assembly is expected to operate under: working stresses, possibility of overloads, environmental conditions leading to weathering or corrosion, etc.

In the case of most well-established metals and alloys, there is much reliable property data available. Also, major initiatives have taken place in recent years to standardise many commercial alloy compositions and specifications internationally. In the case of new alloy compositions, much testing and property evaluation has to take place before the material can enter service. Some of this testing, such as creep and corrosion tests, is very long term. The situation with regard to polymer and ceramic materials is less clear. There is far less standardisation. The amounts and exact nature of fillers and additives in, say, a polypropylene moulding compound may vary significantly from one major manufacturer to another and the user must rely on property data supplied by the specific manufacturer, or else embark on a lengthy and costly programme of testing for himself. In some cases, a new material may be used extensively on the basis of its attractive properties, as determined by short term tests with subsequent problems developing as a result of long-term ageing and weathering problems. An example of this is high-alumina cement. This is a high-strength cement with a very rapid hardening rate. When first developed, it was widely used in the construction industry for structural concrete. Unfortunately, in the long-term, a structural change occurs in one of the cement constituents leading to loss of strength and an increase in porosity which gives a greater susceptibility to weathering. A number of structural failures ensued and the use of high alumina cement in structural concrete is now greatly restricted.

The vast amount of property data available for materials, much of it being quantitative, but some of a qualitative nature, is best handled using the power of a computer. In recent years, there have been a number of developments in the production of materials property data bases, some of which are available commercially either as subscription services or for outright purchase.

The second aspect, a thorough analysis of the service conditions which a component or assembly has to operate under, and the conditions which may cause premature failure, requires the exercise of considerable skill and experience by the design and materials engineers. As stated earlier, the selection of a suitable material to fit a requirement often involves compromise, with a 'trade-off' between properties. As the selection process is one of optimisation, it is of extreme importance for the engineer to be able to determine which is the most important property to satisfy a particular design—is stiffness more important than fatigue strength, or is fracture toughness more critical than yield strength?

A systematic approach to the problem of materials selection is necessary and the following is cited as one possible approach.

Stage 1

Analyse fully the product specification and determine the minimum acceptable values for all the relevant material properties.

Stage 2

Make first selection by eliminating all materials which do not possess all the minimum criteria.

Stage 3

Assess the degree of relative importance of the various required properties from essential through to desirable, and for each property place the potential materials in ranking order.

Stage 4

Evaluate the material and process costs for each material.

Stage 5

On the basis of the decisions made in Stage 3 and the data from Stage 4, optimise to determine the materials which give the best overall combination of properties for the least cost.

It is hoped that the foregoing paragraphs give the reader some awareness of the complexities of materials selection problems and highlight the necessity for considering the possible materials and processing routes from the commencement of any product design study. It is only in this way that effective and efficient utilisation of materials will be achieved.

Answers to Questions

Chapter 2

2.1 73% of Cu–63, 37% of Cu–65

2.2 107.87

2.3 6.324×10^{24}

2.4 (a) $Z = 13$, $M = 27$ (b) $1s^2$, $2s^2$, $2p^6$, $3s^2$, $3p^1$ (c) 3

2.5 Fe II $1s^2$, $2s^2$, $2p^6$, $3s^2$, $3p^6$, $3d^6$, $4s^2$; Fe III $1s^2$, $2s^2$, $2p^6$, $3s^2$, $3p^6$, $3d^5$, $4s^2$, $4p^1$

2.6

Nucleus	Symbol	Emission
Uranium	$^{235}_{92}U$	α
Thorium	$^{231}_{90}Th$	β
Protoactinium	$^{231}_{91}Pa$	α
Actinium	$^{227}_{89}Ac$	β
Thorium	$^{227}_{90}Th$	α
Radium	$^{223}_{88}Ra$	α
Radon	$^{219}_{86}Rn$	α
Polonium	$^{215}_{84}Po$	β
Astatine	$^{215}_{85}At$	α
Bismuth	$^{211}_{83}Bi$	β
Polonium	$^{211}_{84}Po$	α
Lead	$^{207}_{82}Pb$	Stable

2.7 (a) $^{208}_{82}X$ (b) $^{214}_{82}Y$

2.8 (a) 79.2% (b) 62.8% (c) 39.4%

2.9 1570 years

2.10 (a) and (e) metallic (b) (c) and (g) covalent (d) van der Waal's (f) ionic

2.11 Ionic; XY_2

Chapter 4

4.1 (a) Yes, by addition (b) Yes, by self-condensation (c) Yes, by addition (d) Yes, by addition

4.2 0.188 kg S/kg rubber

4.3 $M = 42078$

4.4 $DP = 56.43$

4.5 $DP = 3951$; $M = 34001$ kg

4.6 107.8 g/100 kg

4.7 Ratio styrene/acrylonitrile = 1.46/1

4.8 T_m at $DP = 50$ is 324.7 K; T_m at $DP = 500$ is 405.2 K; T_m at $DP = 1000$ is 410.8 K.

Chapter 5

5.1 A $(00\bar{1})$; B (220); C $(1\bar{1}1)$; D (0001); E $(10\bar{1}0)$

5.2 A $[10\bar{2}]$; B $[121]$; C $[110]$; D $[00\bar{1}0]$; E $[0001]$

5.3

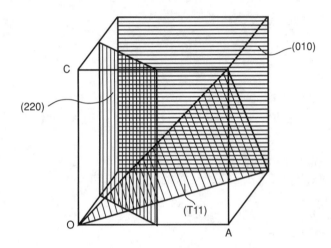

$d = 8^{-\frac{1}{2}}a$ for (220); $d = a$ for (010); $d = 3^{-\frac{1}{2}}a$ for $(\bar{1}11)$

5.4 Packing factor for discs in FCC (111) is 0.48; in BCC (110) is 0.41

5.5 (a) 9035 kg/m³; (b) crystals contain vacancies and dislocations

5.6 (a) 0.039 nm (b) 0.058 nm

5.7 (a) a = 0.556 nm (b) $PF = 0.67$

5.8 a for sample A is 0.367 nm, a for sample B is 0.361, a for pure copper is 0.361 nm so sample B is pure copper while sample A contains foreign atoms in solid solution altering the average atomic diameter

5.9 (a) BCC (b) planes (110), (200), (211), (220), (310), (222), (321) (index numbers 2, 4, 6, 8, 10, 12 and 14); (c) 0.295 nm (d) 0.255 nm

5.10 12 atoms per unit cell (4 C atoms and 8 H atoms)

Chapter 7

7.1 (a) 76.4 MPa (b) 3.67×10^{-4} (c) 208 GPa

7.2 31.8 MPa (double shear, so use 2 × c.s.a.)

7.3 (a) 29.85 GPa (b) 0.32

7.4 (a) 1663 N (b) 70.6 MPa (c) 2.5 MPa

7.5 $\sigma = 846$ MPa; $E_1 = 33$ GPa; $E_t = 1.47$ GPa

7.6 Matrix stress 5.64 MPa (compressive); fibre stress 8.41 MPa (tensile)

7.7 Carbon fibre composite (both composites meet the minimum strength requirement but the glass fibre composite does not meet the minimum E requirement

Chapter 8

8.1 (a) 32.14 MPa (b) 49.24 MPa (c) 43.30 MPa (d) 17.10 MPa

8.2 9.39 MPa

8.3 (a) 6.24 GPa (b) 104.7 MPa (c) theoretical strength assumes a perfect crystal with no dislocations or other defects

8.4 0.297 nm

8.5 1270 km

8.6 (a) 256 grains/mm^2 (b) $\sigma_0 = 56$ MPa—estimated strength of single crystal

8.7 31.2% nickel atoms

8.8 (a) 0.38% volume strain (b) 2.75% volume strain; (c) 0.625% volume strain

Chapter 9

9.1 (a) (i) 0.028, (ii) 0.055 (b) (i) 0.0083, (ii) 0.0099

9.2 (a) 303.4 mm (b) 304.0 mm (c) 300.67 mm

9.3 1.25 MPa

9.4 5.07×10^{16} Pa s

Chapter 10

10.1 160 MPa, 98 MPa

10.2 (a) 367 MPa; (b) 301 MPa

10.3 10.1 mm

10.4 0.28 mm

10.5 Ti gives $\sigma_f = 813$ MPa $\equiv$ design stress $\times$ 1.69, steel gives $\sigma_f = 981$ MPa $\equiv$ design stress $\times$ 1.13, Al gives $\sigma_f = 312$ MPa $\equiv$ design stress $\times$ 1.06

10.6 Thickness = 5.6 mm, diameter = 224 mm

10.7 $-5°$C

Chapter 11

11.1 (a) see diagram below (b) liquidus 850°C, solidus 750°C (c) (i) liquid and β solid solution in proportion L/β = 9/8 (ii) β and γ in proportion β/γ = 25/1 (iii) β and γ in proportion β/γ = 55/10 (d) 63%

11.2 (a) see diagram below

11.3 (a) see diagram below (b) (i) liquid, (ii) liquid + α, (iii) $\alpha + \beta$

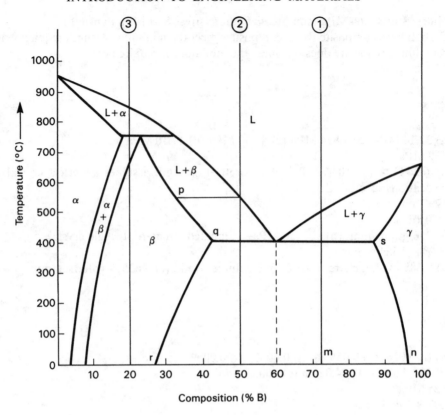

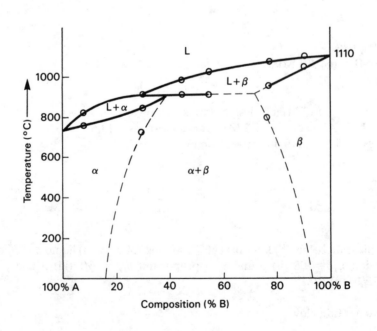

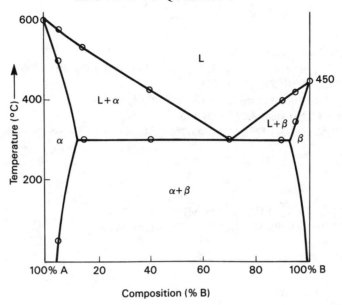

11.4 (a) partial solubility diagram with peritectic, or with eutectic (b) partial solubility diagram with eutectic (c) formation of at least one intermetallic compound (d) complete solid solubility

Chapter 12

12.1 Approximately $\times 4$
12.2 263.7 MJ/kmol
12.3 430°C
12.4 (a) 7.79×10^{-19} J/atom (b) 2.0×10^{-2} mA/mm^2
12.5 22.1 hours, 31.78 MJ/kmol
12.6 (a) 151 MJ/kmol, (b) 3.23×10^{-11} m^{2}s^{-1}
12.7 2 hours 22 minutes
12.8 22170 s (6.16 hours)
12.9 0.33 mm
12.10 -21°C

Chapter 14

14.1 3.736×10^{-19} J (2.33 eV)
14.2 U–V at 4.35×10^{15} Hz
14.3 1.12×10^{19} Hz
14.4 10.5 mm
14.5 3.44°
14.6 (a) 1.62% (b) 0.559 m
14.7 2.025 GJ

Chapter 16

16.1 Steel 1 0.45% C; Steel 2 1.15% C
16.2 (a) (i) 825°C, (ii) 820°C (b) (i) 43%, (ii) 94% (c) (i) 46%, (ii) 94%
16.3 (a) A_1 730°C, A_3 850°C (b) 0.3%C
16.4 (a) 30°C/s, (b) 10–20°C/min, (c) 10–20°C/s, (d) 450 s
16.5 (b) (i) steel C, (ii) steel A, (iii) 40 mm
16.6 (a) 1—3.93% C, 2—4.13% C, 3—4.47% C, 4—4.7% C, 5—3.53% C (b) (i) iron 5, (ii) iron 4

Chapter 27

27.1 (a) 561 MPa (b) 415 MPa (c) 358 MPa
27.2 (a) 2519 MPa (b) 796 MPa (c) 252 MPa (d) 80 MPa
27.3 7.6×10^4 cycles
27.4 1.1×10^5 cycles
27.5 786 K
27.6 4.93×10^5 s (137 hr)
27.7 51 MPa
27.8 (a) $n = 5.5$ (b) 275 MPa

Chapter 28

28.1 (a) 1.59 μm, 19.4 μm, 50.2 μm (b) 62.8 μm, 0.77 mm, 1.99 mm
28.2 P–B ratio = 1.22, oxide likely to be protective

Chapter 29

29.1 H_B for copper = 41.1; H_B for bronze = 87.2
29.2 H_D values: (a) 200.3 (b) 240.4 (c) 225.2, average = 222.0; H_B values: (a) 227.6 (b) 228.2 (c) 226.4, average = 227.4
29.3 $a = 10$, $n = 3.2$
29.4 E (0.2% secant modulus) = 1.28 GPa, $T.S.$ = 29.5 MPa, % elongation = 94%, probably HDPE or PP (proof stress not usually quoted for polymers)
29.5 (a) 1330 MPa (b) 85 GPa (c) 930 MPa (d) 31.1% (e) 1410 MPa
29.6 (a) $k = 732.8$ MPa, $n = 0.2$ (b) 8.8 mm (c) 34.2 kN (d) 435 MPa
29.7 (a) 71.9 GPa (b) 30.85 MPa (c) 73.34 MPa; (d) 44%; pure aluminium
29.8 (a) 29.17 MPa (b) 3.02 MPa

Chapter 30

30.1 4.13 mm; 2.1 MHz
30.2 1.8%
30.3 209.07 GPa

Index